P9-DUC-489

COMPUTATIONAL MATHEMATICS
An Introduction to Numerical Approximation

ELLIS HORWOOD SERIES IN
MATHEMATICS AND ITS APPLICATIONS

Series Editor: Professor G. M. BELL, Chelsea College, University of London

(and within the same series)

Statistics and Operational Research

Editor: B. W. CONOLLY, Chelsea College, University of London

Baldock, G. R. & Bridgeman, T.	Mathematical Theory of Wave Motion
de Barra, G.	Measure Theory and Integration
Beaumont, G. P.	Introductory Applied Probability
Burghes, D. N. & Borrie, M.	Modelling with Differential Equations
Burghes, D. N. & Downs, A. M.	Modern Introduction to Classical Mechanics and Control
Burghes, D. N. & Graham, A.	Introduction to Control Theory, including Optimal Control
Burghes, D. N., Huntley, I. & McDonald, J.	Applying Mathematics
Burghes, D. N. & Wood, A. D.	Mathematical Models in the Social, Management and Life Sciences
Butkovskiy, A. G.	Green's Functions and Transfer Functions Handbook
Butkovskiy, A. G.	Structure of Distributed Systems
Chorlton, F.	Textbook of Dynamics, 2nd Edition
Chorlton, F.	Vector and Tensor Methods
Conolly, B.	Techniques in Operational Research Vol. 1: Queueing Systems Vol. 2: Models, Search, Randomization
Dunning-Davies, J.	Mathematical Methods for Mathematicians, Physical Scientists and Engineers
Eason, G., Coles, C. W., Gettinby, G.	Mathematics and Statistics for the Bio-sciences
Exton, H.	Handbook of Hypergeometric Integrals
Exton, H.	Multiple Hypergeometric Functions and Applications
Faux, I. D. & Pratt, M. J.	Computational Geometry for Design and Manufacture
Goodbody, A. M.	Cartesian Tensors
Goult, R. J.	Applied Linear Algebra
Graham, A.	Kronecker Products and Matrix Calculus: with Applications
Graham, A.	Matrix Theory and Applications for Engineers and Mathematicians
Griffel, D. H.	Applied Functional Analysis
Hoskins, R. F.	Generalised Functions
Hunter, S. C.	Mechanics of Continuous Media, 2nd (Revised) Edition
Huntley, I. & Johnson, R. M.	Linear and Nonlinear Differential Equations
Jaswon, M. A. & Rose, M. A.	Crystal Symmetry: The Theory of Colour Crystallography
Jones, A. J.	Game Theory
Kemp, K. W.	Computational Statistics
Kosinski, W.	Field Singularities and Wave Analysis in Continuum Mechanics
Marichev, O. I.	Integral Transforms of Higher Transcendental Functions
Meek, B. L. & Fairthorne, S.	Using Computers
Muller-Pfeiffer, E.	Spectral Theory of Ordinary Differential Operators
Nonweiler, T. R. F.	Computational Mathematics: An Introduction to Numerical Approximation
Oliviera-Pinto, F.	Simulation Concepts in Mathematical Modelling
Oliviera-Pinto, F. & Conolly, B. W.	Applicable Mathematics of Non-physical Phenomena
Rosser, W. G. V.	An Introduction to Statistical Physics
Scorer, R. S.	Environmental Aerodynamics
Smith, D. K.	Network Optimisation Practice: A Computational Guide
Stoodley, K. D. C., Lewis, T. & Stainton, C. L. S.	Applied Statistical Techniques
Sweet, M. V.	Algebra, Geometry and Trigonometry for Science Students
Temperley, H. N. V. & Trevena, D. H.	Liquids and Their Properties
Temperley, H. N. V.	Graph Theory and Applications
Twizell, E. H.	Computational Methods for Partial Differential Equations in Biomedicine
Wheeler, R. F.	Rethinking Mathematical Concepts
Whitehead, J. R.	The Design and Analysis of Sequential Clinical Trials

COMPUTATIONAL MATHEMATICS
An Introduction to Numerical Approximation

T. R. F. NONWEILER, B.Sc., Ph.D., C.Eng., F.I.M.A., F.R.Ae.S., M.I.P.E.N.Z.

Professor of Applied Mathematics
Victoria University of Wellington

ELLIS HORWOOD LIMITED
Publishers · Chichester

Halsted Press: a division of
JOHN WILEY & SONS
New York · Brisbane · Chichester · Ontario

First published in 1984 by
ELLIS HORWOOD LIMITED
Market Cross House, Cooper Street, Chichester, West Sussex, PO19 1EB, England

The publisher's colophon is reproduced from James Gillison's drawing of the ancient Market Cross, Chichester.

Distributors:

Australia, New Zealand, South-east Asia:
Jacaranda-Wiley Ltd., Jacaranda Press,
JOHN WILEY & SONS INC.,
G.P.O. Box 859, Brisbane, Queensland 40001, Australia

Canada:
JOHN WILEY & SONS CANADA LIMITED
22 Worcester Road, Rexdale, Ontario, Canada.

Europe, Africa:
JOHN WILEY & SONS LIMITED
Baffins Lane, Chichester, West Sussex, England.

North and South America and the rest of the world:
Halsted Press: a division of
JOHN WILEY & SONS
605 Third Avenue, New York, N.Y. 10016, U.S.A.

©1984 T. R. F. Nonweiler/Ellis Horwood Limited

British Library Cataloguing in Publication Data
Nonweiler, Terence R. F.
Computational mathematics. –
(Ellis Horwood series in mathematics and its applications)
1. Numerical analysis
I. Title
519.4   QA297

Library of Congress Card No. 83-12224

ISBN 0-85312-565-1 (Ellis Horwood Limited – Library Edn.)
ISBN 0-85312-586-4 (Ellis Horwood Limited – Student Edn.)
ISBN 0-470-27472-7 (Halsted Press)

Typeset by Ellis Horwood Limited
Printed in Great Britain by Unwin Brothers, Woking.

Table of Contents

8　　　　　　　　　　　　**Table of Contents**

Preface

Numerical approximation is concerned with two related, but contrasted, tasks. It may refer to the replacement of irrational numbers and infinite processes by rational numbers and finite processes, whereby a precise result is rendered less precise but acceptable. It can also refer to the construction of plausible numerical results or relationships founded on imprecise or fragmentary data. Stated in such general terms it could describe virtually the entire range of numerical analysis; at the elementary level of the subject intended in this book, it is taken to exclude only the study of linear systems and of the operations of linear algebra from what might be recognised as the usual scope of an introductory treatment of numerical mathematics. The contrasting roles of approximation, in curtailing the infinite and extending the discrete, provide a common theme pervading the topics which remain.

The book is based upon a course of fifty lectures designed for students whose mathematical preparation has encompassed the use of the Taylor series and its remainder term, and the integral mean-value theorem. The course has been given in two halves, over two consecutive years of the undergraduate curriculum, and the manipulative skills required in dealing with some of the problems demand the more practised second-year attainment; but solutions are provided to all problems, and the text is largely descriptive, having been cleared (so far as seems reasonable) of detailed algebra. For some sections, a knowledge of the elementary algebra of linear spaces, matrices, and vectors is required: in particular, it is assumed that the reader can solve systems of linear equations, and is familiar with the concept of the rank of such a system. The terminology of sets, sequences, and limits is employed where needed, and some statistical measures are occasionally invoked. There is only passing reference to complex variables and differential equations, and none at all to vector calculus or the analysis of functions of more than one variable.

Within these limitations, however, the topics to be discussed are carried considerably further than is usual in an introductory text. The view is taken that numerical work is primarily an experimental branch of mathematics, motivated by the problems and shortcomings involved in its practical application; an approach to any particular technique becomes a matter of exploring its limitations, and continuing the search for the way forward, until it is within sight of the activities of current research. This aim may seem over-ambitious; but it is the intention to reveal the subject, not so much as a realm of applied mathematics where acquired skills find some easy or elegant expression, but as one fruitful in its own concepts, and demanding its own particular mathematical skills — many of which find no place in the mainstream of present-day mathematics.

The extent to which the aim is too ambitious must be a judgment left to the reader. But the intention to provide study in depth accounts for the omission of some topics — most notably perhaps of least-squares approximations — which are certainly basic to the subject and could be penetrated, but which are judged to require a more advanced background of mathematics for their fuller exploration than that assumed.

The discursive nature of the text leaves a lot of essential material to be explored by way of problems. It is of course for the reader to discern whether it is more helpful for these to be studied along with their solutions, as a supplement to the text — as worked examples — or whether they should be tackled with the solution 'unseen'. As has already been stated many are (intentionally) difficult, and belong to the exposition of the subject, while others can be taken as practice. Some call for proofs of assertions or theorems quoted in the text; others take the form of directed discovery of new results, tangential to the text discussion. Numerical work with a calculator or computer is involved in some problems, and the development of simple algorithms is proposed in others; neither of these two essential ingredients of the practice of numerical analysis is given the emphasis that they perhaps deserve, because it is felt that suitable examples of this kind best and most easily arise from the student's own interests and curiosity. There is therefore a considerable flexibility of choice in this supplementary material which can (it is hoped) be adapted to the individual's own aptitudes and inclinations.

In the author's defence, it should be emphasised that the algorithms which appear in the problems and solutions, and here and there in the text, are *not* intended to be examples of efficient, structured programming. Rather the primary intention is that they should be easily read and understood, and readily converted into instruction sequences for a pocket calculator, or programs for a mainframe, or micro-, computer. They are written in a 'loose' form of ALGOL 60, but they do not exploit any sophisticated features of that language; no doubt those interested (as I am myself) will enjoy re-writing them stylishly in a properly-defined language of their choice. Some 50 useful and relevant subroutines written in BASIC are included in an Appendix.

There is, no doubt, much else for which an apology should be due. Fortunately, I have been much assisted by the comments of my colleagues and students at Victoria, and am indebted in particular to John Collins for a number of detailed suggestions and corrections. Not all advice has been heeded – for good or ill – and the responsibility is therefore my own for what remains, a lot of which is new (for a book of this kind) and inevitably idiosyncratic. I am indebted to Bao Nguyen and Shona McLean for the text diagrams, and to Joy Scrimshaw for typing my manuscript. Dr Helen White aided me in proof correction and I wish also to thank the publisher's production staff, including house Editor-in-Chief Fred Sowan, who have been of immense assistance. I owe most of all to the authors of several excellent introductory texts (notably among them Dahlquist & Björck, Carl-Erik Fröberg, Ralston & Rabinowitz, and Stoer & Bulirsch) who have been influential in shaping my own interests, for which I feel no need to apologise. I can only hope they do not regard that omission as unfortunate.

Victoria University of Wellington
December 1983   Terence R. F. Nonweiler

Computational Arithmetic: the limitations and errors in operations with floating-point numbers

In common with many other sophisticated machines which demand a certain skill from their human operators, digital computers and calculators can be regarded as simple to use; their designers often go to considerable trouble to ensure that this is so. A short period of training is enough to allow some quite gratifying results to be achieved. This is certainly true of the operations of simple arithmetic, and of the increasing volume of 'software' – the library programs or modules – which perform the familiar operations of numerical analysis, such as solving equations, or performing interpolation and numerical integration. So much is becoming 'packaged' that it is reasonable to ask if it is really necessary to learn a subject whose techniques must seem secondary to their ability to produce results.

One answer might be that it is necessary to train the next generation of numerical analysts, who in course of time will no doubt be producing enhanced libraries of numerical routines. But a greater number of people than these few devotees will want to study the subject in order to perform some forms of calculation which are sufficiently complex or obscure that the library is, as yet, of no assistance; and a greater number still of students of the subject, at least at its more elementary levels, will be motivated by healthy curiosity, unrelated to any specific need to know.

Even if a library were available to fulfil every need, there is one good reason for seeking to learn something of the subject, despite the fact that it might seem, at first, a negative one; namely, that it is really not quite so simple as it is all made to appear or, more specifically, that the techniques of the subject have their considerable limitations, and the derivation of a numerical answer by their use does not imply necessarily that it is accurate or even realistic in a particular context.

In other words, the very simplicity of the use of computing machinery can be deceptive, and lends itself easily to misuse by anyone unaware of the hidden difficulties. Perhaps the only safe way to approach any non-trivial numerical problem is to regard the process involved as producing the *wrong* answer, and to have an ever watchful eye for the exceptional instance in which no answer can exist. What is meant by 'wrong' needs qualifying, of course: the result need not, indeed should not, be 'very wrong'. The important and frequently most difficult question is 'how far wrong'? What is the likely error? But clearly you never reach such questions if you start by assuming that the answer is precise.

1.1 FIXED-POINT ARITHMETIC

This question of accuracy arises in relation to the simplest of operations of machine arithmetic. There is a widely held view that machine arithmetic is precise, that 2 plus 2 will make 4. So indeed it is, in relation to the operations of addition, subtraction, and multiplication of small integers, so that 2 plus 2 can be relied on to make 4. But the operation of integer division immediately introduces difficulties: the result of $20 \div 3$ can certainly be reduced to the quotient 6, remainder 3, in rational arithmetic, but it cannot be *precisely* expressed in mechanised decimal arithmetic as 'six-point-six recurring', because the number of decimal places that the machine can express is limited. And what would be made of division by zero?

No doubt such an example of imprecision, or of the exception of zero-divide, is familiar already, and not especially alarming. Operations with integer arithmetic, or more generally with fixed-point arithmetic (wherein the numbers are in effect scaled so as to be treated as integers), can with a little care be accomplished with virtually complete precision. This is the circumstance in commercial data processing, dealing with pay-rolls, financial accounts and the like, where the arithmetic obviously needs to be precise.

The precision is made possible because the smallest and largest numbers which occur are well defined by the context. In scientific computing, however, a much wider range of numbers can be encountered, partly because the numerical processes may involve approximations to the abstractions of the infinitesimal and the infinite which are implicit within the operations of calculus. This increased range of magnitude can only be accommodated with economical use of machine memory at the expense of precision.

1.2 FLOATING-POINT NUMBERS

In scientific calculations, a **floating-point representation** of numbers is adopted, in which any number is represented by two (signed) sets of digits, known as the **mantissa** or fixed-point part (m, say), and a signed integer **exponent** or characteristic (e, say). Then the number being represented has the value

$$n = mb^e.$$
(1.2.1)

where b is **base** or radix of the number system in use. For example, $b = 10$ for the decimal system, and this is the base used in most pocket calculators, because a considerable part of their work is concerned with the display of the decimal result of every calculation. Digital computers which only display results on specific demand usually employ binary (base $b = 2$), octal ($b = 8$), or hexadecimal arithmetic ($b = 16$), because systems based on a power of 2 make more efficient use of the machine's memory. Whatever base is adopted, the mantissa and exponent are numbers of that system composed of digits from 0 to ($b - 1$). Thus in the binary system they are made up entirely of 0's and 1's; in the hexadecimal system, they consist of digits 0 to 9 and 'digits', usually written as A–F, representing values of 10 to 15.

Incidentally, the term 'mantissa', although in common use, is a misnomer, its original meaning (derived apparently from the Etruscan) being 'something added'. Clearly m is 'something multiplied'. The corruption has arisen because its magnitude is the *antilog of the mantissa* of $\log_b |n|$.

As it stands, the representation (1.2.1) is not unique: thus in the decimal system 2.5 could be written, consistent with this definition, as 0.25×10, 2.5×10^0, 25×10^{-1}, ... and so on. Such arbitrariness is inconvenient, and one common method of **normalisation** is to define

$$1 > |m| \geqslant b^{-1} \tag{1.2.2}$$

so that, for instance, the **standardised**, or **normalised**, decimal form of 2.5 would be 0.25×10, – consistent, that is, with $m = 0.25$ and $e = 1$. However, this is certainly not a universally accepted convention: many might prefer the normalisation

$$b > |m| \geqslant 1 \tag{1.2.3}$$

whereby 2.5 is simply 2.5×10^0. Clearly, it does not really matter: the important aspect of the standardisation is that it is defined, and in such a manner that the range of the mantissa is between numbers whose ratio is equal to the base, b.

We may note, in passing, that the number zero is exceptional in the sense that it cannot obey such inequalities as (1.2.2) or (1.2.3). A zero number is clearly only representable by $m = 0$, however it is normalised.

The maximum number of digits available to the mantissa is called the **precision**, or number of **significant digits**. Suppose, for example, a decimal floating-point number is to be represented with 3 significant digits (albeit, not a very generous allocation), then the number $\sqrt{2} = 1.414213 \ldots$ would be written as 0.141×10^1 if the standardisation (1.2.2) is used, or as 1.41×10^0 if (1.2.3) is used. The significant digits are the same in either convention. Beware of confusing significant digits with decimal (or more generally, fractional) *places*; the latter term refers to the number of digits behind the decimal (or fractional) point. To avoid this possible confusion, yet a third form of standardisation recommends itself, by which it is required that

$$b^p > |m| \geqslant b^{p-1} \tag{1.2.4}$$

where p is the (integer) precision. This has the convenience of ensuring that the mantissa (like the exponent) is a signed integer – rather than a fractional number – and so encourages the realistic view of a machine's floating-point operations as specialised applications of integer arithmetic. Moreover, integers with one base are always precisely expressible as integers of some other base. In the example previously cited, $\sqrt{2}$ would then appear, subject to (1.2.4), as 141×10^{-2}: there is no decimal point needed, since any mantissa of 3 decimal digit precision appears as a 3-digit integer. In binary with a precision of 8 digits, it becomes $(10110101)_2 \times 2^{-7} = 181 \times 2^{-7} (= 1.414063 \ldots)$. More generally, (1.2.4) requires that the mantissa m is an integer of p digits in any system of precision p.

Most computers allow numbers to be represented in either single precision or **double precision** (usually with *at least* twice the number of significant digits). This is because a limited precision can give rise to considerable errors in some contexts, as we propose to show. On the other hand, of course, double precision working impairs the speed of processing, and uses more memory.

1.3 ROUNDING ERROR OF FLOATING-POINT NUMBERS

Having made the point that normalisation is necessary, but merely a convention open to choice, let us agree to accept the definition (1.2.4) in what follows here, despite the fact that it may appear 'unnatural' or unfamiliar. Also – so as to avoid unnecessary complication – let us continue to illustrate machine arithmetic by using decimal floating-point numbers with 3 significant digits.

Such a low precision serves of course to exaggerate the inaccuracies of more realistic floating-point representations, but does not falsify them. It may have seemed perfectly natural to accept that 141×10^{-2} was the best that could be achieved, with such an imposed limit on precision, in relation to $\sqrt{2}$. The mantissa should be 141.4213 ... but it is truncated or **chopped** to 141, simply by omitting (in this method of normalisation) the fractional part. Likewise, $\sqrt{3}$ would have to be truncated as 173×10^{-2} whereas more precisely the value of the mantissa should be 173.205 .... Similarly, $-\sqrt{3}$ would become -173×10^{-2}, applying the same rule to its absolute value. But what of $\sqrt{5} = 2.23606\ldots$? Should this be represented as 223×10^{-2} or 224×10^{-2}? The latter would clearly be nearer the correct value (with mantissa 223.606 ...), and this process of taking the nearest-integer representation is called **rounding** (or sometimes *proximity* or *correct* rounding). But whether a particular machine employs rounding or chopping may depend upon how the number was generated (for example, was $\sqrt{5}$ generated during a calculation as the square root of 500×10^{-2}, or 'input' – read-in – as 2.23606?).

In order to represent this process algebraically, suppose that the exact (non-zero) value being input is x (for instance, 2.23606 in the above example), and that there exists an integer i (equal to -2 in this instance) such that

$$\left. \begin{array}{l} b^p > |y| \ = \ |x|/b^i \geqslant b^{p-1} \\ \text{i.e.} \qquad b^{p+i} > |x| \ = \ |y| b^i \geqslant b^{p+i-1} \end{array} \right\} \qquad (1.3.1)$$

Comparing this with (1.2.4), y is seen to be the exact, generally non-integer, mantissa of x (equalling 223.606 in the cited example) conforming with the adopted form of normalisation.

The number x is to be represented by the machine in the form of $n = mb^e$, as in (1.2.1), where m is integer and likewise satisfies the inequality (1.2.4). In this representation the sign of m will be the same as the sign of x, and the exponent e will be equal to i (with one possible exception to which we shall return). The only question is, therefore: what is the relation between the magnitudes of the integer mantissa $|m|$ and of the exact mantissa $|y|$?

If the representation is obtained by chopping, then this relation can be written as

$$|m| \ = \ \text{entier}\,(|y|) \qquad (1.3.2)$$

where entier (..) is a function which has as its value the largest integer which does not exceed its argument. Another name given to this useful function is floor(..). In particular, when applied to any positive argument, as in (1.3.2), it represents simply the integer part of the argument (223 in the example) obtained by suppressing the fractional part (0.606), just as in the operation that we have described as chopping.

In the process of rounding, on the other hand, the relation between the input number and its machine representation is defined by

$$|m| \ = \ \text{entier}(|y| + 0.5) \qquad (1.3.3)$$

which makes $|m|$ equal to the integer closest to $|y|$, chopping only if the fractional part of $|y|$ is less than a half, but otherwise rounding-up. In the cited example, rounding therefore implies that $|m| = \text{entier}(224.106) = 224$, and this is an instance therefore of rounding-up. Equation (1.3.3) also implies that there is round-up in the exceptional circumstance that the fractional part of $|y|$ precisely equals a half, in which event of course both integers on either side of $|y|$ are equally close. Why then always round-up in this circumstance? There is no reason other than simplicity: in fact, conventions have been devised which avoid this arbitrary bias, but they are usually reckoned as too complicated to implement on a machine.

There is yet a third method of machine representation which is quite commonly used, called **augmentation**, or round-up, whereby the value of the machine mantissa is invariably rounded-up (rather than chopped) if the exact mantissa is fractional. The word 'augmented' is, like 'truncation', something of a mouthful, and 'round-up' is prone to be confused with rounding, so we here invent a piece of jargon and call this **boosting**. This implies

$$|m| \ = \ -\text{entier}(-|y|) \ = \ \text{ceil}(|y|). \qquad (1.3.4)$$

Applying this rule to the representation of $x = 2.23606$, the value of entier$(-|y|)$ is the largest integer not exceeding -223.606; this integer is -224, so that $m = 224$, and the mantissa is therefore rounded-up or 'boosted'.

The 'ceiling' function ceil(..) used in (1.3.4) is defined to have as its value the smallest integer which is not less than its argument (whatever its sign). Obviously it provides a more straightforward way of expressing the boosted result (224) in terms of the argument (223.606) than the equivalent, but somewhat contrived, used of the double-negative together with entier(..).

Boosting, if used at all, most usually occurs in the arithmetic subtraction of two floating-point magnitudes (e.g. 11.6 minus 0.16 becomes 11.5, not 11.4).

There is in fact one circumstance in which the machine representation is not quite correctly given by the equations (1.3.3) and (1.3.4). Continuing to work with decimal floating-point numbers of precision 3, note what happens if (1.3.3) is applied to an exact mantissa whose magnitude lies in the range $999.5 \leqslant |y| < 1000$. The machine mantissa would then be given by $|m| = 1000$: but this contravenes the normalising requirement (1.2.4), that $1000 > |m| \geqslant 100$. Accordingly, the machine exponent e would have to be increased to $(i + 1)$ and $|m|$ reduced to 100, by way of compensation, in order to conform with the normalisation. The value of n would not thereby be affected, because $1000b^i$ is here the same as $100b^{i+1}$: the modification is to the form, not the value, of the machine representation. Equation (1.3.4) likewise needs the same modification if $999 < |y| < 1000$, but it nonetheless gives the appropriate boosted value of n, if it is assumed that e and i are always equal.

The **error** in the machine representation of a number is simply the difference $(x - n)$ between what the number should be, and how it is approximated: this is also of course the same as the correction to be added to n to obtain the exact result. (Sometimes an expression of the opposite sign is used to express error, but the choice is immaterial, provided that it is consistently maintained.) The name 'absolute error' is also given to this same difference, but this will be avoided here as it is open to confusion with the error magnitude $|x - n|$. A **relative error** (relative, that is, to the number being represented) can be defined by

$$RE(x) = (x - n)/x \qquad (1.3.5)$$

or, since the signs and exponents of x and n are the same, by

$$RE(x) = (|y| - |m|)/|y| . \qquad (1.3.6)$$

It is not difficult to confirm from equations (1.3.2) to (1.3.4) that the numerator of this expression, which is the error in the mantissa introduced by the machine representation, is limited by the inequalities:

$$\left.\begin{array}{ll} 1 > |y| - |m| \geqslant 0 & \text{for chopping,} \\ \tfrac{1}{2} > |y| - |m| \geqslant -\tfrac{1}{2} & \text{for rounding,} \\ 0 \geqslant |y| - |m| > -1 & \text{for boosting} \end{array}\right\} \qquad (1.3.7)$$

Consequently, as these apply whatever the value of $|y|$, the magnitude of the relative error is largest if the denominator in (1.3.6) is least. But (1.3.1) shows that the least value of $|y|$ is b^{p-1}, so that (1.3.7) leads to:

$$
\left.
\begin{aligned}
b^{1-p} &> \mathrm{RE}(x) \geqslant 0 & \text{for chopping,} \\
\tfrac{1}{2}b^{1-p} &> \mathrm{RE}(x) > -\tfrac{1}{2}b^{1-p} & \text{for rounding,} \\
0 &\geqslant \mathrm{RE}(x) > -b^{1-p} & \text{for boosting}
\end{aligned}
\right\},
\qquad (1.3.8)
$$

and it is the existence of these bounds which makes it convenient to use relative, rather than actual, error in describing the accuracy of floating-point arithmetic. In terms of decimal floating-point numbers of precision 3, the largest relative error occurs when an exact mantissa almost equal in magnitude to 101 is truncated to 100, or one of magnitude a little more than 100 is boosted to 101. In rounding, the extreme values of relative error occur when the magnitude of the exact mantissa is 100.5.

The value of $b^{1-p} = \epsilon$, say, which determines the limits to the relative error in (1.3.8), is called the **machine unit** (or round-off unit) of the floating-point representation, and the process of conversion to machine-representation is termed **floating** the number. It is evidently a process by which the continuum of real numbers can be mapped on to a discrete set of rational numbers, and as such it can be regarded as a functional relationship defined by one of the equations (1.3.2) to (1.3.4). If this relationship is written as $fl(..)$, then (1.3.6) and (1.3.8) show that

$$
n = fl(x) = x[1 - \mathrm{RE}(x)] = x(1 - \alpha\epsilon), \text{ say }, \qquad (1.3.9)
$$

where $\alpha = \mathrm{RE}(x)/\epsilon$ is some number which lies in one of the intervals $[0,1)$, $(-\tfrac{1}{2}, \tfrac{1}{2})$, $(-1, 0]$ depending respectively upon whether chopping, rounding, or boosting is in use. The value of $\alpha\epsilon = \mathrm{RE}(x)$ is called the **rounding error**, whatever process is involved.

This is a simple way of expressing the bounds to the error of the floating point representation of any number, but it will be realised that the upper bounds on its magnitude will almost always overestimate the actual rounding error. For instance, the same reasoning by which the inequalities (1.3.8) are derived from (1.3.7) will also show that the range of relative error is substantially less if the mantissa of the number happened to be near the upper limit of the permitted range rather than near the lower limit. Also, of course, particular numbers (like, for example, all integers less than b^p in absolute value) can be represented precisely, with no error at all.

1.4 THE FREQUENCY DISTRIBUTION OF ERROR

The derivation of an 'average' or 'representative' error rather than the upper bounds to that error, is a more difficult problem, particularly so in regard to the input of numbers. This is because input depends on the precision of the data as presented, as well as on whether there has to be a conversion of base between the external representation (presumably expressed in decimal) and that of floating-point arithmetic of the machine. Indeed, many machines use a different method of floating for input (and output) than in other operations.

It is more generally relevant, as well as simpler, to pursue the analysis by considering a machine operation such as finding (say) the square-root of a number already in normalised floating point form. Calling this number n_1, it plays the role of 'input' to a functional calculation, and the 'output' result is also a standardised floating point number, say n_0. If we place $n_1 = 5$, and $n_0 = n$, then we recover the numerical example which was used as an illustration of the analysis in the last section. For present purposes, however, we do not need even to specify the functional operation (as a square-root): let us call it simply $F(..)$, and place $n_0 = \tilde{F}(n_1)$, adding the tilde to the symbol F to denote that the result n_0 is approximate, since it has only a limited precision and might not equal the exact value $F(n_1)$. In fact, we would hope that the computer evaluates the result as

$$n_0 = fl[F(n_1)] = \tilde{F}(n_1) .\qquad(1.4.1)$$

That is, the result n_0 is the exact value reduced in precision merely by the process of floating, and not by any inaccuracy in calculating F. Comparing (1.4.1) with (1.3.9) the exact value $F(n_1)$ takes the place of what was previously called x, and it might well be a transcendental number (like $\sqrt{5} = 2.23606 \ldots$).

Let us go back now to the inequalities (1.3.7) determining the difference between the magnitudes of the mantissa $|m|$ of $n = n_0$ and $|y|$ of $x = F(n_1)$, and agree that we are not just considering here *one* particular functional operation F or *one* particular value of its argument n_1, but the whole range of possible operations upon all possible arguments. That being so, it would seem reasonable to suppose that the fractional part of $|y|$ might, with equal probability, take *any* value between 0 and 1. What amounts to the same thing is to replace (1.3.7) by

$$|m| = |y| - \mu + \delta ,\qquad(1.4.2)$$

where $\mu = \frac{1}{2}, 0$, or $-\frac{1}{2}$ depending upon whether respectively chopping, rounding, or boosting is in use, and where δ could, with equal probability, take any value from the continuum of real numbers between $-\frac{1}{2}$ and $+\frac{1}{2}$. This makes μ equal to the mean value of the error $(|y| - |m|)$ averaged over all possible calculations like (1.4.1), and it represents the bias in calculating the machine mantissa which is only avoided if correct rounding is used. The relative error in representing x is then given from (1.3.6), (1.3.9), and (1.4.2) as

$$\text{RE}(x) = \alpha\epsilon = (\mu - \delta)/(|m| + \mu - \delta)\qquad(1.4.3)$$

where $|m|$ could equally well — so far as it is in general possible to tell — take any integer value between b^{p-1} and $(b^p - 1)$, consistent with (1.2.4). This suggests the point of view that $RE(x)$ is a function of two variables, one real (δ) and the other integer (m), both of which are random in the sense that, with equal probability, they could have any values within their stipulated ranges. In other words, they can be treated as random variables drawn from uniform frequency distributions (one continuous, the other discrete). This implies in turn that values of $RE(x)$ can also be treated as a random variable, with a probability density function describing the frequency of occurrence of particular values of $RE(x)$ or α, which can be calculated from (1.4.3).

   This derivation depends on ideas which are developed within that area of statistics usually called **sampling distribution theory**. The resulting frequency distributions for α (depending on the method of floating used) are shown in Fig. 1.1, its extreme values being of course the same as those already deduced. Some statistics describing the distributions are given in Table 1.1. The distribution derived from the use of boosting is the same as that of chopping, except

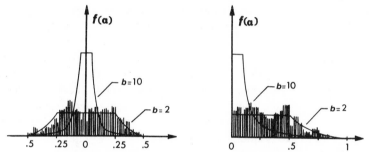

Fig. 1.1 – Theoretical frequency distribution of α resulting from correct (symmetrical) rounding, and from chopping, for numbers with binary and decimal base. Bar diagrams are typical actual distributions (for binary base).

Table 1.1

Statistics relating to the distribution of α

Method of rounding	Base (b)	Mean ($\bar{\alpha}$)	Variance (σ_α^2)	Standard deviation (σ_α)	Lower quartile	Median of α	Upper quartile
Chopping ($\mu = \frac{1}{2}$)	2	0.347	0.0466	0.216	0.167	0.333	0.500
	8	0.149	0.0196	0.140	0.056	0.111	0.188
	10	0.128	0.0170	0.130	0.045	0.091	0.158
	16	0.092	0.0123	0.111	0.029	0.059	0.106
Correct rounding ($\mu = 0$)	2	0	0.0417	0.204	−0.167	0	0.167
	8	0	0.0104	0.102	−0.056	0	0.056
	10	0	0.0083	0.091	−0.045	0	0.045
	16	0	0.0052	0.072	−0.029	0	0.029

that the sign of the error is reversed; correct rounding causes a symmetric distribution of error.

As was anticipated at the outset, these figures confirm that 'average' errors are likely to be substantially smaller than the extremes, particularly if the arithmetic base is large, because then mantissa values close to the lower bound, which are susceptible to the largest relative errors, occur less frequently. If rounding is used, half the sampled values of relative error in an operation like (1.4.1) could be expected to lie in the range between plus and minus $\frac{1}{2}\epsilon/(b + 1)$. On the other hand, if chopping is used in (1.4.1), half of the relative errors to be expected in such an operation are likely to fall between zero and $\epsilon/(b + 1)$; and the bias in forming the machine mantissa shows itself as a mean relative error $\bar{\alpha}\epsilon$, where $\bar{\alpha}$ is a little larger than this median value.

1.5 TRANSMITTED ERROR

Any non-trivial sequential calculation will in general involve several operations like (1.4.1), and the 'input' floating-point number (n_1) will itself generally be in error compared with the exact value which should have been input, had the previous calculation been performed with infinite precision. The 'output' number (n_0) in turn is altered from what it should be, because of this error in input, and we call this particular alteration the **transmitted error** (sometimes called the propagated error). Although any single operation in the sequence cannot be held responsible for the bad information which it is forced to accept, it is nonetheless important to know how it modifies it, before passing it on to the next operation.

Suppose that the correct input value should have been x_1, in infinite precision, and likewise let $x_0 = F(x_1)$ be the corresponding exact result of the operation (1.4.1). Further let us place

$$n_k = x_k(1 - \beta_k\epsilon) \tag{1.5.1}$$

so that $\beta_0\epsilon$ and $\beta_1\epsilon$ are respectively the relative errors of the result x_0 and the input number, x_1. Just as before, we assume that $n_0 = \widetilde{F}(n_1) = F(n_1)(1 - \alpha_1\epsilon)$, where α_1 is some particular value of α. In order to find β_0 in (1.5.1) it is necessary somehow to relate $F(n_1)$ to x_0.

Provided that the relative error of the input number is sufficiently small, the values of n_1 and x_1 will be almost equal and are connected by (1.5.1), taking $k = 1$. We can use a Taylor polynomial approximation to show that

$$F(n_1) = F(x_1 - x_1\beta_1\epsilon) \simeq F(x_1) - F'(x_1)x_1\beta_1\epsilon$$

or, since $x_0 = F(x_1)$,

$$F(n_1) \simeq x_0(1 - \phi_1\beta_1\epsilon) \tag{1.5.2}$$

where $\qquad \phi_1 = x_1 F'(x_1)/F(x_1)$

Substituting this expression for $F(n_1)$ in that for n_0, it follows that

$$n_0 \simeq x_0(1 - \phi_1\beta_1\epsilon)(1 - \alpha_1\epsilon) \ . \tag{1.5.3}$$

Expanding the product of the bracketed terms, and ignoring the term involving the very small quantity ϵ^2, we find that

i.e.
$$\left. \begin{array}{c} n_0 \simeq x_0\,[1 - (\alpha_1 + \phi_1\beta_1)\epsilon] \\[2mm] \beta_0 \simeq \alpha_1 + \phi_1\beta_1 \end{array} \right\}, \tag{1.5.4}$$

the expression for β_0 following by comparison with (1.5.1), taking $k = 0$. This shows therefore that the relative error of the result has two components: one equal to $\alpha_1\epsilon$ representing the rounding error, and another $\phi_1\beta_1\epsilon$ which is the transmitted relative error, and which is equal to ϕ_1 times the input relative error. Although these components are added, they could of course have opposite signs – conceivably they could even cancel one another out, although this would be a pure coincidence.

Of particular significance, however, is whether the magnitude of ϕ_1 is large or small compared with unity. If $|\phi_1|$ is larger than unity, then the evaluation of $\tilde{F}(n_1)$ magnifies the received error before it is passed on; on the other hand, if $|\phi_1|$ is small compared with unity, the evaluation more or less eliminates any prior accumulation of error. These contrasting effects relate to what is called the **condition** of the problem posed by the evaluation, and $|\phi|$ is known as its **condition number**. If $|\phi|$ is large compared with unity, the problem of evaluation is said to be **ill-conditioned**, or ill-posed, because a small error in the data will cause a disproportionately large change in the result; if $|\phi|$ is small, it is said to be **well-conditioned** (or well-posed) because it is relatively insensitive to the data.

The condition number depends on both the function and its argument. Usually functions will have ranges of values of their arguments for which they are ill-conditioned, and other ranges for which they are well-conditioned. The most frequently occurring example of transmitted error arises where a function is evaluated close to a (non-zero) root. For example, the function $(x - 1)$ looks innocuous, but is ill-conditioned in the neighbourhood of $x = 1$; the associated condition number is $|x/(x - 1)|$, and this is (for example) greater than 1000 for any value of x between about 0.999 and 1.001. Thus, if an exact input (x_1) of 1.00098 (say) is input instead as 1.00100, the input error is only 1 part in 50 thousand. The true output of $(x_1 - 1)$ should be $x_0 = 0.98 \times 10^{-3}$, but will at best be represented as $n_0 = 1.00 \times 10^{-3}$, with a much larger error of 1 part in 50. If the exact input ought to be the root $x_1 = 1$ and it is represented by *any* value $n_1 \neq 1$ (no matter how close to unity it may be), the output result will be $n_0 \neq 0$ which, compared with the exact value of $x_0 = 0$, has of course an infinite relative error.

This particular form of error transmission is known as **cancellation error**, or loss of significance. It looks obvious enough, even perhaps trivial and unimportant when presented in such an undisguised form; after all, the *actual* error is small. The trouble is that it often lies hidden within a problem, and the large relative error is propagated through the arithmetic to appear in an end result which is not in itself small. But we will come back to this shortly.

Another form of ill-conditioning arises if the function derivative is large. This happens (for instance) if the function has a rapid oscillation: thus $F(x) = 2 - \sin(2000x)$ is never zero, but its condition number oscillates between 0 and a value of more than $1000|x|$. The value of $F(1)$ is $1.06996\ldots$, but $F(1.001) = 2.72116\ldots$. Similarly, ill-conditioning applies in evaluating large powers: if $F(x) = x^{1000}$, for example, its condition number is equal to 1000 for any positive value of x, and whereas obviously $F(1) = 1$, we can calculate that $F(1.001) = 2.71692\ldots$. Such examples make it clear that ill-conditioning is a property of *what* is being calculated, not of *how* it is calculated.

There is in other words no way of computing an ill-conditioned function which overcomes the inevitable amplification of input error. Suppose, for example, we calculated the value of 1.001^{1000} by the formula $x^a = \exp[a \ln(x)]$. First we would have to calculate $\ln(x)$ for $x = 1.001$. But the condition number of $\ln(x)$ is $|1/\ln(x)|$, and it is large for x close to unity: in other words, this part of the calculation involves large cancellation error. Multiplying the value of $\ln(x)$ by $a = 1000$ carries the resulting loss of significance through to the argument of $\exp(..)$, and on to the final result.

Likewise in the other quoted example, the value of the trigonometric function $\sin(2000x)$ might well be calculated by using the reduction formula:

$$\sin(z) = \sin[z - 2\pi\,\text{entier}(z/2\pi)] \ .$$

If $x = 1.001$, then $z = 2000x = 2002$, and therefore $\text{entier}(z/2\pi) = 318$, so that this reduction formula suggests that we calculate the sine of $2002 - 318 \times 2\pi = 2002 - 1998.05\ldots$ and whoops! we have cancellation error once more.

The problem cannot be evaded, so it is a waste of time trying! One type of transmitted error can be turned into another, and that is all that is achieved. But at least the attempt has shown how cancellation error can lie hidden within a calculation. It can even occur in a well-conditioned calculation, and possibly with rather disasterous consequences.

1.6 GENERATED ERROR

Take, for example, the calculation of $F(x) = \exp(x) - 1$ for a small value of x: the condition number is $|x/[1 - \exp(-x)]|$, and is close to 1 if x is small. The problem is therefore certainly *not* ill-conditioned. However, if we first evaluate $\exp(x)$, it will yield a value close to unity, and subtracting 1 will then introduce a large cancellation error. For example, do this in decimal arithmetic with a precision of 3 and take $x = 0.0123$, say: then $\exp(x) = 1.01$ correct to 3 significant digits, and $\exp(x) - 1 = 0.0100$ with the same precision. The answer should be $F(0.0123) = 0.0124$ rounded to 3 decimal digits, and there is therefore a 19% relative error. What has happened?

We have done it wrong – that is what has happened! We divided the calculation into two operations: the first of which (finding $\exp(x)$) was extremely well-conditioned (with small condition number $|x|$), but the second (differencing the result from 1), extremely ill-conditioned (condition number $1/|1-\exp(-x)|$). The first well-conditioned evaluation avoids transmitted error, but not rounding error, as will be clear from (1.5.4): this rounding error is input into the other badly-conditioned operation, and amplified.

This brings us up against the problem of error propagation in a sequential calculation. Supposing that the exact problem, whatever it may be, is evaluated in a series of stages:

$$x_2 = F_1(x_1), x_3 = F_2(x_2), \ldots, x_k = F_{k-1}(x_{k-1}), x_0 = F_k(x_k)$$

where we have added a subscript to the F's to denote different functions. What would be programmed for machine calculation would then be $n_2 = \tilde{F}_1(n_1), n_3 = \tilde{F}_2(n_2), \ldots$ and so on, up to the kth and last stage $n_0 = \tilde{F}_k(n_k)$. Defining the relative error of n_j by $\beta_j\epsilon$ as in (1.5.1), and the condition number of $F_j(x_j)$ by ϕ_j then as in (1.5.4)

$$\beta_{j+1} \simeq \alpha_j + \phi_j \beta_j$$
$$\phi_j = x_j F_j'(x_j)/F_j(x_j) \tag{1.6.1}$$

for $j = 1, 2, \ldots, k$, where the relative error of the final result is given by $\beta_0 = \beta_{k+1}$. This is an example of a **recurrence formula** for the β's, each β being defined in terms of its predecessor. Some algebra will produce the cumbersome expression:

$$\beta_0 = \beta_{k+1} = \alpha_k + \phi_k\alpha_{k-1} + \phi_k\phi_{k-1}\alpha_{k-2} + \ldots$$
$$+ (\phi_k\phi_{k-1}\ldots\phi_2)\alpha_1 + (\phi_k\phi_{k-1}\ldots\phi_2\phi_1)\beta_1 . \tag{1.6.2}$$

Further, the product of all the ϕ's, which multiplies β_1, can be shown in fact to be the condition number ϕ of the function $x_0 = F(x_1)$ which represents the complete computation (see problem 6, section 1.5).

This will be easier to follow if we revert to the particular example of the two-stage calculation of $F(x) = \exp(x) - 1$. This was divided by first calculating $x_2 = F_1(x_1) = \exp(x_1)$, and then finding the result as $x_0 = F_2(x_2) = x_2 - 1$. Equations (1.5.1) and (1.6.1) show that the notation relating to the numerical calculation is given by

$$n_2 = \tilde{F}_1(n_1) = fl(\exp(n_1)) = x_2(1-\beta_2\epsilon) = x_2[1-(\alpha_1 + \phi_1\beta_1)\epsilon] ,$$
$$n_0 = \tilde{F}_2(n_2) = fl(n_2 - 1) = x_0(1-\beta_0\epsilon) = x_0[1-(\alpha_2 + \phi_2\beta_2)\epsilon] ,$$

where

$$\phi_1 = x_1 F_1'(x_1)/F_1(x_1) = x_1,$$

$$\phi_2 = x_2 F_2'(x_2)/F_2(x_2) = x_2/(x_2 - 1) = 1/[1 - \exp(-x_1)] ,$$

so that the product of the condition numbers is the condition number of the combined function $\exp(x_1) - 1$: that is,

$$\phi_1\phi_2 = x_1/[1 - \exp(-x_1)] = x_1 F'(x_1)/F(x_1) = \phi .$$

If we refer back to the calculation using decimal arithmetic of precision 3 (for which $\epsilon = 10^{-2}$) with input $n_1 = x_1 = 0.0123$, then the following values will be found to apply:

$$n_2 = 1.01, x_2 = 1.01237 \ldots, \beta_2 = 0.23, \alpha_1 = 0.23, \phi_1 = 0.0123, \beta_1 = 0;$$

$$n_0 = 0.01, x_0 = 0.01237 \ldots, \beta_0 = 19, \quad \alpha_2 = 0, \quad \phi_2 = 81, \quad \beta_2 = 0.23 .$$

Note that there was no input error ($\beta_1\epsilon$) since $n_1 = x_1$ precisely, nor any rounding error ($\alpha_2\epsilon$) in subtracting 1 from 1.01 to form n_0. The final error was due entirely to the second-stage transmitted error $\phi_2\beta_2\epsilon$, applying to the input $\beta_2\epsilon$ which was in turn entirely due to first-stage rounding error $\alpha_1\epsilon$. Algebraically, substituting for $\beta_2 = \alpha_1 + \phi_1\beta_1$ in the expression $\beta_0 = \alpha_2 + \phi_2\beta_2$ it follows that

$$\beta_0 = \alpha_2 + \phi_2\alpha_1 + \phi_1\phi_2\beta_1 \qquad (1.6.3)$$

as is consistent with (1.6.2) if $k = 2$. Because α_2 was found to be zero, and β_1 was assumed zero, all the error in this example stems from the middle term which is merely a magnification by the large condition number $|\phi_2| = 81$ of the rounding error of the intermediate result n_2.

We call this **generated error**, because it is generated within the problem by the chosen method of evaluation; it is neither the transmitted input error $\phi_1\phi_2\beta_1\epsilon = \phi\beta_1\epsilon$, nor the rounding error of the final result $\alpha_2\epsilon$. It had not been detected earlier, because its existence contravenes the assumption which had been made when stating that the function $F(..)$ of (1.4.1) was computed *exactly* and then rounded. If the evaluation is broken down into k stages rather than just two, the generated error is represented in (1.6.2) by the contribution to β_0 of the terms

$$\phi_k\alpha_{k-1} + \phi_k\phi_{k-1}\alpha_{k-2} + \ldots + (\phi_k\phi_{k-1} \ldots \phi_2)\alpha_1 \qquad (1.6.4)$$

and it is seen to be the transmitted form of all the rounding errors of the $(k-1)$ intermediate results (Fig. 1.2).

Generated error excludes the transmitted error of the input, which we can do nothing about (since the product of the condition numbers $\phi_1\phi_2 \ldots \phi_k$ is the condition number of the problem), and which for this reason we call the **inherent error** of the problem. It also excludes the contribution to β_0 of the final rounding error component α_k, which again is inescapable, in so far as any result,

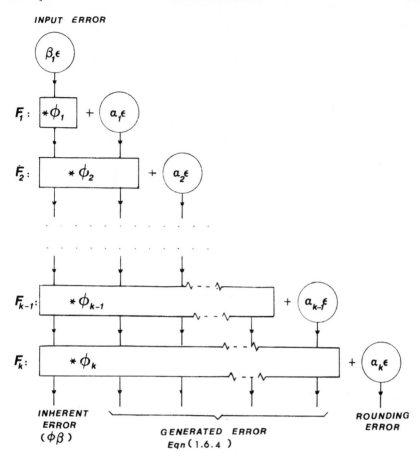

Fig. 1.2 – Generated error.

good or bad, has to be rounded to machine precision. We can, however, do something about this remaining generated error, by a suitable arrangement of the stages of the problem. It may be unduly optimistic to suppose that it can be arranged to be negligible compared with the maximum bounds of the inherent and rounding errors, but ideally it should never be large compared with these other errors.

Admittedly, if the problem as a whole is ill-conditioned, at least one of the component condition numbers is bound to be large. But if it is possible to arrange for this to be the condition number of the *first* stage of the calculation, then the generated error may be kept small despite the ill-conditioning, because ϕ_1 is absent from the expression (1.6.4). In other words, placed at the beginning of a calculation, an ill-conditioned step is harmless, as there is no accumulated rounding error for it to magnify; only the input error, and this will be transmitted through the problem to have the same effect upon the end-result, independently of how the calculation is arranged.

The early insertion of ill-conditioning in a problem might at first sight have seemed likely to degrade the rest of the calculation. But in fact the reverse is true: the conditioning of subsequent stages will thereby be improved (because the product of all the ϕ's is fixed). Often there may not be much choice about where ill-conditioning arises in the sequence of operations; but if there is, then the rule is: *insert it as soon as possible*.

For example, let us evaluate $x^2 - 1$ with $x = 1.07$ and decimal arithmetic of precision 3. The 'obvious' way, perhaps, is to place $n_2 = n_1 * n_1 = 1.14$ and $n_0 = n_2 - 1 = 0.140$, with condition numbers $\phi_1 = 2, \phi_2 = 8$. But better is $n_2 = n_1 - 1 = 0.070$ and $n_0 = n_2 * (2 + n_2)$ which rounds to 0.145. This is the correctly rounded form of the precise answer (0.1449) and is achieved with $\phi_1 = 15, \phi_2 \simeq 1$.

One effective way of viewing the rule of early ill-conditioning is to draw an analogy with the use of a Taylor series expansion in describing function behaviour near some critical value $x = \xi$, say, by a series expanded in terms of $(x - \xi)$. To calculate the series, the small (ill-conditioned) difference $(x - \xi)$ has to be formed first of all. Thus, in the above example, in effect we expanded $(x^2 - 1)$ about its root $x = 1$ as $2(x - 1) + (x - 1)^2$. To remove the stigma that arises from the use of the term 'ill-conditioning' in this entirely beneficial context, we call this approach a **differenced solution**.

The occurrence of generated error in the example of the evaluation of $F(x) = \exp(x) - 1$ for $x = 0.0123$ is due to the fact that ϕ_1 is small, and ϕ_2 large: the wrong way round. Given that large generated error can (usually) be avoided, what can be done about it here? If there happened to be a library function sinh(..) available, then because

$$\sinh(z) = \tfrac{1}{2}[\exp(z) - \exp(-z)] \ ,$$

some algebra shows that

$$\exp(x) - 1 = 2\exp(x/2)\sinh(x/2) \ ,$$

and this formulation should provide a much more accurate result. But the evaluation of $\sinh(x/2)$ embodies the same problem as that involved in evaluating $F(x)$, and so this approach merely relies on the efficacy of someone else's solution to the problem! A do-it-yourself solution would be to replace $F(x)$ by its Taylor series expansion about the root $x = 0$, and evaluate

$$\exp(x) - 1 = x + (x^2/2) + (x^3/6) + \dots \tag{1.6.5}$$

The first two terms are sufficient to give $F(0.0123)$ correctly to 3 significant figures. Of course, if $F(x)$ were to be evaluated for a range of x-values, other questions arise, such as: how many terms must then be used, and for what range of values of x is this formulation preferable to the straightforward method using the library function exp(..)? For the present purposes, however, it is enough to note that the series expansion provides the differenced solution and avoids the generated error: it also (incidentally) helps to show why the condition number $|x/[1 - \exp(-x)]|$ of $F(x)$ is almost equal to unity if x is small.

It will be realised that very few sequential calculations can be arranged in such a simple manner as to allow the propagation of error to be expressed as in (1.6.1). In general, any intermediate result will depend not merely on the immediately preceding result, but rather on some or all of the previous results. Moreover, most problems will depend on a number of input data, not just a single value, x_1. Such generalisations, however, in no way invalidate the general conclusions that emerge from the simplified model.

No matter how many input values – or independent variables – may affect a final result, it remains true that the *inherent error* of that result is *independent of the method or order of calculation*. In such a context, inherent error is the algebraic sum of the errors introduced into that result by the relative error in *each* of the input values. If one's sole interest is in generated error, it is therefore permissible to ignore the input errors of each independent variable and concentrate attention on the transmission of rounding error through the calculation. This, as we have seen, is certainly dependent on the manner of calculation – the algorithm – selected.

1.7 ERROR IN ARITHMETIC OPERATIONS

Looked at in detail, any computation is made up ultimately of the simple arithmetic operations $(+ - * /)$, combined perhaps with some logical discriminations, and this applies equally to the mysterious workings of such library functions as exp(..) as to overt problem-oriented programming. Generated error arises from the rounding errors of each component arithmetic operation of the vast number which make up any non-trivial calculation, and from their transmission through the tangled network of relationships defined by the program to the final result. The task of analysing propagation in such detail would usually be judged too difficult to be tried. The best that can reasonably be attempted is to study groups of operations which frequently occur as part of problem solving, and see what general rules of good practice can be inferred from these.

If we denote any of the operations $+ - *$ or $/$ for the moment by ω, and its performance by a machine by $\tilde{\omega}$, then the best we can hope for is that the machine representation of the result would be

$$n_0 = n_1 \tilde{\omega} n_2 = fl(n_1 \omega n_2) . \tag{1.7.1}$$

In other words, the machine result of (say) multiplying two normalised floating point numbers n_1 and n_2 would be the exact product of these numbers reduced to the required precision by the process of floating. This assumption is similar to that made in relation to (1.4.1), and may likewise prove to a greater or less extent optimistic, depending on the performance of the machine's arithmetic unit. Obviously, no machine can (for example) always perform an 'exact' division preparatory to floating, because the quotient may be a transcendental number. Whether or not it can form a 'sufficiently exact' value which 'almost always' agrees with (1.7.1) is the real issue — and a complicated one to argue.

If we suppose, as before, that the result n_0, and the operands n_1 and n_2, have relative errors $\beta_0 \epsilon$, $\beta_1 \epsilon$ and $\beta_2 \epsilon$ compared with their respective precise values x_0, x_1, and x_0, as in (1.5.1), then for instance in multiplication:

$$n_0 = x_0(1 - \beta_0 \epsilon) = n_1 \tilde{*} n_2 = (n_1 * n_2)(1 - \alpha \epsilon)$$
$$= [x_1(1 - \beta_1 \epsilon) * x_2(1 - \beta_2 \epsilon)](1 - \alpha \epsilon)$$
$$= x_1 x_2 (1 - \beta_1 \epsilon)(1 - \beta_2 \epsilon)(1 - \alpha \epsilon)$$
$$= x_0[1 - (\beta_1 + \beta_2 + \alpha)\epsilon + O(\epsilon^2)] .$$

In the last expression, the term $O(\epsilon^2)$ involves both second and third powers of ϵ; it will usually be quite negligible compared with unity if the machine unit is small, and so β_0 is identified as $(\beta_1 + \beta_2 + \alpha)$. The terms $\beta_1 \epsilon$ and $\beta_2 \epsilon$ then represent the transmitted relative errors of the two input operands (both transmitted with condition number 1), and $\alpha \epsilon$ is the rounding error of the operation.

A similar analysis of each of the four operations yields the following relations:

$$
\left.
\begin{aligned}
\beta_0 &= [x_1 \beta_1 + x_2 \beta_2 + (x_1 + x_2)\alpha]/(x_1 + x_2) && \text{(addition)}, \\
&= [x_1 \beta_1 - x_2 \beta_2 + (x_1 - x_2)\alpha]/(x_1 - x_2) && \text{(subtraction)}, \\
&= \beta_1 + \beta_2 + \alpha && \text{(multiplication)}, \\
&= \beta_1 - \beta_2 + \alpha && \text{(division)}
\end{aligned}
\right\} \quad (1.7.2)
$$

There is no real distinction between addition and subtraction, since the operands are signed numbers. However, the method of floating often depends on whether the magnitude of the result is the sum or the difference of the absolute values of the operands: if it is the sum, then chopping may be used, and should this be so, the difference will almost certainly be boosted.

The elementary operation which is ill-conditioned is an addition (or subtraction) which involves the difference between two nearly equal quantities. This of course we have already recognised as cancellation error, and its being the only ill-conditioned elementary operation explains why all forms of ill-conditioning ultimately can be identified (if we look closely enough) as cancellation error. On the other hand, it can be seen that the rounding error ($\alpha \epsilon$) of such differences would in fact be zero if the operands are floating-point numbers *with the same exponent*, as is of course likely if they are nearly equal in absolute value (e.g. $999 \times 10^e - 100 \times 10^e = 899 \times 10^e$, with no rounding necessary).

A repeated operation may conveniently be regarded as performed by iteration – that is, by repeated application of the same elementary rule. Thus, starting with the value $n_0 = n_1$, and performing the k assignments $n_0 := n_0 \tilde{\omega} n_{j+1}$ for $j = 1, 2, \ldots, k$ implies that $n_0 = n_1 \tilde{\omega} n_2 \tilde{\omega} n_3 \ldots \tilde{\omega} n_{k+1}$, which is the machine representation of $x_0 = x_1 \omega x_2 \omega x_3 \ldots \omega x_{k+1}$ for some particular operation ω. If ω is the additive operator $+$, then x_0 is the sum $(x_1 + x_2 + \ldots + x_{k+1})$, which can be more concisely represented by $\sum\limits_{j=1}^{k+1} x_j$. If ω is the multiplicative operator, then x_0 is the continued product $(x_1 x_2 \ldots x_{k+1})$, which can be represented as $\prod\limits_{j=1}^{k+1} x_j$. Since (1.7.2) provides the relationships for the elementary (iterated) operation $n_0 \tilde{\omega} n_j$, it is therefore possible to write down a relationship for the growth of the relative error in the result n_0.

For instance, consider the formation of a continued product of $k+1$ operands. The operation $n_0 := n_0 \overset{*}{\sim} n_{j+1}$ represents the jth multiplication of the sequence, and after it has been performed, the value of the relative error in n_0 becomes $\beta_0^{(j)} \epsilon$ (say). Referring to (1.7.2), we see that $\beta_0^{(j)}$ is the sum of three terms:

(i) the input relative error of the first operand, which is here n_0, and before the jth operation its relative error is $\beta_0^{(j-1)} \epsilon$;
(ii) the input relative error of the second operand, which is now not n_2 but n_{j+1}, so to be consistent with (1.5.1) we call this $\beta_{j+1} \epsilon$;
(iii) the rounding error of the operation, and since this will be different for each multiplication in the sequence, we call it $\alpha_j \epsilon$.

Hence we see that for each of the k multiplications:

$$\beta_0^{(j)} = \beta_0^{(j-1)} + \beta_{j+1} + \alpha_j, \quad (j = 1, 2, \ldots, k) .\tag{1.7.3}$$

This is a recursion formula describing the relative error $\beta_0^{(j)} \epsilon$ of each newly formed value of n_0 in terms of its previous value $\beta_0^{(j-1)} \epsilon$. Since n_0 is initially given the value n_1, it follows that before the first multiplication, $\beta_0^{(0)} = \beta_1$. This can be confirmed by placing $j = 1$ in (1.7.3), and noting that it is then the same as the relationship $\beta_0 = \beta_1 + \beta_2 + \alpha$ given in (1.7.1), but with the trivial replacements of $\beta_0^{(1)}$ for β_0, and α_1 for α. It may be shown by induction that (1.7.3) leads to the formula:

$$\begin{aligned} \beta_0 = \beta_0^{(k)} &= (\beta_1 + \beta_2 + \beta_3 + \ldots + \beta_{k+1}) + (\alpha_1 + \alpha_2 + \ldots + \alpha_k) \\ &= \sum_{j=1}^{k+1} \beta_j + \sum_{j=1}^{k} \alpha_j \end{aligned}\tag{1.7.4}$$

The sum of the β's represents the inherent error – that is, the transmitted error for all the $k+1$ input values $x_1, x_2, \ldots, x_{k+1}$ of the continued product. The sum of the α's (omitting the last one, α_k) is the generated error. Both components of error can be quite appreciable if there is a large number of terms, unless correct rounding is employed.

For instance, using decimal arithmetic of precision 3 (with the machine unit $\epsilon = 10^{-2}$), and forming the value of 1.23^{100} by the repeated multiplication of 1.23 by itself — not, incidentally, the best way of performing such an operation — we find the results shown in Table 1.2, depending on the method of rounding assumed. The exact result, correctly rounded to three decimal digits, should be 978×10^6. This table also lists the maximum bounds on the generated error, as determined by (1.3.8), but these bear no resemblance to the actual errors. On a probabilistic basis, if the α's in such a calculation were all independent of each other and pseudo-random values from a distribution with mean $\bar{\alpha}$ and standard deviation σ_α, then the value of $\sum_{j=1}^{k} \alpha_j$ in (1.7.4) would be drawn from a distribution with mean value $(k\bar{\alpha})$ and standard deviation $k^{\frac{1}{2}}\sigma_\alpha$. In particular, if the α-distribution were that suggested in section 1.4, values of $99\bar{\alpha}$ and $\sqrt{99}\sigma_\alpha$ can be extracted from Table 1.1, and these are also shown in Table 1.2. The actual generated plus final rounding errors for chopping (or boosting) are seen to be significantly larger than would be expected on this basis — but this is hardly surprising as, for one thing, the operands of the product are not chosen at random.

Table 1.2

Generated error in repeated multiplication

Method of rounding	Result (1.23^{100})	Relative error $(\beta_0 \epsilon)$	From Table 1.1		Max. possible error
			$99\bar{\alpha}\epsilon$	$\sqrt{99}\sigma_\alpha\epsilon$	
Chopping	805×10^6	18%	13%	1.3%	63%
Correct rounding	982×10^6	−0.4%	0	0.9%	±39%
Boosting	123×10^7	−26%	−13%	1.3%	−63%

The important deduction is that since $\bar{\alpha}$ is inevitably non-zero if chopping or boosting is used, the generated error will grow more or less in proportion to the number of multiplications. The effect is to cause **magnitude deflation** if chopping is in use, since the absolute value of the result progressively decreases as the computation proceeds.

On the other hand, if correct rounding is used, the probable generated error may have either sign, and is likely to grow only in proportion to the *square root* of the number of operations, because its magnitude is determined by the standard deviation $k^{\frac{1}{2}}\sigma_\alpha$, the mean $\bar{\alpha}$ being zero. (In fact, the mean value of $|\alpha|$ is $(2/k\pi)^{\frac{1}{2}}\sigma_\alpha$, if k is sufficiently large.)

Repeated summation is a much more common operation than repeated multiplication, and the relative error in forming $\sum_{j=1}^{k+1} x_j$ may be found by a similar argument to be given by

$$\beta_0 = \beta_0^{(k)} = (\sum_{j=1}^{k+1} x_j \beta_j + \sum_{j=1}^{k} s_{j+1}\alpha_j)/s_{k+1} \qquad (1.7.5)$$

where $s_t = \sum\limits_{j=1}^{t} x_j$ is the partial sum of the first t terms, so that s_{k+1} is the completed sum. The sum involving the β's is the inherent transmitted input error of the operands, and the other summation represents the generated error arising from the rounding errors $\alpha_j \epsilon$ of each intermediate summing operation, weighted according to the partial sum achieved by that operation, plus the rounding error $\alpha_k \epsilon$ of the final summing operation.

Note that it does not matter how many of the partial sums s_{j+1} are small compared with s_{k+1} — or even zero — implying the occurrence of earlier cancellation error during the formation of the sum. In fact, this would be beneficial, because the associated values of $s_{j+1}\alpha_j$ will be small, not only because s_{j+1} is small, but because α_j might well be zero (since $\alpha_j \epsilon$ is the rounding error in forming $s_j + x_{j+1} = s_{j+1}$, so that s_j and x_{j+1} may well have opposite signs but the same machine exponent). As has been remarked before, the insertion of an ill-conditioned stage *early* in a calculation is helpful, and the relative error of the total sum depends inevitably on the value of the last summand, x_{k+1}.

This is well-illustrated by considering the problem of the sum of a large number of more or less equal positive data with the value of each x_j relatively close to some non-zero mean value $\bar{x}$, say. Then s_j is approximately $j\bar{x}$, and the generated error of the total sum will be approximately $\sum\limits_{j=1}^{k-1} (j+1)\alpha_j \epsilon/(k+1)$, more or less independent therefore of the mean $\bar{x}$. If the α_j's were all random and independent, then this generated error would be randomly distributed with a mean value of about $k\alpha\epsilon/2$; if chopping is in use, there would be an appreciable magnitude deflation, just as in the evaluation of a repeated product. On the other hand, if it were possible to guess the value of $\bar{x}$ as ξ (say) before summing, even if ξ was only roughly equal to the true mean value, and the differences $(x_j - \xi)$ were then summed, the various partial sums $(s'_j$, say) would all be small compared with s_j, possibly even with fluctuating sign. The value $\sum\limits_{j=1}^{k} s'_{j+1}\alpha_j$ would likewise be small and possibly of either sign. If, finally, the value of $(k+1)\xi$ is added, so that the true sum $s_{k+1} = s'_{k+1} + (k+1)\xi$ of the x_j's is restored, then — because this last term is relatively large — the generated relative error $\sum\limits_{j=1}^{k} s'_{j+1}\alpha_j \epsilon/s_{k+1}$ would become small.

This use of a **provisional mean** (ξ) is another example of the benefit to be derived from a 'differenced solution', and is particularly significant in this context, because the generated (chopping) error $k\alpha\epsilon/2$ of a repeated sum of positive terms is large compared with the inherent error. Thus, returning to (1.7.5), if all the factors $\beta_1, \beta_2, \dots, \beta_{k+1}$ in the input errors of the summands are regarded as random and independent, and drawn from a distribution with mean $\bar{\beta}$, then the inherent error also has a frequency distribution with mean $\bar{\beta}\epsilon$. Since $\bar{\beta}\epsilon$ and $\bar{\alpha}\epsilon$ are likely to be comparable in value, the generated error would be about $k/2$ times as large as the inherent error.

By comparison, the generated error (due to chopping or boosting) of a repeated product will be found by a similar argument to be comparable to the inherent error, since the sum of the β's in (1.7.4) has a mean value equal to $(k + 1)\overline{\beta}$. In the context of total error, therefore, it is not likely to be anywhere near so dominating as that generated in summation. This method of comparing error is called **backward error analysis**.

Some strange and perhaps unexpected effects arise in repeated summation. Suppose, for example, that a decimal floating-point number of precision 3 were used as a counter, to produce $0, 1, 2, \ldots$ and so on quite precisely. Ultimately the counter records $999 \times 10^0, 100 \times 10^1$, but thereafter adding 1 to this result – whether correct rounding or chopping is used – would leave the counter unchanged. This can be termed accumulator **saturation**. If we were adding together values between (say) 0 and 2 whose mean value is 1, then of course long before 1000 of these had been added together, some of the significant digits of each summand would be lost – giving rise to an increasing generated error before the accumulator becomes completely 'saturated'.

This may seem an unlikely hazard when using machines of high precision: but is it? A meteorologist fed about 20000 observations of reduced sea-level barometric pressure into a machine – with values mostly between about 960 and 1020 mb – with the object of adding them together to find a mean, and so on. The results *looked* all right, but were no more than nonsense, because the machine had a precision of about 6 decimal digits. The accumulated sum had reached a value of about $100\,000 \times 10^2$ after ten thousand of these values had been read-in, and the reamining ten thousand were being chopped to either 900 or 1000 before addition to the accumulator. The mistake being at length discovered – somewhat fortunately, because how do you check such a large calculation? – the program was rearranged to accumulate in double-precision, and all then was well. Had this more precise arithmetic been unavailable, the large generated error could have been avoided by subtracting a provisional mean of 1000 mb (or some other *a priori* estimate) from each observation before it was accumulated.

On occasions, as for instance in this last example where numbers are being read-in, they can be sorted into some prearranged order before being summed. If there is such a choice available, it is best to sum in the order of increasing magnitude. Looking at (1.7.5) we see that it can be rearranged as the double sum

$$\beta_0 = \sum_{j=1}^{k+1} x_j (\beta_j + \sum_{i=j-1}^{k} \alpha_i)/s_{k+1} \tag{1.7.6}$$

with α_0 defined as zero; any particular x_j therefore multiplies the accumulated rounding error of all subsequent summations. The earlier values $x_1, x_2, x_3, \ldots$ consequently tend to have more effect on the total generated error than later ones. Similarly, in summing a series of decreasing terms, it is better (from this point of view) to start with the tail-end of the series and work backwards, rather than vice versa. But note that the ordering on this basis is entirely changed if differencing is applied to the data before summation; and note, too, that ordering is at best a *probable*, not an *inevitable*, benefit.

1.8 EXCEPTIONAL ARITHMETIC CONDITIONS

We have so far been concerned with the effects of the limited precision of normalised floating-point numbers. It is of less consequence, but nonetheless not to be forgotten, that the integer exponent e of the representation (1.2.1) has only a limited range. Most pocket calculators working in decimal arithmetic attempt to represent only numbers whose magnitude lies in the range less than 10^{100} and greater than or equal to 10^{-99}. Computers working in binary, octal, or hexadecimal usually provide a more limited range, perhaps from about 2^{-128} to 2^{128} (3×10^{-39} to 3×10^{38}) or thereabouts. Any attempt to form a number whose magnitude would imply an exponent value below the lower integer limit imposed by the machine is said to cause **underflow**; an attempt to form a number of too large a magnitude results in **overflow**.

It is an invariable rule that machines suspend execution upon detection of overflow. Some do the same if underflow occurs; others replace the number which would have caused underflow by zero, with or without issuing a warning to this effect. This replacement is often quite appropriate, and it is a pity perhaps that there is no conventional machine representation of $\pm\infty$ by which an over-flowed number could be represented so as to allow the calculation to proceed.

The exceptional condition of overflow should be distinguished from that of **zero-divide** which results from the division of any floating-point number by zero. The distinction might seem subtle, because zero-divide might well arise when trying to represent a large number, which circumstance could equally well lead to overflow. However, zero-divide also arises where zero is divided by zero, an obvious indeterminacy, but one which, if properly resolved (as, for example, in the limiting value of $\sin(x)/x$ as $x \to 0$), might well provide a value which by no stretch of imagination would be expected to cause overflow.

In a 'one-shot' calculation it may be that, for the particular problem data of interest, no such exceptional conditions are likely to occur. But in writing a program which is to be used many times over, it may seem unduly pernickety, but it is good practice to have in mind the possibility of any of these exceptional conditions arising, however unlikely they may seem. If nothing else, it will help to enforce a good habit which almost certainly will be rewarded sometime by the avoidance of a failure condition.

For instance, every time the division symbol is used, ask: is it guaranteed that the denominator is non-zero? It is not enough to *believe* that it will not be zero, because sometime later you will have forgotten that act of faith. Ask what will be the meaning of zero-divide: if the answer is truly that something is wrong, then leave it, because that is what will be observed in the computation. But

often it will occur in an indeterminacy, and frequently in the context of something like $[\exp(x)-1]/x$ where, as we have seen, the numerator also needs
some attention; such instances should always be the subject of a programmed
exception, which deals correctly with the indeterminate limit rather than being
left as a potential failure condition.

Overflow or underflow could occur anywhere in a calculation, and the best
that can be done is to be aware of the situations in which it is most likely to
occur. The library function $\exp(..)$ is perhaps the most likely situation, as it
overflows or underflows outside a quite modest range of argument values. For
instance the evaluation of $\exp(x)/[1+\exp(x)]$ should never be allowed to cause
overflow, because its standardised floating-point prepresentation would be unity
for all $x>\ln(2/\epsilon)$. Likewise, the equivalent expression $1/[1+\exp(-x)]$ will
overflow for quite modest negative values of x, but maybe its representation by
zero would then be good enough. If underflow is treated by the machine in use
as zero-replacement, this second expression is evidently the way the function
should be evaluated for positive arguments, whilst the first form is then the
appropriate method if x is negative. The operation of exponentiation (finding
the power of a number) is another likely source of overflow or underflow.
It could well happen that overflow occurs in evaluating something like $(x^2+1)^{\frac{1}{2}}$,
yet this could be replaced by $\mathrm{abs}(x)$ if $|x|>\epsilon^{-\frac{1}{2}}$ without any significant loss of
accuracy in the machine representation of the result.

1.9 REPRISE

This chapter has been concerned with sources of arithmetic inaccuracy. This, in
the main, depends on the machine treatment of floating-point arithmetic —
whether rounding, chopping, or what we called boosting is adopted by the
machine designer in preserving the limited precision of floating-point numbers.
We often do not know which is used — ideally, we should not have to know.

In fact, if the machine precision is high enough, only the longest and most
complicated computations may be seriously impaired by the inaccuracies of
limited precision. Many such problems (like the meteorologist's) in which it does
appear can be rendered sufficiently accurate by recourse to double-precision
arithmetic; and there is no harm in taking that way out of the difficulty (if it
is available). Much the most important reason for understanding the origins
of numerical inaccuracy is in order to recognise the situations in which double-
precision working might be justified as the easiest (or only) way out of
the difficulty. Certainly, the conditions of magnitude deflation (or inflation)
in repeated multiplications, or in summing numbers of the same sign, are such
instances on a machine which does not use correct rounding.

The occurrence of cancellation error is another danger point. But here it
may be the preceding calculation which needs to be conducted in high precision,
not the subtraction operation itself. On the other hand, although this form of

error certainly arises in the context of an ill-conditioned calculation, cancellation occurring early enough in a problem, or in association with subsequent well-conditioned stages of the computation, is no particular hazard — in fact sometimes a benefit.

We started with the intention of destroying one popular misconception about computers — that they are inevitably accurate. On the way, perhaps another mistaken belief has been disposed of — that their existence somehow obviates the need for numerate or algebraic skills. Enough has been said already (and more is to follow!) to make it clear that trouble-free computation often requires considerable forethought on the part of the programmer.

The chapter concluded with some cautionary notes on the limited range of arithmetic and the avoidance of exceptional conditions. In these matters, the experienced programmer exercises caution as an habituated response, even if no trouble is envisaged — much in the way that a car-driver may signal his direction to an empty road.

CHAPTER 2

Function evaluation: summing series and working with continued fractions

Chapter 1 was concerned with the numerical error; that is, the source of inaccuracy which can be reduced to as small a value as we please by somehow making the representation of numbers sufficiently precise. In practical terms, it is that source of inaccuracy which can be dramatically reduced by changing from single to double precision number representation and arithmetic, if it is available. Except that, even if it is, it might well be judged to be unduly wasteful of the machine resource to resort to such an option unless there were good reason.

The present chapter is intended to illustrate an entirely separate source of error – an error which would still exist even if the precision were infinite. It will be, moreover, a continuing theme played through the remaining chapters of the book, because it is only these two sources of inaccuracy which have to be considered. This second kind of error arises because, in all but very particular examples, it would be impossible to calculate a result with indefinitely high accuracy, simply because that would require infinitely many operations. It is, in other words, the error due to *algebraic* imprecision, as opposed to numerical imprecision. There is no point in seeking to calculate some result with remarkable algebraic accuracy if the lack of numerical precision destroys that accuracy. That would be merely wasteful of machine resource. But whereas the numerical precision is determined by the machine, the algebraic precision is governed by the user, which is why people write books about it. Yet both kinds of precision should be related, and the algebra – or more strictly the algorithm – tailored to suit the numerical environment or, where possible, 'elasticated' to adapt to any such environment.

It was said that there were two sources of inaccuracy: but in practice there is almost always a third. What is being worked out, however precisely it may apparently be formulated or calculated, will often be merely an approximation to what is really required. This is the **idealisation error** due to modelling the process in a mathematical form. It is present even in applications of the 'exact' sciences. Not knowing the application, there is little more that can be said about it. But the responsibility rests upon the individual who poses the problem also to formulate it in terms compatible with the accuracy required. To be told that a complicated calculation needs 'only two or three figure accuracy' may be useful knowledge, but it does not usually ease the task of writing the algorithm. At best it may shorten the run-time of the computation.

However, the existence of idealisation error certainly modifies the attitude to be taken in regard to algebraic inprecision. The machine environment must be seen as imposing a discernible and overriding limit on the calculation: it may happen that this limit does not need to be approached, but we do not, in general, know this to be so.

2.1 TRUNCATION ERROR

The algebraic limit of which we have spoken is called **truncation error**, which is the same term as that which was at one time frequently used to describe the numerical error caused by 'chopping'. It is as well that the terminology has changed, as it avoids the possible confusion. But it is understandable that the same term might be used, for it refers (in the algebraic context) to the error arising because any strictly infinite mathematical process has to be curtailed — truncated — if it is to be calculated. In the same way as (for example) the number $1/9 = (1/10) + (1/10)^2 + (1/10)^3 + \ldots = 0.\dot{1}$ needs to be truncated, so the calculation of the infinite series $x + x^2 + x^3 + \ldots$ must needs be stopped somewhere. Except, of course, that nobody should want to calculate this particular series term-by-term, as it is one of the few that can be summed and expressed in finite form, as $x/(1 - x)$.

This serves to dispose quickly of one obvious fact: that there is no point in introducing an infinite process, and truncation error, where the precise calculation can be expressed by a finite number of elementary arithmetic operations (involving only $+ - * /$). Instances where an infinite series can be summed in finite terms very often depend upon the binomial theorem in some guise or another, but it sometimes takes an experienced eye to discern this. There are other more obviously finite algebraic processes, such as solving a set of linear equations, or finding the mean or variance of a finite set of numbers; but finding a standard deviation (the square root of the variance) is, in general, an infinite process, as are all processes which evaluate the roots of non-linear equations. Functions which have no finite representation are called **transcendental**, and it is the calculation of such functions which is our present concern.

2.2 INFINITE SERIES

Different processes involve quite different mechanisms of truncation. It is the purpose of what follows to show how the truncation error involved in summing an infinite series affects the algorithm by which the sum is computed. An infinite series provides perhaps the most straightforward application of truncation – obviously one cannot calculate separately an infinity of terms – and moreover, series expansion often seems the most convenient way of expressing, and so evaluating, the transcendental functions that arise in applied mathematics.

The successive terms of a series will be denoted by $t_0, t_1, t_2, \ldots$ and all these terms will be supposed (in general) non-zero. In other words, although one or two may be non-zero, there is no denumerably infinite subset of the terms which are all zero. Quite often, it is convenient to treat t_0 as zero, and start with the term t_1. The **partial sum** of the first $(n+1)$ terms (or the first n terms if $t_0 = 0$) is then written as

$$S_n = \sum_{k=0}^{\infty} t_k = t_0 + t_1 + \ldots + t_n , \qquad (2.2.1)$$

and the infinite series is said to be **convergent** if the limit of S_n as $n \to \infty$ exists and is bounded. This limit is of course what is meant by the sum of the series, and its value can be denoted by S_∞. The part of the infinite series omitted from the partial sum is another infinite series, which we shall denote by τ_n, and which is called the **series tail** or remainder. Evidently, therefore:

$$S_\infty = S_n + \tau_n = S_n + \sum_{k=n-1}^{\infty} t_k = S_n + \sum_{k=0}^{\infty} t_{n+1+k} . \qquad (2.2.2)$$

The terms of the series will obey some systematic rule, and this will be such as to imply that $|t_k| \to 0$ if the infinite series converges. There will be a **term magnitude function** $t(k)$, say, having as its domain the set of non-negative integers $k = 0, 1, 2, \ldots$ such that $t(k) = |t_k|$; and almost invariably there will exist a 'natural' generalisation of this function to at least a limited domain of real numbers, $t(u)$, say, where u is real. Thus if the infinite series is, for example $\sum_{k=1}^{\infty} k^{-2}$ so that $t_k = k^{-2}$ (but t_0 is defined as zero) then $t(u) = u^{-2}$ for all $u \geq 1$. Terms whose expression involves the factorial function $k!$ might seem to be excepted, but the appropriate generalisation of the factorial function is the so-called **gamma function** $\Gamma(u)$ such that $\Gamma(k) = (k-1)!$ for all positive integers k, which also has a consistent definition for all real values of u.

Note that the generalisation of $t(k)$ is not unique. If $t_1(u)$ is one such generalisation, then $t_2(u) = t_1(u) + c \sin(N\pi u)$ is (for example) another for any real c, and any integer N, because $t_2(k) = t_1(k) = |t_k|$. But this is sheer perversity! In practice there is usually no doubt as to what the simplest generalisation might be.

If the partial sum of the term magnitudes $\sum_{k=0}^{n} |t_k|$ has a bounded limit as $n \to \infty$, the infinite series $\sum_{k=0}^{n} t_k$ is said to be **absolutely convergent**. Most of the series likely to be encountered are either composed entirely of positive terms (in which event, if convergent, they will necessarily be absolutely convergent), or they are **alternating series** for which $t_{k+1}/t_k < 0$, so that the signs of successive terms alternate. An alternating series may be convergent, but not necessarily absolutely convergent. The favourite example is the alternating series $\sum_{k=1}^{\infty}(-1)^{k+1}k^{-1}$ which converges to the value $\ln(2)$, whereas $\sum_{k=1}^{\infty}k^{-1}$ diverges: in fact, the partial sum $\sum_{k=1}^{n}k^{-1}$ varies in proportion to $\ln(n)$ as $n \to \infty$.

The ordering of the terms of an alternating series is only significant if it is *not* absolutely convergent. Otherwise we can take liberties with it. For instance, provided it converges absolutely, we can regard it as the difference between the sum of two component infinite series of positive terms $(t_0 + t_2 + t_4 + ...)$ and $(t_1 + t_3 + t_5 + ...)$. On the other hand, provided it is merely convergent – whether or not absolutely so – it can be rearranged as a **Liebnitz sum** of consecutive pairs of terms:

$$(t_0 + t_1) + (t_2 + t_3) + ... \; .$$

Since the terms must – at least at some point – consistently decrease in magnitude, so that $0 > (t_{k+1}/t_k) > -1$, the bracketed pairs become a series of positive terms. Thus, for example, the alternating series $\sum_{k=1}^{\infty}(-1)^{k+1}k^{-1}$ is in this way seen to be equal in sum to $\sum_{k=1}^{\infty}[2k(2k-1)]^{-1}$. However, a *partial* sum may of course always be rearranged in any order one chooses: algebraically there is no difference, although no doubt the numerical error in finite precision will be affected by the ordering. Despite the fact that reverse order summation – from the tail backwards – is likely to be more accurate, in practice partial sums are almost always formed in their natural order because the terms are usually more easily composed that way.

Another important class of infinite series are the power series whose general term $t_k = a_k x^k$, for $k = 0, 1, 2, ...$, where a_k is a numerical coefficient and x is a real variable. To emphasise the dependence on x, the partial sum is written as

$$S_n(x) = \sum_{k=0}^{n} a_k x^k \; , \tag{2.2.3}$$

and similarly $\tau_n(x)$ and $S_\infty(x)$ are the series tail and sum. The term magnitude function is

$$\left. \begin{aligned} t(k) &= a(k)\exp(-pk) \\ p &= \ln|1/x| \end{aligned} \right\} \tag{2.2.4}$$

where

and where $a(k) = |a_k|$ for non-negative integers k. A power series may alternate in sign as written, with x tacitly assumed to be positive, in which event the terms will all have the same sign if x is negative; or conversely it may appear to be the sum of positive terms, but in fact becomes an alternating series if x is negative.

For instance after division by x, the inifinte series (1.6.5) is

$$[\exp(x)-1]/x = 1 + \tfrac{1}{2}x + \tfrac{1}{6}x^2 + \ldots = \sum_{k=0}^{\infty} x^k/(k+1)! \quad (2.2.5)$$

and clearly this of the latter kind, alternating for $x < 0$. It is absolutely convergent for all x, and the magnitude function of its coefficients is $a(k) = 1/(k+1)!$ which generalises to $a(u) = 1/\Gamma(u+2)$ for real $u \geqslant 0$.

Some power series, like $\cos(t) = 1 - (t^2/2) + (t^4/24) - + \ldots$, are essentially alternating, and others like $\cosh(t) = 1 + (t^2/2) + (t^4/24) + \ldots$ are always composed of positive terms. However, written in this way they do not accord with the adopted convention that successive terms of a series shall all be non-zero: clearly all odd powers of t are omitted. The convention could be met by treating them as expansions in $x = t^2$, with the value of x restricted to non-negative values. But in fact the series $1 + (x/2) + (x^2/24) + \ldots$ then represents the value of $\cosh(\sqrt{x})$ if $x \geqslant 0$, whereas it becomes the alternating series for $\cos(\sqrt{-x})$ if $x \leqslant 0$: which gives us two series for the price of one! Similarly, the power series for $\sinh(t) = t + (t^3/6) + (t^5/120) + \ldots$ can be interpreted as the series $1 + (x/6) + (x^2/120) + \ldots$ for $t^{-1}\sinh(t)$ if $x = t^2$ is non-negative, which also equals the series for $t^{-1}\sin(t)$ if $x = -t^2$ is non-positive. The vast majority of power series likely to be encountered in practice can be brought into line with the convention of generally non-vanishing consecutive terms by a similar substitution.

2.3 THE CONTROL OF ERROR IN SERIES SUMMATION

If we denote the *precise* estimated value of the sum of a series by S_e, then the truncation error is E, say, where the true sum is

$$S_\infty = S_e + E . \tag{2.3.1}$$

The simplest way to provide a truncated sum is to choose S_e as the partial sum S_n (with some particular known value of n). In this event, it follows from (2.2.2) and (2.3.1), that the truncation error is simply the value of the (neglected) tail of the series; and so $E \equiv E_n = \tau_n$, where the subscript n is added to E to denote its dependence on the number of terms included in the partial sum.

Because S_e is evaluated with finite precision, what is calculated – however S_e may be formulated – is, say,

$$\tilde{S}_e = fl(S_e) = S_e(1 - \beta\epsilon) \tag{2.3.2}$$

where $\beta\epsilon$ is the relative error due to the effects of finite precision, which can be categorised into components of inherent, generated, and final rounding error. Thus from (2.3.1) in (2.3.2)

$$\left.\begin{aligned} \tilde{S}_e &= S_\infty(1 - \beta\epsilon)(1 - E/S_\infty) \\ &\simeq S_\infty(1 - \beta\epsilon - E/S_\infty) \end{aligned}\right\}, \tag{2.3.3}$$

the second approximate relationship being adequate if $\beta\epsilon$ and the relative truncation error (E/S_∞) are both small.

As has already been remarked, there is no point in making the truncation error smaller than the numerical error, so somehow the relative values of E/S_∞ and $\beta\epsilon$ have to be assessed, in at least a rough and ready way, to ensure that the selected formulation of S_e is neither too coarse, nor yet so accurate as to be unduly wasteful of computation time. Since the assessment is rough, it would usually be preferred to ensure that E/S_∞ is unlikely to exceed $\beta\epsilon$, even if this implies some unnecessary computation. In other words, if S_e were simply a partial sum S_n, and there is doubt about the appropriate value of n, it would be chosen as larger, rather than smaller. This means that the choice would be based on a pessimistic *over*-estimate of E/S_∞, but an *under*-estimate of $\beta\epsilon$.

There are three quantities involved in this assessment: E, S_∞, and $\beta\epsilon$, none of which can be exactly known. At the point of the calculation at which the decision is made, there is of course an acceptable approximation available to S_∞ in the shape of $\widetilde{S}_e$ or $\widetilde{S}_n$, so it can be accepted that it is adequate to compare $E/\widetilde{S}_e$ (rather than E/S_∞) with $\beta\epsilon$. Ways of providing an upper bound on E will be discussed in what immediately follows, but what about the numerical error, $\beta\epsilon$? There would be some hope maybe of providing an *upper* bound on this — with considerable difficulty maybe — but this is not what is wanted. A lower bound to $|\beta\epsilon|$ is zero — which is a possible, if not a probable, value — but this is no help. A 'probable' value of numerical error would be even more difficult to assess: so what can be done?

It seems best to select some known component of $\beta\epsilon$ and ensure that $E/\widetilde{S}_e$ does not greatly exceed some likely value of this. A suitable component is the rounding error, which is an inescapable concomitant of storing the result $\widetilde{S}_e$ in the machine. An 'expected' value $\overline{\alpha}\epsilon$, equal (say) to $\epsilon/(b+1)$, could be pre-computed, but this has the disadvantage that the algorithm embodying the discrimination of a suitable criterion of truncation becomes machine-dependent. It would seem better to use instead a criterion such as

$$fl(\widetilde{S}_e + E_{\max}) = \widetilde{S}_e \qquad (2.3.4)$$

as the condition for truncation: or in a programming context some coding such as

if *sum* + *err* = *sum* **then goto** *SUMMED*

where *sum* represents the current estimate of $\widetilde{S}_e$, and *err* provides the suitably pessimistic overestimate $E_{\max}$ of $|E|$. This addition (*sum* + *err*) would be evaluated as equal to *sum* only if *err* were of such a small size compared with $|S_e|$ as to be comparable with its rounding error $|S_e|\alpha\epsilon$, so that (2.3.4) implies that $|E/\widetilde{S}_e| < \alpha\epsilon$ for some possible α. This can be termed truncation by **rounding error discrimination**, and (2.3.4) is an **adaptive** use of this technique, because it modifies itself to suit the machine environment, whereas (say)

$$|E_{\max}| < |\tilde{S}_e|\bar{\alpha}\epsilon \tag{2.3.5}$$

is non-adaptive because it would need pre-calculation of the number $\bar{\alpha}\epsilon$ for the machine in use.

If it is a power series which is being summed, then alternatively the inherent error could be used, which will be large if the series evaluation is ill-conditioned. The condition number of a power series is given by

$$\phi = xS'_\infty(x)/S_\infty(x) = (\sum_{k=1}^{\infty} ka_k x^k)/S_\infty(x) \ . \tag{2.3.6}$$

A partial sum which provides an estimate D_e (say) of $xS'_\infty(x)$ could be computed along with the partial sum of $S_\infty(x)$, so that ϕ could be estimated (as D_e/S_e) with considerable accuracy. Then an **inherent error discrimination** based on

$$|E_{\max}| < \max(|D_e|, |S_e|)\bar{\alpha}\epsilon \tag{2.3.7}$$

would ensure that rounding-error discrimination is used if $|\phi| < 1$ (that is, if $|D_e| < |S_e|$), or else that the value of $|E/S_e|$ is less than $|\phi|\bar{\alpha}\epsilon$, which would be the inherent error arising from a rounding error $\bar{\alpha}\epsilon$ in the value of x. An adaptive form of (2.3.7), like (2.3.4), can be similarly produced. However, this method of discrimination is rarely if ever adopted: it is usually not at all clear whether it really saves computational time (which is all that is at stake) since, although it may lead to an earlier truncation, the need to sum the value of D_e implies extra computation.

We turn attention now to the methods of evaluating truncation error. What is to be looked for is some upper bound $E_{\max}$, say, where it can be asserted that $0 < E < E_{\max}$, or perhaps $|E| < E_{\max}$, in the expectation that the discrimination of the condition for truncation will ensure that $E_{\max}$, and therefore E, is small. Furthermore, as a start, it will be assumed that the partial sum S_n is to be taken as providing the estimate S_e, and we look first of all at the problem of the accurate summation of series of alternating terms.

2.3.1 Alternating series

Suppose that for all $k > K$, the value of t_{k+1}/t_k lies in the interval $(-1, 0)$, so that the magnitude of successive terms approaches zero monotonically. Then, if $n \geqslant K$, the series tail τ_n has the same sign as the first neglected term, t_{n+1}. This may be clear from Fig. 2.1. It stems from the fact that the Liebnitz sum of the tail

$$(t_{n+1} + t_{n+2}) + (t_{n+3} + t_{n+4})$$

is composed of (the bracketed) terms which all have the same sign, which is also the same sign as that of the first term of each bracketed pair. Hence τ_n and τ_{n+1}

have opposite signs – in fact, the signs of t_{n+1} and t_{n+2} respectively. But this implies that one of S_n, S_{n+1} is too large, the other too small: consequently S_∞ must lie between S_n and S_{n+1} in value, and so with $0 < \mu < 1$

$$S_\infty = S_n + \mu(S_{n+1} - S_n) = S_n + \mu t_{n+1} .$$

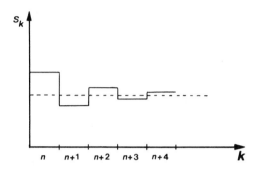

Fig 2.1 – Partial sums of an alternating series.

But, as in (2.2.2), $S_\infty = S_n + \tau_n$, so that $\tau_n = \mu t_{n+1}$; or in other words, if $n \geqslant K$, then

$$0 < (\tau_n / t_{n+1}) < 1 . \tag{2.3.8}$$

Clearly then if an alternating infinite series is to be estimated by a partial sum, and adaptive rounding error discrimination is employed as the criterion of truncation, the value of E_{max} may be replaced in (2.3.4) simply by the next term due to be summed (t_{n+1}). Accordingly, truncation takes place for such a value of n that

$$fl(\tilde{S}_n + \tilde{t}_{n+1}) = \tilde{S}_n \Big\}$$
that is, $$\tilde{S}_{n+1} = \tilde{S}_n \Big\}. \tag{2.3.9}$$

Thus, summing proceeds until it fails any longer to affect the value of the summand, which is therefore saturated. This very convenient technique may be called **summing to saturation**. In a programming context, the previous value of *sum* is temporarily preserved (as *oldsum*, say) and the discriminating statement is

          if *oldsum* = *sum* **then goto** *SUMMED* ; .

Strictly, there may also need to be a test to discriminate that the terms are decreasing in magnitude (i.e., $n > K$). But ulitmately this *must* happen, provided at least that the alternating series converges – and of course it cannot otherwise be summed.

For instance, if x is negative, the power series (2.2.5) is alternating with monotonically decreasing terms for all $0 > (t_{k+1}/t_k) = x/(k+2) > -1$; that is, for all $k > -(x+2)$. If $x = -2$, the first two terms $t_0 = 1$, $t_1 = x/2$ are equal and opposite, but the rest successively decrease in magnitude (and alternate in sign). The series sums to $0.432 \dots$ for this value of x, and (for example) $t_{20} \simeq 2 \times 10^{-14}$, so that 21 terms are sufficient to saturate a summand having a precision of 13 decimal digits (or 43 binary digits).

If the first several terms increase in magnitude, so that $K \gg 1$, this should be taken as a danger signal, for it suggests that terms which are large compared with the final sum are being differenced, and a considerable generated error due to cancellation may arise (see section 2.4).

2.3.2 Taylor's series

Turning attention to series of positive terms, it would appear that if a power series is to be summed, the remainder term of a Taylor series should at once provide bounds upon the magnitude of the tail of a partial sum (that is, of a Taylor polynomial). Thus, using the Lagrange form of this remainder,

$$S_\infty(x) = \sum_{k=0}^{n} x^k S_\infty^{(k)}(0)/k! + \tau_n(x)$$

$$\left. \begin{array}{l} \\ \\ \end{array} \right\} \quad (2.3.10)$$

where $\quad \tau_n(x) = x^{n+1} S_\infty^{(n+1)}(\mu x)/(n+1)!, \quad (0 < \mu < 1)$

Here $S_\infty^{(k)}(\xi)$ is the kth derivative of $S_\infty(x)$ with respect to x at $x = \xi$. If, for instance, $x > 0$, then the fact that the terms are positive implies that $S_\infty^{(k)}(0) > 0$ for all k, so that all the derivatives of $S_\infty^{(n+1)}(\xi)$ at $\xi = 0$ are also positive. Hence, $S_\infty^{(n+1)}(\mu x)$ increases with μ, and bounds upon the value of $\tau_n(x)$ are consequently provided by taking $\mu = 0$ and $\mu = 1$ in (2.3.10), so that

$$t_{n+1} < \tau_n(x) < S_\infty^{(n+1)}(x) x^{n+1}/(n+1)! \ . \qquad (2.3.11)$$

The same inequality also applies if $x < 0$, provided that the terms are all positive.

Now whereas it is easy enough to infer the values of $S^{(k)}(0) = a_k k!$ from the supposedly known terms of the power series $\sum_{k=0}^{\infty} a_k x^k$, the evaluation of $S_\infty^{(n+1)}(x)$ for some general value of x is another matter entirely. It is usually more complicated a task than that of evaluating $S_\infty(x)$ itself. Even where this is not so, the upper bound provided by (2.3.10) often turns out to be very pessimistic.

For example, the exponential series is composed of positive terms if $x > 0$, and moreover every derivative of $\exp(x)$ is equal to $\exp(x)$. Thus the tail of the exponential series is

$$[\exp(x) - \sum_{k=0}^{n} x^k/k!] < \exp(x) x^{n+1}/(n+1)!$$

if $x > 0$. Substituting this inequality in the power series (2.2.5) for

$[\exp(x) - 1]/x$, it follows that its tail, starting with the term t_n, is

$$\tau_{n-1}(x) < \exp(x) x^n/(n + 1)! \ .$$

With some rearrangement, using the fact that $\tau_{n-1} = \tau_n + t_n$, the following inequality is established:

$$\tau_n(x) < (n + 2) S_\infty(x) t_{n+1} \ .$$

Since $S_\infty(x)$ is greater than 1 for $x > 0$, this upper bound is more than $(n + 2)$ times the first neglected term t_{n+1}. But the next method to be described will show that the tail of the series is only slightly larger than t_{n+1}. A great deal of ingenuity has been wasted! In truth, despite its apparent promise, this approach is rarely of any practical use.

2.3.3 Comparison series
A more promising approach to the assessment of an upper bound for the tail of a series of positive terms is provided by the idea that there may exist an easily summable infinite series – such as the geometric series – which is term-by-term larger than the tail, so that this comparison series provides a bound on its value.

This is best shown by example. Consider again the tail $\tau_n(x)$ of the power series (2.2.5). This is $\tau_n(x) = \sum_{k=n+1}^{\infty} x^k/(k + 1)!$, so that

$$\tau_n(x)/t_{n+1} = 1 + [x/(n + 3)] + [x^2/(n + 3)(n + 4)] + \ldots$$
$$= (n + 2)! \sum_{k=0}^{\infty} x^k/(n + k + 2)! \ .$$

Assuming that $x > 0$, so that all the terms are positive:

$$1 < \tau_n(x)/t_{n+1} < 1 + [x/(n + 3)] + [x/(n + 3)]^2 + \ldots$$
$$= \sum_{k=0}^{\infty} x^k/(n + 3)^k \ .$$

Using the result that $\sum_{k=0}^{\infty} x^k = (1 - x)^{-1}$ it follows that

$$1 < \tau_n(x)/t_{n+1} < (n + 3)/(n + 3 - x) \ .$$

The ratio of successive terms is $t_{k+1}/t_k = x/(k + 1)$, so that they do not decrease until $k \geq x$, and in practice the partial sum S_n would not approximate S_∞ unless n is considerably larger than x. This implies that the tail is barely more than the first neglected term, and accordingly summation should proceed until the summand is saturated. For example, with $x = 2$, the series sums to $3.19 \ldots$, and since $t_{19} \simeq 2 \times 10^{-13}$, 19 terms are sufficient to saturate a summand having a precision of 13 decimal digits. Since $\tau_{18}(2) < (21/19) t_{19}$, the difference between the value of the tail and the first neglected term is quite unimportant.

This series is of course quite rapidly convergent. A slowly convergent series of positive terms, would, on the other hand, have a tail τ_n which is large in value compared with the first neglected term, t_{n+1}. Even if such a series were summed to saturation, the truncation error would therefore still be large compared with the rounding error.

In talking of a series of positive terms being 'summed to saturation' – implying that $\tilde{S}_n$ and $\tilde{S}_{n+1}$ are equal – the very unlikely possibility that the machine *boosts* the result of any addition has been disregarded. Saturation in summing such a series occurs only if chopping or correct rounding is in use. Were boosting to be used, *any* positive increment to $\tilde{S}_n$, however small, would cause rounding-up so that $\tilde{S}_{n+1} > \tilde{S}_n$.

It might be possible to find a summable series which is term-by-term *smaller* than that of the tail (but larger than t_{n+1}), in which event it provides a lower bound on τ_n which, if added to S_n, provides an improved estimate of S_∞ whose truncation error is then the difference between the upper and lower bounds on τ_n. This same idea of using a double-sided bound on τ_n to provide an assessment of τ_n without direct summation is exploited in the next method to be described.

2.3.4 An integral approximation

If the series to be summed consists for all $k > K$ (say) of positive terms t_k whose term magnitude function $t(u)$ is a continuous and monotonically decreasing function of the real variable u for $u \geqslant K + 1$, then provided that $n \geqslant K$,

$$\int_{n+1}^{\infty} t(u)\,\mathrm{d}u < \tau_n < t_{n+1} + \int_{n+1}^{\infty} t(u)\,\mathrm{d}u \ . \qquad (2.3.12)$$

For a proof, see Fig. 2.2. Thus if we estimate the sum of the series by

$$S_e = S_n + \int_{n+1}^{\infty} t(u)\,\mathrm{d}u \qquad (2.3.13)$$

the integral is an estimate of the tail of the partial sum S_n, and from (2.3.1) the truncation error is now the error in this estimate: namely,

$$E_n = S_\infty - S_e = \tau_n - \int_{n+1}^{\infty} t(u)\,\mathrm{d}u \ . \qquad (2.3.14)$$

Accordingly (2.3.12) shows that

$$0 < E_n < t_{n+1} \ . \qquad (2.3.15)$$

This implies, in the same way as does (2.3.8), that it is adequate to sum the series to saturation, no matter how slowly it converges, *provided that* the integral in (2.3.13) is added to the saturated partial sum to compensate for the omitted tail. If the series happens to be rapidly convergent, then (presumably) the integral will have little or no significant effect upon the value of S_e.

Consider, for example, the slowly convergent series $S_\infty = \sum_{k=1}^{\infty} k^{-2}$ which is composed of positive terms decreasing successively in magnitude for all $k > 0$. Working, say, to a precision of 3 decimal digits the sum will be found to saturate after adding 15 terms (if rounding is used) with $S_{15} = 1.58$, and since we can take $t(u)$ to be the monotonically decreasing function u^{-2}, the integral term

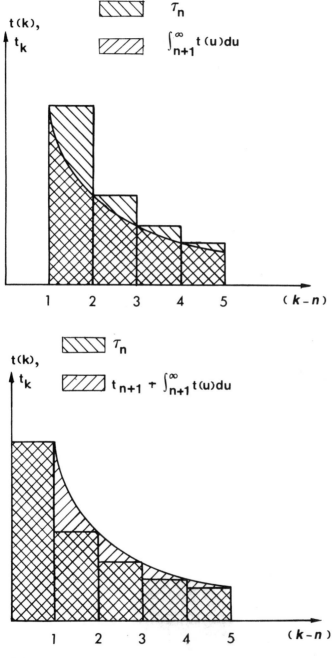

Fig. 2.2 – Approximation of series tail τ_n by relation (2.3.12). Block diagram of t_k and graph of $t(k)$.

in (2.3.13) is correspondingly $\int_{16}^{\infty} u^{-2} du = [-u^{-1}]_{16}^{\infty} = 0.0625$. This provides the estimate $S_e = 1.64$, and from (2.3.15) the error is between zero and $t_{16} = 0.00391$. In fact, the true value is $\pi^2/6 = 1.64493\ldots$ and S_e is therefore the correct sum rounded to 3 decimal digits. But to have achieved this result by the evaluation of a partial sum alone (without the integral approximation to the tail) would have needed reversed summation, or accumulation with enhanced precision, of over 200 terms of the series.

However, it is usually possible to provide an even closer approximation to the series tail. If the curvature of the graph of $t(u)$ versus u is of one sign (i.e., if $t''(u) > 0$) for all $u > n + \frac{1}{2}$, so that $t'(u)$ decreases monotonically in magnitude for all $u > n + \frac{1}{2}$, then as may be clear from the construction of Fig. 2.3 it follows that

$$(t_{n+1}/2) + \int_{n+1}^{\infty} t(u) du < \tau_n < \int_{n+\frac{1}{2}}^{\infty} t(u) du \ , \tag{2.3.16}$$

and these two bounds are usually much tighter than those of (2.3.12). Their difference is

$$\int_{n+\frac{1}{2}}^{n+1} t(u) du - (t_{n+1}/2) = \int_{n+\frac{1}{2}}^{n+1} [t(u) - t(n+1) du$$
$$= \int_{n+\frac{1}{2}}^{n+1} (u - n - \frac{1}{2}) [-t'(u)] du \ ,$$

where the integral involving $t'(u)$ is obtained by an integration by parts. Over the range of the integral from $u = n + \frac{1}{2}$ to $n + 1$, the value of $(u - n - \frac{1}{2})$ is between 0 and $\frac{1}{2}$ and $[-t'(u)]$ is also non-negative and largest at $u = n + \frac{1}{2}$. Thus the difference between the bounds of (2.3.16) is less than

$$-t'(n + \frac{1}{2}) \int_{n+\frac{1}{2}}^{n+1} (u - n - \frac{1}{2}) du = -t'(n + \frac{1}{2})/8 \ ,$$

and this therefore is a limit on the inaccuracy of either the upper or lower bound of (2.3.16), regarded as an estimate of the value of the series tail, τ_n.

In fact the upper bound of (2.3.16) provides a somewhat simpler expression, as well as a rather closer estimate (as might be suggested by the appearance of Fig. 2.3). Thus if the series sum is estimated by

$$S_e = S_n + \int_{n+\frac{1}{2}}^{\infty} t(u) du \tag{2.3.17}$$

in place of (2.3.12), the truncation error E_n of this overestimate is bounded by

$$0 > E_n > t'(n + \frac{1}{2})/8 \tag{2.3.18},$$

provided that $t''(u) > 0$ for all $u > n + \frac{1}{2}$.

Applying this estimate to the slowly convergent series $\sum_{k=1}^{\infty} k^{-2}$, and truncating at $n = 15$ as before, the value of $t'(n + \frac{1}{2})/8$ equals $-2/31^3$, and since $t''(u) = 6/u^4$ is essentially positive, the truncation error predicted by (2.3.18) is between 0 and -0.00007. This suggests that S_e evaluated as in (2.3.17) would be accurate to nearly 5 significant decimal digits. In fact, with $n = 15$, it evaluates as $1.64499\ldots$ instead of the correct value of $1.64493\ldots$. Such an accuracy could only be achieved by summing more than *fifteen-thousand* terms of the original series in enhanced precision!

The snag of this otherwise highly promising method is that it may be diffi-cult to evaluate the integral of the term magnitude function. This is obviously so in regard to series whose terms involve the factorial function. Likewise it is generally difficult for power series, which have a term magnitude function given

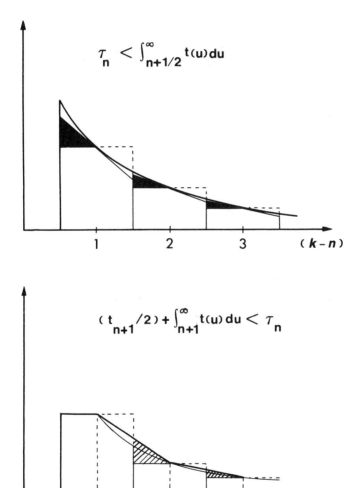

Fig. 2.3 – Approximation of series tail τ_n by relation (2.3.16). Block diagram of t_k and graph of $t(k)$.

In both diagrams, the shaded triangles are equal in area to the cut-off portions of the blocks.

by (2.2.4): integrating this and changing the variable of integration from u to $v = u - n - 1$,

$$\left. \begin{aligned} \int_{n+1}^{\infty} t(u)\,du &= \int_{n+1}^{\infty} a(u)\exp(-pu)\,du \\ &= x^{n+1}\int_{0}^{\infty} \exp(-pv)a(v+n+1)\,dv \end{aligned} \right\} \qquad (2.3.19)$$

where $p = \ln|1/x|$, and integrals of this latter form are called **Laplace transforms** of the function $a(..)$. These are listed in various reference books, but most are not easy to evaluate numerically.

One way round this difficulty is to look for functions $t_1(u)$ and $t_2(u)$ which are easy to integrate and which obey the inequalities $t_1(u) \leqslant t(u)$ and $t_2(u) \geqslant t(u)$, for all $u \geqslant (n+1)$. Then a weaker form of (2.3.16) is

$$(t_{n+1}/2) + \int_{n+1}^{\infty} t_1(u)\,du < \tau_n < \int_{n+\frac{1}{2}}^{\infty} t_2(u)\,du \qquad (2.3.20)$$

which by a similar line of reasoning can lead to an estimate with a truncation error larger than that given by (2.3.18), but still perhaps smaller in magnitude than t_{n+1}.

But it is usually preferable to seek approximate methods of evaluating the integral of $t(u)$, for example by the use of numerical quadrature, or series expansion. For example, when summing the series $\sum_{k=0}^{\infty} (k^5 + 1)^{-1}$, although there is no explicit formula for the integral of $t(u) = (u^5 + 1)^{-1}$, this function can be expanded as a series in inverse powers of u within the integrand and integrated term by term to give:

$$\begin{aligned} \int_{v}^{\infty} t(u)\,du &= \int_{v}^{\infty} (u^5 + 1)^{-1}\,du = \int_{v}^{\infty} u^{-5}(1+u^5)^{-1}\,du \\ &= \int_{v}^{\infty} u^{-5}(1 - u^{-5} + u^{-10} - + \ldots)\,du \\ &= (v^{-4}/4) - (v^{-9}/9) + (v^{-14}/14) - + \ldots \; . \qquad (2.3.21) \end{aligned}$$

At first sight it might appear that all that has been achieved is the replacement of one infinite series (the tail) by another (representing the integral approximation to the tail). But the latter may converge much more rapidly than the former. Thus supposing that $\sum_{k=0}^{\infty} (k^5 + 1)^{-1}$ is to be evaluated correctly to 10 significant decimal digits using (2.3.17), then (2.3.18) suggests that the partial sum S_{30} should be adequate when added to the integral given by (2.3.21) with $v = n + \frac{1}{2} = 30.5$. But working to this precision, only the first term of the series representation of this integral contributes significantly to the value of S_e.

The method as applied to series of positive terms is obviously so successful a means of avoiding extensive summation as to raise the question whether something like it may be applicable to *alternating* series. It would be possible, of course, to represent an alternating series as a Liebnitz sum of positive terms, in which event the method is obviously applicable. Alternatively the terms of an absolutely convergent alternating series could be rearranged as

$$S_\infty = \sum_{k=0}^{\infty} (-1)^k |t_k| = \sum_{k=0}^{\infty} |t_k| - 2 \sum_{k=0}^{\infty} |t_{2k+1}|$$

and the bounds (2.3.16) applied, with appropriate modifications, to the values of the tail of both these component series of positive terms. It can then be shown that the estimate

$$S_e = S_n + t_{n+1}/2 \tag{2.3.22}$$

has a truncation error E_n bounded by

$$0 < \mathrm{sgn}(t_{n+1})E_n < -t'(n+1)/2 \tag{2.3.23}$$

provided that $t''(u) > 0$ for $u > n + 1$. Here sgn(..) is the *signum* function which takes the value ± 1 depending on the sign of its argument, or the value zero if the argument is zero (which of course does not apply here).

This result is certainly generally a considerable improvement on the technique of accepting the saturated partial sum of the alternating series as an estimate (as described in section 2.3.1). Moreover the integral of $t(u)$ has, very conveniently, disappeared. But, by the same token, this result can be deduced much more directly without referring to the concept of an integral approximation to the tail.

It needs only to be observed that the estimate (2.3.22) is simply the partial sum of the first $(n + 1)$ terms of the **averaged series** $\sum_{k=0}^{\infty} (t_{n-1} + t_n)/2$, where we define t_{-1} as zero. If t_k decreases in magnitude for all $k \geqslant n + 1$, then the tail of the averaged series is composed of alternating terms, and so the truncation error of (2.3.22) must be less than the first omitted term of the averaged series: or, in other words,

$$0 < \mathrm{sgn}(t_{n+1})E_n < |t_{n+1} + t_{n+2}|/2 \ . \tag{2.3.24}$$

This implies the somewhat weaker bound of (2.3.23), because if $t''(u) > 0$, then $-t'(n+1) > [t(n+1) - t(n+2)]$. Also, since $|t_{n+1} + t_{n+2}|/2$ is less than $|t_{n+1}|$, the averaged series converges faster than the original alternating series.

The averaged series may itself be averaged, and this recursive technique of **repeated averaging** is often successful as applied to the acceleration of convergence of slowly-convergent alternating series (see problem 5 of section 2.3.1); moreover, it can be conducted numerically without any need for preparatory algebra.

2.4 GENERATED ERROR IN SERIES SUMMATION

There are available, therefore, methods which will usually enable the sum of a slowly-converging series, whether composed of positive or alternating terms, to be evaluated without extending term-by-term summation even so far as would be necessary to saturate the summand, let alone beyond it (with the attendant need for working in enhanced precision). Such avoidance of extensive summation is not merely a convenience simply because it reduces what might otherwise be time-consuming computation. It is also essential if a considerable generated error in the resulting sum is to be avoided. We know that generated error grows in summation with the number of summing operations. It may happen as well that the successive terms of the series are formed recursively by repeated operations, and this also introduces growing error. The rate of growth of error can be particularly serious, and is in direct proportion to the number of operations involved, if the machine arithmetic employs chopping as a means of adjusting precision rather then correct rounding.

Where there appears no alternative to the term-by-term summation of a slowly-converging series of positive terms beyond the point of summand saturation, it is essential either to sum in reversed order — using an *a priori* truncation condition based on (2.3.5), with perhaps a partial sum of the first few terms replacing S_e — or better still, to sum using a double-precision accumulator (to achieve a correctly rounded single-precision result).

However, where double-precision is not available it is possible to employ two single-precision accumulators whereby a second register (*spill*) picks up the pieces, as it were, by accumulating that portion of each term which has not been included in the primary accumulator (*sum*). If it is assumed that *term* has been computed, the sequence of instructions within the iterated loop performing summation then reads:

$$oldsum : = sum ;$$

$$sum : = sum + term ;$$

$$spill : = ((oldsum - sum) + term) + spill \ ;$$

and (*sum* − *oldsum*) here represents how *term* has been stored in *sum*. A single-precision sum is then recovered, when required, by computing (*sum* + *spill*), assuming both are initialised as zero.

A better and more subtle alternative is **Kahan's device**, by which these loop instructions are rearranged as

$$spill : = ((oldsum - sum) + spill) + term ;$$

$$oldsum : = sum ;$$

$$sum : = sum + spill ; .$$

With *oldsum, sum,* and *spill* all initialised as zero, the effect of accumulating the current value of *spill* (rather than of *term* which is here added to *spill*) is to prevent *spill* becoming so large that it too becomes saturated. Moreover, the value of *sum* needs no subsequent correction after leaving the iterated loop.

In summing an alternating series, if the initial terms and partial sums happen to be large compared with the final sum then, as is clear from (1.7.5), a relatively large generated error can result, due to subsequent cancellation. The effect is called 'smearing' (a U.S. slang equivalent of Eng. 'clobbering'). Because successive partial sums provide opposite bounds upon the value of the final sum, this error can usually only occur if the term magnitude first increases before ultimately decreasing. For example, in evaluating the series for $\exp(x)$ at a value of $x \cong -10$, or thereabouts, the sum is known to be about 4.5×10^{-5}, but the terms t_9 and t_{10} are opposite in sign, and both (approximately) equal in magnitude to $10^9/9! \cong 2700$: any relative rounding error in forming these terms is destined to appear multiplied by a factor approaching 10^8 in the relative error of the final result. Of course they may correctly and precisely cancel each other out if x is precisely 10: but in any event, other terms on either side of t_9 and t_{10} are going to have almost as large an effect. This overwhelming error is in no way alleviated by accumulating the result in double-precision.

Since $\exp(-10)$ is not a conspiciously ill-conditioned calculation, it is clear that this is generated error which one ought to be able to avoid. And of course if it were really the exponential function which was being calculated there would be a number of ways of avoiding the need to evaluate this series, by taking advantage instead of the function's special properties. However, other transcendental functions may exhibit the same problem, but not so easily provide a means of escape. The Bessel function $J_n(x)$ forms a case in point, and we shall face up to the difficulties imposed by this evaluation later on (in section 2.10). Some alternative method of formulation has to be sought, which avoids the use of the problem series.

2.5 SEMI-CONVERGENT SERIES

Despite these dire consequences of slow-convergence, there is, strangely, one important context in which certain divergent series may be numerically extremely useful. This most commonly arises in relation to what are called **asymptotic series**. These are series expansions in inverse powers of x (say), which are intended to describe the behaviour of a function for very large values of x. They may be convergent or divergent; (2.3.21) is an example of a convergent series expansion of this kind, with x replaced by the (positive) value of v. We now look at an example of an asymptotic series which diverges, but which none the less can provide a numerical result of significant use and accuracy.

We shall consider the evaluation of the complementary error function erfc(..) used extensively in statistics and probability theory since it describes the area of the 'tail' of the graph of the probability density function for the normal distribution of error. It is defined for $x \geqslant 0$ by the integral:

$$\text{erfc}(x) = 2\pi^{-\frac{1}{2}}\int_x^\infty \exp(-t^2)\,dt = \pi^{-\frac{1}{2}}\int_{x^2}^\infty s^{-1/2}\exp(-s)\,ds \quad (2.5.1)$$

where the substitution $s = t^2$ can be used to transform the one integral into the other (provided $x \geqslant 0$). The function evidently tends to zero as $x \to \infty$ and we look for an expansion in inverse powers of x.

Its derivation depends on successive use of integration by parts: if we place I_n (say) equal to $\int_{x^2}^{\infty} s^{-n-1/2} \exp(-s) \, ds$, so that I_0 is the same as erfc(x), then integrating $\exp(-s)$ and differentiating the power of s within the integrand,

$$I_n = [-s^{-n-1/2} \exp(-s)]_{x^2}^{\infty} - (n + \tfrac{1}{2}) \int_{x^2}^{\infty} s^{-n-3/2} \exp(-s) \, ds$$

$$= x^{-2n-1} \exp(-x^2) - (n + \tfrac{1}{2}) I_{n+1} .$$

In this way we calculate that $I_0 = x^{-1} \exp(-x^2) - (1/2) I_1$, where $I_1 = x^{-3} \exp(-x^2) - (3/2) I_2$, where $I_2 = \ldots$ and so on. Making the relevant substitutions for $I_1, I_2, \ldots$ back into the expression for $I_0 = $ erfc(x), we establish the truncated asymptotic series expansion:

$$\text{erfc}(x) = [\exp(-x^2)/(\pi^{1/2} x)](t_0 + t_1 + \ldots + t_n + E_n) \Bigg)$$

with $\qquad\qquad t_k = (-1)^k (2k)! / [(2x)^{2k} k!] \qquad\qquad\qquad\Bigg\}\qquad$ (2.5.2)

and $\qquad\qquad E_n = t_{n+1} \int_{x^2}^{\infty} (x^2/s)^{n+3/2} \exp(x^2 - s) \, ds \qquad\Bigg) .$

This looks rather formidable, but the ratio of successive terms (t_{k+1}/t_k) is simply equal to $-(2k + 1)/2x^2$, so that the series can be written numerically as:

$$\text{erfc}(x) \sim [\exp(-x^2)/(\pi^{1/2} x)] [1 - (x^{-2}/2) + (3x^{-4}/4) - (15x^{-6}/8) + \ldots] .$$

Notice that an equals sign cannot be used here, because the ratio of magnitudes of successive terms is greater than 1 for all $k > x^2 - 1/2$, so that the *infinite* series is alternating but divergent, and its sum does not exist.

However, taken as a finite truncated series, the first several terms decrease in magnitude quite rapidly if $x \gg 1$. Further, the truncation error E_n in (2.5.2) is bounded because

$$0 < E_n/t_{n+1} < \int_{x^2}^{\infty} \exp(x^2 - s) \, ds = 1 . \qquad\qquad (2.5.3)$$

The upper bound here is established by noting that (x^2/s) within the integrand is less than 1 for all values of s within the limits of integration. In other words, the truncation error is between 0 and the first neglected term, t_{n+1}. Hence, selecting any given partial sum $S_n = \sum_{k=0}^{n} t_k$ of the divergent series, the truncation error of this partial sum will tend to zero as $x \to \infty$. Correspondingly, in numerical work, if x is sufficiently large, summing the series to saturation will provide the correct 'sum', despite the divergent tail.

On the other hand, if x is regarded as fixed, there is no point in increasing n beyond a certain value (Fig. 2.4), because — as n increases — the error E_n will decrease to a minimum at some particular value of n, and then start to increase. Since the value of E_n cannot be precisely known, all that can be done in order to find this 'optimal' value of n is to assess that value of n which minimises the upper bound on $|E_n|$, in the reasonable expectation that this will in effect be the same as minimising the error. If, as in (2.5.3), it is known that $|E_n| < |t_n + 1|$,

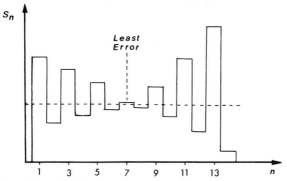

Fig. 2.4 – Partial sums of an alternating semi-convergent series.

then accordingly this *least upper bound* on $|E_n|$ is achieved by selecting n so that $|t_{n+1}|$ is as small as possible. This is most easily accomplished by examining the variation with n of the term-ratio $|t_{n+1}/t_n|$, and selecting the appropriate value of n as the largest value compatible with a non-increasing term magnitude – that is, the largest n for which $|t_{n+1}/t_n| \leqslant 1$.

In the example of the series for $\mathrm{erfc}(x)$, if we place $r = |t_{n+1}/t_n| = (2n + 1)/(2 x^2)$, then $n = rx^2 - \frac{1}{2}$, and the largest value of n comparable with $r \leqslant 1$ is $n = \mathrm{entier}(x^2 - \frac{1}{2})$. Accordingly this is the value of n which makes $|t_{n+1}|$ least and, for instance, t_9 is the term of smallest magnitude if x is between 2.92 and 3.08. Thus if $x = 3$ (say), the use of the partial sum S_8 would ensure that the upper bound to the truncation error is least, and $(2.5.3)$ shows that its truncation error is then somewhere between 0 and -1.74×10^{-4}. More generally, the variation of the least error bound, and of the number of convergent terms, with the value of x is shown in Fig. 2.5.

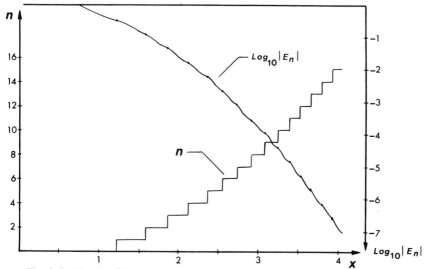

Fig. 2.5 – Number of convergent terms (n) and minimum truncation error E_n of semi-convergent series for $\pi^{1/2} x \exp(x^2) \, \mathrm{erfc}(x)$.

It is the characteristic feature of such **semi-convergent** series that an infinitely precise value of its 'sum' cannot be determined, although bounds on the truncation error are known. If it is an asymptotic series, then the maximum precision obtainable increases as x increases. But of course the absence of infinite precision is of no consequence numerically, since it could not be realised even if it were available, and by placing a lower bound on the value of x for which the series is used, it is possible to ensure that the truncation error is small enough to be compatible with the rounding error of the machine representation.

Perhaps the best-known semi-convergent asymptotic series is **Stirling's formula** for the gamma function:

$$\ln[\Gamma(z)] - \tfrac{1}{2}\ln(2\pi) - (z - \tfrac{1}{2})\ln(z) + z$$
$$= \sum_{j=1}^{n}(-1)^{j+1}B_j/[2j(2j-1)z^{2j-1}] + E_n \qquad (2.5.4)$$

where $z > 0$, and the truncation error E_n has a value between zero and that of the next $(n+1)$th term of the series. The values of B_j are known as **Bernoulli numbers**:

$$B_1 = 1/6, \quad B_2 = 1/30, \quad B_3 = 1/42, \quad B_4 = 1/30, \quad B_5 = 5/66,$$
$$B_6 = 691/2730, \quad B_7 = 7/6, \quad B_8 = 3617/510, \quad B_9 = 43867/798,$$
$$B_{10} = 174611/330, \ \ldots$$

There is no *simple* general formula for these, but for larger values of j they are most easily calculated from the sum of the series $\sum_{k=1}^{\infty} k^{-2j}$, because

$$B_j = 2(2j)!(2\pi)^{-2j} \sum_{k=1}^{\infty} k^{-2j} . \qquad (2.5.5)$$

Because the series sum is not very different from unity if j is large, it will also be seen that

$$B_j \simeq 2(2j)!(2\pi)^{-2j} \ (j \gg 1) . \qquad (2.5.6)$$

A useful approximation to $k!$ can be obtained by setting $n = 0$ and $z = k$ in (2.5.4), then taking the exponential so as to form $\Gamma(k) = (k-1)!$. It follows that

$$k! = k\Gamma(k) = (2\pi)^{1/2}k^{k+1/2}\exp(-k + E_0) \qquad (2.5.7)$$

where E_0 is between 0 and $1/(12k)$. Thus for $k \geqslant 8$ (say), the neglect of E_0 in this formula only causes at most a 1% underestimate of $k!$, which is quite accurate enough for 'back-of-the-envelope' calculations.

For example, using (2.5.7) in (2.5.6), and ignoring E_0, it follows after some algebra that

$$B_j \simeq 4\pi e^{1/2}(j/\pi e)^{2j+1/2} .$$

This gives (for instance) $B_5 \simeq 0.07505$, within 1% of the true value $0.07576\ldots$

Another example of the use to which the approximation (2.5.7) may be put is provided by the determination of the least term of the asymptotic series (2.5.2) for $\mathrm{erfc}(x)$. This will be the term t_k where $k = \mathrm{entier}(x^2 + \frac{1}{2})$. Thus using (2.5.7) and neglecting E_0,

$$|t_k| = (2k)!/[(2x)^{2k}k!] \simeq 2^{1/2}(k/ex^2)^k$$

or since $k \simeq x^2$:

$$|t_k| \simeq 2^{1/2}\exp(-k) \simeq 2^{1/2}\exp(-x^2) .\qquad(2.5.8)$$

It is readily estimated therefore that if (say) $x = 3$, so that $k = 9$, then $|t_9| \simeq 2^{1/2}\exp(-9) = 1.7 \times 10^{-4}$.

2.6 CHOOSING A SERIES EXPANSION

Even if a power series expansion is convergent for all real values of x, it is most unlikely to be computationally suitable for more than some bounded range of x, and of course many series converge only within a bounded range. The evaluation of a particular transcendental function will therefore involve either some recursive formula to extend the range of evaluation, or else different series will need to be employed to cover different ranges of the value of the independent variable, including (possibly) asymptotic series for very large values of $|x|$.

Examples of range-reduction formulae are legion. These include simple periodicity or symmetry properties, like for instance

$$\sin(x) = \sin(x + 2n\pi) = -\sin(-x) .$$

There are addition theorems such as $\exp(x + y) = \exp(x)\exp(y)$, product theorems such as $\ln(xy) = \ln(x) + \ln(y)$, and difference and functional equations like those for the gamma function

$$\Gamma(z + 1) = z\Gamma(z), \qquad \Gamma(z)\Gamma(1 - z) = \pi/\sin(\pi z) ,$$

which can 'stretch' the range of applicability of Stirling's formula (2.5.4) to the domain of all real numbers, not merely those much larger than unity. Even where such formulae are known it may still be necessary or desirable — because (perhaps) a particular formula introduces unnecessary loss of significance by cancellation error — to employ different series expansions in particular parts of the range of argument values. The criteria by which the series and its range of use may be selected are best illustrated by considering in some detail a particular example; they depend primarily on the need to achieve accuracy, or if this still leaves some choice, on the computational work implied by their use.

Let us consider the various series representations of the error function, defined for $x \geqslant 0$ by:

$$
\begin{aligned}
\mathrm{erf}(x) = -\mathrm{erf}(-x) &= 2\pi^{-1/2} \int_0^x \exp(-t^2) \, dt \\
&= \pi^{-1/2} \mathrm{sgn}(x) \int_0^{x^2} s^{-1/2} \exp(s) \, ds \\
&= 1 - \mathrm{erfc}(x)
\end{aligned}
\quad (2.6.1)
$$

and related, as shown, to the complementary error function (2.5.1). It is an odd function of x, so there is no loss of generality in restricting consideration to $x \geqslant 0$. There are three simple series expansions available. Expansion of $\exp(-t^2)$ in terms of t^2, and term-by-term integration of the first integral, yields the power series expansion:

$$
\begin{aligned}
\mathrm{erf}(x) &= 2\pi^{-1/2} x [1 - (x^2/3) + (x^4/10) - (x^6/42) + \ldots] \\
&= 2\pi^{-1/2} x \sum_{k=0}^{\infty} (-x^2)^k / [(2k+1)k!]
\end{aligned}
\quad (2.6.2)
$$

which converges for all x. This is in fact the Taylor series expansion about $x = 0$. Successive integration by parts of the other integral — integrating the half-power of s and differentiating $\exp(-s)$ — produces the result:

$$
\begin{aligned}
\mathrm{erf}(x) &= 2\pi^{-1/2} x \exp(-x^2) [1 + (2x^2/3) + (4x^2/15) + \ldots] \\
&= 2\pi^{-1/2} x \exp(-x^2) \sum_{k=0}^{\infty} (2x)^{2k} k! / (2k+1)!
\end{aligned}
\quad (2.6.3)
$$

which also converges for all x. It is seen to be the Taylor series expansion of the product $\exp(x^2)\mathrm{erf}(x)$ about $x = 0$. Finally, the relationship with $\mathrm{erfc}(x)$ provides a semi-convergent asymptotic series as given in (2.5.2), accurate for sufficiently large x.

To make the criteria of selection definite, suppose that it is desired to employ these series to provide an estimate of $\mathrm{erf}(x)$ based on the use of correctly rounded floating-point arithmetic with 13 significant decimal digits. First consider how large must be the value of x for the asymptotic series (2.5.2) to provide a value of $\mathrm{erf}(x)$ consistent with this precision. For large x, $\mathrm{erf}(x)$ is close to, but a little less than, 1 and so the truncation error of this asymptotic representation must be less than $10^{-13}/2$ (i.e. half of the least significant unit digit) for it to be consistent with the assumed precision. But this value must also therefore be the truncation error $[E_n \exp(-x^2)/(\pi^{1/2}x)]$ of $\mathrm{erfc}(x)$, and Fig. 2.5 shows that this is equal to 5×10^{-14} if $x = 3.81$, with no more than 13 terms of the asymptotic series required to achieve this accuracy. Hence it can be taken (for the time being) that for all $x > 3.81$ the first 13 terms of the asymptotic series will be used to represent $\mathrm{erf}(x)$, although of course the sum of the asymptotic series will be saturated with fewer terms included if x is substantially greater than 3.81.

Both the power series (2.6.2) and (2.6.3) are possibly applicable in the range $0 \leqslant x \leqslant 3.81$. The alternating series (2.6.2) converges slightly more rapidly, and it does not require the evaluation of $\exp(-x^2)$ like the other. It would seem clearly preferable to the other series were it not for the fact that, for $x > \sqrt{3}$, its first few terms increase in magnitude and, as we have pointed out before, this suggests that there may be problems with cancellation error. In fact if it is to be used for all x up to the value of $x = 3.81$ beyond which the asymptotic series takes over, then at $x = 3.81$, the largest term of (2.6.2) is that corresponding to $k = 13$ which contributes a value to the sum equal (approximately) to 3×10^4. Since $\mathrm{erf}(3.81)$ is only a little less than 1, this would imply that the rounding error in forming these large terms would be magnified by a factor of up to 10^4 in the relative error of the estimate of $\mathrm{erf}(3.81)$, and the least significant 4 or 5 digits of the sum would therefore be in error. Thus the alternating series could not be used without unduly large generated error except in the range $0 \leqslant x \leqslant 1$ or thereabouts.

The series (2.6.3) is composed of positive terms, and if it is used in the remaining range $1 \leqslant x \leqslant 3.81$, then towards the top of this range, it too will contain large terms. Indeed it will sum at $x = 3.81$ to a value of about $\pi^{1/2} \exp(-x^2)/(2x) = 4.7 \times 10^5$, since it has to be divided by this value to produce the value of $\mathrm{erf}(3.81) \simeq 1$. Consequently this series will saturate when the terms become less than 10^{-7}. This implies that about 50 terms of the series will be needed at $x = 3.81$, (but only about 16 at $x = 1$).

Such a disproportionately large number of terms (as compared with the 13 required for the asymptotic series) suggests a disproportionately large generated error at, or near to, $x = 3.81$ – particularly if the machine were to employ chopping rather than correct rounding. So it might well be decided to take advantage of the more rapid evaluation of the asymptotic series for $x \leqslant 3.81$, even although its truncation error is then not completely negligible compared with the precision of the machine representation. A change-over at $x = 3.7$ might well be judged to be the right compromise if correct rounding is in use, implying a truncation error less than 3×10^{-13} for the asymptotic series. A somewhat lower changeover value of x would be appropriate if chopping were used.

We see therefore that all three series have their possible range of use. Certainly at least two expansions of any transcendental function are almost always necessary to compute its value over an 'infinite' domain. Whether *both* the power series (2.6.2) and (2.6.3) would be used in any practical application as has been suggested is doubtful. The purpose of introducing the possibility is merely to illustrate the kind of considerations that may determine the choice of series expansions for less familiar functions than $\mathrm{erf}(x)$, for which a library routine – incidentally, almost certainly *not* based on series evaluation – is very likely to be available.

But to pursue the example further in this same spirit, it should be noted that the switch from a power-series to an asymptotic series development of

$\text{erf}(x)$ occurs at a value of x which is machine-dependent. If it were required to make the choice adaptive to any machine environment, the criterion (as to whether x was greater or less than 3.7 or whatever value is chosen) would need to be replaced by an expression evaluating the change-over value. For instance, it could be expressed in the context of a program by a statement sequence such as:

e: = exp($-x * x$);

if $x + e * e = x$ **then goto** ASYMPTOTIC SERIES **else** …

This implies a jump to the instruction sequence for asymptotic series evaluation only if its truncation error, which from (2.5.1) and (2.5.8) roughly equals $(2/\pi)^{1/2} \exp(-2x^2)/x$, is smaller than a possible value of the rounding error of its sum (equal nearly to unity). The value of $\exp(-x^2)$, required by both the power series (2.6.3) and the asymptotic series, is stored and available for use in either summation. In this adaptive approach, the number of terms required in the semi-convergent series is also no longer known beforehand, so that a test would have to be incorporated terminating summation either when the summand is saturated or when the terms diverge.

It is not uncommon for functions to have alternative forms, like $\text{erf}(x)$ and $\text{erfc}(x)$. But it is not necessarily sufficiently accurate to form $\text{erfc}(x)$ as the algebraically equivalent value of $[1 - \text{erf}(x)]$, other than for small values of x, because of the loss of significance arising from cancellation error. If either function value might be required, then both functions need separate development, or else $\text{erf}(x)$ should be calculated if $x \leqslant x_0$, say, and $\text{erfc}(x)$ if $x > x_0$, where x_0 is chosen so that the alternative function can be determined simply by differencing the other from unity without any loss of significance. If the arithmetic base of the machine is b, then this implies that x_0 must be chosen so that both $\text{erf}(x_0)$ and $\text{erfc}(x_0)$ are within the interval $[1/b, (b-1)/b]$. In other words, if a binary base is used, the change-over must be at $x_0 \simeq 0.47$, where $\text{erf}(x_0) = \text{erfc}(x_0) = 0.5$. But with other bases, a wider range of choice is possible – for instance, x_0 could be as large as 1.16 (so that $\text{erf}(x_0) \simeq 0.9$) on a machine using a decimal base.

If this idea were followed up, a new difficulty would arise because there is (apparently) no suitable series expansion for $\text{erfc}(x)$, other of course than its asymptotic series (which is only accurate for $x \geqslant 1$), which avoids the cancellation error inherent in its formation as $1 - \text{erf}(x)$. It might be that some other method of evaluation than by series expansion would need to be enlisted, to fill the gap between the change-over value $x = x_0$ and that at which the asymptotic series 'takes over'. It is at just such a pause for scratching of the head that a numerical analyst warms to the task ahead, and surveys the whole panoply of methods available. The user, of course, might be quite content with the result $[1 - \text{erf}(x)]$, loss of significance and all! After all, it could be that *actual* error rather than *relative* error is relevant to the practical problem. But if not, … ?

2.7 CONTINUED FRACTIONS

Evaluating a transcendental function, like $\text{erfc}(x)$, which is defined by an integral immediately suggests that numerical integration may provide a possible alternative method of evaluation to series summation. Techniques of this kind will be discussed later, but they would generally be reckoned as computationally too slow although they can certainly provide an accurate answer. Alternatively, there will exist expansions in terms of other basis functions than the successive powers of the independent variable (used in a power series). Possible sequences of basis functions include trigonometric functions and orthogonal polynomials, and these again will be referred to later.

However, a very useful technique is based on the idea of representing the function as a *quotient* of two power series. In a simple form, the possible advantages of this have already been demonstrated: the expression for $\text{erf}(x)$ in (2.6.3) could be represented as the quoted series $[1 + (2x^2/3) + (4x^4/15) + ...]$ divided by the series $[1 + x^2 + (x^4/2) + ...]$ for $\exp(x^2)$, the quotient of two series of positive terms replacing the alternating series (2.6.2). Of course, as the series in the denominator is expressible as an exponential there is no point in writing (2.6.3) in this way. But the possibilities opened up by this approach are so wide as to be baffling: after all, the function to be evaluated, $f(x)$ say, could be multiplied by *any* other function $B(x)$ which has a known series expansion, and if the power series $A(x)$ representing the product $B(x)f(x)$ can be formed, then $f(x)$ is the quotient of the two series $A(x)/B(x)$. It is difficult to lay down criteria by which $B(x)$ might best be chosen, although there are approaches which in effect seek to do this. An alternative is to accept what appears to be a convenient and relatively easy way of generating such expansions, in the hope, or maybe expectation, that it will be of assistance. One such method leads to a curious construction called a **continued fraction**. In days gone by, these used to be familiar entities to the algebraist, but dropped from fashion until recently, when their usefulness in computation has reawakened interest. Because they may be unfamiliar, some brief description here will not be amiss.

A *terminated* continued fraction is an expression of the form

$$\cfrac{p_1}{q_1 + \cfrac{p_2}{q_2 + \cfrac{p_3}{q_3 + \cdots + \cfrac{p_{n-1}}{q_{n-1} + \cfrac{p_n}{q_n}}}}}.$$

This representation is typographically very cumbersome: in many books it is replaced by the more compact arrangement

$$\frac{p_1}{q_1+}\;\frac{p_2}{q_2+}\cdots\frac{p_{n-1}}{q_{n-1}+}\;\frac{p_n}{q_n}\;.$$

If written as an expression in algorithmic language it would appear as

$$p_1/(q_1 + p_2/(q_2 + \ldots + p_{n-1}/(q_{n-1} + p_n/q_n)\ldots))$$

with $(n-1)$ closing brackets. Any of these (equivalent) forms will be here represented symbolically by $\underset{j=1}{\overset{n}{K}}\,p_j \downarrow q_j$. The values of p_j and q_j are respectively the **partial numerators** and **partial denominators** of the continued fraction, and with $k \leqslant n$, the value of $\underset{j=1}{\overset{k}{K}}\,p_j \downarrow q_j$ is called the kth **convergent** of the continued fraction. The generalisation to an unterminated, or infinite, continued fraction $\underset{j=1}{\overset{\infty}{K}}\,p_j \downarrow q_j$ is intuitive; it has a value, and is said to converge, if the limit of $\underset{j=1}{\overset{n}{K}}\,p_j \downarrow q_j$ exists as $n \to \infty$. An infinite continued fraction terminates at the kth convergent if $p_j \neq 0$ for $j = 1, 2, \ldots, k$, but $p_{k+1} = 0$.

As so far stated, the representation of any continued fraction is not unique. The jth partial numerator and its complete denominator may both be multiplied by a non-zero value, c_j (say), without altering the value: this is called an equivalence or **similarity transform**, and it implies that

$$
c_0 \underset{j=1}{\overset{\infty}{K}}\,p_j \downarrow q_j = \cfrac{c_0 c_1 p_1}{c_1 q_1 + \cfrac{c_1 c_2 p_2}{c_2 q_2 + \cfrac{c_2 c_3 p_3}{\cdots}}} \cdots
$$

$$
= \underset{j=1}{\overset{\infty}{K}}\, c_{j-1} c_j p_j \downarrow c_j q_j \;. \tag{2.7.1}
$$

Continued fractions, such as these, which have an identical sequence of convergents are said to be **equivalent**, rather than equal — which would merely imply that their convergents had the same final value or limit. For instance, assuming that none of the partial denominators q_j vanishes, and placing $c_j = 1/q_j$ for $j = 0, 1, 2, \ldots$ we see that

$$
\underset{j=1}{\overset{\infty}{K}}\,p_j \downarrow q_j = \underset{j=1}{\overset{\infty}{K}}\,(p_j/q_j q_{j-1}) \downarrow 1 \tag{2.7.2}
$$

with q_0 defined as unity, and here the partial denominators are all equal to unity. This particular standardisation is preferred because it is the form most easily generated.

As will be shown in due course, a continued fraction representation of the complementary error function is given for $x > 0$ by

$$\operatorname{erfc}(x) = \pi^{-1/2} \exp(-x^2) \left[\frac{1}{x+} \frac{1/2}{x+} \frac{2/2}{x+} \frac{3/2}{x+} \cdots\right]$$
$$= \pi^{-1/2} \exp(-x^2) \overset{\infty}{\underset{j=0}{K'}} (j/2) \downarrow x .$$

Note that *the prime applied to K implies that the first partial numerator is to be taken as unity.* In standardised form with unit denominator coefficients, this continued fraction becomes:

$$\left.\begin{array}{l} \operatorname{erfc}(x) = [\exp(-x^2)/(\pi^{1/2}x)] \left[\dfrac{1}{1+} \dfrac{x^{-2}/2}{1+} \dfrac{2x^{-2}/2}{1+} \cdots\right] \\[3mm] \qquad = [\exp(-x^2)/(\pi^{1/2}x)] \overset{\infty}{\underset{j=0}{K'}} (j/2x^2) \downarrow 1 \end{array}\right\} \qquad (2.7.3)$$

The particular form of continued fraction:

$$\frac{1}{1+} \frac{r_1 z}{1+} \frac{r_2 z}{1+} \cdots = \overset{\infty}{\underset{j=0}{K'}} r_j z \downarrow 1 , \qquad (2.7.4)$$

where the r_j's are independent of z, is termed an **S-fraction** in terms of z, and (2.7.3) is an S-fraction representation of $\operatorname{erfc}(x)$ in terms of $(1/x^2)$.

One convenient way of evaluating a continued fraction (there are, as we shall see, others) is by expressing the kth convergent $\overset{k}{\underset{j=1}{K}} p_j \downarrow q_j$ as equal to the quotient (A_k/B_k) where

$$\left.\begin{array}{l} A_{-1} = 1, \quad A_0 = 0, \quad A_j = q_j A_{j-1} + p_j A_{j-2} \\[2mm] B_{-1} = 0, \quad B_0 = 1, \quad B_j = q_j B_{j-1} + p_j B_{j-2} \end{array}\right\} \qquad (2.7.5)$$

for $j = 1, 2, \ldots, k$. In particular, $A_1 = p_1$ and $B_1 = q_1$, consistent with the value p_1/q_1 of the first convergent. The proof that this recurrence leads to all successive convergents is left to the reader (problem 1).

For an S-fraction in terms of z as in (2.7.4), the recurrence (2.7.5) takes the particular form

$$\left.\begin{array}{l} A_0 = 0, \quad A_1 = 1, \quad A_j = A_{j-1} + r_{j-1} z A_{j-2} \\[2mm] B_0 = 1, \quad B_1 = 1, \quad B_j = B_{j-1} + r_{j-1} z B_{j-2} \end{array}\right\} \qquad (2.7.6)$$

for $j = 2, 3, \ldots$, and the initial 3 convergents are $1, 1/(1 + r_1 z), (1 + r_2 z)/[1 + (r_1 + r_2)z]$. It will be seen that A_j and B_j are polynomials in z, and so any convergent of an S-fraction is a ratio of polynomials, or **rational function**, in terms of z. If the degree of the numerator polynomial of a rational function is N (say) whilst its denominator is a polynomial of degree M, then it is said to have **order** N/M. The first few successive convergnets of an S-fraction therefore have orders $0/0, 0/1, 1/1, 1/2, 2/2, 2/3, \ldots$ and an infinite S-fraction in z is a quotient of infinite power series in z — the animal we have been looking for.

2.8 CONVERGENCE OF CONTINUED FRACTIONS

If the continued fraction converges, then successive values of A_j/B_j will tend to a limit, and the computation would have to be broken off when the difference between successive convergents is no longer discernible within the limits imposed either by the machine precision or by some larger prescribed error tolerance. However, as with accumulator saturation in summation, this may or may not be an adequate method of discriminating whether the truncation error is compatible with the machine precision. Moreover, there is one additional hazard of the recurrence formulae (2.7.5) or (2.7.6) in that the values of A_j and B_j, whilst maintaining a more or less fixed proportion to each other, may both rapidly increase (or decrease) in magnitude, and possibly cause overflow (or underflow) as the calculation proceeds. Thus, for instance, if successive convergents of the S-fraction for erfc(x) in (2.7.3) are computed by (2.7.6), then since $r_j = j/2$, it will be found that, whatever the value of $z = x^{-2}$, the magnitudes of both A_j and B_j are ultimately bound to increase as j increases, roughly in proportion to $(j/2)!/x^j$.

This problem of possible overflow or underflow can be overcome either by using a suitable similarity transform to recast the recurrence formula, or by testing the magnitudes of A_j and B_j, and periodically scaling down the currently preserved values of A_j, A_{j-1} and B_j, B_{j-1}, all by the same factor. This latter practice makes no difference to the values of their ratios, nor to the values of successive convergents. However, an alternative method of computing the convergents which avoids this problem, and moreover which provides some insight into convergence properties, is of obvious interest.

If the kth convergent $A_k/B_k = \overset{k}{\underset{j=1}{K}} p_j \downarrow q_j$ is expressed as a sum $\overset{k}{\underset{j=1}{\Sigma}} t_j$, so that the term t_j is simply the difference between successive convergents $(A_j/B_j) - (A_{j-1}/B_{j-1})$, then some algebraic manipulation of (2.7.5) shows that the terms t_j can be generated from the recurrence formulae:

$$\left. \begin{array}{l} \beta_1 = 1/q_1, \ \beta_j = (q_j + p_j \beta_{j-1})^{-1}, \\ t_1 = p_1/q_1, \ t_j = -p_j \beta_{j-1} \beta_j t_{j-1} = (q_j \beta_j - 1) t_{j-1} \end{array} \right\} \qquad (2.8.1)$$

for $j = 2, 3, \ldots$. Here the value of β_j equals the value of the quotient (B_{j-1}/B_j) where B_j is defined by (2.7.5). The convergence of the process would imply that $t_j \to 0$ as $j \to \infty$, and moreover as β_j is a *ratio* of B-values, it diverges in magnitude much less rapidly with increasing j than B_j, if at all. Thus (2.8.1) effectively avoids any danger of overflow or underflow.

If the values of p_j, q_j are all positive, it is seen from (2.8.1) that the β_j's are positive, but that the ratio of successive terms is negative and

$$0 > t_j/t_{j-1} = -p_j\beta_{j-1}/(q_j + p_j\beta_{j-1}) > -1 \; ,$$

so that provided merely that $t_j \to 0$ as $j \to \infty$, the series $\sum\limits_{j=1}^{\infty} t_j$ is convergent and alternating. The kth partial sum of this series being the kth convergent, it follows that the odd-numbered convergents ($k = 1, 3, 5, ...$) form a monotonically decreasing sequence, and the even-numbered ones a monotonically increasing sequence, bounding the value of the continued fraction, and suitable limitation of truncation error is achieved by summing the terms t_j to saturation. In particular, if all the partial denominators q_j are unity (as for an S-fraction) the terms t_j tend to zero as j increases — as is required for convergence — provided at least that bounded positive values of P and J exists such that $0 \leqslant p_j < Pj^2$ for all $j > J$.

It follows that the S-fraction (2.7.3) for erfc(x) is convergent for all non-zero values of x — since $p_j = j/2x^2$ — and the number of convergents required to achieve a specified precision is indicated in Fig. 2.6. Its use enables erfc(x) to be accurately computed in that intermediate interval for which x is too large for the series development about $x = 0$ to be applied, but yet too small for the asymptotic series to be sufficiently precise. The rather slow convergence for smaller values of x can be accelerated by applying the process of repeated averaging to the successive partial sums of the alternating series developed by (2.8.1), or equally to the successive convergents (A_j/B_j) developed by (2.7.6).

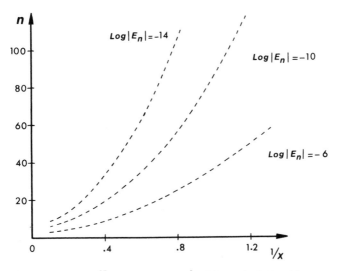

Fig. 2.6 – Number of successive convergents (n) required to achieve specified relative error E_n for the S-fraction development of $\pi^{1/2}\exp(x^2)\,\mathrm{erfc}(x)$.

Confining attention to S-fractions $\overset{\infty'}{\underset{j=0}{K}} r_j z \downarrow 1$, their convergence would be
more rapid, and the terms t_j would decrease roughly in geometric progression,
if $r_j z$ tends to a finite positive limit as $j \to \infty$, rather than (as in the above
example) increasing with the value of j. Indeed, if $r_j z \to 0$ as $j \to \infty$, the convergence is more rapid than a geometric progression. On the other hand, should
it happen that $r_j z$ tends to a finite *negative* limit ($-l$, say) as $j \to \infty$, then the
terms t_j developed by (2.8.1) all have the same sign, at least for sufficiently
large j. However, they decrease asymptotically in geometric progression provided
$0 < l < 1/4$; in particular, successive terms are (approximately) halved if $l = 2/9$.
The general conclusion can be drawn that summing the series developed by
(2.8.1) to saturation is an adequate means of controlling truncation error
provided that, if $r_j z$ tends asymptotically to any finite limit, then its value must
be greater than about $-2/9$. Limits in the interval between $-2/9$ and $-1/4$
imply slow convergence, whilst the S-fraction in fact diverges if $r_j z$ tends to a
value less than $-1/4$ (see, for instance, problem 4 of section 2.7).

Once the number of convergents (N, say) which are necessary in order to
evaluate any continued fraction with required precision is established, it is
possible to recover its value much more conveniently by the reversed recurrence
relation:

$$C_{N+1} = 0, \quad C_k = p_k/(q_k + C_{k+1}) \quad (k = N, N-1, \ldots, 1) \; .$$
$$(2.8.2)$$

If we refer back to the non-abbreviated expression of a terminated continued
fraction in section 2.7, this will be seen to be an obvious 'back-to-front' method
of unwinding it, with C_k representing the value of $\overset{N}{\underset{j=k}{K}} p_j \downarrow q_j$, and C_1 being a
sufficiently precise value of the infinite fraction. It involves considerably less
computation than either (2.7.5) or (2.8.1) and so would generally be preferred,
on grounds both of speed and avoidance of generated error. The only contrary
indication arises where cancellation error may occur late on in the calculation, in
forming $(q_k + C_{k+1})$ for small values of k, but early on — or even not at all —
in forward recurrence. Equally, quite the opposite may happen; but neither
circumstance is ordinarily encountered.

2.9 GENERATION OF S-FRACTIONS

As has been noted before, each convergent to an S-fraction is a rational function,
and if this rational function is developed into a power series, it would be
expected that the power series of successive convergents tends in some way
towards the power series of the function represented by the infinite S-fraction.
The peculiarity of S-fractions, which distinguishes them from the multitude of
other possible rational representations of the function, is the manner in which
these power series converge towards each other.

Take as an example the S-fraction for erfc(x) in (2.7.3). Its first four convergents are $1, (1 + x^{-2}/2)^{-1}, (1 + x^{-2})/(1 + 3x^{-2}/2)$ and $(1 + 5x^{-2}/2)/(1 + 3x^{-2} + 3x^{-4})$. Applying the binomial theorem to expand the denominator into a power series in $z = x^{-2}$, we find that the series expansions of these convergents are respectively:

$$1,$$

$$1 - (x^{-2}/2) + (x^{-4}/4) - (x^{-6}/8) + \ldots,$$

$$1 - (x^{-2}/2) + (3x^{-4}/4) - (9x^{-6}/8) + \ldots,$$

and $\quad 1 - (x^{-2}/2) + (3x^{-4}/4) - (15x^{-6}/8) + \ldots .$

This illustrates what is in general true, that the $(k + 1)$th convergent, when expanded as a power series, has the same initial k terms as the previous $(k$th) convergent. This is an inherent property of S-fractions, and can be proved by (for instance) showing that the term t_{k+1} of (2.8.1), which represents the difference between these two convergents, varies as z^k as $z \to 0$ (see problem 2, section 2.8). Starting with a known power series development of a function, its S-fraction is constructed by matching the kth term of the power series of its kth convergent with that of the function. This agreement will then be maintained in all subsequent convergents.

In the example of the S-fraction representation of erfc(x), it will be found by comparing (2.7.3) with (2.5.2), that the function represented by the S-fraction is the semi-convergent asymptotic series for $\pi^{1/2}x \exp(x^2)$erfc$(x) = 1 - (x^{-2}/2) + (3x^{-4}/4) - (15x^{-6}/8) + \ldots$ and it can be seen that (at least for $k \leqslant 4$) the kth convergent has a power series expansion consistent with the first k terms. Note that although the asymptotic series is only semi-convergent, the derived S-fraction is convergent for all $x > 0$.

It might nonetheless seem likely that the implementation of such a method of generating the S-fraction would be complicated: but if it is accomplished numerically, it transpires to be quite straightforward. The process consists of constructing a triangular array of numbers. One convenient way of referring to any element in this array is by 'row, column' subscripts as will be familiar from matrix notation. Accordingly, suppose that the numbers are denoted by r_{jk} for integers j, k such that $j \geqslant k \geqslant 0$, and imagine them to be arranged as a triangular matrix:

r_{00}

$r_{10} \; r_{11}$

$r_{20} \; r_{21} \; r_{22}$

$r_{30} \; r_{31} \; r_{32} \; r_{33}$

$\ldots \ldots$

$\ldots \ldots$

Further let the function, whose S-fraction representation in terms of z is being sought, have a known power series representation, $S_\infty(z) = \sum\limits_{j=0}^{\infty} a_j z^j$, none of whose coefficients are zero. (Without loss of generality it could be assumed that the function is scaled so that $a_0 = 1$, but this does not produce any particular simplification.) The numbers in the first two columns can then be inserted straight away, for they are expressed in terms of the coefficients of the series $S_\infty(z)$ by:

$$\left.\begin{array}{l} r_{00} = a_0, \quad r_{j0} = 0, \\ r_{j1} = -a_j/a_{j-1} \end{array}\right\} \tag{2.9.1}$$

with $j = 1, 2, 3, \ldots$.

The calculation proceeds by repeatedly adding a new row to the array: assume this to be the jth row, where $j \geqslant 2$ – the first two rows already being completely specified by (2.9.1) with $j = 1$. This same equation also shows r_{j0} is zero and it provides r_{j1}, so that it remains to calculate $r_{j2}, \ldots, r_{jj}$. That is, values are required for r_{jk} for $k = 2, 3, \ldots, j$, filling in the array row from left to right. These values are defined recursively in terms of the elements of the preceding $(j-1)$th row and of the partially filled current row by the relations:

$$\begin{aligned} r_{jk} &= r_{(j-1)(k-2)} - r_{(j-1)(k-1)} + r_{j(k-1)} \quad (k \text{ even}) \\ &= (r_{(j-1)(k-2)}/r_{(j-1)(k-1)}) r_{j(k-1)} \quad (k \text{ odd}) \end{aligned} \tag{2.9.2}$$

for $k = 2, 3, \ldots, j$. It will be seen that different rules of calculation apply to elements of the even-numbered columns $k = 2, 4, 6, \ldots$ and to those in the odd-numbered columns $k = 3, 5, 7 \ldots$. Each new value depends on three preceding elements of the array which have been already calculated, and these elements have a particular spatial disposition (in the triangular matrix) relative to the target element being assigned. This pattern of relationship can be illustrated by 'two steps of a staircase':

Such 4-term recurrence formulae are also called **rhombus rules**, since the elements concerned lie (or can be imagined as lying) at the corners of a rhombus; the arrangement of the array is usually altered to emphasise such a view – but the 'staircase' pattern is an equally acceptable image.

Having completed the jth row, then the last-calculated element is the coefficient of the $(j + 1)$th partial numerator of the S-fraction: that is

$$r_j = r_{jj} \, (j = 0, 1, 2, \ldots) . \qquad (2.9.3)$$

Note that $r_0 = a_0$ is the first partial numerator, but it could be extracted as a factor of the S-fraction (since $a_0 \neq 1$) so as to preserve the convention that r_0 is unity.

This process is known as the **QD-algorithm** (or quotient-difference algorithm) so named because the recursion formula (2.9.2) involves alternately quotients and differences. As applied to the asymptotic series (2.5.2) with $z = x^{-2}$, the following triangular array is developed:

```
1

0    1/2

0    3/2   1

0    5/2   1    3/2

0    7/2   1    5/2   2

0    9/2   1    7/2   2    5/2

0   11/2   1    9/2   2    7/2   3

        ......
```

and it is easy to discern a pattern in the results. Notice (as often happens) that the odd- and even-numbered columns exhibit different patterns. It will also be observed that the numbers r_{jj} along the diagonal extremity are indeed the coefficients of the S-fraction in (2.7.3). The theoretical reason for this is explained in problem 3.

It is worthwhile working through this array of values to verify that the rules of this 'numbers game' are understood. However, it is a computational technique that could be readily mechanised. Even so, if this whole process of calculating the coefficients of the S-fraction had to be performed each time a function was to be evaluated, it would add very considerably to the computation work required. Obviously, it would be prefereable that the values of $r_0, r_1, r_2, \ldots$ were pre-calculated by the QD algorithm and stored for subsequent use. But even this would not be all the preparation necessary: some exploratory calculations would need to be done in order to establish how many of the coefficients would be needed. These chores would seem routine when writing an algorithm for a computational library; they would however provide a strong disincentive in a 'one-shot' calculation. Of course, the same problems may afflict power series evaluation if there is no quick and convenient method of evaluating the coefficients: a number of series, for example, involve the evaluation of the Bernoulli numbers which likewise need to be pre-calculated and stored prior to use.

Hand calculation of the first few terms of a continued fraction using rational arithmetic will occasionally reveal the algebraic law by which the r_k's can be generated, as in the example of the S-fraction for $\text{erfc}(x)$. If this is so, the programmed evaluation of the S-fraction is straightforward, and computationally highly efficient. It can indeed happen that the S-fraction coefficients are given by a simpler relation than those of the series from which they are calculated. A notable example of this is provided by $\tan(x)$, whose power series expansion is

$$x^{-1}\tan(x) = 4 \sum_{k=1}^{\infty} (4^k - 1) B_k \, x^{2k-2}/(2k)!$$

which involves the Bernoulli numbers, whereas its S-fraction expansion is

$$x^{-1}\tan(x) = \frac{1}{1+} \frac{(-x^2/3)}{1+} \frac{(-x^2/15)}{1+} \frac{(-x^2/35)}{1+} \cdots$$

$$= \mathop{K'}_{j=0}^{\infty} x^2/(1-4j^2) \!\downarrow\! 1 \;.$$

Applying a similarity transformation, as in (2.7.1), this can be expressed even more simply as

$$x^{-1}\tan(x) = \frac{1}{1-} \frac{x^2}{3-} \frac{x^2}{5-} \frac{x^2}{7-} \cdots = \mathop{K'}_{j=0}^{\infty} (-x^2) \!\downarrow\! (2j+1) \quad .(2.9.4)$$

Moreover, the continued fraction converges for all x^2, unless $\tan(x)$ is infinite, whereas the series converges only for $|x| < \pi/2$. This again reflects the fact that continued fractions generally have much better convergence properties than the series from which they are derived. The reason why they are so neglected is largely because they are less easily manipulated (algebraically) than series, which can be added together or operated upon in various ways. For numerical work, however, they are almost always superior.

One reason for this superiority is not difficult to discern. The power series expansion of $\tan(x)$ cannot converge for $|x| \geqslant \pi/2$ because it is impossible for a series to represent the pole at $|x| = \pi/2$ where $\tan(x)$ is infinite, and yet remain convergent. A rational approximation is in no way so limited, because it can represent the pole as a zero of the denominator.

2.10 RECURRENCE RELATIONS AND NUMERICAL INSTABILITY

A glance at the continued fraction (2.9.4) shows that it contains no explicit mention of π; it would be correctly inferred that the poles of $\tan(x)$ at, for instance, $|x| = \pi/2$, are not in fact generated by composing a factor $(x^2 - \pi^2/4)$ in the denominator of the rational function. Rather, what should be discovered if successive convergents (A_k/B_k) to the continued fraction are calculated at $x = \pm\pi/2$, using the recurrence formula (2.7.5), is that the magnitude of the numerator A_k diverges as k increases, whereas the denominator B_k

ought to tend to zero, as $k \rightarrow \infty$, producing (in the limit) an infinite quotient. One says 'ought to tend to zero' because if this calculation is performed, what happens in practice will be quite different. The values of A_k *will* steadily increase in magnitude as k increases but, after an initial decline in magnitude, the successive values of B_k will suddenly start to increase, with or without a change of sign, and the ratio A_k/B_k will thereafter soon settle to a fixed value. The magnitude of this value may be quite large – though almost certainly not bigger than $1/\epsilon$, where ϵ is the machine unit – and of either sign.

And what, it might well be asked, is so curious about that? The value of π cannot be *precisely* represented, as it is a transcendental number, so what has been worked out is a value of $(1/x)\tan(x)$ for a value of $|x|$ *close* to $\pi/2$, and a large value of either sign is what consistently should result. In the same way, if the same calculation is performed with $|x|$ 'equals' π, the result will not be precisely zero (nor should it be expected to be). The only curiosity (if there is one at all) is that the numerical process converges in the neighbourhood of a pole, but in the context this is no more to be doubted than that it should converge in the region of a zero.

The reason for mentioning this at all is because many functions satisfy recurrence relations of one kind or another, and in particular some important functions occur in families whose members are connected by a 3-term linear recurrence of the kind used to generate the convergents A_k, B_k. As it happens, the values of A_k and B_k for the continued fraction (2.9.4) at $|x| = \pi/2$, should be proportional to $Y_{k+1/2}$ $(\pi/2)$ and $J_{k+1/2}$ $(\pi/2)$ respectively; these functions are called half-order, or spherical, Bessel functions. (They are of some importance in applied mathematics, because they describe the propagation of spherical waves.) It *should* happen that $J_{k+1/2}(x) \rightarrow 0$ as $k \rightarrow \infty$ for *any* x. As has been implied, numerically it apparently does no such thing: in fact, it will tend to $\pm \infty$ as $k \rightarrow \infty$ not only for $|x| = \pi/2$ but for any x, if it is calculated by the recurrence relation. Plainly, this does not matter in the context of the calculation of the continued fraction: the consequence can even be judged reasonable. However, it would matter quite a lot if the task were simply the evaluation of the function $J_{k+1/2}(x)$: an answer which is liable to overflow is a poor approximation to one that should be nearly zero!

We are dealing here with a rather exceptional, but strikingly awful, example of error due to **numerical instability**. Its relevance to the calculation of Bessel functions is important, but may give the wrong impression that it only occurs in obscure contexts. We can invent a much simpler illustration. The 3-term recurrence like (2.7.5):

$$Z_k = q_k Z_{k-1} + p_k Z_{k-2} \qquad (2.10.1)$$

can be regarded as defining a function Z_k whose domain is the set of integers k. Given initial starting values of Z_0 and Z_1, the recurrence relation (2.10.1) defines Z_k for all integer k. In the event that $p_k = p$ and $q_k = q$ are constants, the recurrence equation has a known algebraic solution:

$$Z_n = A\lambda_1{}^n + B\lambda_2{}^n \tag{2.10.2}$$

where A and B are arbitrary constants and λ_1 and λ_2 are the two roots of the **characteristic equation** of the recurrence, namely:

$$\left.\begin{array}{l} \lambda^2 - q\lambda - p = 0 \\ \text{i.e.} \qquad \lambda_1, \lambda_2 = [q \pm (q^2 + 4p)^{1/2}]/2 \end{array}\right\} \tag{2.10.3}$$

It will be assumed here that $p > -q^2/4$, so that $\lambda_1 \neq \lambda_2$ and both roots are real. Placing $n = 0$ and 1 in (2.10.2), and solving for A and B in terms of the starting values, it can be found that

$$A = (Z_1 - \lambda_2 Z_0)/(\lambda_1 - \lambda_2), \quad B = (\lambda_1 Z_0 - Z_1)/(\lambda_1 - \lambda_2) . \tag{2.10.4}$$

Taking, for example, the recurrence relation

$$Z_k = (100/9)Z_{k-1} - (11/9)Z_{k-2} , \tag{2.10.5}$$

the characteristic equation has the solution $\lambda_1 = 11$, $\lambda_2 = 1/9$, and so (2.10.2) shows that

$$Z_n = 11^n A + 9^{-n} B . \tag{2.10.6}$$

It can easily be verified by substitution that $Z_n = 11^n$ or 9^{-n} are both possible solutions of (2.10.5). The solution $Z_n = 11^n$ corresponding to $A = 1, B = 0$, would result from starting values of $Z_0 = 1, Z_1 = 11$, whilst the decreasing sequence $Z_n = 9^{-n}$ would be the appropriate solution if $Z_0 = 1$ but $Z_1 = 1/9$, so that $A = 0$ and $B = 1$. For general values of Z_0 and Z_1, it would be expected that both A and B would be non-zero.

Pretend now that it was not known that (2.10.5) had such a simple algebraic solution, and that the only way of solving the recurrence relation was by the numerical evaluation of $Z_2, Z_3, Z_4, \ldots$ in turn. With decimal arithmetic of precision 3, and starting values of $Z_0 = 1.00$, $Z_1 = 0.111$, it would be expected that what is calculated would be a numerical approximation to the decreasing sequence of values $Z_n = 9^{-n}$. But look what happens if the recurrence relation (2.10.5), with the correctly rounded values of its coefficients, is applied to calculate this sequence:

$$Z_2 = 11.1 * 0.111 - 1.22 * 1.00 = 1.23 - 1.22 = 0.0100$$

$$Z_3 = 11.1 * 0.0100 - 1.22 * 0.111 = 0.111 - 0.135 = -0.024$$

$$Z_4 = 11.1 * (-0.024) - 1.22 * 0.0100 = -0.266 - 0.0122 = -0.278 ,$$

and already it is clear that something seems to have gone badly wrong: the values have become negative, and their magnitude is increasing, not decreasing.

What has happened is that cancellation has magnified the relative rounding and inherent error of the products used in calculating Z_2 (its value should be 0.0123, not 0.0100). In forming Z_3, the values of Z_2 and Z_1 are used, and owing to the large relative error of the former, they are not in the correct ratio 1:9 (but rather 1:11.1), with the consequence that the **dominant** solution $11^n A$ of (2.10.6) is spuriously excited (with small negative A) and Z_3 is (roughly) 11 times the error in Z_2, whilst Z_4 further increases in magnitude by a factor of about 11.

This dramatic effect of numerical instability depends on three essential conditions:

(i)   the existence of two solutions to the recurrence relation, whose ratio diverges in magnitude as the iteration progresses;

(ii) the intention to isolate the smaller **decrescent** solution; and

(iii) the prior knowledge of the ratio of the starting values which ought to lead to the decrescent solution.

Had any other ratio of starting values for Z_0 and Z_1 been used in the example, the error would have been barely noticeable, because both numerically and precisely the dominant term would correctly be present, and would increase as the iterations proceed, to obscure the decrescent term.

The existence of two such solutions to a recurrence relation like (2.10.1), whether or not p_k and q_k are constants, is to be inferred if $|p_k/q_k q_{k-1}| \ll 1/4$, at least for all $k > K$ (say). In the example of the Bessel functions occurring in the expression for the convergents of the continued fraction (2.9.4) which provoked this discussion, $q_k = (2k-1)$ and $p_k = -x^2$ (a constant), so that the presence of a dominant solution ($Y_{k+1/2}$) and a decrescent solution ($J_{k+1/2}$) could indeed be predicted; although without specialised knowledge, there would be no way of knowing that, if $|x| = \pi/2$, the starting values of $B_0 = 1$, $B_1 = 1$ should identify the values of B_k with the decrescent solution.

Faced with the need to find the decrescent solution of a recurrence like (2.10.1) for say $k = 0, 1, 2, \ldots, n$, it is necessary to turn the recurrence into a *backwards* iteration, starting with some assumed Z_n and working back to Z_0 by recasting (2.10.1) as

$$Z_{k-2} = (-q_k Z_{k-1} + Z_k)/p_k \qquad (2.10.7)$$

for $k = n, (n-1), \ldots, 2$. Working this way round, the decrescent solution becomes the diverging, or dominant solution, and vice versa, with the consequence that the problem of numerical instability is completely overcome. It will not matter what value is given to Z_n, as the object is to determine a sequence of values *proportional* to the decrescent solution. If (for instance) the value of Z_0 is required to be unity by definition, but (as is almost inevitable) this reversed recursion leads to a different value, all the Z_k for $k = 0, 1, 2, \ldots, n$, must be multiplied by that particular factor which makes Z_0 unity consistent with the definition.

However, in order to start the iterative generation of $Z_{n-2}, Z_{n-3}, \ldots, Z_0$ by the reversed recurrence (2.10.7), the value of Z_{n-1} is also needed, as well as Z_n. This can be found from the continued fraction expansion:

$$-Z_n/Z_{n-1} = \frac{p_{n+1}}{q_{n+1}+} \frac{p_{n+2}}{q_{n+2}+} \ldots = \mathop{K}_{j=n+1}^{\infty} p_j \downarrow q_j . \qquad (2.10.8)$$

The proof is left to the reader (problem 3). Although this evaluation involves once more the forward recurrence, no problems of numerical instability arise because a *ratio* of decrescent solutions is being found and not the decrescent solution itself. Moreover the fact that the recurrence relation has a decrescent solution ensures that the continued fraction is rapidly convergent.

Because numerical instability *can* be avoided, it is clearly a form of generated error. It is quite common, incidentally, for *any* form of what we have called 'generated error' to be described as 'numerical instability', although we have specifically reserved the term here to describe the exponentially divergent growth of error in an iterative process – that is, a growth of error by a *factor* (greater than 1) after each iteration. As has been indicated, it is quite dramatic in its effect, and of a quite different character from other forms of numerical error so far encountered. It is met most commonly in 3-term recurrences, but occurs in any other context where divergent and convergent sequences occur as alternative solutions of a particular problem.

2.11 REPRISE

Two related purposes determined the choice of material for this chapter. There was the intention of illustrating methods of accounting for the truncation error involved in attempting to evaluate transcendental functions, and this was quite naturally linked to a study of the techniques of series and continued fraction evaluation. Finally, there was a return to the subject of the first chapter, in the discussion of what was termed numerical instability.

In drawing these threads together into one pattern, maybe there have been some surprises. On the face of it, series evaluation would seem one of the most straightforward of numerical techniques. But without having to concoct any highly involved examples, it has been apparent that there are occasions when it is not good enough merely to sum a series of positive terms until the terms become insignificant; and that even if this is done for an alternating series, when it is demonstrably 'good enough', the result could be largely a nonsense; whilst of course there are many series which are so slowly converging that anyhow it would be quite unrealistic to attempt summation up to the point of saturation. On the other hand, certain kinds of divergent series can be 'summed' very precisely; and stranger still, perhaps, is the cosmetic treatment achieved by recasting an infinite power series as a continued fraction, which not only may extend the range of convergence of the calculation, but may even convert a divergent series into a convergent continued fraction – making a veritable silk purse from a sow's ear.

The unfamiliarity of continued fractions may have caused an emphasis in our treatment of them disproportionate to their utility. But there is a growing realisation that rational expansions of finite order have distinct advantages over polynomials as a means of representing the behaviour of functions, and in any event it is rather pleasing to see discarded parts of classical mathematics resurrected, dusted off, and given a bright new image. It will already be apparent, especially from the problems — and will become the more so as we move on — that computational mathematics abounds in 'classical' references, to judge at least by the name-dropping it occasions. This perhaps reflects the fact that many famous mathematicians of times gone by used to do a lot of patient calculation.

Likewise it will also be apparent by now, and will become increasingly obvious, that recurrence relations, or more broadly iterative procedures, provide the backbone of numerical computation on which all else hangs. Indeed, the subject of series evaluation could have been presented as arising from — rather than leading to — a two-term recurrence relation $S_n = S_{n-1} + t_n$, although that might have seemed unnecessarily obscure. In choosing to illustrate the condition for the termination of the loop of iterated summation by relating truncation to numerical error, it must once again be emphasised that this only arises in practice if maximum precision is sought — as would seem prudent if no specific context for the evaluation is known. But what has been said about the criteria of termination applies equally if a coarser precision is known to be adequate and well within the capabilities of machine arithmetic. The problems are essentially unaltered, except that the superfluous numerical precision then available may obviate the need for some of the more extreme remedies, such as double-precision working.

At a practical level it had better be admitted that, with so many of the more familiar transcendental functions available from numerical libraries, it is possible that a need for the techniques described will only arise in relation to rather uncommon and perhaps complicated functional dependencies. However, as was seen in Chapter 1, instances abound of particular combinations of familiar functions which involve a considerable loss of significance unless evaluated jointly as a single 'corporate' function.

Also it needs to be pointed out that what has been described relates to the *a priori* evaluation of functions having series representations. Such evaluations, from the definition of the function, are often computationally slow and sometimes (as we have seen) quite difficult. If a function has to be evaluated repeatedly — as of course is assumed of the functions provided in a numerical library — then some more rapid and simple *a posteriori* method of evaluation needs to be sought. Its derivation will depend upon the theory of approximation which is to be touched upon in the next chapter. But even in such a context, the *a priori* evaluation is the necessary precursor on which the approximation is developed, and by which its accuracy is confirmed.

CHAPTER 3

Interpolation: curve drawing and function approximation

So far, the existence of error has been regarded as a limitation upon what can be accomplished. Yet it is also possible to introduce error into a numerical process as a benefit — as a means whereby computational speed can be increased at the expense of unnecessary precision. How this might be accomplished in relation to the evaluation of a function is one particular theme among the many which belong to the **theory of approximation**. It is one which will be touched upon later in this chapter, but it is not difficult to anticipate the broad principle involved. A power series in x (for example) is almost certainly rapidly convergent near $x = 0$, where very few terms are needed to provide an accurate sum, but it may be only very slowly convergent for x-values near the limits of its range of application. It would clearly be beneficial if it proved possible to modify the series in such a way that, although it loses its superfluous accuracy near $x = 0$, it becomes more rapidly convergent for larger values of x; and this, as we shall see, does indeed prove possible to arrange.

A much simpler idea would be to employ a large number of different series, each representing the function over a relatively narrow range of values of its argument, and each rapidly convergent in its own selected domain; in this approach, the major problem would be how to generate these series numerically — and once that is decided, the speed of computation would be to a large extent dependent upon the amount of data storage that could be afforded. In an extreme form, this can be likened to the process of tabulating the value of a function at such close intervals that intermediate values can be readily inferred with sufficient accuracy by a very simple rule — the process of **tabular interpolation**. A lot of what might be called the classical theory of approximation alludes to this approach, which was of course the primary method of enabling calculations involving transcendental functions to be made, before the advent of the computer.

The word 'interpolation' is from a Latin root which means 'to polish between' or 'to give a new appearance to'; and that is quite an apt description of the bulk of the work of approximation. Specifically the word is also applied to the construction of a continuous relation connecting two variables, the one depending on the other, when only discrete pairs of data values of these variables are known. In mind here are the results of some practical experiments — like, for example, the records of plant growth over a period of time — where there may not even be a mathematical model of the dependency involved, let alone a known functional relationship, but where nonetheless a continuous variation can be expected, and needs to be constructed. This is what will be called here the process of curve-fitting or **curve-drawing**, more akin perhaps to art than science, since there is no exact basis for judging how true the interpretation may be.

The same term 'curve-fitting' is usually applied to a similar problem which arises where the data values are known to involve observational error, and where the constructed curve or graph does not therefore need to pass through all data values: there is even more room for artistic license available in this problem, which will be distinguished here as **data smoothing**. Or it may be that the data can be assumed approximately to obey one of a family of known relationships, and it is desired to select that particular relation which provides the best fit with the data. This is a process known as **regression** within the subject of statistics, but it is also part and parcel of approximation theory.

Approximation is therefore a widely diversified topic, sometimes entirely cosmetic in intent, sometimes more purposefully remedial. There is considerable overlap in 'approach between the component topics, and no generally accepted terminology by which these components are usually identified.

Despite the undoubted importance of the problem of data smoothing and of the process of regression, involving the construction (for example) of 'least-squares' approximations with which the reader may already be familiar, and despite the fact that such problems and processes are what is often meant by 'approximation', nonetheless it is not the part of the subject to be followed here. Throughout the rest of the book, the data (whether derived from a functional relationship or discrete observation) are taken as precise, and the inprecision — the 'approximation' — results from the process applied.

We start with an account of one of the more cosmetic treatments: that of finding a systematic method of constructing a 'fair curve' through a number of data points. Our concern is with the appearance of the curve — its shape, or more exactly its Cartesian geometry — and how it may be constructed and computed.

3.1 CURVE-DRAWING BY INTERPOLATION

Suppose that there are $(n + 1)$ discrete data pairs relating values of a dependent variable y to another (independent) variable x: say (x_0, y_0), (x_1, y_1), ..., (x_n, y_n). Unless otherwise stated these data pairs will all be assumed distinct — no two data values of x are the same. Also it will be assumed that they are ordered so that $x_0 < x_1 < \ldots < x_n$, unless it is specifically stated that this is immaterial (as not infrequently happens). The values of y_k are the data **ordinates**, and x_k are the corresponding data **abscissae**. These x-values are also referred to by various other names. Collectively, they are sometimes called the net of x-values, or the **support** of the interpolative scheme; individually, they are referred to by such names as grid points, breakpoints, nodes, knots and so on – the terminology usually seems to depend on the context.

It is required to find a method by which a relationship $y = y(x)$ can be constructed which describes the shape of a curve passing through, or **inter-polating**, each data point. Thus it is assumed that $y(x_k) = y_k$, and further it will be required that $y(x)$ is single-valued and continuous over an interval $a \leqslant x \leqslant b$, which embraces all the data points (so that $a \leqslant x_0 < x_n \leqslant b$). If the interval of interest extends outside $[x_0, x_n]$ then the process of **extrapolation** of the data outside the bounds for which values are given is involved, as well as interpolation between the given values. Unless otherwise stated, however, it will be assumed that extrapolation is not the primary interest.

Obviously something more needs to be said about the required function $y(x)$ if one method of constructing it is to be preferred to another. Given a particular set of data (Fig. 3.1) there are of course an infinity of ways in which a functional relationship might be interpolated – a few are shown in Figs 3.1a – f. Here (a) can be dismissed, as $y(x)$ is assumed single-valued, and (b) and (c) do not appear to meet the requirement that the curve should be continuous; but why should (d) and (e) appear less acceptable than (f)? The eye may well reject the possibility of the interpolated oscillations in (e) because the data provides no positive evidence for their existence. The sequence of straight lines joining consecutive points, as in (d), is hardly the 'smooth' curve provided by (f); so evidently continuity of function value *and* of derivatives of $y(x)$ enters into one's judgement of acceptability. This can be expressed by asserting that $y(x)$ should belong to the function space $C^m[a, b]$ where m is a positive integer, meaning that $y(x)$ and its first m derivatives are to be continuous for all x in $[a, b]$. Then the curve (d) is rejected (since m must be positive), but the eye is also to some extent 'offended' by discontinuities in curvature as well as slope, suggesting that m should not be less than 2. But note that even if m were required to be infinite, this would not eliminate a curve like that shown in Fig. 3.1e.

We are not, for the moment at least, going to be able to state what charac-terises an 'acceptable' method of constructing $y(x)$ beyond the observation that in the example of Fig. 3.1, the curve (f) is of the kind we are looking for. However, the reader will in all probability be in no doubt what is the broad intent, for no-one finds it difficult to draw a 'smooth' curve through a set of points. The brain solves the problem with very little fuss, although admittedly never twice in the same way – which is reasonable enough, as it is not suggested that there is a uniquely correct 'solution'. Yet, perforce, in mechanised com-putation, a particular set of rules has to be set down. About all that can be said at the outset is that if these are to embody the intent of hand-draughting, then continuity of slope and (perhaps) curvature, as well as function value, must be implied, but (for example) seemingly unnecessary oscillatory excursions between data points would not be acceptable. We try to find the *simplest* method of constructing $y(x)$ which is unobjectionable as judged by such broad considerations. This approach will itself provide some insight into what in fact underlies the subjective judgements involved in hand-draughting.

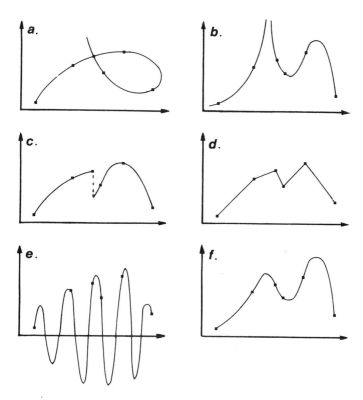

Fig. 3.1 — A set of data: six curve-fittings.

3.2 POLYNOMIAL INTERPOLATION FORMULAE

One simple and 'obvious' way of constructing the function $y(x)$ which ensures continuity of *all* its derivatives, not just the first and second (which determine slope and curvature), is to define it as the polynomial $p_n(x)$ of least degree n which intersects all the $(n + 1)$ data points. If there were only two data points, then this polynomial would be the straight line described by $y = p_1(x)$, joining (x_0, y_0) to (x_1, y_1), given formally by

$$p_1(x) \equiv \frac{(x - x_1)}{(x_0 - x_1)} y_0 + \frac{(x - x_0)}{(x_1 - x_0)} y_1$$

$$= [(x - x_1)y_0 - (x - x_0)y_1]/(x_0 - x_1) \; . \qquad (3.2.1)$$

Similarly, three data pairs would be interpolated by a parabola $y = p_2(x)$, unless it happened that the three data points were collinear, in which event the curve $y = p_2(x)$ would degenerate into a straight line. The general expression for the quadratic can be written as

$$p_2(x) \equiv \frac{(x - x_1)(x - x_2)}{(x_0 - x_1)(x_0 - x_2)} y_0 + \frac{(x - x_2)(x - x_0)}{(x_1 - x_2)(x_1 - x_0)} y_1$$

$$+ \frac{(x - x_0)(x - x_1)}{(x_2 - x_0)(x_2 - x_1)} y_2 \; , \qquad (3.2.2)$$

and it can easily be checked that $p_2(x_0) = y_0$, $p_2(x_1) = y_1$ and $p_2(x_2) = y_2$ as is clearly required if the parabola is to intersect all three data points. Also it can be seen that, if the three points are collinear, then the multiple of x^2 in the expression for $p_2(x)$ must be zero, so that this exception arises if

$$0 = y_0/[(x_0 - x_1)(x_0 - x_2)] + y_1/[(x_1 - x_2)(x_1 - x_0)]$$

$$+ y_2/[(x_2 - x_0)(x_2 - x_1)] \; .$$

This can be rearranged as the simpler, and perhaps more familiar, condition that three data points are collinear, namely:

$$(x_2 - x_1)y_0 + (x_0 - x_2)y_1 + (x_1 - x_0)y_2 = 0 \; .$$

But of course such a condition will not usually be satisfied.

This process of polynomial interpolation can be extended to any number of data points: in general, the polynomial $p_n(x)$ of least degree which intersects $(n + 1)$ data points will be of degree not greater than n — because such a polynomial has $(n + 1)$ adjustable coefficients, matching the number of constraints. Exceptionally, for particular sets of data, it may degenerate to a polynomial of degree less than n. Also, just as in (3.2.1) and (3.2.2), this polynomial can be expressed as the sum of $(n + 1)$ terms each linearly dependent on *one* of the data ordinates and independent of all others (though not independent of the data abscissae). This can be expressed formally by writing

$$p_n(x) \equiv \sum_{k=0}^{n} y_k \, s_{nk}(x_0, x_1, \ldots, x_n; x) \; , \qquad (3.2.3)$$

where the $s_{nk}(..)$ are called **shape functions**.

It will be of importance to find out how to express these functions, and a glance back at (3.2.1) and (3.2.2) might well correctly suggest that they are going to be polynomials in x, whose coefficients depend on the data values $x_0, x_1, \ldots, x_n$. However, before following up this suggestion, it is worthwhile setting what is going to be discussed in a rather wider context. If a curve is being fitted which — as implied by (3.2.3) — has the form $y = \sum\limits_{k=0}^{n} y_k s_{nk}(x_0, x_1, \ldots, x_n; x)$, where what we have termed the shape functions s_{nk} are independent of the data ordinates, then irrespective of how these functions are expressed — whether or not they are polynomials — the process is said to be an example of **linear interpolation**. The term 'linear' obviously applies to the dependence of y on the data *ordinates*, and not the abscissae. Most forms of interpolation which will be discussed in this chapter fall into this general category.

If a function $p_n(x)$ is defined as a polynomial of degree $\leqslant n$, and $p_n(x)$ is expressed as a linear combination of $(n+1)$ terms, say as $\sum\limits_{k=0}^{n} a_k B_k(x)$, where the a_k's are arbitrary coefficients independent of x, and the B_k's are linearly independent functions of x, then the set of $(n+1)$ functions $B_k(x)$ for $k = 0, 1, \ldots, (n+1)$ is said to form a **basis** for the polynomial. The most familiar basis would be the set $\{1, x, \ldots, x^n\}$ of ascending powers of x, and such a selection would lead to the usual method of writing down a polynomial in its **power form** as

$$p_n(x) \equiv \sum_{k=0}^{n} a_k x^k = a_0 + a_1 x + \ldots + a_n x^n$$

However, there are infinitely many ways of composing a basis for a polynomial. Incidentally, the condition that the basis functions $B_k(x)$ should be linearly independent requires that the equation $\sum\limits_{k=0}^{n} a_k B_k(x) = 0$ is true if, and *only* if, all the coefficients a_k are zero.

Linking this definition to (3.2.3), it can be inferred that the shape functions s_{nk} for $k = 0, 1, \ldots, n$, must form a basis for any polynomial of degree $\leqslant n$. Because of this, they are sometimes referred to as *cardinal* basis functions with support $x_0, x_1, \ldots, x_n$. Since shape functions (or cardinal basis functions) will be mentioned in relation to various examples of linear interpolation in what follows, it will be valuable to try to develop at the outset a feeling for their meaning.

As a consequence of the assertion that $y = \sum\limits_{k=0}^{n} y_k s_{nk}(..)$ interpolates the data, and that the shape functions are independent of the data ordinates, it follows that for each value of $k = 0, 1, \ldots, n$, each of these shape functions must therefore obey the condition that

$$s_{nk}(x_0, x_1, \ldots, x_n; x_j) = \delta_{jk} \qquad (j = 0, 1, \ldots, n) \ . \qquad (3.2.4)$$

Here we have used a useful symbol called the **Kronecker delta**, δ_{jk}, which is defined to be equal to 1 if the two subscripts are the same (that is, if $j = k$), but to have the value 0 if the subscripts are different. For example, referring back to polynomial interpolation and equation (3.2.1), and accordingly considering n to be 1, there are then just two shape functions, s_{10} and s_{11}, to be considered, which can be identified respectively as equal to $(x - x_1)/(x_0 - x_1)$ and $(x - x_0)/(x_1 - x_0)$. Evidently, (3.2.4) states that $s_{10}(x_0, x_1; x_j) = \delta_{j0}$ and it is indeed evident that s_{10} is 1 at $x = x_0$ (that is, at $x = x_j$ for $j = 0$), and zero at $x = x_1$, consistent with the fact that $\delta_{10} = 0$. Likewise, s_{11} is zero at $x = x_0$, and unity at $x = x_1$, consistent with the statement that $s_{11}(x_0, x_1; x_j) = \delta_{j1}$ for $j = 0, 1$.

Suppose now that one of the data ordinates, say y_m at $x = x_m$, is either in error, or is for some reason changed, by an increment δy_m. Then the ordinate y of the interpolating curve $y = \sum_{k=0}^{n} y_k s_{nk}(x_0, x_1, \ldots, x_n; x)$ will evidently be changed by an increment $\delta y_m s_{nm}(x_0, x_1, \ldots, x_n; x)$ for all values of x. In other words, the function $s_{nm}(..)$ can be regarded as determining for any value of x, the incremental change in shape of the curve being fitted per unit change (or error) in the particular data ordinate y_m. This applies of course to any data ordinate, and so each shape function s_{nk} denotes the influence on the shape of the interpolating curve of a change or error in the ordinate y_k with which it is said to be **associated** (as identified by its second subscript). It is this important property which gives 'shape' functions their name.

It can also be observed that the curve $y = s_{nk}(x_0, x_1, \ldots, x_n; x)$ for particular values of n and k intersects the x-axis at all but one of the support points $x = x_j$ for $j = 0, 1, \ldots, n$, the exception being the point $x = x_k$ corresponding to $j = k$; and at this particular value of x, $y = \delta_{jk} = 1$. In other words, this particular curve would interpolate the set of $(n + 1)$ data pairs (x_j, δ_{jk}) for $j = 0, 1, \ldots, n$: all of these data ordinates are zero, except that belonging to the associated pair $(x_k, 1)$. This interpretation does not seem to throw any extra light on the meaning to be associated with shape functions, but it provides a simple conceptual principle by which the form of the shape function can be deduced.

Returning now to consider polynomial interpolation, it follows that $s_{nk}(..)$ in (3.2.3) must be selected so that $y = s_{nk}(..)$ is the particular example of a polynomial $p_n(x)$ of degree at most n, which interpolates the $(n + 1)$ data pairs (x_j, δ_{jk}) for $j = 0, 1, \ldots, n$. But this will be the polynomial of genuine degree n with the n real zeros $x = x_j$ for all values of $j = 0, 1, \ldots, n$, save $j = k$. Further, since $s_{nk}(..) = 1$ at $x = x_k$ it follows therefore that this shape function can be written as

$$s_{nk}(x_0, x_1, \ldots, x_n; x) = \prod_{\substack{j=0 \\ j \neq k}}^{n} [(x - x_j)/(x_k - x_j)] . \qquad (3.2.5)$$

This states that $s_{nk}(..)$ is the product of n factors $(x - x_j)$, omitting only the factor $(x - x_k)$ from the total given by $j = 0, 1, \ldots, n$, and the form of the denominator ensures its unit value at $x = x_k$. It is not difficult to see that this expression is consistent with the required condition (3.2.4) for a shape function. Further, since (3.2.3) shows that $p_n(x)$ is a linear sum of these functions, it too becomes a polynomial (as of course is required) of degree *not greater than n*. One says 'not greater than' since there exists the possibility that terms of highest degree in x in each component $y_k s_{nk}(..)$ of the sum may cancel each other out when they are added together. Trivially, for example, if it happened that all the y_k were equal to one another, then the polynomial through these points is the straight line $y = y_0$ parallel to the x-axis, and $p_n(x)$ would be constant – that is, it would have zero degree.

The relation $y = p_n(x)$, where $p_n(x)$ is given by (3.2.3), and (3.2.5) is termed **Lagrange's interpolation formula**; if $n = 1$ or 2, it can be verified that it simplifies to the expressions (3.2.1) and (3.2.2) for interpolation by linear and quadratic polynomials. There are numerous other and perhaps more convenient ways in which $p_n(x)$ could be written down – one might want to sort out the multiples of various powers of x and write it as, for example, $p_n(x) = a_0 + a_1 x + \ldots + a_n x^n$ – but in whatever way written, it is the unique polynomial of degree not greater than n which interpolates the data. This uniqueness can most easily be proved by *reductio ad absurdum*. For suppose that there were a different polynomial $q_n(x) \neq p_n(x)$ also of degree not greater than n which interpolated the same points: then $[p_n(x) - q_n(x)]$ would be a (non-vanishing) polynomial of degree not greater than n, and yet it would have $(n + 1)$ zeros at the distinct points $x = x_0, x_1, \ldots, x_n$, because $p_n(x_k) = q_n(x_k) = y_k$ for $k = 0, 1, \ldots, n$. But a polynomial with $(n + 1)$ distinct zeros at $x = x_k$, which is not identically zero, must have $(n + 1)$ different linear factors, $(x - x_k)$, and therefore be at least of degree $(n + 1)$: which is a contradiction of our assumption. In fact, therefore, q_n and p_n must be one and the same polynomial, and their difference is accordingly identically zero.

There are infinitely many polynomials of degree *greater* than n which interpolate the same data. Any polynomial of degree $(n + 1)$ or more, which has zeros at $x_0, x_1, \ldots, x_n$, can be added to $p_n(x)$ and still provide an expression for y which accords with all the data. One might therefore ask, why choose the polynomial of least degree? or, indeed, for that matter, why choose a polynomial? The answer to such questions has to be, why not? For there is no unique solution to the problem posed, and any 'solution' can be considered possible and reasonable unless it can be faulted. Experience shows that Lagrangian interpolation can indeed be faulted.

To see why this should be, it is necessary to consider the behaviour of the shape functions $s_{nk}(..)$ given by (3.2.5). These, as we have seen, describe the influence of each associated data value y_k upon the shape of the curve, by determining the increment to the value of $p_n(x)$ arising from a change, or error, in y_k. If there are a lot of points, then intuitively one would imagine that a particular $s_{nk}(..)$ should contribute dominantly to the shape of the curve near $x = x_k$. Obviously precisely at $x = x_k$ it is the *only* contributor to the value of $p_n(x)$, since all the other shape functions s_{nj} with $j \neq k$ are zero at $x = x_k$. But its contribution to the value of $p_n(x)$ should progressively

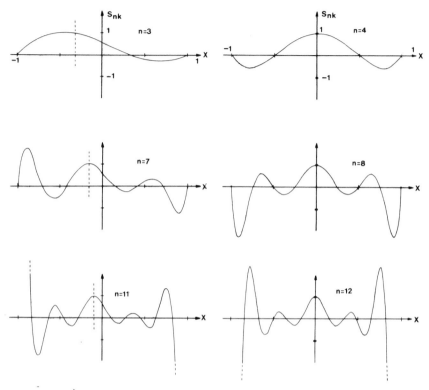

Fig. 3.2 – Shape functions $s_{nk}(x)$ for Lagrangian interpolation on equispaced support, $x_j = (2j/n) - 1$, of various degrees n and with $k = $ entier $(n/2)$.

diminish as the value of $|x - x_k|$ increases. Rather more precisely, because $s_{nk}(..)$ becomes zero at all data points other than $x = x_k$, it could be anticipated that it should oscillate about these zeros with decreasing amplitude as $|x - x_k|$ increases. But Fig. 3.2 shows what happens in the quite common condition of equidistant data values (such that $x_k - x_{k-1} = h$, a constant). Considering a value of k corresponding to a data point at or near the middle of the set of data, and allowing n to increase, it is clear that, far from its contribution decreasing as $|x - x_k|$ increases, the opposite happens. It provides a large contribution to the value of $p_n(x)$ close to the end points of the interval – much larger in fact than those near $x = x_k$ if $n > 6$. In other words, a small change to a data point in the middle of the set of values can produce a very large excursion in the curve near the ends – quite an unnatural consequence which is called **Runge's phenomenon** (or the Runge–Meray syndrome). The outcome is that, as applied to sets of a dozen or so data with more or less equally spaced support, the curve described by the Lagrangian polynomial often oscillates even although the data ordinates decrease monotonically with x (Fig. 3.3).

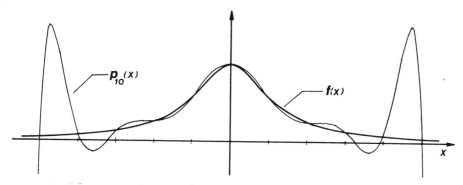

Fig. 3.3 – Lagrangian interpolation on an equispaced 11-point support of
$$f(x) \equiv (1 + x^2)^{-1}.$$

It will be demonstrated later that there are particular arrangements of x-values – particular supports – for which high-degree Lagrangian interpolation is exempt from such undesirable behaviour. Moreover the Runge phenomenon is absent if there are no more than, say, half-a-dozen data pairs to be interpolated. So it certainly cannot be dismissed as an uninteresting technique. But in the context of curve-drawing, the number and arrangement of the data are not open to choice, and a numerous and more-or-less equi-spaced support is so common a requirement of interpolation that any defect it reveals becomes an overriding deficiency. There are other methods of – successfully – adapting polynomial interpolation to the problem, which will be discussed later and which are much to be preferred.

3.2.1 Techniques of polynomial evaluation

Nonetheless, because there are important applications of polynomial interpolation, it is relevant to consider how it can be most efficiently used in computational work. Although (3.2.3) and (3.2.5) together provide an algebraically succinct way of stating the formula for $p_n(x)$, they do not provide a very suitable algorithm for calculating its particular value for some given x. A variety of methods of doing this have been suggested.

Both methods to be discussed in this section arrange the computation as the generation of a triangular array of numbers, with the required value of $p_n(x)$ at its apex. The elements of this array can be referenced by the usual matrix notation as $a_{j,k}$ to denote the element of the array in row j and column k, including row and column 0:

a_{00}

$a_{10} \quad a_{11}$

$a_{20} \quad a_{21} \quad a_{22}$

$a_{30} \quad a_{31} \quad a_{32} \quad a_{33}$.

The table extends during the course of calculation up to the row numbered n, and finishes with the calculation of a_{nn}. New rows are added, one-by-one, and in **Aitken's method** the elements of row j (say) are computed from left to right using the recursion formula:

$$a_{j,0} = y_j, \quad a_{j,k+1} = [a_{jk}(x - x_k) - a_{kk}(x - x_j)]/(x_j - x_k) \quad (3.2.6)$$

for $k = 0, 1, \ldots, (j - 1)$. Thus $a_{00} = y_0$ in the zeroth row, and $a_{10} = y_1$, $a_{11} = [y_1(x - x_0) - y_0(x - x_1)]/(x_1 - x_0)$ in the first row, and so on. The last value calculated, a_{nn}, will then be the value of $p_n(x)$.

In performing this calculation by hand, it is useful to append a column of values of $(x - x_j)$ to the left of the triangular array: then the recursion formula of (3.2.6) assigns the value

of the determinant ratio

$$\begin{vmatrix} A & D \\ B & C \end{vmatrix} \div \begin{vmatrix} A & 1 \\ B & 1 \end{vmatrix}$$

$$= (AC - BD)/(A - B)$$

$(x - x_j)$	a_{j0}	a_{j1}	a_{j2}	a_{j3}	$\cdots$
	$\cdot$	$\cdot\,\cdot$			
A	$\cdot$		D		
$\vdots$	$\vdots$	$\vdots$	$\vdots$		
B	$\cdot$		C	$*$	$\cdot$

to the (typical) element asterisked in the boxed diagram, the element D always being on the diagonal of the array in the previous column.

Although the triangular arrangement is descriptively convenient, and useful in hand computation, there is in fact no need to preserve all the values of this array: only the diagonal elements of previous rows are subsequently referenced. So for machine calculation, if these are stored in a vector d_k (say) for $k = 0, 1, \ldots, n$, the effect of this calculation is achieved by the instruction sequence:

> **for** $j: = 0$ **step** 1 **until** n **do**
>
>> **begin** $d_j: = y_j$; $s: = x - x_j$;
>>
>>> **if** $j > 0$ **then for** $k: = 0$ **step** 1 **until** $j - 1$ **do**
>>>
>>>> $d_j: = d_j + (d_j - d_k) * s/(x_j - x_k)$
>>
>> **end**; .

The values assigned in this way to d_j are equal temporarily to $a_{j,k+1}$, and finally to a_{jj}, so that d_n is the required value of $p_n(x)$. Evidently the scheme adapts neatly and economically to iterated (as well as tabulated) calculation.

The reason why the method produces the correct value of $p_n(x)$ is unlikely to be immediately obvious: it depends on a simple and general recursive proposition. In order to exhibit this, it is useful to introduce the notation of set theory. Let us denote a particular set of (say) m data pairs by **S**: this *could* be the set of data pairs (x_k, y_k) for $k = 1, 2, \ldots, m$, but equally well it can be *any* combination of data pairs from the $(n + 1) \geqslant m$ available in the specific problem, not necessarily those successive in subscript value. Further, let $p_{m-1}(S;x)$ be the polynomial which interpolates these m data pairs, and which therefore has degree not greater than $(m-1)$, one less than the number of elements (or cardinality $|S|$) of the set S. Note that this polynomial will depend upon the *set* S but not the *sequence* of the data pairs within the set: in other words, it is completely immaterial to the value of $p_{m-1}(S;x)$ for any given x how the data pairs making up S may be ordered if the elements of the set are written out.

Now consider two sets of data pairs, both different, call them S_j and S_k, which both have cardinality $(m+1)$ and whose intersection is the set S: this means that both sets S_j and S_k are composed of the m data pairs of set S, plus one more. To be specific, suppose that the difference set $S_j \backslash S$ has as its single element the data pair (x_j, y_j); likewise, let $S_k \backslash S$ be the set with the one element (x_k, y_k) where $k \neq j$. In the notation established, the polynomials $p_m(S_j;x)$ and $p_m(S_k;x)$ interpolate the $(m+1)$ data pairs of S_j and S_k respectively, and so both are polynomials of degree $\leqslant m$. Then the recurrence on which Aitken's method is based can be written as:

$$p_{m+1}(S_j \cup S_k;x) = [(x-x_k)p_m(S_j;x)-(x-x_j)p_m(S_k;x)]/(x_j-x_k) .$$
(3.2.7)

Evidently the form of this relation ensures that $p_{m+1}(..)$ represents a polynomial of degree not greater than $(m+1)$: and it is not difficult to verify that – as implied by the notation – it does indeed interpolate all the $(m+2)$ data pairs belonging to either S_j or S_k. Clearly (x_j, y_j) and (x_k, y_k) are two elements of the union of S_j and S_k: and the form of (3.2.7) ensures, for example, that $p_{m+1} = y_j$ at $x = x_j$, because $p_m(S_j;x_j) = y_j$; and likewise $p_{m+1} = y_k$ at $x = x_k$. In particular, if S is an empty set (that is, $m = 0$) then (3.2.7) becomes identical with the formula (3.2.1) for linear interpolation (replacing j, k by 0, 1). If $m \neq 0$, and if the data pair (x_i, y_i) is any one of the m elements of S, then by definition

$$p_m(S_j;x_i) = p_m(S_k;x_i) = y_i ,$$

and so (3.2.7) ensures that $p_{m+1}(S_j \cup S_k;x_i) = y_i$, implying interpolation of each data pair of S. Hence it interpolates all data pairs of $S_j \cup S_k$.

The form of (3.2.7) is known as a **linear cross-mean**, and it is also to be seen in the expression (3.2.1) for linear interpolation. Starting by placing $m = 0$ in (3.2.7) – which produces linear interpolating polynomials – the same formula can be applied (with $m = 1$) to obtain quadratic polynomials interpolating three points, then reapplied to these (with $m = 2$) to provide cubic interpolates, and so on. Thus by the application of a succession of linear cross-means (3.2.7) allows a polynomial of least degree to be progressively built up which interpolates any prescribed set of $(n+1)$ data pairs – and this construction can be effected either algebraically or numerically. There are many different ways in which this building-up process could be conducted. To observe how the recursion (3.2.7) is applied in Aitken's method, some other notation is necessary which explicitly refers to the data pairs which each polynomial interpolates.

This can be accomplished by multiple subscripting of y: y_i is then consistently the constant 'polynomial' of zero degree interpolating (x_i, y_i); y_{ij} the linear polynomial interpolating (x_i, y_i), (x_j, y_j); and so on. Since the ordering of the data is immaterial, the subscripts can be permuted in any way without effect upon the meaning (so that y_{ij} and y_{ji} represent the same polynomial); conventionally, they will be ordered in increasing value. Then comparison of (3.2.6) with (3.2.7) shows that the triangular array of Aitken's method is really made up of the (numerical) values at the given x of a corresponding array of polynomials:

y_0

$y_1 \quad y_{01}$

$y_2 \quad y_{02} \quad y_{012}$

$y_3 \quad y_{03} \quad y_{013} \quad y_{0123}$

$y_4 \quad y_{04} \quad y_{014} \quad y_{0124} \quad y_{01234}$

or in formal terms

$$a_{jk} = y_{0,1,\ldots,(k-1),j} \cdot \qquad\qquad (3.2.8)$$

Hence $a_{nn} = y_{0,1,\ldots,n}$ is the value of the polynomial interpolating all $(n+1)$ points — which (it needs to be recalled) was what was to be proved.

But this analysis associates a meaning with *every* a_{jk} of the scheme, as equal to the value of a polynomial which interpolates some subset of the data.

Moreover the proof makes it clear that the ordering of the data pairs (x_k, y_k) is quite immaterial to the validity of Aitken's method — the fact that the x_k for $k = 0, 1, \ldots, n$ were originally supposed numerically increasing has not been used. The only significance of the subscripts appended to the data coordinates (x, y) is in determining the order in which the data pairs are incorporated in the tabular layout, and this can be a matter of choice.

It has been noted that the linear cross-means of (3.2.7) can be recurrently applied in many different ways to build up the final value of $p_n(x) = y_{0,1,\ldots,n}$. Another procedure is **Neville's algorithm**, by which the jth row is formed by the recursion:

$$a_{j0} = y_j, \quad a_{j,k+1} = [a_{j,k}(x - x_{j-k-1}) - a_{j-1,k}(x - x_j)]/(x_j - x_{j-k-1})$$
$$(3.2.9)$$

with $k = 0, 1, \ldots, (n-1)$. Again the final value of the last row (a_{nn}) equals the required value of $p_n(x)$. The pattern by which this value is built can be indicated as

$$y_1$$

$$y_1 \quad y_{01}$$

$$y_2 \quad y_{12} \quad y_{012}$$

$$y_3 \quad y_{23} \quad y_{123} \quad y_{0123}$$

$$y_4 \quad y_{34} \quad y_{234} \quad y_{1234} \quad y_{01234}$$

$$\cdots\cdots$$

which implies the identification:

$$a_{jk} = y_{(j-k),(j-k+1),\ldots,j} \ . \tag{3.2.10}$$

The values (a_{jj}) on the diagonal are evidently the same as those of the Aitken method.

In evaluating this triangular array by hand, if a left-hand column of values of $(x - x_j)$ is included, then the spatial distribution of the elements referenced by the recursion (3.2.9) is indicated by the appended diagram where, as before, the asterisked element is assigned equal to the ratio of the determinants

$$\begin{vmatrix} A & D \\ B & C \end{vmatrix} \div \begin{vmatrix} A & 1 \\ B & 1 \end{vmatrix}$$

$$= (AC - BD)/(A - B).$$

But note that now D is in the previous row of the previous column – that is on the diagonal from $*$ in the previous column – and that the element $A = (x - x_{j-k-1})$ matches the term $a_{j-k-1,0} = y_{j-k-1}$, which is also on this same diagonal from $*$.

The economical arrangement of Neville's algorithm for machine computation is not entirely straightforward. Only the last row calculated needs to be preserved for further reference. If each element a_{jk} is assigned to a vector element r_{n-j+k}, then the new values of the jth row can be overwritten on those of the previous row preserved in one and the same vector. Then (3.2.9) shows that $a_{j,k+1}$ depends on $a_{j,k}$ (just assigned to r_{n-j+k}) and upon $a_{j-1,k}$, which has been previously recorded in the vector in the next storage position $r_{n-j+k+1}$ to which the value of $a_{j,k+1}$ (when calculated) is due to assigned. An instruction sequence which economises on storage in this way is

for j: $= 0$ **step** 1 **until** n **do**

    begin r_{n-j}: $= y_j$; s: $= x - x_j$;

        if $j > 0$ **then for** m: $= n - j + 1$ **step** 1 **until** n **do**

$$r_m: = r_{m-1} + (r_{m-1} - r_m) * s/(x_j - x_{n-m})$$

    end; .

In this algorithm, the values of $(n - j + k)$ are assigned to m for $k = 1, 2, \ldots, j$ and the value of $p_n(x) = y_{0,1,\ldots,n}$ is finally assigned to the element r_n. However, the vector r_k for $k = 0, 1, \ldots, n$, leaves the outer loop with its elements equal to those of the last row, a_{nk}, and not those of the diagonal as in Aitken's method.

3.2.2 Newton's form of the interpolating polynomial

Although the ideas embodied in Aitken's and Neville's algorithms are of importance — because they illustrate particular applications of the idea of the recursive generation of polynomials by linear cross-means, as given in general by the formula (3.2.7) — they do not, in fact, provide the best method of calculating the value of the interpolating polynomial. Certainly if several values of the same polynomial are required for different values of x, as for example would be necessary in constructing a curve through discrete data, it would seem likely to be more efficient to pre-calculate the coefficients of the polynomial, as there are then far fewer operations required to provide its value than in constructing the triangular tableaux of Aitken's or Neville's scheme. One obvious and relatively quick way of arriving at a value would result from a knowledge of the coefficients of $a_0, a_1, \ldots, a_n$ in the rearrangement of the Lagrange polynomial in its power form $a_0 + a_1 x + \ldots + a_n x^n$. But there is another similar formulation which is almost as quickly evaluated, and whose coefficients are much more readily computed. This is called **Newton's form** of the interpolating polynomial:

$$p_n(x) = d_0 + d_1(x - x_0) + d_2(x - x_0)(x - x_1) + \ldots$$
$$+ d_n[(x - x_0)(x - x_1) \ldots (x - x_{n-1})]$$
$$= \sum_{j=0}^{n} [d_j \prod_{k=0}^{j-1} (x - x_k)] , \qquad (3.2.11)$$

where the values of $d_0, d_1, \ldots, d_n$ are numerical coefficients, and the $x_0, x_1, \ldots, x_n$ are the data abscissae. In the symbolic representation, note that the repeated product is to be replaced by unity if its upper limit $(j - 1)$ is negative.

A glance at this polynomial is enough to show that its value for any given x would be much more quickly calculated than the Lagrangian form (3.2.3), which is the sum of $(n + 1)$ shape functions each with n factors involving x. Supposing that the coefficients of (3.2.11) are known, then $p_n(x)$ could be efficiently computed using nested multiplication:

$$p_n(x) = ((\ldots (d_n \times (x - x_{n-1}) + d_{n-1}) \times (x - x_{n-2}) + \ldots$$
$$+ d_2) \times (x - x_1) + d_1) \times (x - x_0) + d_0$$

the expression starting with $(n - 1)$ opening brackets. Alternatively, arranged as a sequential calculation rather than as an expression, this becomes

$$y: = d_n;$$

for $j: = n - 1$ **step** -1 **until** 0 **do** $y: = y * (x - x_j) + d_j$;

The value finally assigned to y is then equal to $p_n(x)$. Either way, there are just n multiplications, subtractions and additions.

It can also be immediately observed that the coefficients must to some extent be dependent on the ordering of the data: obviously if x_0 and x_1 were merely interchanged everywhere in (3.2.11) but there was no other change in the ordering of the data, and we assume that there are no consequential changes in the values of any of the coefficients, then the value of $p_n(x)$ would be incremented by the amount $d_1(x_0 - x_1)$. This cannot in general be correct (since the polynomial value must be unaffected), and so the assumption that the coefficient values are unaltered cannot be correct. We return to this point again later, but for the moment it is important to state that, although the ordering of the data is therefore significant, no *particular* ordering need be assumed (such as, for example, $x_0 < x_1 < ... < x_n$). The sequence of the data is, in other words, fixed but arbitrary.

To see why the coefficients are easy to compute, let us suppose that for any non-negative integer $m < n$, the polynomial $p_m(x)$ is defined merely by replacing n by m in (3.2.11), but leaving all the coefficients unchanged: that is, we write

$$p_m(x) = \sum_{j=0}^{m} [d_j \prod_{k=0}^{j-1} (x - x_k)] \qquad (m = 0, 1, ..., n) \ .$$

Then in particular (with $m = n - 1$) it would follow that

$$p_{n-1}(x) = p_n(x) - d_n[(x - x_0)(x - x_1) ... (x - x_{n-1})] \ ,$$

and so the values of $p_n(x)$ and $p_{n-1}(x)$ coincide at all the points $x = x_0, x_1, ..., x_{n-1}$. But, by definition, $p_n(x)$ interpolates all the $(n + 1)$ data points: so therefore $p_{n-1}(x)$ interpolates the n data pairs (x_k, y_k) for $k = 0, 1, ..., (n-1)$. Proceeding in this way, we see inductively that, for all $0 \leqslant m \leqslant n$, the polynomial $p_m(x)$ so defined interpolates the $(m + 1)$ data pairs (x_k, y_k) for $k = 0, 1, ..., m$; or in terms of the notation of the previous section

$$p_m(x) = y_{0,1,...,m} \ .$$

The values of $p_m(x)$ for particular values of x are therefore the values a_{mm} of the elements on the diagonal of the triangular tableaux generated by Aitken's or Neville's method.

Recalling that each $p_m(x)$ has as its coefficients a subset $\{d_0, d_1, ..., d_m\}$ of those belonging to $p_n(x)$, it can be observed that, for any given m, any particular one, say d_k, has the property of being the coefficient of the highest power x^k in the expression for $p_k(x)$. To confirm this, it is merely necessary to observe that d_k multiplies the term $(x - x_0)(x - x_1) ... (x - x_{k-1})$ which is the only one of degree k in the expression for $p_k(x)$, just as d_n multiplies the term $(x - x_0)$ $(x - x_1) ... (x - x_{n-1})$ which is the only one of degree n in the expression (3.2.11) for $p_n(x)$. Let us denote the leading coefficient (that is, of the highest power, x^m) of any polynomial interpolating $(m + 1)$ points – whatever the

value of m — by appending a caret to the relevant symbol: then it follows that
the coefficients of the Newton form (3.2.11) are given by

$$d_m = \hat{p}_m(x) = \hat{y}_{0,1,...,m} \qquad (m = 0, 1, ..., n) \ .$$

The significance of this particular coefficient $\hat{p}_m(x)$ is that $p_m(x)/x^m \to \hat{p}_m(x)$
as $x \to \infty$; correspondingly, $y_{0,1,...,m}/x^m \to \hat{y}_{0,1,...,m}$ or, referring back to the
triangular arrays, $a_{mm}/x^m \to d_m$ as $x \to \infty$. A rough and ready way of estimating
d_m which might at once suggest itself would therefore be to calculate the
triangular array for some 'very large' x, and then to divide each diagonal element
a_{mm} by x^m. However, one can do very much better than this.

In either Aitken's or Neville's scheme, (3.2.8) and (3.2.10) show that the
elements a_{jk} in the kth column are values of polymomials of degree $\leqslant k$ interpolating $(k + 1)$ points. Hence the values of a_{jk}/x^k tend to the coefficients $\hat{a}_{jk}$
of the highest (kth) power of x in each such polynomial. Thus in the Aitken
scheme, replacing a_{jk} by $\hat{a}_{jk} x^k$ in (3.2.6) and allowing x to tend to infinity, it is
seen that values of a_{jk} are determined by

$$\hat{a}_{j0} = y_j, \ \hat{a}_{j,k+1} = (\hat{a}_{jk} - \hat{a}_{kk})/(x_j - x_k) \ , \qquad (3.2.12)$$

for $k = 0, 1, ..., (j - 1)$ and $j = 0, 1, ..., n$. This recurrence leads, just as in
Aitken's method, to a triangular array of values of $\hat{a}_{jk}$, and so to the terms on its
diagonal which have been identified with the values

$$\hat{a}_{mm} = \lim_{x \to \infty} (a_{mm}/x^m) = \hat{y}_{0,1,...,m} = d_m \ . \qquad (3.2.13)$$

The recurrence therefore provides a way of calculating the coefficients of
Newton's interpolating polynomial.

The formula (3.2.12) — which incidentally can be deduced without appealing to the idea of a limit (see problem 4, section 3.2.1) — is clearly a great deal
simpler than that of (3.2.6) on which it is based: the two multiplications by
linear factors in x in each application of the latter recurrence are omitted. This
not only avoids a possible source of generated error, but saves a total of $n(n + 1)$
arithmetic operations in the formation of the array, plus the additional $(n + 1)$
subtractions in forming the linear factors — that is, a total of $(n + 1)^2$ operations
in all. Thus even if the polynomial is to be evaluated for only one of x, it is
still quicker to form the divided difference table, and then to perform the $3n$
operations required to evaluate Newton's interpolating polynomial, than to
execute Aitken's (or Neville's) scheme. One omits $(n + 1)^2$ operations, and gains
only $3n$. If there are a number of evaluations of the same polynomial, the benefit
is even more marked. Moreover, this economy of effort involves no additional
susceptibility to numerical error — quite the reverse, in fact.

The expressions $(\hat{a}_{jk} - \hat{a}_{kk})/(x_j - x_k)$, which are real numbers independent
of x, are called **divided differences** (referring to their formal appearance), and
so the same name is applied to the elements $\hat{a}_{jk}$ of the triangular array, as well as
the coefficients of the Newton form (3.2.11). Many books denote explicit

references to the data values on which any particular divided difference is based by a notation such as, for example, $y[x_0, x_1, x_2 \ldots]$ rather than $\hat{y}_{012}\ldots$ which is the form of expression adopted here. For instance, using (3.2.8), we would refer to the divided difference $\hat{a}_{jk}$ derived from the recursion (3.2.12) as equal to $\hat{y}_{0,1,\ldots,k-1,j}$, meaning by this that $\hat{a}_{jk}$ is the coefficient of the highest power (x^k) of x in the polynomial interpolating the $(k+1)$ data pairs (x_i, y_i) for $i = 0, 1, \ldots, (k-1)$ and for $i = j$.

The basic property of the divided differences, whatever data values are used in their formation, is given by a relation between the leading coefficients of the general interpolating polynomials defined in (3.2.7). Thus on identifying and comparing the coefficients of powers of x^{m+1} on both sides of (3.2.7), it is seen that

$$\hat{p}_{m+1}(S_j \cup S_k; x) = [\hat{p}_m(S_j; x) - \hat{p}_m(S_k; x)]/(x_j - x_k) \quad (3.2.14)$$

and (3.2.12) is just a special case of this recursion, with the intersection $S_j \cap S_k$ being the set of data pairs $\{(x_i, y_i): i = 0, 1, \ldots, (k-1)\}$; that is, (3.2.12) states, as is consistent with (3.2.14), that

$$\hat{y}_{0,1,\ldots,k,j} = (\hat{y}_{0,1,\ldots,(k-1),j} - \hat{y}_{0,1,\ldots,k})/(x_j - x_k) .$$

Furthermore, the notation of (3.2.14) makes it clear that, like the values of the polynomials of which they are the leading coefficients, the divided differences are *symmetric* functions of the data pairs on which they are based. They do not depend on any particular ordering of these data, but only on the totality of the set of data involved. The same goes, therefore, for the coefficients of the Newton interpolating polynomial, so that from (3.2.13) $d_m = \hat{y}_{0,1,\ldots,m}$ depends on the $(m+1)$ data pairs (x_i, y_i) for $i = 0, 1, \ldots, m$, but not their particular sequence. This subset of the data pairs may be permuted at will, maintaining of course the association of each y_k ordinate with its corresponding abscissa x_k, and d_m will be unaltered. On the other hand, such a permutation is bound to affect the constituent data pairs, and so the values, of other coefficients d_k of the Newton form having smaller subscript values $k < m$.

To clarify this, consider as an example the table of divided differences arising from the interpolation of the four data pairs $(0, 1)$, $(1, 3)$, $(2, 11)$, $(3, 49)$. First of all, arranging them as written, in order of increasing x, the triangular array $\hat{a}_{jk}$ given by (3.2.12) is

$$
\begin{bmatrix}
y_0 \\
y_1 & \hat{y}_{01} \\
y_2 & \hat{y}_{02} & \hat{y}_{012} \\
y_3 & \hat{y}_{03} & \hat{y}_{013} & \hat{y}_{0123}
\end{bmatrix}
=
\begin{bmatrix}
1 = d_0 \\
3 & 2 = d_1 \\
11 & 5 & 3 = d_2 \\
49 & 16 & 7 & 4 = d_3
\end{bmatrix} .
$$

and the Newton form of interpolating polynomial is accordingly

$$p_3(x) = 1 + 2x + 3x(x-1) + 4x(x-1)(x-2)$$
$$= 1 + 7x - 9x^2 + 4x^3 .$$

If the data pairs at $x = 0$ and $x = 2$ are interchanged, the table becomes

$$\begin{bmatrix} y_0 \\ y_1 & \hat{y}_{01} \\ y_2 & \hat{y}_{02} & \hat{y}_{012} \\ y_3 & \hat{y}_{03} & \hat{y}_{013} & \hat{y}_{0123} \end{bmatrix} = \begin{bmatrix} 11 = d_0 \\ 3 & 8 = d_1 \\ 1 & 5 & 3 = d_2 \\ 49 & 16 & 7 & 4 = d_2 \end{bmatrix}$$

leading to the rearranged, but equivalent, polynomial

$$p_3(x) = 11 + 8(x-2) + 3(x-2)(x-1) + 4(x-2)(x-1)x$$
$$= 1 + 7x - 9x^2 + 4x^3 .$$

Note that d_2 and d_3 are unchanged because they depend on a permuted but unchanged set of data: however, d_0 and d_1 are different because they each depend on a changed set of data.

The coefficients of the Newton polynomial can just as well be provided by evaluating a triangular array of divided differences arranged as in Neville's (rather than Aitken's) method. Placing $a_{jk} = \hat{a}_{jk} x^k$ in (3.2.9) and, taking the limit $x \to \infty$, one obtains the recursion

$$\hat{a}_{j0} = y_j, \quad \hat{a}_{j,k+1} = (\hat{a}_{jk} - \hat{a}_{j-1,k})/(x_j - x_{j-k-1}) \qquad (3.2.15)$$

for $k = 0, 1, \ldots, (j-1)$ and $j = 0, 1, \ldots, n$. Further, equation (3.2.10) shows that here $\hat{a}_{jk} = \hat{y}_{(j-k),(j-k+1),\ldots,j}$. The triangular array of $\hat{a}_{jk}$-values derived from (3.2.15) starts with the same values in column $k = 0$, and leads by a different progression to the same diagonal elements $\hat{a}_{mm} = d_m$, as in the adaptation of Aitken's scheme given by (3.2.12). Applied to the example set of data $(0, 1)$, $(1, 3)$, $(2, 11)$, $(3, 49)$, taken in the order written, the triangular array for $\hat{a}_{jk}$ generated by (3.2.15) is then

$$\begin{bmatrix} y_0 \\ y_1 & \hat{y}_{01} \\ y_2 & \hat{y}_{12} & \hat{y}_{012} \\ y_3 & \hat{y}_{23} & \hat{y}_{123} & \hat{y}_{0123} \end{bmatrix} = \begin{bmatrix} 1 = d_0 \\ 3 & 2 = d_1 \\ 11 & 8 & 3 = d_2 \\ 49 & 38 & 15 & 4 = d_3 \end{bmatrix} .$$

In hand-computation, this arrangement is often preferred to that of Aitken's Modifying the shape of the triangular pattern a little:

$$
\begin{array}{llll}
y_0 & & & \\
 & \hat{y}_{01} & & \\
y_1 & & \hat{y}_{012} & \\
 & \hat{y}_{12} & & \hat{y}_{0123} \\
y_2 & & \hat{y}_{123} & \\
 & \hat{y}_{23} & & \\
y_3 & & &
\end{array}
$$

it has an easily recognised symmetry. Each divided difference is then the difference of those two immediately to its left, divided by the difference in the x-values each subscripted with one of the extreme values from the range of subscript dependence: thus for example $\hat{y}_{123}$, which is the element $\hat{a}_{32}$, is equal to $(\hat{y}_{23} - \hat{y}_{12})$ divided by $(x_3 - x_1)$, or if one prefers it, differencing in the other direction, $y_{123} = (\hat{y}_{12} - \hat{y}_{23})/(x_1 - x_3)$. Such visual cues to the pattern of relationships are obviously irrelevant in machine computation, for which Aitken's arrangement is in fact the more convenient.

This is because when generating new terms of the divided difference table, the Aitken scheme references previously calculated diagonal elements of the triangular array, and those elements in any event have to be preserved, since they are the required coefficients of the Newton polynomial. This is clearly a convenience, but especially so in 'interactive' implementations which allow one to test the effect of adding a new data pair to those already included in the Newton polynomial. This form of the interpolating polynomial is ideally suited to just such a requirement, as all that is then entailed is the addition of an extra term $d_{n+1}(x - x_0)(x - x_1)\ldots(x - x_n)$ to (3.2.11), the previously calculated coefficients $d_0, \ldots, d_n$ remaining unchanged – they display the property of **permanence**. If the 'new' data pair is called (x_{n+1}, y_{n+1}) then (3.2.12) shows that the new coefficient d_{n+1} is calculated by the instruction sequence

$$d_{n+1} := y_{n+1};$$

for $k := 0$ **step** 1 **until** n **do** $d_{n+1} := (d_{n+1} - d_k)/(x_{n+1} - x_k)$;

the value of d_{n+1} being assigned, in turn, the values of $\hat{a}_{n+1, 0}, \hat{a}_{n+1, 1}, \ldots$ and finally $\hat{a}_{n+1, n+1}$ as required.

However, in the adaptation of Neville's scheme in (3.2.15), if a new row is to be added, the values of the previous row are needed, and this means that they would have had to have been preserved in a separate vector against such future use; or else the whole triangular array would need to be recalculated from scratch. This objection is overcome if one uses what is sometimes called the **reversed Newton form** of the interpolation polynomial:

$$p_n(x) = c_n^{(n)} + c_{n-1}^{(n)}(x - x_n) + c_{n-2}^{(n)}(x - x_n)(x - x_{n-1}) + \ldots$$

$$+ c_0^{(n)}[(x - x_n)(x - x_{n-1}) \ldots (x - x_1)]$$

$$= \sum_{j=0}^{n} [c_{n-j}^{(n)} \prod_{k=0}^{j-1} (x - x_{n-k})] \ . \qquad (3.2.16)$$

At first sight, this seems different only at a trivial level from (3.2.11): all that has happened is that the order of subscripts has been reversed. Correspondingly, it follows immediately by comparison with (3.2.13) that the coefficients of (3.2.16) are given by:

$$c_m^{(n)} = \hat{y}_{m, m+1, \ldots, n} \qquad (m = 0, 1, \ldots, n) \ . \qquad (3.2.17)$$

However, if it is taken that the subscript enumeration is defined by the order in which the data are presented in the divided difference table, then the modification is not so trivial, because the vector $(c_n^{(n)}, c_{n-1}^{(n)}, \ldots, c_0^{(n)})$ is equal to $(\hat{y}_n, \hat{y}_{n-1, n}, \ldots, \hat{y}_{0,1,\ldots,n})$, and this in turn is the vector formed by the last row $(\hat{a}_{n0}, \hat{a}_{n1}, \ldots, \hat{a}_{nn})$ of Neville's scheme of divided differences derived from (3.2.15), reading from left to right. Referring to the *symmetric* arrangement of this triangular array (indicated on page 97), the consequence of reversing the order of the subscripts is that the vector of coefficients has been simply interchanged from the top (downward sloping) diagonal to the bottom (upward sloping).

This means that the scheme of divided differences derived from Neville's algorithm has the same advantage *vis-a-vis* the reversed Newton form as has Aitken's in relation to the normal form. The effect of adding a new data element (x_{n+1}, y_{n+1}) is to add an extra term to (3.2.16), and the new coefficients $c_m^{(n+1)}$ are immediately calculable from the existing values $c_m^{(n)}$ by the instruction sequence:

$$c_{n+1} := y_{n+1};$$

for $l := n$ **step**-1 **until** 0 **do** $c_l := (c_{l+1} - c_l)/(x_{n+1} - x_l)$.

Here $c_{n+1}, c_n, \ldots, c_0$ are simply the values of $\hat{a}_{n+1,0}, \hat{a}_{n+1,1}, \ldots, \hat{a}_{n+1,n+1}$ given by (3.2.15). But note now that it is not merely one new coefficient being calculated: *all* the coefficients are changed — they do not possess permanence like the d_m's — which is why we have added a superscript to these coefficients to distinguish the new $c_m^{(n+1)}$ from its previous value $c_m^{(n)}$.

Thus to determine the effect on the polynomial value (at some given x) due to the addition of a new data pair, *all* the terms of (3.2.16) have to be recalculated, along with the new addition. Using the 'direct' Newton form (3.2.11), only the newly introduced term $d_{n+1}(x - x_0)(x - x_1) \ldots (x - x_n)$ needs to be calculated and added to the previous polynomial value. In fact, the extra computational work involved by the use of the reversed form is relatively small: the offence is more to one's sense of what is neat and tidy (as evinced in the property of permanence). However, as will be shown, there are other forceful reasons for preferring Neville's arrangement of the divided difference table, and as a consequence the reversed Newton form will be found to be frequently used in library implementations of polynomial interpolation, including those designed for interactive use.

3.2.3 Tabular interpolation

One particular advantage of the arrangement of the divided difference table by Neville's scheme is that it leads to a considerable simplification in the event that the support values x_k are equispaced. That of course is the circumstance encountered when interpolating a table of data, and it is a problem which used to attract a lot of attention – and still remains of some interest, at least in relation to hand computation. In conformity with the basic assumption, it is supposed that $x_k = x_0 + kh$, where h is the (constant) increment between successive data points; and the advantage of which we spoke in relation to Neville's scheme arises because the denominator of the expression for $\hat{a}_{j,k+1}$ in (3.2.15) is then simply equal to $(k+1)h$. This has the same value for each divided difference in the same column, and it is convenient therefore to compose the coefficients of the Newton interpolating polynomial from a **table of differences**, rather than *divided* differences, since the latter can be formed merely by applying the appropriate factor to each column of the difference table.

The algebra of finite differences is most conveniently represented by **difference operators**. Where (as here) it is supposed that one knows values of $y(x)$ at discrete equispaced values of $x = x_0 + kh$, such that $y(x_k) = y_k$, then the **shift operator** E applied to the value of (say) y_i is defined by $Ey_i = y_{i+1}$. This definition is extended inductively to yield the interpretation $E^k y_i = y_{i+k}$ by considering the sequence

$$E^k y_i = E^{k-1}(Ey_i) = E^{k-1}y_{i+1} = E^{k-2}y_{i+2} = \dots$$

$$= Ey_{i+k-1} = y_{i+k} \ .$$

This leads quite naturally to the definition $E^v y_i = v[x_0 + (i + v)h]$ for any v, whether or not integer, and (in particular) whether or not v is positive.

The **forward difference** operator Δ applied to y is defined so that $\Delta y_i = y_{i+1} - y_i$, and this is called a finite difference **of first order**. Differences of general order k can again be defined inductively so that

$$\Delta^k y_i = \Delta^{k-1}(\Delta y_i) = \Delta^{k-1}(y_{i+1} - y_i) = \Delta^{k-1}y_{i+1} - \Delta^{k-1}y_i \ ,$$

and for example

$$\Delta^2 y_i = \Delta y_{i+1} - \Delta y_i = (y_{i+2} - y_{i+1}) - (y_{i+1} - y_i) = y_{i+2} - 2y_{i+1} + y_i \ .$$

In terms of the shift operator, $\Delta y_i = Ey_i - y_i = (E-1)y_i$, so that Δ can be replaced by $(E-1)$, and similarly Δ^k can be replaced by $(E-1)^k$. Thus, as is consistent with the previous result,

$$\Delta^2 y_i = (E-1)^2 y_i = (E^2 - 2E + 1)y_i = y_{i+2} - 2y_{i+1} + y_i ,$$

and this makes it clear that the operators can be manipulated algebraically – because they obey the usual associative and distributive laws. Accordingly, an operation like $\Delta^k y_i = (E-1)^k y_i$ can be interpreted algebraically, if required, by applying the binomial theorem to the expansion of $(E-1)^k$, to produce a linear combination of repeated shift operations.

It will also be convenient to introduce the **backward difference** operator ∇ (read as 'del', or 'nabla') which is defined so that $\nabla y_i = y_i - y_{i-1} = (1 - E^{-1})y_i$; and elsewhere in this chapter we refer to the **central difference** operator $\delta y_i = (E^{1/2} - E^{-1/2})y_i$. These definitions can also be extended inductively, so that for example:

$$\nabla^2 y_i = (1 - E^{-1})^2 y_i = (1 - 2E^{-1} + E^{-2})y_i = y_i - 2y_{i-1} + y_{i-2} ;$$

and $\delta^2 y_i = (E^{1/2} - E^{-1/2})^2 y_i = (E - 2 + E^{-1})y_i = y_{i+1} - 2y_i + y_{i-1}$.

Obviously there are various relationships between these operators, and one could as easily manage with just one, rather than three or four.

Returning to the Neville scheme for the divided difference table, if the support is equispaced so that $x_k = x_0 + kh$, and one places

$$\hat{a}_{jk} = \nabla^k y_j / (h^k k!) , \tag{3.2.18}$$

then substituting in (3.2.15) yields the recurrence

$$\nabla^0 y_j = y_j, \ \nabla^{k+1} y_j = \nabla^k y_j - \nabla^k y_{j-1} \tag{3.2.19}$$

for $k = 0, 1, \ldots, j-1$ and $j = 0, 1, \ldots, n$. This is at once seen to be satisfied by the properties of the backward difference operator: for (3.2.19) is merely the recursive definition of that operator, starting with the definition $\Delta^0 y_j \equiv y_j$ for all j, and proceeding with

$$\nabla^1 y_j = \nabla^0 y_j - \nabla^0 y_{j-1} = y_j - y_{j-1} ,$$
$$\nabla^2 y_j = \nabla^1 y_j - \nabla^1 y_{j-1} = (y_j - y_{j-1}) - (y_{j-1} - y_{j-2})$$
$$= y_j - 2y_{j-1} + y_{j-2} ,$$

and so on. Equally well, (3.2.18) could be equivalently replaced by $\hat{a}_{jk} = \Delta^k y_{j-k} / (h^k k!)$ in terms of the forward difference operator, in which event (3.2.15) would be found to be consistent with its recursive definition.

In terms of either of these operators, the entries in a table of differences are identified by one or other of the equivalent arrangements:

$$
\begin{bmatrix}
y_0 & & & \\
& \nabla y_1 & & \\
y_1 & & \nabla^2 y_2 & \\
& \nabla y_2 & & \nabla^3 y_3 \\
y_2 & & \nabla^2 y_3 & \\
& \nabla y_3 & & \\
y_3 & & & \\
& \cdots & &
\end{bmatrix}
=
\begin{bmatrix}
y_0 & & & \\
& \Delta y_0 & & \\
y_1 & & \Delta^2 y_0 & \\
& \Delta y_1 & & \Delta^3 y_0 \\
y_2 & & \Delta^2 y_1 & \\
& \Delta y_2 & & \\
y_3 & & & \\
& \cdots & &
\end{bmatrix}
$$

which can of course be extended indefinitely to include more data and higher order differences. In either arrangement, each entry is simply the difference between the two immediately to its left (the lower minus the upper). Moreover, it is clear from (3.2.18) that if each value in the column of kth differences is divided by $(h^k k!)$ they are equal to appropriate columns of divided differences in the symmetric arrangement (p. 97) of Neville's scheme.

Thus, using the same example set of data as before, $(0, 1)$, $(1, 3)$, $(2, 11)$ and $(3, 49)$, the difference table is easily calculated as

$$
\begin{bmatrix}
y_0 & & & \\
& h\hat{y}_{01} & & \\
y_1 & & 2h^2\hat{y}_{0,1,2} & \\
& h\hat{y}_{12} & & 6h^3\hat{y}_{0,1,2,3} \\
y_2 & & 2h^2\hat{y}_{1,2,3} & \\
& h\hat{y}_{23} & & \\
y_3 & & &
\end{bmatrix}
=
\begin{bmatrix}
1 & & & \\
& 2 & & \\
3 & & 6 & \\
& 8 & & 24 \\
11 & & 30 & \\
& 38 & & \\
49 & & &
\end{bmatrix}.
$$

But here $h = 1$, and so to recover the values of the Neville table, it is merely necessary to divide the 2nd differences by 2, and the (single) 3rd difference by $3! = 6$, in accord with the relation (3.2.18), and as indicated in the left-hand table above.

However, since the purpose of the divided difference table is to evaluate the coefficients of the Newton form, it is only necessary to pick out the relevant differences (along the top or bottom of the table) and scale these by the appropriate divisor. Evidently from (3.2.13) and (3.2.17):

$$
\left.
\begin{aligned}
d_m &= \hat{y}_{0,1,\ldots,m} = \nabla^m y_m/(h^m m!) = \Delta^m y_0/(h^m m!) \\
c^{(n)}_{n-m} &= \hat{y}_{(n-m),\ldots,n} = \nabla^m y_n/(h^m m!) = \Delta^m y_{n-m}/(h^m m!)
\end{aligned}
\right\} \quad (3.2.20)
$$

for $m = 0, 1, \ldots, n$.

The historical importance of the interpolation of function tables has also led to a considerable body of work devoted to various expressions for the value of the Lagrangian interpolate using difference operators, but the few occasions when tabular interpolation is required nowadays hardly justifies their treatment here. The evaluation of either Newton form (3.2.11) or (3.2.16), once its coefficients are known, presents little difficulty, particularly bearing in mind that Lagrangian interpolation is not recommended for large sets of equispaced data. What would be recommended in such a context is a question that needs an answer, but before we come to that, one other form of polynomial interpolation remains to be considered where again the arrangement of the divided difference table in accord with Neville's scheme has very particular advantages.

3.2.4 Osculatory or Hermitian Interpolation

Quite often it may happen that a curve is required to be fitted to data which includes values of slope as well as positional coordinates. The tangent of the slope would, of course, be defined by a value of the first derivative $y'(x)$ of the dependent variable y, and more generally it may be that for some, or all, of the support points the data provide not only the values of y but of a sequence of its lowest order derivatives. The form of interpolation which incorporates this data is termed **osculatory** – a word derived from the Latin verb 'to kiss', referring to the close contact impressed upon the curve by the data. It arises most frequently, perhaps, in relation to the interpolation of discrete data obtained from the numerical solution of differential equations.

The application of continuous polynomial interpolation to this problem gives rise to various relationships which are usually named after the French 19th century mathematician, Charles Hermite. Sometimes a distinction is made between the **Hermitian interpolation** of data which provides slope and ordinate data at every support point, and **modified Hermitian interpolation** of a rather more general kind. Broadly speaking, the improved matching provided by the extra data does not of itself inhibit the existence of the wild excursions between more or less equally-spaced support points which typify Runge's phenomenon in Lagrangian interpolation (for example, see Fig. 3.4). However, the computational methods of Lagrangian interpolation extend quite simply to the

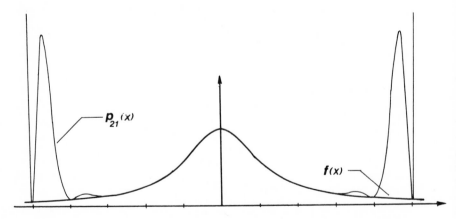

Fig. 3.4 – Hermitian interpolation on an equispaced 11-point support of $f(x) \equiv (1 + x^2)^{-1}$.

problem, which is certainly not without interest or relevance, despite this common limitation.

The notation needs to be changed a little to suit the different data. A convenient convention, and one which is usually adopted for library implementations of Hermitian interpolation, is to suppose that if in the sequence of data two successive data pairs have identical x-values, then the y-value of the first data pair is the positional value (that is, the ordinate corresponding to the x-value), but the next y-value relates to the first derivative of the required interpolating polynomial at that value of x. Since the data are all supposed to be distinct, this is the only valid context in which identical values of x can arise. This is a particularly useful convention, as it does not demand that derivative information is supplied for every support: indeed it allows for the possibility that none is supplied – in which event, of course, the data merely require the application of Lagrangian interpolation.

It is simplest first of all to restrict consideration to the method of dealing with slope and value data at one particular support point, say $x = x_i$. According to the convention adopted, there will then be two data pairs (x_i, y_i), (x_{i+1}, y_{i+1}) such that $x_{i+1} = x_i$, which are to be taken as implying that the interpolating polynomial $y = y(x)$ meets the conditions: $y(x_i) = y_i$ and $y'(x_i) = y_{i+1}$. However, in order to apply the methods developed for Lagrangian interpolation, what is assumed to be implied by the convention is that the second data pair should be replaced by $(x_i + \epsilon_i, y_i + \epsilon_i y_{i+1})$ where ϵ_i is infinitesimal but non-zero. Since in fact $y_{i+1} = y'(x_i)$, this 'fictitious' point is close to (x_i, y_i), but distinct from it, and on the tangent at $x = x_i$ to the curve being constructed. The idea is that, performing Lagrangian interpolation on what has now become a complete set of distinct data, real and fictitious, we then allow the value of ϵ_i to tend to zero, converting the ith and $(i+1)$th data pairs into a 'double point'. The interpolating polynomial then not only intersects (x_i, y_i) but also an almost coincident point on the required tangent at $x = x_i$, and in the limit as $\epsilon_i \to 0$, it is constrained to have the correct value *and slope* at this double point.

Applying this idea to Neville's algorithm, there is only one element of the triangular array which is not defined if the limit is treated merely by substituting $\epsilon_i = 0$, which would imply that the two data pairs were taken as coincident: and that one element is $a_{i+1,1} = y_{i,i+1}$ which is the formula for the linear polynomial joining these two points. However, using the 'fictitious' data pair $(x_i + \epsilon_i, y_i + \epsilon_i y_{i+1})$ in place of (x_{i+1}, y_{i+1}), then (3.2.9) shows that

$$a_{i+1,1} = [a_{i+1,0}(x - x_i) - a_{i,0}(x - x_i - \epsilon_i)]/\epsilon_i$$
$$= [(y_i + \epsilon_i y_{i+1})(x - x_i) - y_i(x - x_i - \epsilon_i)]/\epsilon_i$$
$$= y_i + (x - x_i)y_{i+1} \ .$$

Now regarding x as variable, and recalling that $y_{i+1} = y'(x_i)$, this is seen to be the value of y on the required tangent to the curve of $x = x_i$. This is true, whatever the value of ϵ_i. On the other hand, the element $a_{i+1,0}$, which by (3.2.9) should be equal to the fictitious ordinate, obviously tends simply to y_i as $\epsilon_i \to 0$ – the same value as for the element $a_{i,0}$. This is what would be expected: in the limit, the double point becomes two coincident points, and the line between them becomes the tangent at the point.

The idea can easily be extended to deal with a sequence of data providing the y-value and the successive low order derivatives. Let the convention be extended by supposing that if $(m + 1)$ successive data pairs have the same x-value (so that $x_{i+p} = x_i$ for $p = 1, 2, \ldots, m$), then the associated y-values are to be interpreted as implying that $y(x_i) = y_i$, and that the pth derivative of the required interpolating polynomial is to be taken as $y^{(p)}(x_i) = y_{i+p}$ at $x = x_i$ for $p = 1, 2, \ldots, m$. The data then define a 'multiple point' which can be treated as the limit of $(m + 1)$ distinct points, coalescing in the limit so as to provide the appropriate ordinate and derivatives. Thus the values $a_{j,0}$ in column $k = 0$ of the Neville algorithm are all equal to the y-value y_i of the coalesced point for $j = i, i + 1, \ldots, i + m$. The entries $a_{j,1} = y_{j-1,j}$ are the linear polynomials representing the straight line joining the $(j - 1)$th and jth points; for $j = i + 1, i + 2, \ldots, i + m$, these points coalesce and define this line (in the limit) to be the tangent at $x = x_i$. The entries $a_{i+2,2}, \ldots, a_{i+m,2}$ are similarly y-values on parabolic arcs having 3-point contact, and the table builds up in the following manner:

$$(x = x_i) \quad y_i \quad \ldots \qquad\qquad\qquad \ldots$$

$$(x = x_i) \quad y_i \quad y_i + (x - x_i)y_{i+1} \quad \ldots$$

$$(x = x_i) \quad y_i \quad y_i + (x - x_i)y_{i+1} \quad y_i + (x - x_i)y_{i+1} + \tfrac{1}{2}(x - x_i)^2 y_{i+2}$$

$$\ldots \qquad\qquad \ldots$$

with otherwise indeterminate entries in the kth column replaced by the Taylor polynomial of degree k expanded about $x = x_i$, it being remembered that $y_{i+p} = y^{(p)}(x_i)$. Formally, we express the modification required to (3.2.9) in the event that $x_{i+p} = x_i$ for $p = 1, 2, \ldots, m$, by

$$\left. \begin{aligned} a_{i+p,0} &= y_i \\ a_{i+p,k} &= a_{i+p,k-1} + (x - x_i)^k y_{i+k}/k! \\ &= a_{i+p,k-1} + (x - x_i)^k y^{(k)}(x_i)/k! \end{aligned} \right\} \qquad (3.2.21)$$

for $k = 1, 2, \ldots, p$. In the event that the data consist of only one support point $x = x_0$ but $(m + 1)$ data pairs, all with identical x-value, defining the value (y_0) and the first m derivatives of the required interpolating polynomial at $x = x_0$, this scheme simply generates the Taylor polynomial

$$a_{m,m} = y_{0,1,\ldots,m} = y_0 + (x - x_0)y'(x_0) + \ldots + [(x - x_0)^m y^{(m)}(x_0)/m!],$$

which has the required properties matching the data.

Computationally, we know that it is easier and quicker to use Newton's form of the interpolating polynomial and a divided difference table. Neville's arrangement of this table corresponding to the recurrence (3.2.12) can be modified, subject again to the same convention, merely by noting that, if $x_{i+p} = x_i$ for $p = 1, 2, \ldots, m$, then from (3.2.21)

$$\left.\begin{aligned} \hat{a}_{i+p,0} &= y_i \\ \hat{a}_{i+p,k} &= y_{i+k}/k! = y^{(k)}(x_i)/k! \quad (k = 1, 2, \ldots, p) \end{aligned}\right\} \quad (3.2.22)$$

Further, the identity between these x-values will now of course imply that powers $(x - x_i)^p$ will replace the group of factors $(x - x_i)(x - x_{i+1}) \cdots (x - x_{i+p})$ in the Newton form of the interpolating polynomial. Again, if there is only one support point $x = x_0$, but $(m + 1)$ data pairs all with this x-value, the Newton polynomial becomes the Taylor polynomial of degree m developed about $x = x_0$.

As an example of the divided difference scheme, consider that corresponding to the data $(0, 1)$, $(0, 2)$, $(1, 6)$, $(1, 12)$, $(1, 32)$. According to the convention this implies that

$$y(0) = 1, \ y'(0) = 2, \ y(1) = 6, \ y'(1) = 12, \ y''(1) = 32 \ .$$

The table of $\hat{a}_{jk}$ values according to the recurrence (3.2.12), as modified by (3.2.22), then leads to

$$
\begin{array}{llllll}
(x_0 = 0) & 1 & & & & \\
(x_1 = 0) & 1 & 2^* & & & \\
(x_2 = 1) & 6 & 5 & 3 & & \\
(x_3 = 1) & 6 & 12^* & 7 & 4 & \\
(x_4 = 1) & 6 & 12^* & 16^* & 9 & 5 \ ,
\end{array}
$$

the asterisked values being those whose indeterminacy is resolved by (3.2.22). The Newton form (3.2.11) and its reverse (3.2.16) are then the equivalent quartics:

$$p_4(x) = 1 + 2x + 3x^2 + 4x^2(x - 1) + 5x^2(x - 1)^2 \ ,$$
$$= 6 + 12(x - 1) + 16(x - 1)^2 + 9(x - 1)^3 + 5(x - 1)^3 x \ .$$

Arranged for machine computation, the derivative values would be initially entered into the first column ($k = 0$) just as if they were ordinates. However, x-values are tested for equality before a divided difference is formed. If there is equality, the elements are moved, and divided by the column number (so as to develop the factor $1/k!$ applied to the kth derivative), as indicated in the following progressive construction of the example table given above. Here the bracketed entries to the left of the table are the x-values, and asterisks denote detection of an incipient zero-divide, signalling an equality of the x-values:

```
(0)   1           (0)   1                 (0)   1
(0)   2    *      (0) ↓ 1  (2/1)          (0)   1    2
                           ────→
                  (1)   6    5    3       (1)   6    5    3
                  (1)   12   *            (1) ↓ 6  (12/1) 7    4
                                                  ────→
                                          (1)   32   *


(0)   1                            (0)   1
(0)   1    2                       (0)   1    2
(1)   6    5    3                  (1)   6    5    3
(1)   6    12   7    4             (1)   6    12   7    4
(1) ↓ 6  (32/1)  *                 (1)   6  ↓ 12  (32/2)  9    5
        ────→                                 ────→
```

It would be possible also to provide a difference table which correctly incorporates derivative information and is arranged according to Aitken's scheme, but the rules to be followed are considerably more difficult to express and implement. By contrast, it is the very easy way in which Neville's arrangement of the divided difference table provides this powerful generalisation which causes it to be preferred for general use in computational libraries.

It would seem a useful generalisation to be able to drop the requirement that data giving a value of $y(x)$ and of its *successive* derivatives, starting with the first, are required. However this would give rise to very considerable problems. Consider for example the construction of a quadratic polynomial such that $p_2(1) = p_2(-1) = 1$, and $p_2'(0) = 0$, but without a prescribed value being given at $x = 0$: there are an infinity of possible solutions. Or if $p_2(-1)$ were -1, instead of $+1$, there would be no solution. Clearly an alogirthm without any form of logical discrimination is not going to provide this kind of information: it must derive from the solution of a set of linear equations which may not have a unique solution. The provision of *successive* derivative values, however, is sufficient to ensure both uniqueness and existence of the interpolating polynomial.

3.3 PIECEWISE POLYNOMIAL INTERPOLATION

The difficulties associated with polynomial interpolation, which give rise to the Runge phenomenon, are avoided if the curve being fitted to the data is pieced together by a number of arcs of different polynomials of low degree, rather than by one single polynomial of high degree. This of course is to sacrifice some measure of continuity of the interpolating function in order to achieve the rather less well defined property of what could be called conformability.

Thus in **piecewise Hermitian interpolation** of a set of discrete data providing (say) values of y_k and first derivatives y_k' at a number of **nodes** $x = x_k$ for $k = 0, 1, \ldots, n$, ordered so that $x_{k-1} < x_k$ for $k = 0, 1, \ldots, n$, it is possible to describe the variation of $y(x)$ over any interval $[x_{k-1}, x_k]$ between successive nodes by the particular cubic polynomial $\pi_k(x)$, say, which meets the four data constraints:

$$\pi_k(x_j) = y_j, \quad \pi_k'(x_j) = y_j': \quad j = k - 1, k . \tag{3.3.1}$$

Clearly, these match the given value and derivative at the end points of $[x_{k-1}, x_k]$. The interpolating function is then made up of n such cubic polynomial **pieces**:

$$\left. \begin{aligned} y(x) &= \pi_1(x) \quad a \leqslant x \leqslant x_1 \\ &= \pi_k(x) \quad x_{k-1} \leqslant x \leqslant x_k: \quad k = 2, 3, \ldots, (n-1) \\ &= \pi_n(x) \quad x_{n-1} \leqslant x \leqslant b \end{aligned} \right\} \tag{3.3.2}$$

where the values $a \leqslant x_0$ and $b \geqslant x_n$ define the total interval $[a, b]$ of interest. This function belongs to the space $C^1[a, b]$ of continuously differentiable functions: that is, it has continuous value and first derivative over $[a, b]$, whereas the unique Hermitian polynomial of degree $2n + 1$ which interpolates *all* the data belongs of course to the space C^∞ of infinitely differentiable functions. However, the eye is not very sensitive to the discontinuity of curvature implied by the existence of a discontinuity in $y''(x)$ across the nodes. Further, because each component polynomial $\pi_k(x)$ is entirely dependent on *local* data there is no 'effect at a distance' due to the global influence of complete continuity which underlies the Runge phenomenon.

Of course, if it happened that the data provided first and second derivates at each node, then there would be six constraints on each polynomial piece $\pi_k(x)$, three at each end of the interval $[x_{k-1}, x_k]$, and these polynomials must then be taken to be quintics – in order that all the data be incorporated – and the composite interpolating function $y(x) \in C^2[a, b]$, having continuity of slope and curvature. The discontinuity in the third derivative at each node would be almost impossible for the eye to discern.

But of course much the most common form of curve drawing requires an interpolative scheme based entirely on data pairs (x_k, y_k) without any information on derivatives. The lowest degree polynomial $\pi_k(x)$ depending only on the localised data values y_{k-1}, y_k at the ends of the interval in which it is defined is then linear, and the interpolating function is composed of a series of straight line segments joining successive data coordinates. Of course this would not be judged acceptable to the eye, unless the number of data was extremely large and everywhere dense. It is called **Piecewise lineal interpolation**.

In the absence of derivative information, it is therefore necessary to reintroduce some measure of 'effect at a distance' so as to provide a higher degree of continuity to the interpolative scheme. For instance, in **piecewise Bessel interpolation** the polynomials $\pi_k(x)$ are assumed to be cubics having the correct values at the end points (as provided by the data), but the derivative values at the ends of the interval are constructed from neighbouring data. The derivative at $x = x_k$ is assumed to be the same as that of the quadratic arc $y_{k-1,k,k+1}$ which interpolates the three data pairs $(x_{k-1}, y_{k-1}), (x_k, y_k), (x_{k+1}, y_{k+1})$. This gives the two constraints

$$y(x_j) = y_j, \quad y'(x_j) = (\hat{y}_{j-1,j} h_{j+1} + \hat{y}_{j,j+1} h_j)/(h_j + h_{j+1}) \quad (3.3.3)$$

at each interior node x_j for $j = 1, 2, \ldots, (n-1)$. Here we have defined $h_j = x_j - x_{j-1}$ and the formula for $y'(x_j)$ is a generalisation of the result given in problem 7 of section 3.2.2. Evidently something needs to be done about the end pieces $\pi_1(x)$ and $\pi_n(x)$, but all the other cubic pieces $\pi_k(x)$ with $k = 2, 3, \ldots, (n-1)$ each satisfy both the constraints (3.3.3) at both ends of the interval $[x_{k-1}, x_k]$ over which they are defined. This is evidently a form of piecewise Hermitian interpolation in which (at the interior nodes, at least) the derivative is constructed, or in some sense 'invented', rather than provided by data. In particular, if the nodes are equally spaced (so that $h_j = h$, a constant), the derivative at $x = x_k$ is therefore taken to be $(y_{k+1} - y_{k-1})/2h$, which can be interpreted as the tangent of the slope of the straight line joining the data coordinates (x_{k-1}, y_{k-1}) and (x_{k+1}, y_{k+1}).

As in piecewise Hermitian interpolation, the composite interpolating function $y(x)$, as given by (3.3.2), belongs to the space $C^1[a, b]$ of functions with continuous value and slope, Evidently, the same idea could be extended to invent a value of curvature at each interior node $x = x_k$ (equal to that of the quadratic interpolating the same three data pairs): this would provide the extra constraints

$$y''(x_j) = 2\hat{y}_{j-1,j,j+1} = 2(\hat{y}_{j,j+1} - \hat{y}_{j-1,j})/(h_j + h_{j+1}) \quad (3.3.4)$$

at each node $j = 1, 2, \ldots, (n-1)$, which could be met by taking $\pi_k(x)$ to be a quintic, for $k = 2, 3, \ldots, (n-1)$, satisfying six constraints, three at each end of the interval of definition $[x_{k-1}, x_k]$; and then $y(x) \in C^2[a, b]$.

The algebraic expressions for $\pi_k(x)$ as determined by Bessel interpolation are rather involved, but the principle is fairly clear, that the behaviour of $\pi_k(x)$, and so of $y(x)$ within $[x_{k-1}, x_k]$, depends upon the 4 data pairs (x_j, y_j) for $j = k-2, k-1, k$ and $k+1$. Although each of the non-local data (with $j = k-2$ and $k+1$) determine only derivative values at each respective end point of $[x_{k-1}, x_k]$, they nonetheless affect the variation of $\pi_k(x)$ over this entire interval.

This can be looked at another way: any particular data pair (x_j, y_j) affects the form of interpolation over four adjacent segments from $x = x_{j-2}$ to $x = x_{j+2}$; on the other hand this data pair has no effect whatsoever on the form of $y(x)$ outside $[x_{j-2}, x_{j+2}]$. Reverting to the idea of the representation of $y(x)$ by shape functions as in (3.2.3), we could represent the interpolation by, say,

$$y(x) = \sum_{k=0}^{n} y_k \, s_k(x_{k-2}, \ldots, x_{k+2} \, ; \, x) \tag{3.3.5}$$

where, as in (3.2.4), the value of $s_k(..)$ is equal to δ_{jk} at $x = x_j$ (that is, it is unity at $x = x_k$, but zero at the other nodes). This shape function $s_k(..)$ is then zero everywhere except within $[x_{j-2}, x_{j+2}]$; within this interval, it is composed of polynomial segments, either cubics consistent with the conditions imposed on $y(x)$ by (3.3.3), when $s_k(..)$ would belong to the function space C^1, or quintic pieces together belonging to C^2 in the event that the additional constraints of (3.3.4) are imposed. The expressions for these polynomial pieces of $s_k(..)$ depend only upon the local support – that is, the values of $x_{k-2}, \ldots, x_{k+2}$ – which is why we have omitted the subscript n on the s_k, since the totality of the data has no influence upon their value. The variation of these shape functions is shown in Fig. 3.5 for an equally-spaced support. Also shown is the shape function corresponding to piecewise lineal interpolation between data coordinates which

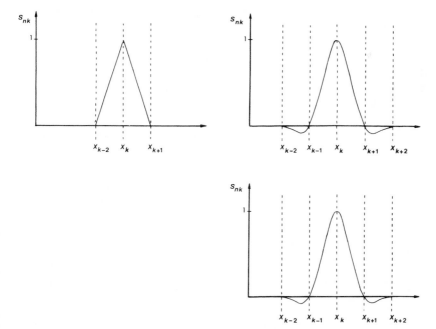

Fig. 3.5 – Shape functions for piecewise straight-line interpolation, and for Bessel-interpolation using piecewise cubic or quintic polynomials (upper and lower diagrams respectively).

obviously provides a much less smooth way of incorporating the influence of any data ordinate.

But we have one problem left over: how does one deal with the first and last segments $[x_0, x_1]$ or $[x_{n-1}, x_n]$? There are no data for $x < x_0$ or $x > x_n$ by which the derivative at $x = x_0$ or $x = x_n$ may be constructed: the values of (x_{-1}, y_{-1}) or (x_{n+1}, y_{n+1}) required in (3.3.3) and (3.3.4) for $\pi_1(x)$ or $\pi_n(x)$ do not exist. There is always a difficulty with the end intervals in any form of piecewise interpolation which incorporates more than strictly local data. It might of course happen that derivative values at $x = x_0$ and $x = x_n$ are supplied, although this is not usually so; otherwise, it is probably most generally satisfactory simply to omit the constraints on $\pi_1(x)$ and $\pi_n(x)$ which would be needed to provide these polynomials with derivatives at $x = x_0$ and $x = x_n$ respectively, and to lower the degree of these end-pieces accordingly. If this is done, whether cubic or quintic forms of polynomial pieces are employed, both end intervals $[x_0, x_1]$ and $[x_{n-1}, x_n]$ can be interpolated by segments of the quadratic polynomials $y_{0,1,2}$ and $y_{n-2, n-1, n}$ which interpolate the first and last three data pairs respectively. This ensures continuity of value and derivatives at $x = x_1$ and $x = x_{n-1}$ so that the continuity of $y(x)$ is not disturbed. These same quadratics can, if required, provide extrapolation for $x < x_0$ or $x > x_n$, as implied by (3.3.2).

Such an exception at the end intervals affects the definition of the shape functions s_0, s_1 for $x < x_1$, and s_{n-1}, s_n if $x > x_{n-1}$. However, provided the formula (3.3.5) were used for interpolation only in the interval (x_1, x_{n-1}), no modification is necessary. Thus if, for example, the end pieces $\pi_1(x)$ and $\pi_n(x)$ are taken to be quadratics (as suggested), then any required interpolation or extrapolation in the range $x \leqslant x_1$ or $x \geqslant x_{n-1}$ can be achieved by evaluating the relevant Lagrangian polynomial $y_{0,1,2}$ or $y_{n-2, n-1, n}$, without reference to the shape functions.

Piecewise Bessel interpolation deserves a better reputation than it seems to enjoy, as it is both serviceable and simple, and moreover it adapts reasonably to irregular supports. On the other hand it is an entirely pragmatic piece of numerical engineering. Modern practice favours an involvement of some measure of influence of *all* the data pairs on the shape of the interpolating polynomial pieces – that is, a global basis of interpolation rather than a purely localised basis. In other words, it is viewed as unreasonable that the shape functions s_k in (3.3.5) should vanish outside the interval (x_{k-2}, x_{k+2}); rather one would expect them to dwindle away *gradually*, and so produce some influence over the whole interpolating range (a, b), without however producing a large 'effect at a distance' as in Lagrangian interpolation.

3.4 INTERPOLATION BY SPLINES

An idea which leads to a graduated form of influence of neighbouring data on the interpolating polynomial pieces can best be illustrated by an example.

Suppose as before that the data consist of $(n+1)$ discrete data pairs (x_k, y_k) with $k = 0, 1, \ldots, n$, ordered so that $a \leqslant x_0 < x_1 < \ldots < x_n \leqslant b$, and that $y = y(x)$ is the equation of the interpolating function, with $y(x)$ defined over the interval $[a, b]$ in terms of n polynomial pieces $\pi_k(x)$ with $k = 1, 2, \ldots, n$ as in (3.3.2). As it is the intention to improve on linear piecewise interpolation, suppose further that the $\pi_k(x)$ are quadratic polynomials. Each such polynomial piece has three adjustable coefficients, but the conditions that $\pi_k(x_{k-1}) = y_{k-1}$ and $\pi_k(x_k) = y_k$ — in other words, that $y(x)$ has a continuous value and interpolates all the data coordinates — provide two constraints on the coefficients of each $\pi_k(x)$. This leaves the system with n adjustable degrees of freedom, one for each polynomial piece. Suppose now — and this is the innovation — that it is to be arranged that $y(x)$ is continuously differentiable, so that it belongs to the space $C^1[a, b]$. Each quadratic piece has continuous derivative, and consequently discontinuities could only occur at the interior nodes $x_1, x_2, \ldots, x_{n-1}$ where the pieces join: our assumption can therefore be restated as

$$\pi_k'(x_k) = \pi_{k+1}'(x_k) \quad k = 1, 2, \ldots, (n-1) \ . \tag{3.4.1}$$

Here $\pi_k(x)$ and $\pi_{k+1}(x)$ describe the interpolation in the intervals $[x_{k-1}, x_k]$ and $[x_k, x_{k+1}]$ respectively, and their derivatives are required to be equal where they join. But (3.4.1) provides $(n-1)$ equations for the n undetermined coefficients, so that it is clear that $y(x)$ *can* be made continuous in value *and* slope, and there is still one condition to spare. It will turn out that lack of this one extra condition is rather like the want of the nail that lost the shoe, the horse, and the rider; but that is not yet apparent. Let us merely assume for the time being that a value of $\pi_1'(x_0) = y'(x_0) = y_0'$ is provided as data, and then the method of interpolation is fully defined.

To see what it implies let us write the equation for each polynomial piece as:

$$\pi_k(x) = [(x - x_{k-1})y_k + (x_k - x)y_{k-1} - \mu_k(x - x_{k-1})(x_k - x)]/h_k, \tag{3.4.2}$$

where $h_k = x_k - x_{k-1}$ and $k = 1, 2, \ldots, n$. The form of the equation ensures that $\pi_k(x)$ interpolates the data (x_{k-1}, y_{k-1}) and (x_k, y_k); the n values of μ_k are, however, adjustable coefficients, determining the curvature of each quadratic piece. A little algebra shows that the known value of $\pi_1'(x_0)$ and equation (3.4.1) together lead to the following n equations for these n unknowns:

$$\left.\begin{array}{l} \mu_1 = \hat{y}_{0,1} - y_0' \\[2mm] \mu_k + \mu_{k+1} = \hat{y}_{k,k+1} - \hat{y}_{k-1,k} \quad k = 1, 2, \ldots, (n-1) \end{array}\right\} , \tag{3.4.3}$$

where $\hat{y}_{j,k} = (y_j - y_k)/(x_j - x_k)$ is a divided difference whose value is determined by the data. Equation (3.4.3) is a simple recurrence relationship for each successive unknown $\mu_1, \mu_2, \ldots, \mu_n$. It can be verified that it leads to the general expression:

$$\mu_k = \hat{y}_{k-1,k} + (-1)^k [2 \sum_{j=1}^{k-1} (-1)^j \hat{y}_{j-1,j} + y_0'] , \tag{3.4.4}$$

although for numerical work the recurrence relation (3.4.3) would provide an easier method of calculation.

At first sight, the derivation of an interpolating function $y(x) \in C^1[a, b]$ by piecewise quadratic interpolation would seem a worthwhile accomplishment, bearing in mind that Bessel interpolation needs cubic pieces to achieve the same continuity. But note from (3.4.4) that μ_k depends only on the values of y_0' and the divided differences $\hat{y}_{0,1}, \hat{y}_{1,2}, \ldots, \hat{y}_{k-1,k}$, and these latter in turn depend only on the first $k + 1$ data pairs. Remembering that, as in (3.4.2), μ_k affects the value of $\pi_k(x)$, this must mean that any particular data pair (x_j, y_j) only affects the interpolation for $x \geqslant x_{j-1}$. If the value of $y(x)$ is expressed in terms of shape functions, as in (3.2.3), so that $s_{nj}(..)$ describes the incremental change in $y(x)$ due to unit displacement of the data value y_j, then s_{nj} would be zero for all $x < x_j$. The shape function is lop-sided: imagining x is a time-variable, the interpolation is influenced only by the past, and unaffected by the future. Of course in particular applications this might be quite appropriate; but it is not so in general. This brings us back to the assumption that the value y_0' is 'known', for the only way to avoid this effect is to replace y_0' by a quantity which has a global significance — which depends on *all* the data.

The definition of μ_k in (3.4.2) shows that it equals the constant value of $\frac{1}{2} h_k \pi_k''(x)$, which is also the value of $\frac{1}{2} h_k y''(x)$ over (x_{k-1}, x_k). Looking for some condition of global significance to replace the assumption that $\pi_1'(x_0) = y_0'$ is known, one might accept the suggestion that y_0' be chosen so as to minimise some overall measure of curvature such as:

$$\int_{x_0}^{x_n} [y''(x)]^2 \mathrm{d}x = \sum_{k=1}^{n} \int_{x_{k-1}}^{x_k} [\pi_k''(x)]^2 \mathrm{d}x = 4 \sum_{k=1}^{n} (\mu_k^2/h_k) \ .$$

There is no need to follow up the algebra: the effect is (as required) simply to derive a value of y_0' which depends on all the data, and which is a linear combination of the ordinates — say $y_0' = \sum_{m=0}^{n} c_m y_m$, where the c_m's depend on the support. Substituting this in (3.4.4), it will be seen that any particular ordinate y_m with $m > k$ will now provide an increment $(-1)^k c_m y_m$ to the value of μ_k. Clearly this increment has the same magnitude but an alternating sign in each of the expressions for $\mu_0, \mu_1, \ldots, \mu_{m-1}$. Thus the ordinate y_m will now, as intended, exercise an influence upon the interpolation in *both* directions, different on the two sides $x < x_m$ and $x > x_m$. But even so, its influence does not attenuate with distance from $x = x_m$ (Fig. 3.6), as should happen if it were to be the kind of interpolation method we are seeking.

It may seem as if a false trail has been laid; but this is not so. Go through the same development as before, but this time try using cubic, rather than quadratic, polynomial pieces. Each has then four adjustable coefficients, but the essential interpolation of the data at each end point provides only two constraints, leaving two spare for each polynomial piece, and so $2n$ unknowns.

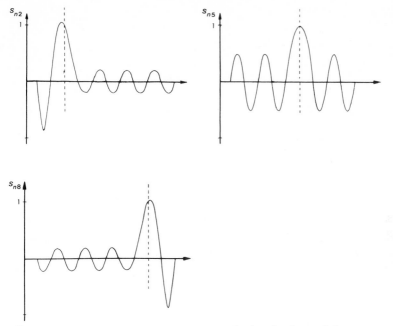

Fig. 3.6 – Shape functions for quadratic spline interpolation.

Suppose then that the piecewise polynomial $y(x)$ is required to belong to the function space $C^2[a, b]$ with continuous value, slope and curvature. It follows that (3.4.1) must now be replaced by

$$\pi'_k(x_k) = \pi'_{k+1}(x_k), \ \pi''_k(x_k) = \pi''_{k+1}(x_k): \ k = 1, 2, \ldots, (n-1) \ ,$$
$$(3.4.5)$$

giving $(2n-2)$ equations for the $2n$ unknowns. As for the remaining two conditions, assume (for the time being) that $\pi'_1(x_0) = y'_0$ and $\pi'_n(x_n) = y'_n$, where y'_0 and y'_n are prescribed as data. Notice that the partiality of choosing just one end at which to ascribe a boundary condition has been avoided.

There being sufficient conditions to determine the unknowns, set down the general equation for the cubic pieces as

$$h_k \pi_k(x) = (x - x_{k-1})y_k + (x_k - x)y_{k-1}$$
$$- (x - x_{k-1})(x_k - x)[(h_k + x_k - x)M_{k-1} + (h_k + x - x_{k-1})M_k]/6 \ ,$$
$$(3.4.6)$$

for $k = 1, 2, \ldots, n$. This form ensures not only that $\pi_k(x)$ interpolates the end points of x_{k-1}, x_k, but that $\pi''_k(x_{k-1}) = M_{k-1}$ and $\pi''(x_k) = M_k$. Hence the condition in (3.4.5) that the second derivative is continuous across the nodes is automatically satisfied, because $y''(x_k) = M_k$ for $k = 0, 1, \ldots, n$. These $(n + 1)$ unknowns M_k are then determined by the two end-conditions at $x = x_0$ and x_n, and by the requirement of (3.4.5) that the slope $y'(x)$ be continuous at each node. After some algebra directed towards evaluating the derivatives $\pi'_k(x)$, this leads to the system of $(n + 1)$ simultaneous linear equations:

$$(2M_0 + M_1)h_1 = 6(\hat{y}_{0,1} - y'_0)$$

$$h_k M_{k-1} + 2(h_k + h_{k+1})M_k + h_{k+1}M_{k+1} = 6(\hat{y}_{k,k+1} - \hat{y}_{k-1,k}):$$

$$k = 1, 2, \ldots, (n-1)$$

$$(M_{n-1} + 2M_n)h_n = 6(y'_n - \hat{y}_{n-1,n}) \tag{3.4.7}$$

A further notational simplification results from defining $x_{-1} = x_0$ and $x_{n+1} = x_n$, and writing $\hat{y}_{-1,0}$ and $\hat{y}_{n,n+1}$ in place of y'_0 and y'_n respectively. (It will be recognised that this is consistent with the conventionalised treatment of derivative data in osculatory interpolation.) Dividing each equation through by $(h_k + h_{k+1}) = (x_{k+1} - x_{k-1})$, the system (3.4.7) can then be expressed in matrix form as:

$$\begin{bmatrix} 2 & (1-\lambda_0) & 0 & 0 & & \\ \lambda_1 & 2 & (1-\lambda_1) & 0 & & \mathbf{O} \\ 0 & \lambda_2 & 2 & (1-\lambda_2) & & \\ & & \cdot & \cdot & \cdot & \cdot & \cdot \\ & & \cdot & \cdot & \cdot & \cdot & \cdot \\ \mathbf{O} & & & \lambda_{n-1} & 2 & (1-\lambda_{n-1}) \\ & & & 0 & \lambda_n & 2 \end{bmatrix} \begin{bmatrix} M_0 \\ M_1 \\ M_2 \\ \cdot \\ \cdot \\ M_{n-1} \\ M_n \end{bmatrix} = 6 \begin{bmatrix} \hat{y}_{-1,0,1} \\ \hat{y}_{0,1,2} \\ \hat{y}_{1,2,3} \\ \cdot \\ \cdot \\ \hat{y}_{n-2,n-1,n} \\ \hat{y}_{n-1,n,n+1} \end{bmatrix},$$

$$\tag{3.4.8}$$

where $\lambda_k = h_k/(h_k + h_{k+1})$ and for instance $\hat{y}_{1,2,3} = (\hat{y}_{2,3} - \hat{y}_{1,2})/(x_3 - x_1)$ is a divided difference. Note that $h_0 = x_0 - x_{-1}$ and $h_{n+1} = x_{n+1} - x_n$ are both zero, so that $\lambda_0 = 0$ and $\lambda_n = 1$.

    This linear system does not yield a simple solution in closed form like that of (3.4.4) derived from the recurrence relation (3.4.3). However, it does have a unique solution, because the tri-diagonal coefficient matrix has diagonal dominance, and is therefore non-singular. This is because $0 \leqslant \lambda_k \leqslant 1$ for all k, so that none of the matrix elements is negative, and the diagonal elements are twice the sum of the magnitudes of the two off-diagonal elements of any row. In particular, if the nodes are equispaced, so that $h_j = h$ is constant for $j = 1, 2, \ldots, n$, then the coefficient matrix has the simple form:

$$\begin{bmatrix} 2 & 1 & 0 & 0 & & & \\ \frac{1}{2} & 2 & \frac{1}{2} & 0 & & 0 & \\ 0 & \frac{1}{2} & 2 & \frac{1}{2} & & & \\ & & & \cdots\cdots & & & \\ & & & \cdots\cdots & & & \\ & 0 & & & \frac{1}{2} & 2 & \frac{1}{2} \\ & & & & 0 & 1 & 2 \end{bmatrix}.$$

and the terms $6\hat{y}_{k-1,\,k,\,k+1}$ for $k = 1, 2, \ldots, (n-1)$, on the right-hand side of (3.4.8), are simply equal to

$$3\delta^2 y_k/h^2 = 3(y_{k+1} - 2y_k + y_{k-1})/h^2$$

where δ is the central difference operator.

The shape function corresponding to a data ordinate in the middle of a large number of equally-spaced points is shown in Fig. 3.7, and it displays the kind of variation that we have been seeking. Any particular ordinate extends only a limited influence on the interpolation of non-local intervals, and its effect attenuates rapidly with distance: each successive lobe between support points of the shape function shown in Fig. 3.7 has an amplitude of only about one quarter of that of its predecessor. On the other hand, the influence of any ordinate on the interpolation is not localised in a hard and fast way as it is in Bessel interpolation.

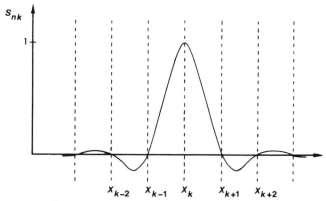

Fig. 3.7 – Shape functions for cubic spline interpolation.

The curve produced by this form of piecewise polynomial interpolation is called a **cubic spline,** and the (less successful) application of the same idea using quadratic instead of cubic pieces is called interpolation by a **quadratic**

spline. In general, an mth degree spline is an interpolating function belonging to $C^{m-1}[a, b]$ which reduces to polynomials of degree $\leqslant m$ on each interval between nodes.

The origin of the word 'spline' is of special significance: it means a narrow strip (of wood, or metal, etc.) but refers in particular to the flexible rod used by some designers as an aid in drawing a smooth contour between points – a mechanical analogue of interpolation. This rod, or *spline*, is pinned to the drawing board at the given data points, and the elastic properties of the material of the spline determine how it curves to take up the strain between the pins. Provided it is thin and uniformly flexible, the shape which it adopts approximates to the particular curve that we have called a cubic spline.

There is one further and interesting consequence of the fact that the cubic spline is a model of a physical process. In reality, the shape adopted by the designer's spline is such as to minimise the strain energy consistent with the constraints imposed by the pins. For a thin, uniformly flexible spline, the strain energy per unit length is proportional to $1/r^2$, where r is its radius of curvature: hence the integral of $1/r^2$ taken along the length of the spline is the minimum consistent with the constraints on its shape imposed by pinning. In the mathematical model, $(1/r)$ is replaced by the second derivative of the interpolating curve, $y''(x)$, and this integral is approximated by $\int_{x_0}^{x_n} [y''(x)]^2 \mathrm{d}x$. Accordingly, the cubic spline would then be defined as that function $y = y(x)$ which interpolates the $(n + 1)$ data, has continuous value and slope in $[x_0, x_n]$, has a prescribed slope at $x = x_0$ and $x = x_n$, and which minimises this integral. This is called the **variational definition** of the spline, and it leads to just the same formulation as already obtained (though by a rather more complicated path – see problem 4). However, it is to be noted that the continuity of the second derivative is then a *consequence* of the definition, not an assumption: the assumption is, instead, that the mean-square value of $y''(x)$, as measured by the integral, is to be as small as possible, implying that the curve flexes no more than is necessary.

Both the descriptive definition which has been used to lead us to the idea of a spline, and the equivalent variational definition, can be generalised to provide interpolating splines composed of polynomial pieces of *odd* degree $(2m + 1)$, say, which are piecewise continuous functions belonging to $C^{2m}[a, b]$, and which minimise the value of $\int_{x_0}^{x_n} [y^{(m+1)}(x)]^2 \mathrm{d}x$. The lineal spline $(m = 0)$ is simply the curve formed by piecewise lineal interpolation between data coordinates, and is of limited interest. A certain body of opinion favours the use of quintic splines $(m = 2)$ over the cubic $(m = 1)$, but this is about as far as practical application (as distinct from theoretical interest) usually extends.

The end-conditions for a cubic spline need a little further attention. It is of course unlikely in most examples of curve drawing that the values of the end derivatives y_0', y_n' would be available. In their absence, one possibility is to produce what is called a **natural cubic spline** for which $M_0 = M_n = 0$ – that is, the curvature at the end points $x = x_0$, x_n is assumed to vanish. This is called 'natural' because the curvature of the designer's spline vanishes at the end pins. Likewise if there is no constraint imposed on the value of slope at $x = x_0$ and $x = x_n$, then the minimisation of the integral $\int_{x_0}^{x_n} [y''(x)]^2 \, dx$ implies that $y''(x_0) = y''(x_n) = 0$, so that it is also a 'natural' consequence of the variational definition. If such an end-condition is assumed, the first and last equations of (3.4.7) and (3.4.8) are no longer needed. For example, the latter provides a system of $(n-1)$ equations for $M_1, M_2, \ldots, M_{n-1}$ given by

$$\lambda_k M_{k-1} + 2M_k + (1 - \lambda_k) M_{k+1} = 6\hat{y}_{k-1, k, k+1} \qquad (3.4.9)$$

with $k = 1, 2, \ldots, (n-1)$ and $M_0 = M_n = 0$.

However, there is one other feature of the interpolation — or rather the extrapolation — of the cubic spline which would consistently follow from such an approach. The free ends of the physical spline extend uncurved from the position of the end pins, since there are no constraints in those regions to induce the stress implied by curvature. Likewise in the variational definition of a cubic spline, if the limits of the integral are changed so that $\int_a^b [y''(x)]^2 \, dx$ is to be minimised, then indeed it would be a clear consequence that this implies that $y''(x) = 0$ for all $x \in [a, x_0]$ or $x \in [x_n, b]$. The condition that $y''(x_0) = y''(x_n) = 0$ must be seen therefore merely as a 'natural' way of blending the constrained part of the spline (between the first and last pins at $x = x_0$ and $x = x_n$) with the straight-line extensions in the extrapolated regions, and for this reason it is termed the **free-end** condition.

Now of course up to this moment it has been somewhat uncritically assumed that, as in (3.3.2), extrapolation is achieved in the intervals $[a, x_0]$ and $[x_n, b]$, if indeed these exist (that is, if $a < x_0$ and $x_n < b$), by a simple **analytic extension** of the polynomial pieces $\pi_1(x)$ and $\pi_n(x)$. What is now being implied is that (perhaps) there should be extra polynomial pieces $\pi_0(x)$ and $\pi_{n+1}(x)$, say, representing $y(x)$ in the extrapolated regions; and that consistent with the 'natural' cubic spline, these extrapolation pieces should be straight lines.

This may indeed be 'natural' in the sense that it replicates the behaviour of the physical spline, but such lineal extrapolation would seem quite unnatural to the eye. Surely it would be more 'natural' if the extrapolated regions 'borrowed' some curvature from the region of interpolation, as might seem the more likely to happen if $\pi_1(x)$ and $\pi_n(x)$ were used as extrapolation pieces. In the present context, however, such an expectation would be misplaced: these cubic pieces have been given points of inflexion at the end points $x = x_0$, x_n and consequently would have the opposite curvature in the extrapolated and inter-polated regions on either side of each end point. In this instance, therefore, the use of lineal extrapolation is a better alternative, even if still an unsatisfactory one.

In more general contexts than this, it would readily be conceded that analytic extension provides the best method, if not merely the only reasonable basis, by which extrapolation may be performed. The dissatisfaction with the natural spline stems essentially from the imposition of the free-end condition, and not from overlooking the possibility that the extrapolating pieces might differ analytically from the first and last interpolating pieces. What is needed is a different form of end-condition.

The generally preferred approach is first of all to construct the 'unnatural' cubic spline which interpolates the $(n-1)$ data corresponding to the internal nodes $x_1, x_2, \ldots, x_{n-1}$, which will be in need of two end-conditions to be properly defined. Next one agrees on some reasonable way by which this spline shall be 'extrapolated' over the intervals $[a, x_1)$ and $(x_{n-1}, b]$, and then introduces the two missing conditions by requiring that these 'extrapolations' shall interpolate the data (x_0, y_0), (x_n, y_n) which have been otherwise ignored. The word 'extrapolation' is here placed in quotes because these intervals include both regions of pseudo-extrapolation $[x_0, x_1)$, $(x_{n-1}, x_n]$ as well as of true extrapolation $[a, x_0)$, $(x_n, b]$; by combining them, one is of course simply treating them as single polynomial pieces, $\pi_1(x)$ and $\pi_n(x)$, just as in (3.3.2).

One reasonable basis of extrapolating a cubic spline is to accept that $\pi_1(x)$ and $\pi_n(x)$ are **quadratic end pieces** continuous in value, slope and curvature with the cubic spline at $x = x_1$ and $x = x_{n-1}$ respectively. The form assumed for these polynomials in (3.4.6) will ensure that they provide this continuity, as well as also respectively interpolating the end nodes at $x = x_0, x_n$, so that it is merely necessary to impose the condition that these pieces are quadratics: this must mean that their second derivatives are constant, and that

$$M_0 = M_1, \qquad M_n = M_{n-1} , \qquad (3.4.10)$$

it being remembered that $y''(x_k) = M_k$. It is easy to check from (3.4.6) that this condition (3.4.10) implies that the term x^3 vanishes in the expressions for $\pi_1(x)$ and $\pi_n(x)$. It will be recalled that dropping the degree of the polynomial end pieces by one was also the procedure suggested in relation to cubic Bessel interpolation.

However, it being usually accepted that extrapolation is best accomplished by analytic extension, another reasonable suggestion is to apply this principle of extension to the *pseudo*-extrapolation, and therefore require that $\pi_1(x) = \pi_2(x)$ and $\pi_{n-1}(x) = \pi_n(x)$. Since these pairs of cubic pieces would differ only because the piecewise constant third derivative $y'''(x)$ may be discontinuous, this requirement is met simply by requiring that $\pi_1'''(x) = \pi_2'''(x)$ and $\pi_n'''(x) = \pi_{n-1}'''(x)$. But (3.4.6) shows that $\pi_k'''(x) = (M_k - M_{k-1})/h_k$; so that this requirement leads, after some algebra, to the conditions (valid for $n > 2$) that:

$$(1 - \lambda_1)M_0 = M_1 - \lambda_1 M_2, \qquad \lambda_{n-1}M_n = M_{n-1} - (1 - \lambda_{n-1})M_{n-2} . \qquad (3.4.11)$$

If these conditions are imposed, then $y(x)$ is given by the same cubic piece over the entire interval (a, x_2), and by another single cubic piece over (x_{n-2}, b). All derivatives are continuous at $x = x_1$ and $x = x_{n-1}$, not merely the first two.

Another way of looking at this latter end-condition is to regard $[x_0, x_2]$ and $[x_{n-2}, x_n]$ as the intervals of definition of the end-pieces of the spline, and to provide the two end-conditions of this spline by requiring that these two end pieces respectively intersect the data coordinates (x_1, y_1) and (x_{n-1}, y_{n-1}). This is effectively to take the view that the nodes (or knots) at $x = x_1$ and $x = x_{n-1}$ are absent, because in contradistinction to all others, they cannot mark discontinuities in the third derivative of $y(x)$. With the adoption of this point of view, (3.4.11) has been catchily christened the **not-a-knot** condition.

Either of the pairs of conditions (3.4.10) and (3.4.11) can be incorporated in the interpolation scheme as replacements for the first and last equations of the systems (3.4.7) and (3.4.8), but in hand computation at least it is more convenient to use them to replace the values of M_0 and M_n in the first and last of the $(n-1)$ equations (3.4.9). In particular, if the nodes are equispaced (so that $\lambda_1 = \lambda_2 = \ldots = \lambda_{n-1} = \frac{1}{2}$), the not-a-knot condition (3.4.11) simplifies to $\Delta^2 M_0 = \Delta^2 M_n = 0$, and accordingly substituting $M_0 + M_2 = 2M_1$ and $M_{n-2} + M_n = 2M_{n-1}$ in the first and last equations of (3.4.9), these are found to be converted into the simple identities:

$$M_1 = 2\hat{y}_{0,1,2} = \delta^2 y_1/h^2, \quad M_{n-1} = 2\hat{y}_{n-2,n-1,n} = \delta^2 y_{n-1}/h^2 \ ,$$

implying that the curvature at $x = x_1, x_{n-1}$ is that of the quadratic interpolating the three end points. Thus only a system of $(n-3)$ equations for $M_2, M_3, \ldots, M_{n-2}$ remain to be solved, and when this is accomplished the values of M_0 and M_n can be found from (3.4.11).

3.5 INTERPOLATION AND EXTRAPOLATION BY RATIONAL FUNCTIONS

Although the use of piecewise continuous polynomials is successful as a means of interpolating data, it is much less so as a means of extrapolation. Indeed, as we have seen, it leaves the precise method of extrapolation as a matter of choice, and although the interpolation of interior points may not be much affected, the shape of the curve being fitted to the data close to and beyond the end points can be grossly altered by that choice. Of course, extrapolation is always a stab in the dark: but it seems particularly unguided in schemes of piecewise-continuous interpolation. It is natural, therefore, to look for another method of representation which does not involve discontinuities, so that the whole of the data can be enlisted – without interruption, as it were – in shaping the analytic extension beyond the interval of support, and yet which is exempt from the problems that attend the use of Lagrangian interpolation.

A favoured method involves the use of rational functions: that is, the representation of $y(x)$ as a quotient of polynomials:

$$y(x) = A_\mu(x)/B_\nu(x) = \left(\sum_{k=0}^{\mu} a_k x^k\right)\bigg/\left(\sum_{k=0}^{\nu} b_k x^k\right) . \tag{3.5.1}$$

The degrees of the numerator and denominator polynomials are here (at most) μ and ν respectively, so that in general the rational function has order μ/ν. It is at once obvious that by suitable choice of $(\mu - \nu)$ it is possible to control the behaviour of $y(x)$ for large values of $|x|$, and so avoid the wild excursions of value that are inevitable if $y(x)$ is a high degree polynomial (which is clearly a particular form of rational function with $\nu = 0$). Moreover, $(\mu - \nu)$ can be so chosen no matter how many data are provided, since the number of degrees of freedom (that is, undetermined coefficients) of the rational function depends upon the sum of μ and ν. Also, by way of encouragement, it will be recalled that rational functions (in the shape of continued fractions) are usually more successful in achieving a rapidly convergent representation of the behaviour of transcendental functions than series, and moreover the interval of convergence of the infinite continued fraction is usually much wider than that of the infinite series.

On the other side of the coin, the use of rational functions implies that the method of interpolation is not linear (because the coefficients b_k in the denominator of (3.5.1) depend on the data ordinates, as well as those in the numerator). One is thereby robbed of the means of insight into the method of interpolation provided by the use of shape functions. But more signigicantly, the very strength of rational approximation in accommodating (for example) a few unusually large data ordinates, simply by contriving that the magnitude of the denominator polynomial locally reaches a small minimum, by which one might judge that the rational function has excellent properties of 'conformability' – is also its weakness. The denominator polynomial may have zeros in the interval of interpolation, and if Lagrangian interpolation is rejected on the grounds of the Runge phenomenon, what is one to say of a scheme which possibly allows an *unbounded* excursion of value between data points?

For example, if the data are ordered in increasing value of x, and the data ordinates y_k change sign m times (say) – which might imply that the curve $y = y(x)$ should be oscillating about $y = 0$, not necessarily regularly or with large amplitude – then if $\mu < m$, any rational function of order μ/ν fitted to these data will have at least one pole in the interval of support. This is because $y(x)$ can only change sign at a zero of either the numerator or denominator polynomial, and the numerator $A_\mu(x)$ has at most μ zeros in the interval. Moreover, this is merely a sufficient condition, not a necessary one, for the existence of such a pole.

It is not difficult therefore to envisage conditions in which rational interpolation would be quite inappropriate, and it cannot be recommended for general use. There are, as will be demonstrated, other difficulties – of indeterminancy and lack of uniqueness – which may beset its use in particular applications which are of a more subtle and even less predictable nature. However, in common with Lagrangian interpolation, the disqualification from general application is not to be interpreted as dismissing the possibility of its successful use, and it is of importance to realise that it *can* be used as a method of interpolation or extrapolation without the need for a lot of laborious calculation.

In regard to extrapolation, there is particular interest (for the reason already stated) in representation by rational functions of order μ/ν where $|\mu - \nu| \leqslant 1$. Also it will be recalled that S-fraction developments of transcendental functions proceed by successive convergents of order $0/0, 0/1, 1/1, 1/2, \ldots$ and so on. Furthermore, there is no loss in generality by restricting consideration here to such rational functions. For instance, if one wanted for some reason to construct an interpolating rational function of order μ/ν where $\mu = m + \nu + 1 > \nu$, then Lagrangian interpolation could be applied to find the polynomial $p_m(x)$ which interpolates the $(m + 1)$ data pairs (x_k, y_k) for $k = 0, 1, \ldots, m$ (say) with the consequence that

$$y(x) = p_m(x) + (x - x_0)(x - x_1) \ldots (x - x_m)\overline{A}_\nu(x)/\overline{B}_\nu(x)$$

is a rational function of the required order interpolating these $(m + 1)$ data, and $\overline{A}_\nu(x)/\overline{B}_\nu(x)$ is another rational function (of order ν/ν) which can be arranged to interpolate the remaining data – or more precisely it would need to interpolate not the data ordinates y_k, but rather the values of

$$[y_k - p_m(x_k)]/\prod_{j=0}^{m} (x_k - x_j)$$

for all $k > m$.

It can also be observed that if $A_\mu(x)/B_\nu(x)$ interpolates some given set of data pairs (x_k, y_k), then $B_\nu(x)/A_\mu(x)$ will interpolate the corresponding set of data pairs (x_k, y_k^{-1}). This is a **reciprocity principle**. It indicates how the problem of interpolation by a rational function of order $\nu/(\nu + m + 1)$ could be transformed into one like that just described, of interpolation by a rational function of order $(\nu + m + 1)/\nu$, provided at least that any zero data ordinates are excluded from those treated by Lagrangian interpolation (since they would transform into infinite values of y_k^{-1}). For example, if it was wished to construct the rational function of order $0/n$ interpolating the $(n + 1)$ data pairs (x_k, y_k) for $k = 0, 1, \ldots, n$, where all the ordinates y_k are (say) positive, then this will be given by $y(x) = 1/p_n(x)$, where $p_n(x)$ is the Lagrangian polynomial interpolating (x_k, y_k^{-1}) for $k = 0, 1, \ldots, n$.

It will be seen from (3.5.1) that there are $(\mu + \nu + 2)$ coefficients $a_0, \ldots, a_\mu$, $b_0, \ldots, b_\nu$ in the description of $y(x)$, which are determinate except for a constant factor. Any particular non-zero coefficient could be taken as unity without loss of generality, but no particular coefficient can be guaranteed to be non-zero, so they must be left as they are. If this rational function interpolates $(\mu + \nu + 1)$ distinct data pairs (x_j, y_j) for $j = 0, 1, \ldots, (\mu + \nu)$, then (3.5.1) shows that the coefficients must satisfy the system of $(\mu + \nu + 1)$ homogeneous linear equations

$$\sum_{k=0}^{\mu} x_j{}^k a_k - \sum_{k=0}^{\nu} y_j x_j{}^k b_k = 0: \quad j = 0, 1, \dots, (\mu + \nu) \ . \qquad (3.5.2)$$

Of course there is a trivial solution of these equations in which all the coefficients are zero, but a homogeneous system always has an infinity of non-trivial solutions, and it is not difficult to show that the b_k's in these solutions cannot all be zero (which would imply that $B_\nu(x)$ was identically zero). If the system (3.5.2) is non-singular, and so of rank $(\mu + \nu + 1)$ — which is neither obvious, nor necessary, but is in fact generally the case — then the non-trivial solutions all differ from each other in that each coefficient is proportional to one and the same arbitrary non-zero constant. This constant will then be a common factor of $A_\mu(x)$ and $B_\nu(x)$, and clearly its value does not affect $y(x)$, which is therefore uniquely determined.

The solution of the system (3.5.2) would afford one way of constructing $y(x)$. But if $0 \leqslant (\mu - \nu) \leqslant 1$, then there is an easier way, by which $y(x)$ is represented as the terminated continued fraction:

$$y(x) = q_0 + \mathop{K}_{k=0}^{\mu+\nu} (x - x_{k-1}) \downarrow q_k \ , \qquad (3.5.3)$$

where on placing $q_k = q_{kk}$, these partial denominators are formed for $j = 0, 1, \dots,$ $(\mu + \nu)$ by the recurrence:

$$q_{j,0} = y_j, \quad q_{j,k+1} = (x_j - x_k)/(q_{jk} - q_{kk}): \quad k = 0, 1, \dots, (j-1) \ . \qquad (3.5.4)$$

The order of convergents of the rational function $y(x)$ in (3.5.3) increases in the sequence $0/0, 1/0, 1/1, 2/1, 2/2, \dots$ as successive data pairs are added, and the data can be taken in any order without affecting the value of $y(x)$, provided at least that the effects of finite arithmetic precision are ignored (see problem 2).

The elements q_{jk} given by (3.5.4) are called **inverted differences**: and they can be composed in a triangular array, just like divided differences, with the partial denominators $q_k = q_{kk}$ of the continued fraction lying along the diagonal. However, unlike divided differences, they are not symmetrical functions of the data on which they depend, so that the recursion (3.5.4), which is seen to be patterned as in Aitken's method (3.2.12), *cannot* be rearranged in a form similar to Neville's scheme of divided differences (3.2.15).

It follows from the reciprocity principle that the continued fraction

$$y(x) = \mathop{K'}_{k=0}^{\mu+\nu} (x - x_{k-1}) \downarrow \bar{q}_k \qquad (3.5.5)$$

with $\bar{q}_k = \bar{q}_{kk}$ as before and the $\bar{q}_{jk}$ given by the same recurrence (3.5.4) as q_{jk} except that $\bar{q}_{j,0} = y_j^{-1}$, also interpolates the data, and has successive convergents which are rational functions of order $0/0, 0/1, 1/1, 1/2, 2/2, \dots$. Thus this approach extends the development by continued fractions to all rational functions of order $|\mu - \nu| \leqslant 1$. If rational interpolation is unique, the even numbered convergents (with $\mu = \nu$) must be the same in both approaches.

Evidently the continued fraction development of an interpolating rational function parallels the Newton form of interpolating polynomial, and its reverse order evaluation by (2.8.2) is in effect a form of nested division, analogous to nested multiplication. However, aside from the fact that $y(x)$ may be unbounded, which might reveal itself as a zero-divide error during the computation of $y(x)$, there are other difficulties, both numerical and algebraic.

Dealing first with the algebra, there exists the possibility that the interpolating rational function may be **reducible**. This means that both numerator and denominator polynomials have at least one common zero, and therefore a common factor which is a polynomial in x of positive degree; conversely, if they have no common polynomial factor, the rational function is said to be **irreducible**. Two rational functions, P_1/Q_1 and P_2/Q_2 say, are defined as **equivalent** if $P_1 Q_2 \equiv P_2 Q_1$, and were they both also irreducible this would imply that $P_2 = cP_1$, $Q_2 = cQ_1$ for some non-zero constant c; on the other hand, if one or both should be reducible, then their equivalence would imply that, although they may differ to the extent of having different polynomials common to their respective numerators and denominators, they are both reducible to equivalent irreducible forms by cancelling all such common factors.

There are, as we know, infinitely many non-trivial solutions of the linear system (3.5.2) leading to different expressions of the numerator and denominator polynomials of the interpolating rational function: however, whether or not reducibility is implied, all such solutions represent equivalent rational functions. For if $P_1(x)/Q_1(x)$, $P_2(x)/Q_2(x)$ are two such solutions, then the equations of the linear system ensure that $P_1(x_j) = y_j Q_1(x_j)$ for all $j = 0, 1, \ldots (\mu + \nu)$, and similarly $P_2(x_j) = y_j Q_2(x_j)$ for the same set of points $x = x_j$ of interpolation; consequently $(P_1 Q_2 - P_2 Q_1)$, which in this context is a polynomial of at most degree $(\mu + \nu)$, is zero at these $(\mu + \nu + 1)$ points of interpolation, and must therefore vanish identically, implying that P_1/Q_1 and P_2/Q_2 are (as asserted) equivalent.

In general the interpolating rational function will be irreducible, and this equivalence merely implies that any two expressions of the function may differ to the extent of an arbitrary numerical factor common to both numerator and denominator. But in particular if the rank of the linear system (3.5.2) is less than full, then the interpolating function will be reducible. In this event, the coefficients of the numerator and denominator polynomials are all expressible as linear combinations of $(\mu + \nu + 2 - r)$ arbitrary parameters, where r is the rank. Consequently the numerator polynomial $A_\mu(x)$ can be written as $c_1 P_1(x) + c_2 P_2(x) + \ldots$, say, where $c_1, c_2, \ldots$ are the arbitrary parameters and $P_1, P_2, \ldots$ are linearly independent polynomials each of degree $\leqq \mu$; and similarly the denominator can be expressed as $c_1 Q_1(x) + c_2 Q_2(x) + \ldots$ where likewise

$Q_1, Q_2, \ldots$ are linearly independent polynomials of degree $\leq v$. It then follows that $P_1/Q_1, P_2/Q_2, \ldots$ represent different but equivalent solutions of the interpolation problem, and because of the linear independence of the component polynomials, they differ not by a common constant factor, but rather by a common polynomial factor, in the numerator and denominator. In fact, without loss of generality, it can be arranged for the polynomial sequences $P_1, P_2, \ldots$ and $Q_1, Q_2, \ldots$ to have monotonically decreasing degree, and then the equivalence of the functions of the sequence $P_1/Q_1, P_2/Q_2, \ldots$ implies that they are formed by successive cancellation of factors $(x - \xi_1), (x - \xi_2), \ldots$ common to the numerator and denominator. If we write $m = \mu + v + 1 - r$, it follows that the general solution of the interpolating problem is then given by

$$A_\mu(x) = C_m(x)A_{\mu-m}(x), \quad B_v(x) = C_m(x)B_{v-m}(x)$$

where $C_m(x)$ is an arbitrary polynomial of degree m. Where the rank is full $C_m(x)$ is simply c_0 – an arbitrary constant – but where it is deficient $(m > 0)$, then all solutions must be reducible to $A_{\mu-m}(x)/B_{v-m}(x)$, and this rational function of reduced order also interpolates the data.

It is therefore characteristic of the lack of uniqueness implied by deficient rank of the system that all the data happen to be able to be interpolated by a rational function of order $(\mu - m)/(v - m)$, with $m > 0$. Taking as a simple example the construction of a rational function of order $1/1$ interpolating 3 data, the rank of the system is deficient only if these data are interpolated by a function of order $0/0$, or in other words only if all the three ordinate values y_0, y_1, y_2 are equal. In this event, any rational function $(c_0 + c_1 x)y_0/(c_0 + c_1 x)$, which has the required order $1/1$, interpolates the data for arbitrary c_0, c_1. The converse is also true: thus if y_0, y_1, y_2 are equal, then (for instance) $y_0/1$ and xy_0/x are equivalent solutions of the interpolation problem, implying the existence of two linearly independent solutions $(a_0 \, a_1 \, b_0 \, b_1) = (y_0 \, 0 \, 1 \, 0), (0 \, y_0 \, 0 \, 1)$, for the vector of coefficients given by the linear system (3.5.2), which consequently cannot have full rank.

Although deficiency of rank is one circumstance leading to a reducible form for the interpolating function, it is not the only circumstance. For consider the possibility that *all but one* of the $(\mu + v + 1)$ data happen to be interpolated by a rational function $A_{\mu-1}(x)/B_{v-1}(x)$, say, of order $(\mu - 1)/(v - 1)$, and let this 'odd' datum not so interpolated be (x_i, y_i). Then the rational function $A_\mu(x)/B_v(x)$ interpolating *all* the data (the 'odd' one included) is given by

$$A_\mu(x) = c_0(x - x_i)A_{\mu-1}(x), \quad B_v(x) = c_0(x - x_i)B_{v-1}(x)$$

where c_0 is an arbitrary constant. It may readily be checked that the equations of the linear system (3.5.2) are all satisfied, because it is asserted that $A_{\mu-1}(x_j) = y_j B_{v-1}(x_j)$ for all $j \neq i$; and since (by assumption) the rank is not deficient, the indeterminacy of any solution can only be due to the presence of an arbitrary common factor (c_0) of the numerator and denominator polynomials.

These also involve a common polynomial factor $(x - x_i)$, but this is not arbitrary (as in the context of deficient rank); specifically, it is one of the data abscissae (the 'odd' one, in fact) which is here the common zero.

In practice it is of course more likely that all but one of the $(\mu + \nu + 1)$ data are interpolated by a rational function of order $(\mu - 1)/(\nu - 1)$ – as latterly asserted – than that they *all* happen to be so arranged – as would imply rank deficiency; and it is not difficult to show that the interpolating rational function is reducible only in one or the other of these eventualities. Returning to the example of three data fitted by a rational function of order $1/1$, if just two of the data ordinates – say y_0 and y_1 – are equal, then both are interpolated by the line $y = y_0$, which is of course a rational function of order $0/0$, and the three data are interpolated by $y = y_0(x - x_2)/(x - x_2)$, irrespective of the value of y_2. But is the third datum really interpolated? The linear system of equations (3.5.2) is satisfied, but the structure of this system presupposes that interpolation is possible, and one point of view is that in this context this supposition is false. Thus upon cancelling the common factor, the 'interpolating' function is $y = y_0$, which clearly interpolates (x_0, y_0) and (x_1, y_1) since $y_1 = y_0$, but *not* the datum (x_2, y_2) because $y_2 \neq y_0$. In the jargon of the subject, the 'odd' point $(x = x_2)$ is called **inaccessible**. This might appear to place some significance on the value x_2, but rather we see that *any* third datum would be inaccessible, whatever its position, because of the particular arrangement of the rest.

That is one interpretation of the apparent paradox; a preferable view is that it is invalid to cancel the common factor $(x - x_2)$, because x may equal x_2, and if it does, then the interpolated function has an indeterminate value $0/0$. From this point of view, *any* datum ordinate $y = y_2$ is interpolated at $x = x_2$, irrespective of the value of y_2, because the interpolating 'curve' is made up of the two straignt lines $y = y_0$ and $x = x_2$. Two data are interpolated by the former (horizontal) line, and the remaining one by the latter (vertical) line. What has happened (according to this interpolation) is that the interpolating curve, which is in general a rectangular hyperbola with horizontal and vertical asymptotes

$$y - y_0 = (x_1 - x_2)(y_1 - y_0)(y_2 - y_0)/[(x_1 - x_0)(y_2 - y_0) - (x_2 - x_0)(y_1 - y_0)]$$

$$x - x_2 = (y_1 - y_0)(x_1 - x_2)(x_0 - x_2)/[(x_1 - x_0)(y_2 - y_0) - (x_2 - x_0)(y_1 - y_0)]$$

degenerates, if any two data ordinates are equal, into those two asymptotes. As y_1 tends to y_0, both the interpolating curve and asymptotes tend towards the lines $y = y_0$ and $x = x_2$. In the limit, the vertical asymptote $(x = x_2)$ is as much a part of the interpolating curve as the horizontal part. The 'odd' datum (x_2, y_2) *is* therefore interpolated, but the interpolating curve is **degenerate**. If all three data ordinates are equal, the curve remains degenerate; and the three data are all interpolated by the horizontal asymptote $y = y_0$; the position of the vertical asymptote is therefore arbitrary, since it is not fixed by the need to interpolate a datum point. In this event (when the rank of the linear system is deficient) the interpolation is therefore non-unique. It will be shown that the ideas of indeterminacy and the associated degeneracy of the interpolating curve provide a closer description of the numerical phenomena associated with the reducibility of an interpolating rational function, because the precise coincidences invoked in the algebraic treatment become blurred by numerical inaccuracies: 'common' zeros of the numerator and denominator will not be calculated as precisely equal – only nearly the same.

Turning attention now to the numerical evaluation of the continued fractions (3.5.3) or (3.5.5), it is at once obvious that there can be computational difficulties even in contexts which are algebraically innocuous. For example, the computation of the values of $q_{j,0} = y_j^{-1}$ in the table of inverted differences, by which one constructs the partial demoninators of (3.5.5), would encounter a zero-divide error if any of the data ordinates were zero. Similarly, the computation of any of the inverted differences $q_{j,1} = (x_j - x_0)/(y_j - y_0)$ of (3.5.4) would produce the same error condition if $y_0 = y_j$. Such eventualities are by no means uncommon, nor in general is there any reason to regard them as catastrophic: two measures are essential to allow the numerical working to proceed.

Firstly, one must introduce some conventional representation of an (unsigned) infinity into the computation – let us denote it by ∞ – and $(1/\infty)$ must be defined by zero. Secondly, in generating the triangular arrays of values of q_{jk}, if any diagonal element q_{mm} (say) becomes infinite, so that the partial denominator $q_m = 0$, then it is necessary to re-order the data by interchanging the offending data pair (x_m, y_m) with another, (x_n, y_n), say, with $n > m$, which avoids this occurrence. This is essential because if $q_{mm} = \infty$ it follows inevitably from (3.5.3) that all the values of $q_{j,m+1}$ in the next column will be zero, and consequently all the elements $q_{j,m+2}$ will be infinite. If $(\infty \pm \infty)$ is defined as ∞, this succession becomes repetitive, and $q_m = q_{m+2} = \ldots = \infty$, $q_{m+1} = q_{m+3} = \ldots = 0$. The continued fraction is then in effect terminated at the $(m-1)$th convergent, with the partial denominator q_{m-1}, and none of the data (x_k, y_k) for $k \geqslant m$ affect its value.

This would suggest that if the interpolating rational function is reducible, it may not be possible to make any interchange which avoids this phenomenon, because just such a reduction in order of the rational function is then properly to be implied. In any event, nothing can (or need) be done if the last partial denominator $q_{\mu+\nu}$ is infinite: this merely (and correctly) implies that the last data pair is interpolated by the same rational function as the rest of the data. Generally, however, in other instances an interchange in the data order will resolve the apparent difficulty.

This may seem curious, bearing in mind that the ordering of the data should be (in theory) inconsequential to the construction of the continued fraction. But it will be noted that both (3.5.3) and (3.5.5) contain only $(\mu + \nu + 1)$ adjustable parameters (the partial denominators, q_k), whereas in the general rational form

(3.5.1) there are $(\mu + \nu + 2)$ – the coefficients a_k and b_k of the polynomials. In the same way that *usually* one of these coefficients could be chosen as equal to (say) unity without giving rise to any difficulties – but certainly not invariably, because for a particular choice of data that particular coefficient may in fact be zero – so the assumed form of the continued fractions (3.5.3) and (3.5.5) will usually, *but not invariably*, be compatible with the data as presented. In the event of such an incompatibility, a re-ordering of the data may remove the inconsistency because the structure (though not the theoretical value) of the continued fraction depends on the sequence of data values.

To see what happens in the event that the interpolate is degenerate, consider first the interpolation of two data (x_0, y_0), (x_1, y_1) by a rational function of order 0/1 in the event that one of the data ordinates is zero. The second convergent of the continued fraction (3.5.5) is needed: its partial denominators are

$$q_0 = y_0^{-1}, \; q_1 = (x_1 - x_0)/(y_1^{-1} - y_0^{-1}) \; .$$

To avoid $q_0 = \infty$, it would be necessary to arrange the data so that it is the ordinate y_1 which is zero, and then $q_1 = 0$ and the continued fraction is

$$y(x) = 1/[y_0^{-1} + (x - x_0)/0] \; .$$

This, quite appropriately, shows that $y(x) = 0$ except (possibly) at $x = x_0$, where the value of $y(x)$ is undefined (but where of course the value of $y(x_0) = y_0$ is available as data).

Again, if the continued fraction (3.5.3) is used to construct a rational function of order 1/1 interpolating the three data (x_k, y_k) for $k = 0, 1, 2$, its partial denominators are obtained from the recurrence (3.5.4) as the diagonal elements of the array of inverted differences:

$$y_0 = q_0$$

$$y_1 \qquad (x_1 - x_0)/(y_1 - y_0) = q_1$$

$$y_2 \qquad (x_2 - x_0)/(y_2 - y_0) = q_{21} \quad (x_2 - x_1)/(q_{21} - q_1) = q_2 \; .$$

This rational function is reducible if two of the ordinates are equal; if y_0 is one of the two, then y_1 and y_2 would need to be arranged so that q_1 is bounded, and therefore $y_2 = y_0 \neq y_1$ must be taken as the other 'twin' ordinate, giving $q_{21} = \infty$ and $q_2 = 0$. Consequently the interpolating function is

$$y(x) = y_0 + (x - x_0)/[q_1 + (x - x_1)/0]$$

giving $y(x) = y_0$ except (possibly) at $x = x_1$. This is once again seen to be an entirely appropriate result, since an indeterminacy like 0/0 is just what might reasonably be expected numerically to occur at an inaccessible point (where y is multi-valued if the interpolate is regarded as degenerate).

On the other hand, in this latter example if it had happened that $y_1 = y_2 \neq y_0$, so that the point (x_0, y_0) is inaccessible, then none of the partial

denominators is infinite, and the algebra shows that the interpolating function
should be

$$y(x) = y_0 + (x - x_0)/[(x - x_0)/(y_1 - y_0)]$$

which once more correctly implies that $y = y_1$, except possibly at $x = x_0$ where
the 0/0 indeterminacy reappears. Note that such indeterminacies can occur only
where the partial numerators are zero, which means they can occur only at some
support point $x = x_k$ (other than the last one, $x = x_{\mu+\nu}$).

However, in the numerical working of this last example, quite a different
kind of error is likely to be revealed. The quantity in square brackets would be
computed as $[q_1 + (x - x_1)/q_2]$ and the inevitable lack of numerical precision,
first in forming q_1 and q_2 and then in evaluating the square bracketed quantity
for any given value of x, will generally cause its value to be zero at some value
of x close to, but *not precisely* equal to, x_0. The indeterminacy may well be
'missed', and the degenerate interpolate will then be approximated by a
rectangular hyperbola which is almost, but not quite, degenerate.

Indeed as the number of data and the order of the rational function
increases, this becomes much the most likely way in which the reducibility of
the rational function would manifest itself in machine arithmetic. Furthermore,
the exceptional (and, in practice, very unlikely) condition in which the interpolation is not unique will generally reveal itself by producing a 'spike' in the
interpolating curve at a quite arbitrary point, dependent upon the random
nature of numerical error, but correctly mimicking the arbitrary indeterminacy
of the algebraic representation. The higher the arithmetic precision of the
machine, the more localised will be this disturbance to the shape of the interpolating curve, and to all intents and purposes it may well be 'inaccessible' to
discovery.

Enough has been said to explain why interpolation by rational functions
should always be approached with caution. There are many applications in
which the structure of the problem data may be such that the kinds of difficulties
mentioned do not arise, and the interpolated curve is entirely picturesque. On the
other hand, it is the kind of technique by which selective results can be obtained
from uncritical exercises in 'number crunching' which may fail to reveal the
nonsense on which they are based.

3.6 PARAMETRIC AND PERIODIC INTERPOLATION

It has been assumed so far that the curve to be fitted to the data is single-valued,
but there are contexts in which this assumption may be inappropriate. Perhaps
the most common instance is that where the data points describe a closed curve.
However, the closure of a curve is a sufficient, but far from necessary, condition
for the value of y to be a multi-valued function of x, and in *any* such context
where the data imply that y is not single-valued, the methods of curve drawing
so far described cannot be immediately applied.

One obvious remedy for this situation would involve segmenting the interpolating curve into component x- and y- **branches** which can each respectively be represented by $y = \varphi(x)$ and $x = \psi(y)$, where $\varphi(x)$ and $\psi(y)$ are single-valued functions of their arguments. Any curve, however complicated its shape, can be composed as a set of such branches. However, to apply this idea to an interpolation algorithm is far from straightforward. The set of data has to be partitioned into subsets describing each individual branch, and the regulation of such a choice proves in practice to involve quite subtle complications. Then each of the individual branches must be matched to its neighbour, where they join, to preserve an appropriate continuity. It is of course an acceptable method where the whole curve can be described by a single branch (where for example x is a single-valued function of y, but not *vice versa*), but otherwise it cannot be recommended.

It is much neater, computationally, to use **parametric interpolation**: that is, to generate the curve by separately interpolating the data values of x and y against some parameter t (say) which increases monotonically with distance along the curve. In this way, two continuous single-valued functions $x(t)$ and $y(t)$ can be obtained, by any of the methods of interpolation already discussed, for some interval $t \in I$. The interpolating curve C (say) is then the set of points (x, y) in the Elucidean plane such that

$$x = x(t), \quad y = y(t), \quad t \in I . \tag{3.6.1}$$

For a start, this approach simplifies the description of the curve's geometry. If there is a point on C which is given by two (or more) distinct elements t_1, t_2 of I then it is said to be a **multiple point** of C (where the curve crosses over itself). A curve without a multiple point is called a **simple arc**, and a curve with a multiple point corresponding to the end-points of I is evidently a closed curve; if this is the only multiple point it is called a **simple closed curve**. Further, if each of the functions $x(t), y(t)$ belongs to C^k on the interval I, then the curve C is said to be of **class** C^k: thus, for instance, if $x(t)$ and $y(t)$ are cubic splines, the generated curve C will be of class C^2 and have continuous tangent and curvature.

The only real difficulty of this approach concerns the appropriate choice of the parameter t. Of course, it may be that the parameter is part of the data — for example $x(t), y(t)$ might describe spatial positions at different times t, and it is required to interpolate the trajectory in the Euclidean plane from given data triads (x_k, y_k, t_k) — and then clearly there is no difficulty. But otherwise, supposing that the data coordinate pairs are ordered sequentially at successive positions along the curve starting at one end or the other, then all that can be asserted is merely that the parametric value t_k corresponding to any data pair (x_k, y_k) must vary strictly monotonically with k. The precise form of this variation remains to be constructed — virtually in any way that seems convenient.

In the theoretical development of the differential geometry of curves, it is often found convenient to define the parameter as the distance measured along the curve, so that t is defined by the relation $x'(t)^2 + y'(t)^2 = 1$; but computationally such a definition would lead to considerable complications. However, it may be adequate in numerical work to accept a rather crude approximation to this interpretation, by taking $t_0 = 0$ and placing

$$t_k = \sum_{j=1}^{k} [(x - x_{j-1})^2 + (y_j - y_{j-1})^2]^{1/2} , \qquad (3.6.2)$$

so that it is equal to the progressive sum of the lengths of the straight lines (the **chords**) joining successive data points on the curve. In fact, there is no need even to take the square root in this expression: the sum of the squares of the lengths might do just as well. It would be still simpler to place

$$t_k = \sum_{j=1}^{k} (|x_j - x_{j-1}| + |y_j - y_{j-1}|) ,$$

but unlike the former expression (3.6.2), this would cause the values of t_k, and so the shape of the interpolating curve, to depend on the angular orientation of the (x, y) axes relative to the curve. This would seem a clear irrelevance.

The simplest idea of all, and one which assures that the shape of the inter-polated curve is invariant under any transformation of the Cartesian coordinate system, is to suppose that there is some basis of regularity in the positioning of the data, however remote, and to place $t_k = k$ (or equally well, $t_k = a + bk$, where a and b are any fixed constants). This can be called **cardinal parametric interpolation**, and it proves a satisfactory method even if the underlying assumption is not strictly justified. It has the convenience of permitting the method of interpolation to be linear, since the vector equation of the curve $\mathbf{r} = \mathbf{r}(t)$, where $\mathbf{r}$ is the 2-vector (x, y), can be expressed as a linear combination of the data vectors $\mathbf{r}_k = (x_k, y_k)$. Moreover it involves the advantage of an equi-spaced support (of t_k values), which of course implies a considerable simplification of the arithmetic. Thus if $\mathbf{r}(t)$ is to be constructed as a polynomial in t, then this vector function could be expressed as a Newton form (in t), whose coefficients were each 2-vectors calculable from (3.2.20) as $\mathbf{d}_m = \Delta^m \mathbf{r}_0 / m!$, — assuming that the parameter increment h is taken as unity.

Yet another approach is to replace the parameter t by some suitable angular coordinate; obviously the polar equation of a curve $r = r(\theta)$ may be such that $r(\theta)$ is single-valued, in which event it could be directly constructed by inter-polating r against θ. Such a choice of parameter might be eminently suitable in particular contexts, but it cannot be relied upon as the basis of a general method applicable to *any* curve. For instance, it may be likely to succeed if the interpolate is a simple closed curve, but inevitably it fails if the curve is not simple.

In the application of interpolation to the generation of closed curves, whether or not simple, it is clearly possible to treat $x(t)$ and $y(t)$ in (3.6.2) as periodic functions of t, and this introduces the possibility of both simplifications to, and extensions of, the methods of interpolation. Taking the fundamental period as equal to ω, so that any arbitrary interval I of t having length ω is mapped once (and only once) on to the closed curve by (3.6.2), and using (for example) cubic spline, or quintic Bessel, interpolation to construct the functions $x(t)$ and $y(t)$, it follows that the end-conditions can be determined from the fact that $\mathbf{r}'(t) = \mathbf{r}'(t + \omega)$ and $\mathbf{r}''(t) = \mathbf{r}''(t + \omega)$, where $\mathbf{r}(t)$ is the 2-vector $(x(t), y(t))$. This not only resolves an aspect of uncertainty in the use of such methods, but of course it serves to ensure that the interpolated curve has continuous slope and curvature at any and every point (and not merely at all points except the multiple point corresponding to the end values of the interval I).

Interpolation by a single-valued periodic function is a problem which arises in fitting a curve to data pairs (x_k, y_k) which are known to describe a single-valued periodic variation of y in terms of x, and not just in the context of the parametric interpolation of closed curves. This is the important and highly developed subject of **harmonic analysis**, which however is beyond the bounds of adequate treatment here. Methods of generating interpolating curves $y = y(x)$, where $y(x)$ is a periodic function belonging to C^∞ for all real x, employ truncated trigonometric series, or **trigonometric polynomials**: thus for example if the n distinct data pairs (x_k, y_k) for $k = 0, 1, \ldots, (n-1)$ with $x_k \in [0, 2\pi)$ describe a periodic function of known period 2π, then the trigonometric interpolating function would be written as

$$\begin{aligned} y(x) &= (A_0/2) + S_{m-1} + (A_m/2)\cos(mx), &&\text{if } n = 2m, \\ &= (A_0/2) + S_m &&\text{, if } n = 2m + 1, \end{aligned} \right\} \ (3.6.3)$$

where $\qquad S_m = \sum_{k=1}^{m} [A_k \cos(kx) + B_k \sin(kx)]$.

The n coefficients $A_0, A_1, B_1, \ldots$ can be determined uniquely in terms of the data, and depend linearly on the data ordinates. For instance, if the support is equally-spaced so that $x_k = 2k\pi/n$, then it can be shown that (3.6.3) interpolates the data if, and only if,

$$A_j = (2/n) \sum_{r=0}^{n-1} y_r \cos(2\pi jr/n), \quad B_j = (2/n) \sum_{r=1}^{n-1} y_r \sin(2\pi jr/n) .$$
$$(3.6.4)$$

The use of trigonometric polynomials to interpolate a large number of equi-spaced data is exempt from the difficulties that attend the use of algebraic polynomials in similar circumstances. But special consideration has to be given to methods for the rapid evaluation of the sequence of trigonometric functions occurring in (3.6.3), if the computation is not to be unduly time-consuming. This is the problem of **Fourier synthesis**, whilst that of determining the coefficients of (3.6.3) is part of the subject of **Fourier analysis**; the successful development of numerical methods for both analysis and synthesis involves considerable subtleties.

The similar pairs of summations like those occurring both in (3.6.3) and (3.6.4) can for instance be simultaneously determined by defining the recursion

$$U_{n+1} = U_n = 0, \ U_j = c_j + 2U_{j+1}\cos\xi - U_{j+2}:$$

$$j = n-1, n-2, \ldots, 0 \ , \qquad (3.6.5)$$

which leads to

$$\sum_{k=1}^{n-1} c_k \sin(k\xi) = U_1 \sin\xi, \ \sum_{k=0}^{n-1} c_k \cos(k\xi) = U_0 - U_1 \cos\xi \ .$$

$$(3.6.6)$$

This is known as **Goertzel's algorithm**. However, although the three-term non-homogeneous recursion for U_j does not suffer from the catastrophic numerical instability discussed in section 2.10, neither is it numerically strongly stable, unless it is suitably modified. The problem here of loss of precision is probably no worse than that which attends the numerical implementation of the methods for interpolating a large number of data by algebraic polynomials or rational functions; it is merely that trigonometric polynomials are more commonly used in association with large sets of data, and so the problem assumes a greater importance in that context.

3.7 INTERPOLATION AND EXTRAPOLATION OF FUNCTIONS

We turn attention now to the interpolation of discrete data which are obtained as evaluations of a known transcendental function. The data ordinates, which before have been represented merely by values y_k, can now better be regarded as equal to $f(x_k)$, where $f(x)$ represents a function whose definition and algebraic properties are known, and whose domain is some (real) interval of x. In this context, interpolation and extrapolation are no longer simply exercises in producing a graph of pleasing appearance, simulating the hand construction of a curve which fits given data. They can be judged by the extent to which they borrow the functional characteristics of $f(x)$. It becomes meaningful to speak of the error involved – the possible difference between the value of $f(x)$ and the interpolating function.

Broadly speaking, the need to interpolate functions arises from two different origins. Firstly, it may be desired to approximate a function which is complicated to calculate, but which nonetheless *can* be evaluated precisely for any x in the range of interest, the intention being to simplify calculation presumably at some expense in precision. The required result of such function approximation is a formula (usually a polynomial or rational function), its numerical coefficients, and some measure of its accuracy. Without that last piece of information the approximation is meaningless: there is no way of knowing whether it is good or bad.

Alternatively, a function may need to be interpolated because it is only evaluated at discrete values of the independent variable; this arises for example in the numerical solution of differential equations. Or it may need to be extrapolated because a numerical limit value is required which can only be approached but never achieved, in which event extrapolation becomes a means of accelerating the rate of convergence of a sequence of values to this limit. Examples of such acceleration have been encountered already in methods which seek to estimate the tail of an infinite series. These applications are typified by the fact that they arise as a means to an end in numerical experiments, and the arrangement of data abscissae is determined by the context. On the other hand, function approximation is an end in itself, and the support points at which the function is evaluated can be specially chosen to suit that purpose.

Because the applications to numerical experiments can more easily be discussed in the associated context, we shall consider in what follows only the application of interpolation to function approximation. Here the major interest lies in determining an optimal support.

3.7.1 The error in Lagrangian interpolation

Let the function being interpolated at $(n + 1)$ support points, ordered so that $x_0 \geqslant x_1 \geqslant \ldots \geqslant x_n$, be denoted by $f(x)$, and suppose that its first $(n + 1)$ derivatives are continuous in (a, b), where $a \leqslant x_n$ and $b \geqslant x_0$. (Note that this assumed ordering of the support is the *reverse* of previous practice, in order to simplify some later notation.) If $p_n(x)$ is the polynomial of degree not greater than n which interpolates this function at these points, then

$$\left. \begin{aligned} f(x) &= p_n(x) + f^{(n+1)}(\xi_x)\pi_n(x)/(n + 1)! \ , \\ \text{where} \quad \pi_n(x) &= (x - x_0)(x - x_1) \ldots (x - x_n) \end{aligned} \right\} \tag{3.7.1}$$

for some value of $\xi_x \in (a, b)$. This relation is a generalisation of Taylor's theorem: in fact, if all the support points coalesce on $x = x_0$, say, then (as shown in section 3.2.4) the interpolating polynomial becomes the Taylor polynomial expanded about $x = x_0$, $\pi_n(x)$ becomes simply $(x - x_0)^{n+1}$, and the term involving $\pi_n(x)$ is then the Lagrange form of the remainder term of the Taylor series. However, if the support points are distinct, then it is clear that the factor $\pi_n(x)$ in the remainder term merely reflects the fact that $p_n(x_k)$ equals $f(x_k)$ at $x = x_k$ for $k = 0, 1, \ldots, n$, because the polynomial interpolates the function. The dependence of the remainder on the $(n + 1)$th derivative of f is plausible, since the remainder would be zero if $f(x)$ happened itself to be any polynomial of degree $\leqslant n$, as would be implied if $f^{(n+1)}(x)$ were identically zero. Again, if $f(x)$ happened to be a polynomial of degree $(n + 1)$, so that $f^{(n+1)}(x)$ is a constant, then (3.7.1) is simply another way of writing the Newton form (3.2.11) of the Lagrangian interpolating polynomial of degree $(n + 1)$.

This fact provides the clue as to how (3.7.1) may be proved for an arbitrary function f. For let

$$p_{n+1}(x) = p_n(x) + b_{n+1}\pi_n(x)$$

be the polynomial which interpolates $f(x)$ at all the given support points, and also at some arbitrary but fixed point $x = \alpha$, say, where $\alpha \in (a, b)$. Then $[f(x) - p_{n+1}(x)]$ has $(n + 2)$ zeros in (a, b), and from repeated applications of Rolle's theorem, it follows that there is some point $x = \xi_\alpha$, say, in (a, b) where the $(n + 1)$th derivative of $[f(x) - p_{n+1}(x)]$ is zero. But the $(n + 1)$th derivative of $p_{n+1}(x)$ is $(n + 1)! \, b_{n+1}$, so that $b_{n+1} = f^{(n+1)}(\xi_\alpha)/(n + 1)!$; and since $f(\alpha) = p_{n+1}(\alpha) = p_n(\alpha) + b_{n+1}\pi_n(\alpha)$, the relation (3.7.1) is recovered with the (arbitrary) value of α replacing x, and ξ_α replacing ξ_x. The subscript α has been added to ξ to denote its dependence on α – and correspondingly in (3.7.1), ξ_x depends on x – although this proof makes it clear that the dependence is a rather obscure one!

In the context of approximation of the function $f(x)$ by a polynomial $p_n(x)$, the remainder term is the truncation error of the approximation, and one needs to be able to reduce this error in some methodical way. Fortunately, some useful insight into this problem can be obtained by concentrating attention on the factor $\pi_n(x) = (x - x_0)(x - x_1)...(x - x_n)$ applied to $f^{(n+1)}(\xi_x)/(n + 1)!$. This factor is independent of f, and is determined only by the choice of support points within the interval (a, b) of interest. It would seem reasonable to expect that if these points were chosen in some way so as to limit the magnitude of $\pi_n(x)$, then that choice of support might also limit the truncation error. There is no certainty attached to this, because not only does ξ_x depend on x, but undoubtedly it also depends in some way on the position of all the support points as well: thus a change in the support which reduces the factor $\pi_n(x)$ applied to $f^{(n+1)}(\xi_x)$ could nonetheless, for some functions f, actually increase this derivative value and also increase the truncation error represented by the product. But at least this would not happen if $f(x)$ were a polynomial of degree $(n + 1)$, since the $(n + 1)$th derivative is then a constant. One could argue more generally that the magnitude of $\pi_n(x)$ would be likely strongly to influence the error if $f(x)$ has a rapidly convergent power series expansion, since then a good approximation of that error would be obtained simply by truncating the power series to include only the terms up to x^{n+1}.

To make the problem definite we can suppose without loss of generality that the interval of interest is $[-1, 1]$, since any other interval $[a, b]$ may be mapped on to this by a linear transformation, and we seek to minimise some measure of the magnitude of $|\pi_n(x)|$ in $[-1, 1]$ with respect to the distribution of the support points $x_0 \geqslant x_1 \geqslant ... \geqslant x_n$. In particular, from what has just been said, it is known that this measure refers to the magnitude of the truncation error involved by using the interpolating polynomial $p_n(x)$ to approximate x^{n+1}, or indeed any other polynomial of genuine degree $(n + 1)$ whose leading coefficient is unity.

If, for instance, we choose $x_0 = x_1 = \ldots = x_n = 0$, then $p_n(x)$ is the Taylor polynomial $f(0) + f'(0)x + \ldots + f^{(n)}(0)x^n/n!$, and $|\pi_n(x)| = |x|^{n+1}$, so that its maximum value in $[-1, 1]$ is unity. If the support points are equally spaced, so that $x_k = (n - 2k)/n$, for $k = 0, 1, \ldots, n$, then the maximum error occurs close to the end points $x = \pm 1$ if n is large, where $|\pi_n(x)|$ is a little smaller than $(2/e)^{n+1}$. Thus the interpolating polynomial on an equispaced support can, in this sense, be said to be a better approximation than the Taylor polynomial, particularly as n increases. But is there an optimum distribution of support points?

3.7.2 Tchebyshev interpolation

If by an 'optimum' choice of support is meant one which minimises the extreme or maximum value reached by $\pi_n(x)$ over the interval $[-1, 1]$, there is indeed such a unique choice. The values of $x_0, x_1, \ldots, x_n$ must be taken to be the $(n + 1)$ zeros of a particular polynomial $T_{n+1}(x)$, called a **Tchebyshev polynomial**. The value of this polynomial in $[-1, 1]$ is most easily appreciated if one replaces x by a new independent variable $\theta = \arccos(x)$, where $0 \leqslant \theta \leqslant \pi$, for then

$$T_\nu(\cos\theta) = \cos(\nu\theta): \quad \nu = 0, 1, 2, \ldots . \tag{3.7.2}$$

However, written this way, it is not immediately apparent that this is a polynomial in x $(= \cos\theta)$. But if one recalls the trigonometric identity

$$\cos[(\nu + 1)\theta] + \cos[(\nu - 1)\theta] = 2\cos(\nu\theta)\cos\theta ,$$

and substitutes for the cosines from (3.7.2), placing $\cos\theta = x$, it follows that the functions $T_\nu(x)$ obey the recurrence relation:

$$T_{\nu+1}(x) = 2xT_\nu(x) - T_{\nu-1}(x) . \tag{3.7.3}$$

But (3.7.2) shows that $T_0(x) = 1$, since the cosine of zero is unity, and $T_1(x) = x$. Hence, from the recurrence relation (3.7.3) we obtain

$$T_2(x) = 2x^2 - 1, \qquad T_3(x) = 4x^3 - 3x,$$

$$T_4(x) = 8x^4 - 8x^2 + 1, \quad T_5(x) = 16x^5 - 20x^3 + 5x, \ldots$$

and it is clear that $T_\nu(x)$ is a polynomial of genuine degree ν, whose leading coefficient (multiplying x^ν) is $2^{\nu-1}$. This could be proved from (3.7.3) by induction on the value of ν.

The zeros of $T_\nu(\cos\theta)$ in $[-1, 1]$ are given from (3.7.2) by equating $\cos(\nu\theta)$ to zero: it is easily verified that there are ν zeros given by $\theta = (2k + 1)\pi/(2\nu)$ for $k = 0, 1, \ldots, (\nu - 1)$, and since any polynomial of degree ν has at most ν zeros, these are *all* the zeros of $T_\nu(\cos\theta)$. Changing from θ to x, and replacing ν by $(n + 1)$, it therefore follows that what has been asserted is that the choice of the so-called **Tchebyshev abscissae**

$$x_k = \cos[(k + 1/2)\pi/(n + 1)] : k = 0, 1, \ldots, n \tag{3.7.4}$$

provides the **minimax value** of $|\pi_n|$ in $[-1, 1]$ – that is, the minimum value of $\max_{|x| \leq 1} |\pi_n(x)|$ for any possible set $\{x_k \mid x_k \in [-1, 1] : k = 0, 1, \ldots, n\}$.

To prepare the way for a proof of this assertion, note that such a choice of x-values also implies that $\pi_n(x)$ is some multiple of $T_{n+1}(x)$, since it is a polynomial in x of the same degree and with the same zeros. In fact, because the leading coefficient of $T_{n+1}(x)$ is known to equal 2^n, it follows accordingly that

$$\pi_n(x) = T_{n+1}(x)/2^n . \tag{3.7.5}$$

But (3.7.2) shows that $|T_{n+1}(x)| \leq 1$ for all $x \in [-1, 1]$. Moreover these extreme values of $T_{n+1}(x) = (-1)^k$, say, are achieved at the $(n + 1)$ points given by $x = \cos[k\pi/(n + 1)]$ for $k = 0, 1, \ldots, (n + 1)$, which include n interior stationary values where $T_{n+1}(x)$ alternately reaches a maximum $(+1)$ or minimum (-1), in addition to the two extreme values at the ends of the interval ± 1. These alternating extremes are interlaced by the $(n + 1)$ zeros of $T_{n+1}(x)$.

Evidently, then, (3.7.5) shows that the choice of Tchebyshev abscissae leads to a value of $|\pi_n(x)|$ reaching but not exceeding 2^{-n} over the interval $[-1, 1]$; we can now show (as already anticipated) that no other arrangement of abscissae can better this. For it there were such a choice of the x-values which produced a smaller upper bound on $|\pi_n(x)|$, then we could define

$$q_n(x) = 2^{-n} T_{n+1}(x) - \pi_n(x)$$

with the values of x_k in $\pi_n(x)$ so chosen. Clearly $q_n(x)$ is then a non-vanishing polynomial of degree $\leq n$, since the terms in x^{n+1} on the right-hand side cancel out. Further, since $|\pi_n(x)|$ is asserted to be less than 2^{-n} in $[-1, 1]$ with this choice of x_k-values, then $q_n(x)$ must have the sign of $T_{n+1}(x)$ at each of the $(n + 2)$ x-values in $[-1, 1]$ where $|T_{n+1}(x)| = 1$. But we have already noted that this sign alternates between each successive extreme or stationary value of $T_{n+1}(x)$, so that $q_n(x)$ – like $T_{n+1}(x)$ – suffers $(n + 1)$ changes of sign in $[-1, 1]$, and would have to have $(n + 1)$ zeros. However, this is a contradiction, since $q_n(x)$ has degree n; it can only be resolved by deducing that $q_n(x)$ is identically zero. Thus there is no other arrangement of the zeros of $\pi_n(x)$ which can provide a lower upper bound on its magnitude in the interval $[-1, 1]$.

Interpolation using the Tchebyshev abscissae as support is called **Tchebyshev interpolation**. These abscissae are arranged closer together near the ends of the interval (where an equispaced support causes most error): the shape functions (3.7.3) for such a support attenuate with distance away from the particular ordinate whose influence they represent (Fig. 3.8), and the Runge phenomenon is thereby avoided (compare Fig. 3.8 with Fig. 3.2). Tchebyshev interpolation therefore represents a particular form of Lagrangian interpolation which is entirely trustworthy, even when there are a large number of points being interpolated.

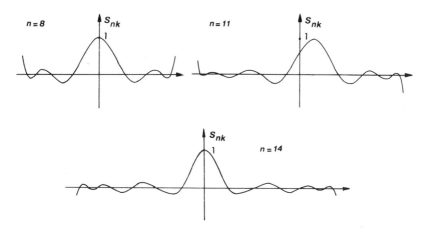

Fig. 3.8 – Shape functions $s_{nk}(x)$ for Tchebyshev interpolation of various degrees n and with $k = $ entier $(n/2)$.

3.8 MINIMAX POLYNOMIALS

It was an expectation that there would be some relevance to function approximation that led us to the formulation of Tchebyshev interpolation: it will now be shown that this form of interpolation is closely related to the derivation of what are called **minimax polynomial** approximations (or, sometimes, 'best uniform' approximations). If we denote the maximum or extreme value of the error magnitude $|f(x) - p_n(x)|$ over some interval $x \in [a, b]$ of interest by $E_n(f)$, then the particular polynomial $p_n(x)$ belonging to the space S_n (say) of all polynomials of degree $\leqslant n$, which minimises the value of $E_n(f)$, is called the minimax polynomial approximation to $f(x)$ over $[a, b]$ of degree n. It can be shown that there is one and only one such polynomial (although its degree may be less than n), and its maximum error magnitude will be denoted by $E_n^*(f)$.

For instance, if x^{n+1} is approximated over $[-1, 1]$ by a polynomial of degree n, then placing $f(x) = x^{n+1}$ in (3.7.1) and noting that $f^{(n+1)}(x) = (n + 1)!$, it follows that the error magnitude is simply the value of $|\pi_n(x)|$. But it was shown in the last section that the maximum of $|\pi_n(x)|$ is least if $p_n(x)$ is the Tchebyshev interpolate of $f(x)$; further, its magnitude is then 2^{-n}, so that $E_n^*(x^{n+1}) = 2^{-n}$. It follows from (3.7.5) that this interpolate must be $x^{n+1} - \pi_n(x)$ or, say,

$$t_{n-1}(x) = x^{n+1} - 2^{-n}T_{n+1}(x) , \qquad (3.8.1)$$

and accordingly, $t_{n-1}(x)$ is the minimax polynomial approximation to x^{n+1} over $[-1, 1]$ of degree n. Because $T_{n+1}(x)$ is a polynomial consisting entirely of multiples of either even or odd powers of x (depending on whether $(n + 1)$ is even or odd), equation (3.8.1) shows that $t_{n-1}(x)$ is in fact a polynomial of genuine degree $(n-1)$. Thus $t_{n-1}(x)$ is also the minimax polynomial approximation to x^{n+1} over $[-1, 1]$ of degree $(n-1)$, and so also $E_{n-1}^*(x^{n+1}) = 2^{-n}$. Nothing can be inferred about minimax polynomial approximations to x^{n+1} of lower degree, but of course $E_m^*(x^{n+1}) = 0$ for all integers $m > n$, since the polynomial x^{n+1} can be precisely represented by a polynomial of degree $\leq m$, if $m > n$. Trivially, therefore, $p_m(x) \equiv x^{n+1}$ is the minimax 'approximation' to x^{n+1} over any interval, if $m > n$.

The most important characteristics of the minimax approximation are contained in the **Tchebyshev alternation theorem**, which states that a necessary and sufficient condition for $p_n(x)$ to be the minimax polynomial approximation of degree n to a continuous function $f(x)$ over $[a, b]$ is that there exists a number ϵ and $(n + 2)$ distinct points $a \leq z_0 < z_1 < \ldots < z_{n+1} \leq b$ such that the error at these points is

$$f(z_k) - p_n(z_k) = (-1)^k \epsilon: \quad k = 0, 1, \ldots, (n+1), \qquad (3.8.2)$$

and that $E_n^*(f) = |\epsilon|$. This is also known as the equi-oscillation or ripple theorem, and the values $x = z_k$ are called the **nodes** or points of equi-oscillation of the minimax approximation. In particular, if $f \in C^{n+1}[a, b]$ and $f^{(n+1)}(x)$ does not change sign in (a, b), then these nodes specifically include the endpoints $x = z_0 = a$ and $x = z_{n+1} = b$ and all the n stationary values of the error $(f - p_n)$ within (a, b). It will be left to the reader to verify that $p_n(x) \equiv t_{n-1}(x)$ satisfies such conditions if $f(x) \equiv x^{n+1}$ (problem 2).

The proof of the necessity of the conditions in this theorem is in parts quite complicated, and will not be given (sufficiency is readily established – see problem 3). But it embodies the notion that if for instance the maximum error magnitude were reached at only one point, then there would be an increment which could be applied to the polynomial which would reduce the error at this point without causing the error elsewhere to reach a larger magnitude.

The usual method of computing a minimax polynomial is an iterative method (called the **Remes algorithm**) which depends on the theorem. Starting with a 'guessed' set of $(n + 2)$ nodes, a polynomial is constructed whose error in relation to $f(x)$ alternates between $\pm e$ at these points, where e is supposed fixed but unknown. This involves solving a system of $(n + 2)$ linear equations for the $(n + 1)$ coefficients of the polynomial and the unknown value of e. The $(n + 2)$ points at which this polynomial involves extreme or stationary values of the error, alternating in sign (and all exceeding $|e|$ in magnitude, to a greater or lesser extent), are then determined. If these $(n + 2)$ local alternating errors differ in magnitude from each other by more than some prescribed tolerance, the process is repeated with this new set of $(n + 2)$ points of nearly-equal oscillation as the next 'guessed' nodal positions.

3.8.1 'Nearly minimax' polynomials
Because the Remes algorithm involves a quite complicated instruction sequence (although the process, when programmed, is rapidly convergent) there is some

considerable interest in **nearly minimax** approximations, not merely because these are needed to start the iterations of the Remes algorithm, but also in their own right. One might 'define' them as approximations leading to maximum error magnitudes $E_n(f) \leqslant K E_n^*(f)$ where $K > 1$ is 'not too large'; but obviously the spirit of the exercise is observable without any hard and fast limit on K.

There are a number of methods by which these may be found, and it will be useful to illustrate them in relation to the approximation of a particular function. For this purpose, consider the approximation of $f(x) = \ln(1 + x/3)$ over $[-1, 1]$ by a quadratic $p_2(x) = a_0 + a_1 x + a_2 x^2$. These coefficients, as determined by the different methods to be discussed, are listed in Table 3.1, along with the positions $x = z_1$, z_2 where the error $[f(x) - p_2(x)]$ reaches a local maximum or minimum, and the respective values of the error e_1, e_2 at these positions. The values e_0 and e_3, which are also tabulated, refer respectively to the extremes of error at the end points of the interval, $x = -1, 1$.

Table 3.1
Some 'nearly minimax' polynomials

	$10^4 a_0$	a_1	a_2	z_1	z_2	$10^3 e_0$	$10^3 e_1$	$10^3 e_2$	$10^3 e_3$	$10^3 E_2(f)$
A	0	0.33333	−0.05556	0	0	−16.58	0	0	9.90	16.58
B	0	0.34309	−0.05801	−0.5240	0.4812	− 4.37	3.74	−2.89	2.61	4.37
C	3.9	0.34311	−0.05865	−0.5389	0.4640	− 4.10	3.54	−3.15	2.83	4.10
D	4.3	0.34315	−0.05888	−0.5447	0.4589	− 3.87	3.59	−3.16	2.99	3.87
E	8.5	0.34321	−0.05974	−0.5657	0.4382	− 3.37	3.47	−3.44	3.37	3.37
F	8.7	0.34315	−0.05976	−0.5650	0.4361	− 3.42	3.42	−3.42	3.42	3.42

The polynomial A is simply the truncated Taylor series expansion about $x = 0$; it is included only because this is an 'easy' way of generating an approximating polynomial. It has an extreme error magnitude of 0.0166 at the ends of the interval.

Polynomial B is the Tchebyshev interpolating quadratic (see problem 4, section 3.7.2), which we have argued should be a 'good' approximation; certainly with a maximum error of 0.0044 it is much better than the Taylor polynomial. The Tchebyshev abscissae forming the support of this quadratic are shown by (3.7.4) to be at $x = -\sqrt{3}/2$, 0 and $+\sqrt{3}/2$.

Another way of employing Tchebyshev polynomials so as to improve any power series approximation is to take the latter some few terms further than the degree actually required, and then to replace the higher order terms by their individual minimax polynomials as given by (3.8.1). Suppose therefore that the power series expansion of

$$\ln(1 + x/3) = (x/3) - (x^2/18) + (x^3/81) - (x^4/324) + (x^5/1215) - \dots,$$

which converges over $[-1, 1]$, is truncated after the x^5-term, consistent with an error magnitude of less than 0.32×10^{-3}. Now if the term $(x^5/1215)$ is replaced by $t_3(x)/1215$, the error introduced does not exceed $2^{-4}/1215 \doteq 0.05 \times 10^{-3}$ in magnitude, giving a total error not greater than 0.37×10^{-3}. Note from their definition in (3.8.1) that the minimax polynomials approximating x^4 and x^5 are respectively

$$t_2(x) = x^2 - 1/8, \quad t_3(x) = 5x(4x^2 - 1)/16 ,$$

so that upon substituting these for x^4 and x^5 in the truncated series, one finds that

$$\ln(1 + x/3) = (x/3) - (x^2/18) + (x^3/81) - [(x^2 - 1/8)/324] + [x(4x^2 - 1)/3888]$$
$$= 0.00039 + 0.33307x - 0.05865x^2 + 0.01339x^3 .$$

To reduce the polynomial to a quadratic (as required) by this process of **power series economisation**, it remains only to replace x^3 by $t_1(x) = 3x/4$, producing a further increment in the maximum error magnitude of $0.01339/4 = 3.35 \times 10^{-3}$, and making the total not greater than 4.10×10^{-3}. This leads to the quadratic C of Table 3.1, which does indeed have a maximum error magnitude of 0.0041.

A rather smaller error would have been achieved had the process of economisation started at a higher power than x^5 – at the expense of some additional arithmetic. However, this process becomes very simple if it is possible to express the function as an infinite **Tchebyshev series**:

$$f(x) = \tfrac{1}{2}c_0 + \sum_{k=1}^{\infty} c_k T_k(x) . \tag{3.8.3}$$

Since each $T_n(x)$ is a polynomial of genuine degree n, a sequence of Tchebyshev polynomials spans the same function space as a sequence of corresponding powers of x, and it is to be anticipated that such infinite expansions as (3.8.3) may exist for some functions $f(x)$, and be convergent in some interval of x. It would take us too far afield for the moment to discuss how they can be derived; it will suffice for present purposes merely to quote the result that

$$\ln(1 + x/3) = \ln(\alpha/6) + 2\sum_{k=1}^{\infty} (-1)^{k+1} T_k(x)/(k\alpha^k) , \tag{3.8.4}$$

where $\alpha = 3 + \sqrt{8}$. Since each $|T_k(x)| \leqslant 1$ for $x \in [-1, 1]$, this series converges much more rapidly than the power series for $\ln(1 + x/3)$. Moreover, if it is truncated to N terms, and the highest power of x (namely, x^N), which occurs only in $T_N(x)$, is replaced by the appropriate multiple 2^{N-1} of $t_{N-2}(x)$ then (3.8.1) shows that the term $T_N(x)$ is cancelled out. Similarly, replacing x^{N-1} by $t_{N-3}(x)$ will cancel out the term involving $T_{N-1}(x)$ and so on. In other words, the process of economisation is effected merely by further truncating the Tchebyshev series. Polynomial D in Table 3.1 is simply the truncated series

$$\ln(\alpha/6) + [2T_1(x)/\alpha] - [T_2(x)/\alpha^2] ,$$

and its maximum absolute error is 0.0039, somewhat less (as we had anticipated) than that of polynomial C.

Such an improvement (though small) is what one would expect. Indeed, there are two ways of reducing the partial sum $S_N = \sum_{k=0}^{N} a_k x^k$ of any convergent power series expansion down to an 'economised' form $p_n(x) = S_{N,n}$ (say) of degree $n < N$. The first is the iterative method, already described, of successively eliminating the highest power of x until x^n is reached. The other would be to convert S_N into a truncated Tchebyshev series C_N (say) of terms up to $T_N(x)$ using a formula relating x^k to $T_k(x), T_{k-1}(x), \ldots, T_0(x)$, and then to economise this by truncating it, so as to ignore all the terms involving $T_j(x)$ with $j > n$. A little thought will suggest that these two methods must be equivalent, and if we denote the truncated Tchebyshev series by $C_{N,n}$, then $C_{N,n} = S_{N,n}$. Subject to appropriate criteria of convergence, the truncation $C_{\infty,n}$ of the infinite Tchebyshev series C_∞ should therefore correspond in the limit to the economised form $S_{\infty,n}$ of the infinite power series S_∞. Our expectation (it should fall short of certainty) is that, if economisation is effective in reducing the maximum absolute error of approximation, then that of $S_{\infty,n}$ should be smaller than the maximum error of $S_{N,n}$ whatever the value of N.

If the infinite Tchebyshev series (3.8.3) is rapidly convergent, then the error involved by truncating it at the term $T_n(x)$ will be roughly equal to the first neglected term, $c_{n+1}T_{n+1}(x)$. But it has been seen that $T_{n+1}(x)$ equi-oscillates over $[-1, 1]$, and reaches its extreme magnitude of 1 at the $(n + 2)$ values $x = \cos[k\pi/(n + 1)]$ for $k = 0, 1, \ldots, (n + 1)$. Thus the truncation error will also (roughly) equi-oscillate at these $(n + 2)$ points and hence the approximation is (roughly) minimax over $[-1, 1]$. This is therefore the reason why the truncated Tchebyshev series can be expected to be 'almost minimax'. To return to our example (in which $n = 2$), the first neglected term of the series in (3.8.4) works out to be $0.0034T_3(x)$, whereas the truncated series alternates over $[-1, 1]$ between extremes whose magnitudes lie between 0.0030 and 0.0039; so it is indeed almost equi-oscillating.

Polynomial E is an approximation with even smaller ripples of error, of magnitude between 0.0034 and 0.0035. It was obtained by the method of **forced oscillation**: that is, by constructing a polynomial $p_n(x)$ which equi-oscillates at the $(n + 2)$ points $x = \cos[k\pi/(n + 1)]$ for $k = 0, 1, \ldots, (n + 1)$. In effect, this applies the first part of the Remes algorithm to a set of nodes $z_0, z_1, \ldots, z_{n+1}$, which are 'guessed' to be the points of equi-oscillation of the first neglected term of the Tchebyshev series, $T_{n+1}(x)$. Here $n = 2$, and accordingly the nodes are 'guessed' as $x = \cos(k\pi/3)$, that is $x \in \{-1, -\frac{1}{2}, \frac{1}{2}, 1\}$, and a polynomial is constructed which is forced to have equal but alternating error at each of these nodes, in succession, as in (3.8.2). This leads to a system of four linear equations

$$f(-1) = \ln(2/3) = p_2(-1) + \epsilon = a_0 - a_1 + a_2 + \epsilon ,$$
$$f(-\tfrac{1}{2}) = \ln(5/6) = p_2(-\tfrac{1}{2}) - \epsilon = a_0 - (a_1/2) + (a_2/4) - \epsilon ,$$
$$f(\tfrac{1}{2}) = \ln(7/6) = p_2(\tfrac{1}{2}) + \epsilon = a_0 + (a_1/2) + (a_2/4) + \epsilon ,$$
$$f(1) = \ln(4/3) = p_2(1) - \epsilon = a_0 + a_1 + a_2 - \epsilon ,$$

which can be solved to give a_0, a_1, a_2 and $\epsilon = -0.0034$. The corresponding polynomial does not in fact have stationary values at $x = \pm\frac{1}{2}$, and consequently is in error by more than the value of ϵ. However, if the tabulated values of z_0, z_1, z_2 and z_3 for this polynomial E are fed back as a new approximation to the points of equi-oscillation, and the process repteated, one deduces polynomial F, which is indistinguishable (to the accuracy quoted) from the true minimax polynomial, for which the maximum error magnitude $E_2^*(f) = 0.00342$.

There are close connections between all the polynomials B through to F. The common link is the truncation of the Tchebyshev series (3.8.3), and the assumption that the first neglected term is virtually the truncation error. If it *were* the total truncation error, then the truncated Tchebyshev series D would be the same as the minimax polynomial F, and economisation of the power series for the function (as in C), by approximating D, would approximate F. If it were the total error, then the error would vanish at the zeros of the first neglected term, and Tchebyshev interpolation, in reproducing this feature by interpolating the function at these zeros, should also lead to the minimax polynomial F. Like-wise, if it were the total error, then by constructing a polynomial (like E) which is forced to have equal but alternating errors at the points of equi-oscillation of the first neglected Tchebyshev polynomial, one should also be constructing the minimax F. Of course, the first neglected term is *not* the total truncation error of the Tchebyshev series; but often (as in our example) it is not far off the total.

As some further and more general indication of how 'nearly minimax' these approximations are, **Powell's theorem** states that, for an arbitrary continuous function on the interval $[-1, 1]$, the magnitude of the error $E_n(f)$ of Tchebyshev interpolation through $(n + 1)$ points is less than $4E_n^*(f)$ if $n \leqslant 20$, and less than $5E_n^*(f)$ if $n \leqslant 100$. And, as in the example we have explored, it is to be noted that the Tchebyshev interpolate is usually the most easily constructed, but not necessarily the closest of the 'nearly minimax' approximations. The value of the minimax error itself, $E_n^*(f)$, tends to zero as $n \to \infty$ provided merely that f is continuous and the interval of approximation is closed and bounded; this is known as **Weierstrasse's approximation theorem**. This general statement is refined in **Jackson's second theorem** which states that, if $f(x)$ also possesses continuous derivatives up to and including $f^{(m)}(x)$, then $n^m E_n^*(f) \to 0$ as $n \to \infty$. Thus for any function which is analytic in the interval (that is, all its derivatives are continuous), $E_n^*(f)$ tends to zero more rapidly than any inverse power of n: a theorem of Bernstein implies in fact that, in this circumstance, $E_n^*(f)$ decreases in geometrical progression at least as fast as λ^n (where λ is positive and less than 1). For example, Table 3.2 shows that the minimax approximation of $\ln(1 + x/3)$ over $[-1, 1]$ has a maximum error $E_n^*(f)$ which decreases (for $n < 8$) by a factor of roughly 7 each time the degree is incremented. Typically, therefore, Powell's theorem can be taken to mean that a Tchebyshev interpolate of degree n is likely to be at least as good an approximation to an analytic function as a true minimax polynomial of degree $(n - 1)$ or $(n - 2)$.

Table 3.2

Variation of $E_n^*(f)$ with degree, n

n	1	2	3	4	5	6	7
$10^n E_n^*(f)$	0.30	0.34	0.45	0.62	0.87	1.3	1.9
$7^n E_n^*(f)$	0.21	0.17	0.15	0.15	0.15	0.15	0.16

3.8.2 Techniques of minimax approximation

It will be realised that, although for convenience the interval of approximation has been standardised as $[-1, 1]$, the accuracy of approximation will also be affected by the actual interval on to which the standardised range will map. To make this clear, suppose that the actual approximation is of a function $g(z)$ for z in the interval $[a, b]$. Mapping this interval on $-1 \le x \le 1$ by placing $z = [(b + a) + (b - a)x]/2$, the function $f(x)$ to be approximated in the standardised interval $[-1, 1]$ is seen to equal $g[\frac{1}{2}(b + a) + \frac{1}{2}(b - a)x]$. Differentiating $(n + 1)$ times with respect to x, it follows that

$$f^{(n+1)}(x) = [(b - a)/2]^{n+1} g^{(n+1)}[\frac{1}{2}(b + a) + \frac{1}{2}(b - a)x] .$$

Hence the term $f^{(n+1)}(\xi)$, which determines the error of interpolation as in (3.7.1), is $[(b - a)/2]^{n+1} g^{(n+1)}(\xi)$ where $\xi \in (a, b)$. This would suggest that with a fixed arrangement of support points, as in Tchebyshev interpolation, the error might be reduced by a factor of round about 2^{n+1} if the interval is halved. It seems reasonable therefore to suppose that the error $E_n^*(g)$ of minimax approximation will be similarly affected by the width of the interval $(b - a)$.

It also suggests the possibility of segmentation as an aid in producing simpler approximations: that is, dividing the required range of approximation into a number of sub-intervals and applying a *different* (lower-degree) approximating polynomial to each. However, provided the function is analytic over the complete range, the minimax error $E_n^*(f)$ usually decreases so rapidly with increase of n that, to achieve the same error, the segmented polynomials are not of much lower degree (and therefore not much 'simpler') than that of the simple polynomial which is minimax over the unsegmented range.

For instance, if $E_n^*(f) = 8^{-n}$ for an approximation over the complete range, then it would be expected that, if that range were divided into two, minimax polynomials of degree m (say) over each sub-interval would have an error $E_m^*(f) \simeq 2^{-m-1} 8^{-m} = 2^{-4m-1}$. For the error to be unaltered, therefore, m would need to be about the same as $(3n - 1)/4$. The choice might then be between (say) a single polynomial approximation of degree $n = 7$ over the complete interval, and two quintics ($m = 5$) over each subinterval.

On the other hand, if $f(x)$ is not analytic over the range of approximation, then segmentation is essential to ensure that the degree of minimax polynomial needed to achieve any desired accuracy is not unduly large. For instance, a minimax polynomial for $f(x) \equiv \arcsin(x)$ over $[-1,1]$ has an error which decreases extremely slowly with n, because its derivative $f'(x) = (1 - x^2)^{-\frac{1}{2}}$ is singular at $x = \pm 1$. On the other hand, using symmetry and range reduction formulae such as

$$\arcsin(x) = -\arcsin(-x) = \pi/2 - 2\arcsin\{[(1 - x)/2]^{\frac{1}{2}}\} ,$$

it will be seen to be adequate merely to approximate $\arcsin(x)$ over $[0, \frac{1}{2}]$, in which interval it is analytic. We have met this approach before, of course, both in relation to the avoidance of ill-conditioning, and as a means of simplifying function evaluation.

It will be appreciated that minimax approximations which minimise the variation of *actual* error, as opposed to *relative* error, are not usually appropriate for computer or calculator approximations. Thus all the polynomials marked C through to F in Table 3.1 have infinite relative error at $x = 0$, because the function, but not the approximating polynomial, is zero at $x = 0$. This particular deficiency would have been overcome if the function to be approximated had been taken to be $f(x) = x^{-1} \ln(1 + x/3)$. As a general rule, zeros (and certainly of course poles or other singularities) should *always* be extracted in this manner. Had this been done, and had a minimax approximation then been constructed, it would still have had rather larger *relative* error at $x = 1$, where $|f(x)|$ is least, than elsewhere. To achieve equi-oscillation of relative error, the term $(-1)^k \epsilon$ on the right-hand side of (3.8.2) has to be multiplied by $f(z_k)$: in other words, $p_n(x)/f(x)$ must equi-oscillate about 1. Provided that the relative variation of $f(x)$ over the interval of approximation is not large, then Tchebyshev interpolation will still provide a polynomial such that $p_n(x)/f(x)$ is 'nearly minimax'; and a suitably modified form of the Remes algorithm, again with ϵ replaced by $\epsilon f(x_k)$, can be applied to improve it if this is judged necessary.

3.9 MINIMAX RATIONAL FUNCTIONS

Generally speaking, much more economical minimax approximation is obtained by the use of rational functions than by polynomials. A theorem due to Walsh states that there is one, and only one, rational function $R_{\mu/\nu}(x) = A_\mu(x)/B_\nu(x)$ of given order μ/ν which minimises the maximum value of $|f(x) - R_{\mu/\nu}(x)|$ over any bounded and closed interval. However, the *form* in which $R_{\mu/\nu}(x)$ can be written is not unique unless we require that it is irreducible, and standardised by (say) taking $B_\nu(0) = 1$. Even then it cannot necessarily be guaranteed of course that the polynomials $A_\mu(x)$ and $B_\nu(x)$ have genuine degrees μ and ν; it is known merely that their degrees do not exceed μ and ν, respectively. The integer by which their degree may be less than this upper limit is called the **defect** of $A_\mu(x)$, or $B_\nu(x)$.

Provided that $R_{\mu/\nu}(x)$ is irreducible, and that numerator and denominator polynomials do not *both* have defects, then the Tchebyshev alternation theorem may be extended to show that the rational minimax approximation is characterised by $(\mu + \nu + 2)$ points of equi-oscillation. Similarly, rational interpolation of the function $f(x)$ which is being approximated over $[-1, 1]$, say, at the zeros of the Tchebyshev polynomial $T_{\mu+\nu+1}(x)$ will usually be found to give a 'nearly minimax' rational approximation, although there is no error formula like (3.7.1) on which to base such a conjecture.

However, the improvement of the 'nearly minimax' rational Tchebyshev interpolate is a much more difficult task than that for the polynomial approximation. The Remes algorithm now produces a *non-linear* system of equations for the ripple magnitude ϵ and the coefficients of the polynomials $A_\mu(x)$ and $B_\nu(x)$; the unsuspected existence of defects in both these polynomials may alter the number of points of equi-oscillation; the rational approximation may be degenerate; and of course the initial 'nearly-minimax' approximation may be too coarse to allow the process to converge.

Perhaps the simplest and most effective approach is to envisage a sequence of iterates $A_{\mu,i}(x)$, $B_{\nu,i}(x)$ with $i = 0, 1, 2, \ldots$, and to seek to minimise the values of

$$\max_{|x| \leqslant 1} \left[|f(x) B_{\nu,i}(x) - A_{\mu,i}(x)| / B_{\nu,i-1}(x) \right]$$

for $i = 1, 2, \ldots$, in the expectation that after a sufficient number of iterations each of the coefficients of $A_{\mu,i}(x)$ and $B_{\nu,i}(x)$ will individually tend towards those of the required minimax approximation, and so become virtually independent of i. Since the coefficients of $B_{\nu,i-1}(x)$ are already known, this now is a *linear* minimax problem to which the Remes algorithm may be adapted and applied, but the conditions necessary for the convergence of the process are not well understood.

Although the derivation of 'best' rational approximations is *not* an elementary or straightforward calculation, the results – when accomplished – usually put polynomial approximation in the shade. For instance, the rational function of order $1/1$ interpolating $f(x) \equiv \ln(1 + x/3)$ at the zeros of $T_3(x)$ gives the expression $1.9857x/(5.9145 + x)$, which has a maximum error of 1.42×10^{-3} over $[-1, 1]$; this is less than half that of the minimax quadratic (Table 3.1). The true rational minimax approximation of order $1/1$ has an error of only 0.88×10^{-3} over $[-1, 1]$, practically a quarter of that of the minimax quadratic. This advantage becomes even more pronounced as more terms are included: the sequence of rational minimax approximations of order $1/1, 2/1, 2/2, 3/2, \ldots$ has successive errors decreasing by a factor of over 20; this compares with a reduction factor of about 7 which was found to apply to minimax polynomials each time their degree was increased by one. So minimax rational approximations of higher order and high accuracy usually involve far fewer terms – and are accordingly more quickly computed – than high degree minimax polynomials of the same accuracy.

3.10 OTHER CRITERIA OF 'BEST' APPROXIMATION

Although it seems natural to measure the accuracy of an approximation by its maximum error, there are other possbile measures, and it is worth enquiring if any of these provide an easier path to the 'best' approximation. If the error is denoted by $e(x)$, and the interval of approximation is considered to be $[-1, 1]$, as before, then

$$\|e\|_p = \{ \tfrac{1}{2} \int_{-1}^{+1} w(x)|e(x)|^p dx \}^{1/p}: \quad p \geqslant 1 \qquad (3.10.1)$$

is a measure of the 'average' value of $e(x)$ over $[-1, 1]$, which is called a **Holder norm**. The symbol $\|e\|_p$ is read as the L_p-**norm** of e. The $w(x)$ in (3.10.1) is any integrable, positive weight function standardised by assuming that its integral $\tfrac{1}{2} \int_{-1}^{+1} w(x) dx$ is unity (so that if $e(x)$ is a constant, then the norm is equal to the same constant). For instance, if the average *relative* error were of importance, then one would place $w(x)$ proportional to $|f(x)|^{-p}$, where $f(x)$ is the function being approximated. Often, no weight is specified, and one takes $w(x)$ to be unity.

As p increases without limit, it may be shown that if $w(x)$ is independent of p then

$$\lim_{p \to \infty} \{\|e\|_p\} = \max_{|x| \leqslant 1} |e(x)| = \|e\|_\infty , \qquad (3.10.2)$$

so that the maximum error magnitude is the L_∞-norm (also known as the uniform or Tchebyshev norm). The L_2-norm is also much used; it is sometimes referred to as the **Euclidean norm**, and it may be recognised from (3.10.1) in more familiar terms as the RMS or root-mean-square value of $e(x)$ over $[-1, 1]$. It is particularly amenable to the processes of analysis, performing an important role in the theory of approximation and in particular in the development of Fourier series and orthogonal polynomials. An approximation which renders $\|e\|_2$ a minimum is commonly called a **least-squares approximation**.

It is not the intention here to explore the rich field of theory opened up by a study of the least-squares function approximation. However, there is a very simple result relevant to the 'best' polynomial approximation of a function as judged by the so-called **absolute norm** or L_1-norm. Suppose the function is continuous over $[-1, 1]$ and that $p_n(x)$ is polynomial of degree $\leqslant n$ which interpolates $f(x)$ at the $(n + 1)$ points:

$$x_k = \cos[(k + 1)\pi/(n + 2)] \quad \text{for} \quad k = 0, 1, \ldots, n . \qquad (3.10.3)$$

Provided that the error $e(x) = f(x) - p_n(x)$ does not change sign in $[-1, 1]$ except at these points – or in other words, provided there are no other 'accidental' points of interpolation – then the interpolate $p_n(x)$ is that particular polynomial of degree $\leqslant n$ which minimises $\|e\|_1 = \tfrac{1}{2} \int_{-1}^{+1} |f(x) - p_n(x)| dx$. A sufficient condition that the error does not change sign except at the points of interpolation is that $f^{(n+1)}(x)$ exists and has no zeros in $[-1, 1]$, as will be clear from a consideration of (3.7.1).

These particular points of interpolation $x_0, x_1, \ldots, x_n$ given by (3.10.3) are similarly distributed to the support points used in Tchebyshev interpolation as in (3.7.4), although they are not quite so concentrated towards the ends of the interval $[-1, 1]$. For instance, with $n = 2$, Tchebyshev interpolation fits a quadratic to the function values at $x = -\sqrt{3}/2, 0, \sqrt{3}/2$, whereas the unweighted absolute error norm is (possibly) minimised by that quadratic which interpolates the function at $x = -1/\sqrt{2}, 0, 1/\sqrt{2}$. The L_1-norm would certainly be minimised by this quadratic interpolate if $f''(x)$ is monotonic in $[-1, 1]$, because this would imply that $f'''(x)$ is non-zero in $[-1, 1]$, and therefore that the error $e(x)$ vanishes in this interval only at the three points of interpolation.

The simplicity of this rather remarkable result is in marked contrast to the complication of finding the true minimax polynomial. But of course the benefit of minimising the L_1-norm is less easily discerned than the limitation of error achieved by minimising the L_∞-norm, as in the derivation of minimax approximation.

3.11 REPRISE

Throughout this chapter, we have regarded the data – in whatever way they are derived – as precise, and references to 'curve-drawing' or 'fitting' have been intended to imply the description of a smooth plane curve which intersects all data coordinates. The construction of such a curve by expressing its ordinate as a polynomial in terms of the independent variable is the commonest, probably the simplest and possibly the most reliable of the approaches available. However, we have seen that a completely different application of polynomial interpolation applies to the construction of a curve through a given set of data pairs whose functional representation is non-existent, from that which applies to the approximation of a transcendental function. In the former application, preference is given to the representation of the ordinates by a piecewise polynomial, and fitting more than (say) half a dozen such discrete data by a single continuous polynomial is regarded as a hazardous operation. In the other application, almost the reverse is true: the use of segementation in function approximation, whilst it is not in any sense hazardous, is not usually judged to be a worthwhile alternative to the use of a single polynomial arc – provided, at least, that the function is analytic over the range of approximation.

Of course there are very good reasons for this completely different emphasis. In the one application the data support is fixed (and unchangeable) and might be quite arbitrarily arranged; in relation to the commonly occurring example of an equi-spaced support, the Runge phenomenon has been frequently cited as the reason for eschewing continuous high-degree polynomial interpolation, but no doubt one could concoct even more pathological examples of 'bad' supports. In function approximation, on the other hand, one is at liberty to select the support as one pleases. The advantage of particularly 'good' arrangements of support points, such as the Tchebyshev abscissae, can be exploited – an advantage which derives from the higher density of information near the ends of the interval of interpolation, where it best compensates for the absence of any from outside the interval. Moreover, as in the iterative approach to minimax approximation, one can (if need be) call upon innumerable function evaluations at any points one likes in the quest for the 'best' support.

Again, the construction of a curve through a fixed but arbitrary set of data is what has been described as an 'artistic' exercise, in which much emphasis has been placed on 'eye appeal'. No doubt such anthropomorphism is unjustifiable: if anyone ever uncovers the set of rules which a person uses to guide his hand in curve-sketching, it seems doubtful that they will resemble those which govern the shape of a spline, whether it be made of plastic or pure philosophy. Nonetheless, there is a consensus view about the success of spline interpolation which is difficult to justify by more robust reasoning. In function approximation, on the other hand, there is no need for heuristic argument: an 'error' can be found, and a number – a norm – attached to it by which one may judge what is best; and that is science, not art. Some element of the heuristic approach might still be engaged in selecting the appropriate 'norm' to use; latterly, we had begun to open up the options that are available. But the aptness of the maximum norm and minimax approximation is at once recognisable, without the need to appeal to the force of other people's opinion.

Aside from a few brief remarks about the use of trigonometric polynomials in interpolating periodic data, the only other form of interpolating function discussed has been the rational function, mainly in its guise as a continued fraction. Such a formulation can be – and frequently is – much more effective in all applications than a polynomial, piecewise or continuous; but the fact that it can – and does – produce poles, even without good cause (in those exceptional conditions in which the rational function is reducible or nearly so), makes it an unreliable form for general use. Its use will certainly be considered where a problem of function approximation is worth the expenditure of time and effort, or where extrapolation rather than interpolation is a dominant concern. But rational function interpolation and approximation is always to be approached with caution, even if the subtleties of the computation do not stand in the way.

Some effort has been spent in describing the techniques of interpolation because these are examples of manipulative algebra which fall outside the ambit of mainstream mathematics: some limited skill in their performance is also of value in hand computation. Arguably, however, one can use interpolation 'packages' in a computational library without knowing how to perform the techniques involved. Less contentious is the fact that the ideas involved in interpolation and approximation, as they have been set out here, are important to understand because they provide an essential background to many of the topics that remain to be considered in both the following chapters.

CHAPTER 4

Solving equations: finding the real zeros of a function of one variable

---

The need to find the real roots of a non-linear equation arises in many forms of numerical work. It may be provoked by the requirement to invert a known functional relationship: given a function $y = F(x)$ of a real variable x, then the search for the value of x corresponding to some particular value of $y = c$, is simply the process of evaluating the inverse function $x = G(y)$, say, at $y = c$. An obvious example is provided by the positive root of $x^2 = c$ which of course is the value of the function $x = \text{sqrt}(c)$. But equally root-finding can arise from some application of interpolation: given values of $y = F(x)$ corresponding to particular values of x, it is required to predict the value of x which will make y equal to some given value c. Whether or not that happens to be the source of the problem, almost all methods of root-finding are based on this interpolative approach, incorporated into some iterative scheme of successive approximation.

The problem therefore has points of contact with both these aspects of the previous discussion; it certainly also is a process in which a knowledge of numerical error plays an important role, but little can be said about this without reference to the particular example under consideration. If the equation were written as $F(x) = c$, then one might be tempted to suppose that the error in the numerical representation $\tilde{F}(x)$ of $F(x)$ at or close to its precise root $x = \xi$, say, would be related to the rounding error of c. However, this could be a gross underestimate of the error, because $F(x)$ might be the difference or quotient of numerical magnitudes much larger than $|c|$. In any event, it is unlikely that there is just on 'parameter' c on which the solution depends. Generally there would be several, and the equation whose roots are required will for this reason usually be expressed as $f(x) = 0$, without reference to any particular parameter value.

Only when the actual form of the function is known can anything be asserted therefore about the value of $F(x) - \tilde{F}(x)$, or equivalently of $f(x) - \tilde{f}(x)$. However, suppose that this error — whatever its value may be — is denoted by $M\epsilon$, where ϵ is the machine unit. Then the numerical root $\tilde{\xi}$ which causes $\tilde{f}(\tilde{\xi})$ to be computed as zero will be the value of x such that $f(\tilde{\xi}) = M\epsilon$. But the value of $f(\tilde{\xi})$ is approximately $(\tilde{\xi} - \xi)f'(\xi)$, where ξ is the precise root, and consequently

$$\tilde{\xi} \simeq \xi + M\epsilon/f'(\xi) = \xi\{1 + M\epsilon/[\xi f'(\xi)]\} .$$

If $|M/\xi f'(\xi)|$ is large compared with unity — and there is no particular reason to reject that possibility — then the root would necessarily have an error which is large compared with its rounding error. Any attempt to find its value to the full machine precision available would be a waste of computational effort, and the accuracy attained would be completely illusory. A knowledge of bounds on the value of $|M\epsilon|$ can therefore be of considerable significance.

It is also clear that the magnitude of the derivative $f'(\xi)$ is important in determining the accuracy of determination of the root, and this is nowhere more apparent than in the determination of multiple roots for which both function and derivative values simultaneously vanish. This particular problem will receive special attention in a later section. But it is worth realising that there may be many different forms of the equation having the same roots, and correspondingly many possible forms for $f(x)$. There will be forms which serve to reduce the error; these may or may not be those which serve to increase the rate of convergence of the iterative process by which the root is obtained — for as we shall see, this also depends on the chosen form of $f(x)$.

For instance, where there is more than one root to be found, it is a matter of choice whether or not to employ the technique of **zero suppression** by which the equation to be solved, $f(x) = 0$, is treated as $[f(x)/p(x)] = 0$, where $p(x)$ is a polynomial product $(x - \xi_1)(x - \xi_2)\ldots$ whose zeros $\xi_1, \xi_2, \ldots$ correspond to these already found for $f(x)$. This may be very effective in separating closely-spaced roots, but in other contexts it might be an extremely complicated matter to determine whether or not it is, in some sense, of benefit. Very often it will not be worth the investment of time required to find out. Certainly, though the emphasis in the development of the subject is in finding efficient and fast methods of locating a root, it is more important for the user of these methods to be aware that the accuracy of the process should be a dominant concern.

4.1 BRACKETED METHODS OF ROOT FINDING

We suppose that it is required to estimate a root $x = \xi$ (say) in (a, b) of the equation $f(x) = 0$, given that $f(x)$ is single-valued and continuous in $[a, b]$, and that $f(a)f(b) < 0$. This inequality implies that $f(x)$ is of opposite sign at the two ends of the interval, and accordingly therefore at least one root must exist within that interval. In the **bracketed methods** of root finding, one seeks systematically to adjust the values of the interval bounds a and b by some iterative process in such a way that the span $(b - a)$ of the interval containing the root is continually reduced, until ultimately it becomes smaller than some prescribed error tolerance. Although exceptionally it might happen that there is just one root at, or very close to, one or other of the initially provided interval bounds, this will not of course generally be true, and so the iterative process must seek to revise *both* interval bounds.

4.1.1 The method of interval bisection

In the **method of bisection**, the span $(b - a)$ of the interval is continually halved, by calculating the value of the function at the middle point (or median) of the existing interval, and then replacing one or the other of the bounds to the interval by the median so as to maintain the property of bracketing. The appropriate bound (to be replaced) is discerned by matching its sign with that of the function at the mid-point. The process is repeated iteratively, and is terminated either when a zero value of $f(x)$ is determined – not a very likely eventuality, perhaps, when using floating point arithmetic – or when the span of the interval has been shrunk to a sufficiently small value. The only subtlety of the process stems from deciding what constitutes a 'sufficiently small' span.

This becomes clearer if a typical algorithm is considered. In the following instruction sequence for the interative loop which progressively reduces the interval span, the variable xi represents the best current estimate of the root, which is the mid-point of the current interval, and $xtol$ represents a prescribed error tolerance; the sign of the function at the lower limit of the interval is recorded in $asgn$, and this would be initialised (as plus or minus unity) before the loop is entered, whilst the sign of the function at the mid-point of the current interval is assigned to $msgn$:

do begin $xi := (b - a)/2 + a$;

   $QFIN$:   **if** $b - xi \leqslant xtol$ **or** $xi = a$ **then** $msgn := 0$

      else begin $msgn := \mathrm{sgn}(f(xi))$;

         if $msgn = asgn$ **then** $a := xi$ **else** $b := xi$

      end

end   **until** $msgn = 0$; .

Discrimination of termination is provided at the label $QFIN$ by examining the value of $(b - xi)$, which is the half-length of the current interval; this represents the maximum magnitude of the error of the current root estimate xi. If it becomes so small that it is less than (or equal to) the prescribed error tolerance $xtol$, the termination condition (a zero value of $msgn$) is planted, and the rest of the loop is skipped. Otherwise the function is evaluated, and its sign determined, allowing updating of the appropriate interval bound, a or b. Since the function $\mathrm{sgn}(..)$ can have the value zero if its argument – in this instance, the value of $f(xi)$ – is zero, a 'direct hit' on a root would be recorded as a zero value of $msgn$, thereby flagging the termination condition.

The condition $xi = a$, which is also tested at the label $QFIN$, is included as a precaution against creating an 'endless loop' in the event that $xtol$ has been given 'too small' a magnitude. Suppose, for example, that $xtol$ were zero (and such a possibility is not precluded): then the iterations could evidently proceed uninterrupted until a and b were successive, or **adjacent**, values of the ordered set of all floating point numbers representable by the machine. There is then no computable value between a and b, so that xi must be calculated as either a or b (depending on the way the machine rounds in addition/subtraction). If $xi = b$, then the first of the two tests would be satisfied (this merely requires that $0 \leqslant xtol$). But on the other hand, if $xi = a$, then since $(b - a) > 0$ the first test would fail, and this second condition would be needed to trigger an escape from the loop. Escape is essential because further iteration cannot change the then existing values of $a, b,$ or xi.

Thus a zero value of $xtol$ serves to force as precise an indication of the root as can be computed. But note that the 'accuracy' thus implied might be purely illusory. As has already been mentioned, the evaluation of $f(x)$ inevitably involves numerical error and – particularly if $|f'(\xi)|$ is small – its randomised nature might cause several changes of sign in the function value in the vicinity of the root (Fig. 4.1). All that would happen if $xtol$ were prescribed as zero is that one of these random changes of sign would be isolated.

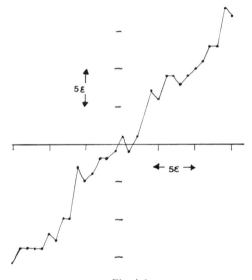

Fig. 4.1.

As written, the bisection algorithm has still one rare though possible 'bug'. If $f(x)$ is positive (say) for all $x > 0$, and negative for all $x < 0$, but is not zero at $x = 0$, then with zero $xtol$, the bisection process will ultimately cause underflow in forming xi, as it searches for the representable floating point number 'adjacent' to zero. Underflow may merely cause zero replacement, or it might be a condition able to be trapped. Alternatively, the precaution could be imposed that $xtol$ always exceeds some positive lower bound related to the smallest representable non-zero magnitude. In any event, its value would have checked to be non-negative before entering the iterated loop of the algorithm. Other necessary preliminaries would involve confirming that b is greater than a, and that $f(a)$ and $f(b)$ are indeed of opposite sign, as well as assigning the correct value to *asgn*.

Although very simple, the bisection method is therefore not entirely without its subtleties. With care it can be made completely reliable, in that it necessarily converges to an answer in a finite number of iterations. Its major disadvantage is that it is slow — or more specifically, that it may involve a large number of function evaluations, which makes it computationally expensive for other than 'easily computed' functions. This is because it relies on only a small though important detail of the function evaluation (the sign), and discards all the information about the position of the root discernible from the function's magnitude.

The fact that it uses so little information enables basically the same process to be applied to such tasks as searching for an item in an ordered list (see, for example, problem 1, section 3.3). Essentially this is the same as the (integer) root of a function defined on an integer domain, and the technique — often referred to as a **logarithmic search** in such applications — can as a consequence be much simplified.

4.1.2 The method of *regula falsi*
It would speed up the process of interval reduction if the information implicit in the function magnitude could be used to partition the interval, in such a way as to make one or other of the interval bounds — ultimately both — closer to the position of the root than the median point of the interval might happen to be. With two function values available, lineal interpolation immediately suggests itself as a means of indicating such a partition, and the estimate of the position of the root of $f(x) = 0$ would then be the intersection with the x-axis of the straight line joining the points $(a, f(a))$, $(b, f(b))$. This intersection is at $x = c$ (say) where

$$c = [af(b) - bf(a)]/[f(b) - f(a)] , \qquad (4.1.1)$$

and, since $f(a)f(b) < 0$, it is easily demonstrated that $a < c < b$. The value of $f(c)$ can then be calculated, and one or other of the bounds a and b replaced by c, depending on the sign of $f(c)$, just as in the bisection method. The process continues iteratively, using the new bounds and calculating a new partition. It is known as the method of false position, or **regula falsi**, meaning 'the artificial ruler' and referring to the formula (4.1.1) as a replacement for geometric construction (using a straight-edge). It is false as in falsies, not falsities.

Unfortunately, the method does not necessarily improve convergence compared with interval bisection — at least not in the latter stages of the process — because it serves to bring only one bound close to the root, and not both. To see why this should be, it can be observed from a graphical construction (Fig. 4.2) that, if the sign of $f''(x)$ does not change over (a, b), then $f(c)$ has the opposite sign to that of $f''(x)$, being negative if the curve $y = f(x)$ is curvex, and positive if it is concave. Since almost all functions become ultimately either concave or convex in the vicinity of a root (unless the root is a point of inflection), the

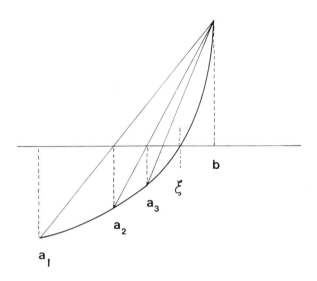

Fig. 4.2 – *Regula falsi* with successive replacements of bound $a = a_1, a_2, a_3, \ldots$

repeated application of (4.1.1) to the replacement of the interval bounds will ultimately leave one bound fixed. The other will converge towards the position of the root, but Fig. 4.3 shows that the rate at which it does so *can* be very slow — much slower in fact than in the bisection method.

To be more precise, if a is the interval bound which is being changed, b remaining fixed, and if $x = \xi$ is the precise root being sought, then for sufficiently small values of the error $(\xi - a)$, it can be shown that each application of (4.1.1) reduces the error by a fraction

$$(c - a)/(\xi - a) \simeq [f'(\xi)/\tan\beta] \qquad (4.1.2)$$

where $\tan\beta = f(b)/(b - \xi)$ is the tangent of the angle between the x-axis and the line joining $(\xi, 0)$ to the fixed point $(b, f(b))$. If, as in Fig. 4.3, the curve $y = f(x)$ is 'flat' in the neighbourhood of the root (that is, $f'(\xi)$ is small), but highly curved, so that it climbs away steeply to the fixed bound at $x = b$ (implying therefore that $\tan\beta$ is large), then the error is reduced by only a small fraction on each iteration, However, if the curve $y = f(x)$ is nearly straight over

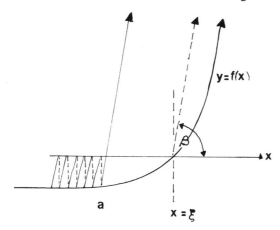

Fig. 4.3 – The slow convergence of *regula falsi*, with arrowed lines intersecting at
distant 'fixed' bounding datum $(b, f(b))$.

(a, b), then $f'(\xi)/\tan\beta$ is only a little short of unity, and (4.1.2) would then
indicate a rapid convergence of the value of a towards the position of the root
$x = \xi$ each time a newly calculated value of c replaces a.

By way of numerical example, suppose the method is applied to locating
the (positive) root of $[\exp(x^2) - 50] = 0$, starting with values of $a = 1, b = 3$
(the precise root is at $x = \sqrt{(\ln 50)} \simeq 1.98$). The value of c calculated from
(4.1.1) is 1.0117; if this is used to replace the bound a, and then c is recalculated,
we find successive values of $1.0233, 1.0348, 1.0462, \ldots$ by repeated iterations
of the calcuation, and clearly the rate of convergence towards the root is very
poor — nothing like as rapid as in bisection. It would still remain slow even close
to the root — the trouble originates from the fact that the other bound $(b = 3)$
remains fixed throughout. The method proves false, after all.

4.1.3 Modified *regula falsi*

An effective remedy is provided by a modification known as the **Illinois method**.
This institutes one significant change to the iteration process: if two successive
iterations cause a change in the same bound, then the (stored) value of the
ordinate at the other bound is halved. This amendment is repeated at each
successive iteration until an interpolated estimate of the root transpires to
be a replacement of the other 'fixed' bound. The effect of this modification
is illustrated in Fig. 4.4: it will be clear that it must ultimately force a change in
the upper bound. In the later stages of iteration, one application of this ordinate-
halving is usually enough — in other words, if two successive iterations cause a
change in the same bound, the institution of this 'fudge' will nearly always cause
the other bound to be changed in the next iteration. If one calls successive
estimates of the root resulting from successive iterations $x_0, x_1, x_2, \ldots$ then the

sequence of bounds ultimately settles down into a cyclic pattern of bound replacement

$$\dots (x_n, x_{n+1}), (x_n, x_{n+2}), (x_{n+3}, x_{n+2}), (x_{n+3}, x_{n+4}), (x_{n+3}, x_{n+5}), \dots$$

which repeats itself after every three iterations, and the process converges in such a way that

$$x_{n+3} - \xi = K(x_n - \xi)^3 , \qquad (4.1.3)$$

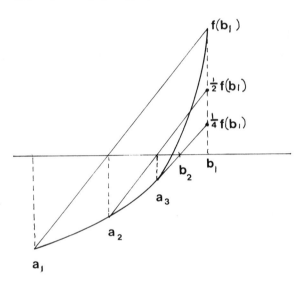

Fig. 4.4 – The Illinois modification of *regula falsi*: upper bound b_1 replaced (by b_2) at second attempt.

where K is a positive constant.

An instruction sequence of the iterative loop of unmodified *regula falsi*, excluding any looping instruction or termination discrimination, can be written in the following form.

LOOP: $x := (a*fb - b*fa)/(fb - fa)$; $fx := f(x)$;
      if $\text{sgn}(fx) = \text{sgn}(fb)$ **then**
          begin   **comment** replace bound b by x because fx and fb
                    have same sign;

     UBMOD:

               $fb := fx$; $b := x$
        end

else if $fx \neq 0$ then

   begin   comment fx and fa have same sign so replace a by x;

   LBMOD:

   $$fa: = fx; \ a: = x$$

   end

Here the variables fa and fb are used to record the values of $f(a)$ and $f(b)$ respectively. In changing these instructions to suit the Illinois method, it is necessary first of all to mark which bound is being replaced in the current iteration. One could agree to do this, for instance, by setting a variable *mark* as (say) $+1$ at label *UBMOD*, and -1 at *LBMOD*. In order to discern whether the bound a was replaced in the previous iteration, it is then merely necessary to examine the value of *mark* before it is reassigned in the current loop. Thus the effect of the modification to *regula falsi* in the Illinois method could be achieved by inserting, at the label *UBMOD*, the conditional statement:

if $mark > 0$ **then** $fa: = fa/2$ **else** $mark: = 1$;

meaning that, if this bound (b) was also replaced in the last iteration (and so $mark = 1$), then halve the recorded value of the ordinate at the opposite bound ($fa = fa/2$), otherwise merely record that this iteration is replacing bound b (by setting $mark : = 1$). Whichever of these alternatives applies, the statement is completed with *mark* assigned as $+1$. In the other branch, at the label *LBMOD*, there would be a precisely analogous instruction:

if $mark < 0$ **then** $fb: = fb/2$ **else** $mark: = -1$;

Initially, before the iterations are started, *mark* could be assigned as zero, so that no modification takes place on the first iteration in either branch (since this might be premature).

   It can be objected that the modification of the Illinois method, though commendably simple, is rather crude. If the successive root estimates $x_n, x_{n+1}, x_{n+2}, \ldots$ are converging rapidly, so that $|f_n| \gg |f_{n+1}| \gg |f_{n+2}| \ldots$ where we have placed $f_k = f(x_k)$, and if these ordinates all have the same sign, it could be reasonably anticipated that a small change in the ordinate at the other bound would suffice to cause the 'overshoot' in the next root estimate, necessary to obtain a replacement of that bound. In the **Pegasus method**, instead of halving the ordinate at the other bound, it is multiplied by β where

$$\beta = f_{n+1}/(f_{n+1} + f_{n+2}) .   (4.1.4)$$

In this context, f_{n+1} and f_{n+2} will have the same sign, since they refer to ordinates of successive replacements of one and the same interval bound. Therefore β will be between 0 and 1, as is necessary to ensure that the next linearly interpolated root estimate lies in the interval (a, b). To implement this modification, the required instruction at the label *UBMOD* now becomes

if $mark > 0$ **then** $fa := fa * fb/(fb + fx)$ **else** $mark := 1$;

with the value of β evaluated as $fb/(fb + fx)$. Similarly at *LBMOD*, fb is assigned as $fa * fb/(fa + fx)$ if $mark < 0$. The effect is that if the last ordinate f_{n+2} (or fx in the algorithm) is much less than f_{n+1}, then β is only slightly less than unity, and the ordinate at the other bound is only slightly reduced. If on the other hand the convergence is very slow, so that f_{n+1} and f_{n+2} are almost equal, then $\beta \simeq 1/2$ as in the Illinois method.

There is an interesting dilemma here. One wants to achieve a root estimate on the 'other side' of the true root position, but also it would be best if this estimate were close to the true root. If an overshoot is not achieved, the new root estimate will nonetheless be an improvement on what would have been calculated had the overshoot not been attempted, so that caution is not without its own reward. Thus, at that stage of the iteration where the sequence of root estimates x_n, x_{n+1}, x_{n+2} is such that only the last two replace the same bound (so that the signs of f_{n+2}, and f_{n+1} are the same, but opposite to that of f_n), there is much to be said for adjusting β so that the next root estimate is as accurate as possible – and there are three points on the curve $y = f(x)$ available on which to base this estimate, namely (x_n, f_n), (x_{n+1}, f_{n+1}) and (x_{n+2}, f_{n+2}).

If this next root estimate x_{n+3}, say, still has an ordinate f_{n+3} of the same sign as f_{n+1} and f_{n+2} – in other words, an overshoot has not been achieved – and if the value of β which was used is denoted by β_n, then on the next iteration, the values of the coordinates of the three points $(x_n, \beta_n f_n)$, (x_{n+2}, f_{n+2}), (x_{n+3}, f_{n+3}) will be available. Of course the first of these points does not lie on the curve $y = f(x)$, because its ordinate has been reduced in magnitude. But if the same method of calculating β is applied once again to these three points, including the one with this false ordinate, there is this time a greater likelihood of deriving an overshoot – that is, of calculating a root estimate x_{n+4} which can replace the bound x_n – because the reduced magnitude of the ordinate $\beta_n f_n$ will cause the estimated root to be closer to $x = x_n$. If not this time, then it would be reasonable to expect that the process will at least succeed if applied often enough.

The essence of this approach is therefore to formulate β so that, as first applied to the points (x_n, f_n), (x_{n+1}, f_{n+1}), (x_{n+2}, f_{n+2}), an *accurate* root estimate is attempted, taking into account the value of f_n, which one might hope will provide a replacement for the fixed bound (x_n), but which benefits the process even if it fails to do so. This suggests the use of a **three-point interpolation** scheme. Thus, if these points are interpolated by a rational function of order $1/1$, and β is accordingly derived by requiring that this interpolate intersects the x-axis at the same point as the straight line joining $(x_n, \beta f_n)$ and

(x_{n+2}, f_{n+2}) – since this is how the root is to be estimated in the next iteration – then it can be shown, on taking account of the way in which x_{n+2} has been derived, that

$$\beta = 1 - (f_{n+2}/f_{n+1}) . \qquad (4.1.5)$$

Note that, although f_n does not appear explicitly, its value is used in forming x_{n+1}, and x_{n+2}, and so its value is implicit in f_{n+1} and f_{n+2}. If $|f_{n+2}/f_{n+1}|$ is small compared with unity – implying rapid convergence – (4.1.5) yields almost the same value of β as that given by the Pegasus method in equation (4.1.4). However, in other circumstances it yields a smaller value than (4.1.4), and therefore an estimate of the root is closer to the interval bound which is in need of being changed. Its use is therefore more likely to force a new value for this bound than the Pegasus method, but if the latter *does* produce a replacement for this bound, it will be closer to the true root. Pegasus, in other words, is the more cautious approach of the two.

Evidently use of (4.1.5) estimates the position of the root of $f(x) = 0$ as that of the rectangular hyperbola interpolating the three existing points (x_k, f_k) for $k = n, n + 1, n + 2$. However, if $(f_{n+2}/f_{n+1}) > 1$, the curve $y = f(x)$ must have a stationary value between x_n and x_{n+2}, whereas the arc of a rectangular hyperbola has no turning value. In fact, if $(f_{n+2}/f_{n+1}) > 1$, the interpolating hyperbola has a pole (but no root) in the interval between x_n and x_{n+2}. In this circumstance, (4.1.5) shows that β would be negative, which would place the intersection of the hyperbola with the x-axis, and so the position of the estimated root, outside the interval in which the true root of $f(x) = 0$ is known to exist. Thus, in using (4.1.5), some positive lower bound on β would have to be imposed to avoid such a misconstruction.

An obvious alternative approach, which avoids this fallacy, is to interpolate a parabolic arc through the three points (x_n, f_n), (x_{n+1}, f_{n+1}) and (x_{n+2}, f_{n+2}), corresponding to interpolation by quadratic polynomial. Such a quadratic has a zero within the interval between x_n and x_{n+2}, because f_n and f_{n+2} have opposite signs, and the position of this zero can be expressed, as before, as the intersection with the x-axis of the straight line joining $(x_n, \beta f_n)$ with (x_{n+2}, f_{n+2}), where

$$\beta = \alpha + \sqrt{(\alpha^2 - \gamma_0)}$$

with $\qquad \alpha = (1 + \gamma_0 - \gamma_1) \qquad\qquad\qquad (4.1.6)$

and $\qquad \gamma_k = f_{n+2}/f_{n+k} .$

Here, $\gamma_0 < 0$ and so β is necessarily positive, provided that the square root is always taken as positive. Also, the fact that $\gamma_1 > 0$ means that β must be less than unity, although this is not immediately obvious (see problem 3). Moreover, it is always a larger value of β than that derived from the use of (4.1.5), and is usually (but not necessarily) smaller than that given by (4.1.4). It is therefore a more cautious estimate of the upper bound than that of the other three-point method, though likewise less cautious than the Pegasus method.

The results of each of these various methods applied to the previously quoted example of the estimation of the root of $[\exp(x^2) - 50] = 0$, starting with interval bounds of $a = 1$ and $b = 3$, are shown in Table 4.1. Each method, as we have seen, uses a different 'fudge factor' β, the unmodified *regula falsi* method, in effect, using $\beta = 1$ – that is, leaving the ordinate of the 'fixed' bound unchanged. All methods start with the same two root estimates before any modification of the 'fixed' bound (in this example, the upper bound) is attempted. Use of the expression (4.1.6) to provide the modification, derived from interpolation by a parabola, provides a sequence of root estimates and bounds which have converged to within seven significant decimal digits in 10 iterations. The use of values of β given by (4.1.4) or (4.1.5) – the former in the Pegasus method, the latter derived from interpolation by a hyperbolic arc – both converge to this same accuracy in 14 iterations (though by quite different sequences of estimates). The simplest assumption, that β is to be taken as equal to 1/2, as in the Illinois method, provides convergence in 16 iterations.

Table 4.1

Successive root estimates (and interval bounds) of equation $\exp(x^2) = 50$, using modified *regula falsi*, given $x \in (1, 3)$.

(*Asterisks denote replacements of the upper bound)

Iteration number	Unmodified method	Illinois method	Pegasus method	Three-point interpolation: Hyperbolic	Parabolic
1	1.011674	1.011674	1.011674	1.011674	1.011674
2	1.023264	1.023264	1.023264	1.023264	1.023264
3	1.034770	1.046144	1.046128	2.617104*	1.167610
4	1.046193	1.090728	1.090618	1.103186	1.597046
5	1.056533	1.175400	1.174865	1.178292	2.079756*
6	1.068790	1.328250	1.325996	2.317565*	1.882963
7	1.079964	1.577823	1.569176	1.426510	1.956714
8	1.091055	1.907325	1.877406	1.608409	1.977173
9	1.102064	2.082212*	2.035208*	2.051982*	1.977912*
10	1.112991	1.961982	1.964812	1.909350	1.977883
11	1.123835	1.974396	1.976260	1.966680	,,
12	1.134598	1.979716*	1.977859	1.977961*	,,
13	1.145279	1.977869	1.977884*	1.977882	,,
14	1.155879	1.977883	1.977883	1.977883	,,
15	1.166386	1.977884*	,,	,,	,,
16	1.176834	1.977883	,,	,,	,,

Comparing the various results of this example, it would appear that convergence improves in conjunction with increased complexity in the evaluation of β. The unmodified method of *regula falsi* would need over 600 iterations before achieving 7 digit accuracy in its lower bound, and its upper bound remains fixed throughout as $b = 3$. On the other hand, in interval bisection, both bounds would reach this accuracy in 25 iterations.

The modified forms of *regula falsi* provide as good a combination of simplicity, speed, and reliability as one can reasonably expect to achieve. However, although they all will ultimately converge to a root within the given interval, there can be no guarantee about the speed of the process, as in the method of bisection. For this reason, it is regarded as good practice for a limit to be imposed upon the number of iterations which are to be permitted before termination is forced.

The iterations would also be terminated if a zero function value is in fact computed, or if the current interval span $(b - a)$ is less than a prescribed tolerance. The limit upon the number of iterations also protects against the consequences of this error tolerance being too small, as well as against those of programming errors in the evaluation of the function – which of course could have quite unforeseeable effects upon the convergence of the process. The instruction sequence previously listed might therefore be placed within the following loop:

$n: = nmax$; **comment** a limit on the number of iterations;
do begin $n: = n - 1$;
LOOP:
    . . .
    . . .

end until $n = 0$ **or** $fx = 0$ **or** $\mathrm{abs}(b - a) \leqslant xtol$;

This, in turn, would need to be 'top-and-tailed', firstly to check input data and initialise the value of fa, fb and *mark*, and finally to assemble output. This latter would include information on the reason for termination and the possible error in the value of the root.

Arguably, it might be better to discriminate termination by reference to function value rather than interval span, using $\mathrm{abs}(fb - fa) \leqslant ftol$, say, as the criterion. As has already been pointed out, a value of $xtol$ could only be assessed from a knowledge of the likely numerical error in evaluating the function at, or close to, its root, so that an assessment of the 'tolerable' error in function value, $ftol$, is in any event implicit – or at least should be so.

4.2 UNBRACKETED METHODS OF ROOT FINDING

Inevitably there are occasions when establishing bounds to an interval isolating a root may be difficult, or even impossible. Most commonly this would arise where an interval is known to contain two roots, or indeed any even number of roots – so that although interval bounds could be asserted, the function value at these bounds has the same sign. A certain amount of exploration would then be necessary to separate them into two odd-numbered groups, and one could envisage that finding one root is a reasonable way of doing this. Of course, it could happen that the pair of roots coincide – or in other words, that there is a double or, more generally, a multiple root, to be found. This kind of problem presents special difficulties, and in our initial discussion of unbracketed methods it will be assumed that the roots are **simple**: that is, that the derivative $f'(x)$ is not zero at the root. But whatever the reason for selecting an unbracketed method, it needs to be kept in mind that there can be no guarantee that the method will in fact locate a root – if for no better reason than that one may not exist. On the other hand, if it fails to locate a root, one cannot be sure that roots do not exist – the methods are not intended as, nor can they act as, efficient means of *searching* for roots.

There are convergence criteria available for the unbracketed methods which generally state that a root will be located if the required initial estimate of its position is 'sufficiently accurate'. However, it is usually impossible to put a figure on what is 'sufficiently' accurate unless interval bounds isolating the root are known, and consequently these criteria are of theroretical rather than practical interest. Also, as will be shown, the very circumstances which may make it difficult to supply bracketing values are those which render the convergence insecure.

4.2.1 The secant method

Supposing that (x_k, f_k) for $k = 0, 1, 2, \ldots$ provide a sequence of root estimates and associated function values $f(x_k) = f_k$, then the **secant method** generates such a sequence recursively by defining $x = x_{n+1}$ to be the intersection with the x-axis of the straight line interpolating the two previously determined points (x_{n-1}, f_{n-1}), (x_n, f_n) lying on the curve $y = f(x)$. This implies that

$$x_{n+1} = (x_{n-1}f_n - x_nf_{n-1})/(f_n - f_{n-1}) = x_n - (x_n - x_{n-1})f_n/(f_n - f_{n-1})$$
$$(4.2.1)$$

for $n = 1, 2, \ldots$, and two initial estimates of the root x_0, x_1 are required to start the sequence. The method is evidently rather like *regula falsi*, and the relation (4.2.1) is the same as that of (4.1.1) but with x_{n-1}, x_n, x_{n+1} replacing a, b, and c; however, note that there is no attempt to ensure that f_n and f_{n-1} are of opposite sign. Because of this, the second form of expression for x_{n+1}, as an increment to x_n, is to be preferred in numerical work to the linear cross-mean, as the effect of cancellation error is thereby much reduced if f_n and f_{n-1} are nearly equal and of the same sign.

The lineal interpolate used in generating x_{n+1} is the straight line

$$y = p_1(x) \equiv (x - x_{n+1})\hat{f}_{n-1, n},$$

where $\hat{f}_{n-1,n}$ is the dived difference $(f_n - f_{n-1})/(x_n - x_{n-1})$. It follows from the expression (3.7.1) giving the error in polynomial interpolation, that if ξ is a precise root of $f(x) = 0$, then on placing $x = \xi$, the error $(\xi - x_{n+1})$ is given by

$$p_1(\xi) = (\xi - x_{n+1})\hat{f}_{n-1,n} = -(\xi - x_n)(\xi - x_{n-1})f''(\bar{\xi})/2 \,,$$

for some (unknown) value of $\bar{\xi}$ in the interval bounding x_{n-1} and ξ. Writing $\xi - x_k = \epsilon_k$, this can be expressed more concisely as

$$\epsilon_{n+1} = C_n \epsilon_n \epsilon_{n-1} \,. \tag{4.2.2}$$

Assuming that the method converges to a root, so that $x_k \to \xi$ as $k \to \infty$, and the divided difference $\hat{f}_{n-1,n}$ tends towards $f'(\xi)$ in this limit, then C_n tends to the value

$$C_\infty = \lim_{n \to \infty} C_n = -\tfrac{1}{2}f''(\xi)/f'(\xi) \,. \tag{4.2.3}$$

Sooner or later in the (convergent) iterative process, C_n will assume the same sign as C_∞ (provided that the root is not also a point of inflection, so that $C_\infty = 0$). When this happens, (4.2.2) shows that if (for instance) $C_\infty > 0$, then either the errors ϵ_n are all positive, or else successive values of the sequence $\{\epsilon_n\}$ are signed $-+--+--+-$ ... with just one positive error in any sequential group of 3. This behaviour is rather picturesquely termed 'waltzing' about the root (Fig. 4.5), and in this circumstance the secant rule adopts its own version of

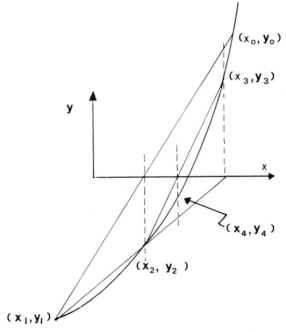

Fig. 4.5 – The waltzing behaviour of the secant rule: $\xi < x_0, x_3, x_6, \dots,$
$\xi > x_1, x_2, x_4, x_5, \dots.$

root-bracketing without any overt effort being made to impose it. Similarly, if C_∞ is negative, root estimates will ultimately overestimate the root (so that the errors are all negative), or else they will 'waltz' about the root with just one negative error in any sequential group of three.

Not only does the relation (4.2.2) enable the sign of the errors to be determined, it also enables a prediction to be made concerning the rapidity with which the sequence of absolute error magnitudes $\{|\epsilon_n|\}$ tends to zero — if the process does indeed converge. In the event that there exist positive numbers p and q such that

$$\lim_{n \to \infty} (|\epsilon_{n+1}|/|\epsilon_n|^p) = q \ , \tag{4.2.4}$$

then p is called the **order of convergence** of the sequence $\{|\epsilon_n|\}$, and the larger the value of $p > 1$, the more rapidly will $|\epsilon_n|$ tend to zero as n is increased — at least in the final iterations of the process. It is not difficult to confirm that values of $q = |C_\infty|^{1/p}$ and $p = (1 + \sqrt{5})/2 = 1.618$ are appropriate to a converging sequence of errors produced by the secant rule.

To give some indication of what such an index means, it can be noted that the *un*modified method of *regula falsi* provides successive root estimates which, according to (4.1.2), have errors which ultimately decrease in geometric progression with each new iteration: this implies $p = 1$ and $q < 1$ (and this is termed **linear convergence**). Although the actual rapidity of convergence will depend on the value of q, it can generally be reckoned as a relatively slow process compared with that achieved by one with a higher order of convergence. Again, accepting the assertion of (4.1.3) about the error of the Illinois method, it is not difficult to determine that its average order of convergence is $p = 3^{1/3} = 1.442$, so again the secant method converges rather more rapidly than this. It is in fact more nearly comparable with the Pegasus method (for which the average order is to determine that its average order of convergence is $p = 3^{1/3} = 1.442$. Thus the secant method converges rather more rapidly than this. It is in fact more nearly comparable with the Pegasus method (for which the order may average about 1.64), or the modifications of *regula falsi* involving three-point interpolation, which can have an order of convergence as high as $5^{1/3} = 1.710$.

However, it needs to be stressed that the order of convergence is a property of those estimates **local** to the neighbourhood of the root, and so it only measures how good the method is when it is near termination; and of course it cannot be assumed for any unbracketed method that this neighbourhood is ever reached. We look at the problem of **global convergence** in section 4.2.3.

4.2.2 Newton-Raphson's method

The term 'secant' applied to the method described in the last section relates not to the trigonometric function of that name, but to its less familiar definition as 'a line which intersects a curve'. If there is no particular difficulty in computing a derivative value $f'(x)$ of the function whose zero is being sought, then one can use the tangent to the curve in place of the secant as a means of constructing a straight-line interpolant — replacing the two distinct points of contact of the secant by one double point contact — so deriving a single-point rather than a two-point iteration formula. What is in mind here is the well-known **Newton-Raphson method** (sometimes called simply Newton's method or the tangent rule), by which an estimate of the root of $f(x) = 0$ is constructed as the intersection with the x-axis of the tangent $y = f(x_n) + (x - x_n)f'(x_n)$ to the curve $y = f(x)$ at the point $x = x_n$, which is the previously obtained estimate of the position of the root. The intercept with $y = 0$ is evidently at $x = x_{n+1}$ where

$$x_{n+1} = x_n - f(x_n)/f'(x_n) = x_n - f_n/f_n', \text{ say }, \qquad (4.2.5)$$

for $n = 0, 1, 2, \ldots$, given some starting value $x = x_0$.

This is a recurrence relation connecting successive root-estimates. As compared with (4.2.1), the divided difference $f_{n-1, n} = (f_n - f_{n-1})/(x_n - x_{n-1})$ is replaced by the derivative f_n', and if x_n is close to x_{n-1}, the two values are nearly equal to each other.

In just the same way as the error formula (4.2.2) is derived, it can be shown that the error $\epsilon_n = \xi - x_n$ now obeys the recurrence

$$\left.\begin{array}{l} \epsilon_{n+1} = -C_n \epsilon_n^2 \\[2mm] C_n = \tfrac{1}{2} f''(\xi_n)/f'(x_n) , \end{array}\right\} \qquad (4.2.6)$$

for some ξ_n between x_n and ξ, and C_n has the same limit C_∞ as $n \to \infty$ as that given by (4.2.3) for the secant method, provided that the iterations converge. The converging errors therefore ultimately all have the same sign, namely that of $-f''(\xi)/f'(\xi)$. Clearly the convergence is of second order, and is said to be **quadratic**.

This is often interpreted as implying that the method is therefore 'better' than the secant rule. The ultimate convergence is certainly faster: quadratic convergence can be interpreted as implying that the number of accurate digits in the floating point representation of the root estimate is doubled by each iteration (as shown by problem 2 of section 4.2.1). But one must also take into account the work required in computing the function *and* its derivative. If these involve roughly equal computational effort, then it would be fairer to compare the use of (4.2.5) with *two* applications of the secant rule (4.2.1) which — regarded as a single joint iteration — would result in an order of convergence equal to $1.618^2 = 2.618$. The Newton-Raphson method is therefore 'worse', not 'better'.

Certainly the fact that the Newton-Raphson method needs only one starting value is always an advantage, and there are a number of instances in which the derivative value can be obtained so simply that its superiority compared with the secant method is not then in doubt. Its best known application is to the estimation of the square root of a positive number c (say) obtained as the positive root of the equation $f(x) \equiv x^2 - c = 0$. Using (4.2.5) the root is the limit of the sequence $\{x_n\}$ where

$$x_{n+1} = \tfrac{1}{2}(x_n + c/x_n) = x_n - \tfrac{1}{2}(x_n - c/x_n) \ . \tag{4.2.7}$$

This is known as **Heron's rule**, and is readily extended to generate nth roots of positive numbers.

This raises another point, which equally applies to all methods of root-finding, but which becomes particularly apparent in relation to this simple example. There are innumerable forms of the function $f(x)$ which will have identical zeros: for instance, the positive root of $f(x) \equiv x^{3/2} - cx^{-1/2} = 0$ is also $x = \sqrt{c}$, and if this form of $f(x)$ is used in (4.2.5), one obtains

$$x_{n+1} = (x_n^2 + 3c)x_n/(3x_n^2 + c) = x_n[1 - (x_n^2 - c)/(3x_n^2 + c)] \ . \tag{4.2.8}$$

This recurrence seems to offer no advantage compared with Heron's rule (4.2.7) – quite the reverse, in fact – until one notes that

$$f''(x) = (3/4)(x^{-1/2} - cx^{-5/2})$$

is also zero when $x^2 = c$, so that the zero of $f(x)$ is now also a point of inflection of the curve $y = f(x)$. This implies that, from (4.2.6), C_n is now proportional to ϵ_n, and the convergence of (4.2.8) is of third order, or **cubic**. On the other hand, the simpler Heron's rule involves only about half the number of arithmetic operations entailed by (4.2.8), so that, in terms of similar computational effort, two iterations of (4.2.7) are comparable to one of (4.2.8), and these two (regarded as a single iterative step) would produce 4th order convergence. Therefore, computationally, Heron's formula is in fact preferable, and once again it is apparent that it is necessary to consider not only the iterative rate of convergence, but the amount of work involved per iteration. What is gained on the swings may be lost on the roundabouts.

4.2.3 Global convergence of secant and Newton-Raphson methods
Whereas the *local* convergence in the neighbourhood of a root of the unbracketed methods so far considered is easy to establish – from the relations (4.2.2) and (4.2.6) – the question whether a particular starting value will allow the iterations to converge to the neighbourhood of a root is much more difficult to answer.

In relation to the Newton-Raphson method, for example, it can be shown that a *sufficient* condition of global convergence is that x_0 should be chosen so that $2|f'(y)/f''(x)| > |\xi - x_0|$ for all values of x, y in the interval bounded by x_0 and $2\xi - x_0$. In relation to the equation $f(x) \equiv x^2 - c = 0$, it follows accordingly that Heron's iteration (4.2.7) will converge to the positive square root $\sqrt{c}$ if $x_0 \in (\sqrt{c}/3, 5\sqrt{c}/3)$. In fact, Heron's iteration will converge to $\sqrt{c}$ for *any* (positive) starting value of x_0, so that clearly this sufficient condition can be far too stringent; but setting this aside, such a condition can be seen to be of little value as a means of forecasting whether a particular starting value will provide convergence, unless close upper and lower bounds are known for the value of the root being sought. Of course if one knew such bounds, there would in general be no good reason for using an unbracketed method.

A more frequently useful criterion is provided by the following proposition: if there exists an interval I, such that $f(\xi) = 0$ for some $\xi \in I$, and $f'(x)$ is bounded, continuous, monotone and non-zero in I, then the root $x = \xi$ in I is unique, and Newton-Raphson iteration will converge to this root for any starting value $x = x_0$ belonging to I such that

$$f'(x_0)/f'(\xi) \geqslant 1 \ . \tag{4.2.9}$$

Essentially this places the starting value $x = x_0$ on the side of the root of $f(x) = 0$ where the magnitude of the slope $|f'(x)|$ is higher – and one might therefore call this the **steep-side condition**.

For example, in relation to Heron's iteration (4.2.7), the interval I can be taken to be $(0, x_0]$ and this condition asserts that the iterations will converge to the positive root $\xi = \sqrt{c}$ from any starting value $x_0 \geqslant \sqrt{c}$ which over-estimates the positive root. Likewise it will converge to the negative root $x = -\sqrt{c}$ of $f(x) \equiv x^2 - c = 0$, for any starting value $x_0 < -\sqrt{c}$ which under-estimates this negative root. More generally, consider the common problem of finding one or other of two zeros $\xi_1 < \xi_2$ of a function $f(x)$ known to be in $[a, b]$, where $f(x)$ is convex or concave over $[a, b]$, so that $f'(x)$ is monotone throughout this interval. Then by taking I as $[a, \xi_1]$, it follows from the condition that any starting value $x_0 \in I$ causes the Newton-Raphson process to converge to $x = \xi_1$. Similarly, placing $I = [\xi_2, b]$ shows that $x_0 \in I$ will lead to the evaluation of ξ_2. In other words, convergence is assured, and the pair of roots may be distinguished, provided at least that the starting value is not between them. That *is* useful knowledge, as it does not require the two roots to be individually bracketed to obtain that assurance.

The proof of the proposition leading to (4.2.9) is instructive. Provided that $x_0 \neq \xi$, then as is at once clear from the graphical construction of Fig. 4.6, the successive root estimates $x_0, x_1, x_2, \ldots$ all lie on the same 'steep' side of the root $x = \xi$, so that the errors all have the same sign, and their magnitudes $|\epsilon_0|, |\epsilon_1|, |\epsilon_2|, \ldots$ form a decreasing sequence. To prove this algebraically, observe from the iteration formula (4.2.5) for Newton-Raphson iteration that

$$1 - \epsilon_{n+1}/\epsilon_n = (x_n - x_{n+1})/(x_n - \xi) = f(x_n)/[f'(x_n)(x_n - \xi)]$$

$$= f'(\xi_n)/f'(x_n) \ .$$

where, by the mean value theorem, ξ_n is some value between x_n and ξ. But if I is taken to be the closed interval between ξ and x_0, and it is assumed that $x_n \in I$ and $f'(x_n)/f'(\xi) > 1$, as is certainly true by assumption for $n = 0$ (provided

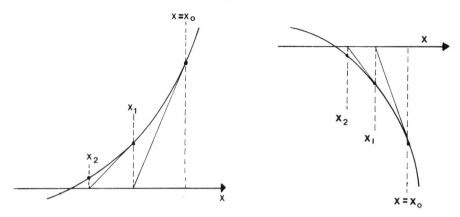

Fig. 4.6 — Monotonic decrease of error in the steep-sided approach to the root of the Newton-Raphson method.

that $x_0 \neq \xi$), then since $f'(x)$ is monotonic in I, $f'(\xi_n)/f'(x_n)$ must be between 1 and $f'(\xi)/f'(x_n)$ which is a value in $(0, 1)$. It follows that

$$0 < \epsilon_{n+1}/\epsilon_n < 1 - f'(\xi)/f'(x_n) < 1 \ .$$

Thus ϵ_{n+1} has the same sign as ϵ_n but a smaller (non-zero) magnitude. Accordingly, $x_{n+1} \in I$ and $f'(x_{n+1})/f'(\xi) > 1$. Hence, by induction, this is true for all n, provided that $x_0 \neq \xi$.

In particular, since $f'(x)$ is monotone, therefore the sequence $\{f'(x_k)/f'(\xi):$ $k = 0, 1, 2, ...\}$, like $\{|\epsilon_k|\}$, is also a monotonically decreasing sequence of positive values, as too is $\{\epsilon_{k+1}/\epsilon_k : k = 0, 1, 2, ...\}$; thus

$$0 < \epsilon_{n+1}/\epsilon_n < 1 - f'(\xi)/f'(x_0)$$

so that the errors decrease more rapidly than some geometric progression, and the global convergence is at least linear. In fact, since there *is* convergence, we know that it will ultimately become quadratic as $n \to \infty$. The proof of the convergence condition (4.2.9) is completed by considering the trivial possibility that $x_0 = \xi$, which of course presents no difficulty.

For a function $f(x)$ fulfilling the same conditions in I, a starting value $x_0 \in I$ such that $f'(x_0)/f'(\xi) < 1$ will lead to a first root estimate $x = x_1$ on the opposite side of $x = \xi$, so that $\epsilon_0 \epsilon_1 < 0$. Clearly this eventuality could be included in the sufficient condition of global convergence by alternatively requiring that either x_1 or x_0 should satisfy (4.2.9), which would then ensure convergence. But whether or not $x_1 \in I$, it would be more generally true that $f(x_0)$ and $f(x_1)$ may then have opposite signs, and therefore that a bracketed method could be employed to locate the root, starting with x_0 and x_1 as interval bounds.

Analogous arguments can be applied to the secant method for finding the zeros of a function $f(x)$ possessing these same properties. The sufficient condition of convergence is that $x_k \in I$ and $f'(x_k)/f'(\xi) \geqslant 1$ for *both* initially supplied estimates (with $k = 0, 1$). However, if x_0 and $x_1 \in I$ but these estimates straddle the root (so that only one satisfies the inequality), then $f(x_0)f(x_1) < 0$, and these estimates can be used to institute a bracketed method of root finding (which necessarily converges). If neither x_0 nor x_1 is on the appropriate side of the root to satisfy the inequality, then ϵ_2 will be of opposite sign to ϵ_0 and ϵ_1, and it is likely (but not certain) that x_1 and x_2 will therefore bracket the root.

Lest these references to the bracketed methods appear merely to evade a difficult question of convergence (which of course they do), it should be remarked that **hybrid** methods of root location – employing more than one particular technique – have much to recommend them, and are often employed in sophisticated library versions of root-finding algorithms. It is arguably always a good idea to escape from an unbracketed method to a bracketed one if, during the former process, suitable interval bounds are fortuitously thrown up for a root.

4.2.4 Fixed-point iteration

The Newton-Raphson method can be viewed as a special example of a class of **one-point iteration methods** which, in their simplest form, involve rewriting the equation to be solved as $\varphi(x) - x = 0$, and then developing the sequence $\{x_k : k = 1, 2, ...\}$ of successive approximations to the root by the recurrence relation $x_{n+1} = \varphi(x_n)$, given some starting value x_0. A geometric interpretation of the process is provided by Fig. 4.7, in which the required root $x = \xi$ is shown as the intersection of the curve $y = \varphi(x)$ with the straight line $y = x$ through the origin. Starting from a point $(x_0, \varphi(x_0))$, one constructs the horizontal line $y = \varphi(x_0)$ until it intersects $y = x$ at the point (x_1, x_1); from there, the vertical line $x = x_1$ intersects the curve $y = \varphi(x)$ at $(x_1, \varphi(x_1))$, and so on. If $\varphi(x)$ increases in the neighbourhood of the root, the sequence $\{x_k\}$ is monotone, as in Fig. 4.7(a); alternatively, if it decreases, then the root estimates successively over and under-estimate its value, as in Fig. 4.7(b). The process may converge as shown, or it may diverge as indicated in parts (c) and (d) of the figure. It will be apparent from the geometry that it is the slope of the curve $y = \varphi(x)$ close to the root which is of relevance to the question of convergence.

In fact, the mean value theorem shows that the ratio of successive increments to the root estimate is

$$(x_{n+1} - x_n)/(x_n - x_{n-1}) = [\varphi(x_n) - \varphi(x_{n-1})]/(x_n - x_{n-1}) = \varphi'(\xi_n) ,$$

where ξ_n lies between x_n and x_{n-1}. Convergence is assured if there exists a neighbourhood N of $x = \xi$, say $|x - \xi| < \delta$, such that $x_0 \in N$, and $|\varphi'(x)| \leqslant m < 1$ for all $x \in N$: that is, provided that the slope of $y = \varphi(x)$ close to $x = \xi$ is small enough. But if $|\varphi'(\xi)| > 1$, the sequence $\{x_n\}$ could converge only in very special circumstances: generally, it will diverge.

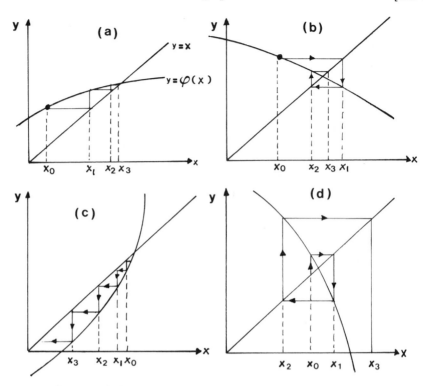

Fig. 4.7.

If convergence does occur, then $x = \xi$ is said to be a **fixed-point** (or point of attraction) of the mapping $x = \varphi(x)$. But the rate of convergence will (in general) only be linear, with the error $\epsilon_n = \xi - x_n$ decreasing ultimately in geometric progression by the factor $\varphi'(\xi)$ at each iteration. Thus the technique, although very simple (and of considerable theoretical interest) is not usually judged computationally efficient. In the special example of Newton-Raphson iteration, the function $\varphi(x)$ is $[x - f(x)/f'(x)]$, which has the derivative

$$\varphi'(x) = f(x)f''(x)/[f'(x)]^2 .$$

This is evidently zero at the root, ensuring that if convergence is achieved, the rate of convergence is super-linear: in fact, we know it is in general quadratic. By enlisting the evaluation of higher derivatives, it is possible to devise iterations that reach even higher-order rates of convergence, but we leave this as something to be explored (by way of problem 3). For the reasons already mentioned in the preceding section, it does not necessarily follow that they succeed in making good use of the computational effort involved in evaluating the function and its derivatives — assuming that this is where the bulk of the computational work lies.

Rather one looks for multi-point iteration functions $x_{n+1} = \varphi(x_n, x_{n-1}, \ldots, x_{n-m+1})$, which make better use of information already obtained as part of the iterative scheme. The secant rule is an example of the form $x_{n+1} = \varphi(x_n, x_{n-1})$ and is a two-point iteration method. We look next at some three-point formulae.

4.2.5 Three-point iteration formulae

There are a variety of methods available which seek to improve on the speed of local convergence of either the secant or Newton-Raphson methods. In **Muller's method**, instead of using lineal interpolation of the last two points to find a new root estimate $x = x_{n+1}$, one fits a quadratic polynomial to the last three data pairs (x_k, f_k) for $k = n - 2, n - 1, n$. The algebra associated with this idea is left as problem 1: the result is that the new root estimate is given by

$$\left.\begin{aligned} x_{n+1} &= x_n - 2f_n \mathrm{sgn}(\omega)/[|\omega| + \sqrt{(\omega^2 - 4f_n \hat{f}_{n-2,n-1,n})}] \\ \text{where} \qquad \omega &= \hat{f}_{n-1,n} + (x_n - x_{n-1})\hat{f}_{n-2,n-1,n} \, . \end{aligned}\right\} \tag{4.2.10}$$

The order of convergence is 1.84, and experience with the method suggests that it is well tolerant of poor starting values. It needs three initial estimates to set the iteration in train, but of course the secant method could be used to produce the third if only two are supplied.

The practical difficulty with this method is that the sign of the square-rooted quantity $(\omega^2 - 4f_n \hat{f}_{n-2,n-1,n})$ is not necessarily positive, and so the iteration may not be able to be continued entirely in terms of real arithmetic. If it is negative, it implies that f_{n-2}, f_{n-1}, and f_n have all the same sign, and that the interpolating quadratic has a maximum or minimum value (according to whether f_n is negative or positive respectively) and so does not intersect the x-axis. This may or may not mean that the equation $f(x) = 0$ likewise has no real roots: but if it has a pair of real roots close together, it must be reckoned as highly likely that Muller's method may miss them both in this way, particularly if the initial root estimates are in error by an amount which is large compared with the true root separation. If such an impasse is reached, and $|x_{n+1} - x_n|$ is still large compared with the accuracy sought, it would seem reasonable either to revert to the secant rule, or to suppress the imaginary component of x_{n+1}, which is equivalent to accepting the stationary value of the interpolating quadratic, namely $x_{n+1} = x_n - \omega/(2\hat{f}_{n-2,n-1,n})$. The latter would certainly be preferable if $f(x) = 0$ has a double root.

To speak of this as a 'difficulty' is to adhere to the limited view that only real roots are of interest. If complex arithmetic is available, the difficulty vanishes, and it is then the particular strength of Muller's method that it allows complex roots to be located from initially real root estimates.

The interpolation by a quadratic polynomial is equivalent to fitting a parabola to the three previously derived points on the curve $y = f(x)$. If instead one fits a rectangular hyperbola, this is equivalent to interpolating by a rational function of order 1/1, and then

$$
\begin{aligned}
x_{n+1} &= x_n - (x_n - x_{n-2})f_n/(f_n - f_{n-2}\,\hat{f}_{n-1,n}/\hat{f}_{n-2,n-1}) \\
&= x_n - f_n/(\hat{f}_{n-1,n} - f_{n-1}\,\hat{f}_{n-2,n-1,n}/\hat{f}_{n-2,n-1}) \ .
\end{aligned} \Bigg\} \quad (4.2.11)
$$

In the final stages of a convergent iteration, this is almost equivalent to the iteration formula (4.2.10), and its order of local convergence is the same. The avoidance of the square root which occurs in (4.2.10) would seem a convenience, but (as was noted in relation to this type of interpolation in modified *regula falsi*) the root estimates given by (4.2.11) can become quite unrealistic if the function $f(x)$ has a stationary value in the neighbourhood of the root. However, none of the unbracketed methods is exempt from problems in that context.

If one can be sure that $f'(x)$ does not change sign in the interval in which the root estimates are being made, it is possible to use inverse interpolation. That is, to construct an approximation to the inverse function $x = g(y)$, say, corresponding to $y = f(x)$, which will be single-valued provided that $f(x)$ is a monotone function of x. Interpolation by a linear polynomial or by a rational function of order 1/1 provides the same root estimates whether the function be in terms of x or y. However, inverse interpolation by a quadratic polynomial provides a different iterative method whose convergence properties and limitations are similar to those of (4.2.11).

Perhaps the most valuable application of inverse interpolation is directed towards an improvement of the Newton-Raphson method, by developing an iteration which depends on f_{n-1} and f'_{n-1} as well as f_n and f'_n. Since the derivative values $g'(y)$ of the inverse function are simply equal to $1/f'(x)$, a cubic Hermitian interpolate $x = p_3(y)$ which has osculatory contact with $x = g(y)$ at $y = f_{n-1}, f_n$ must satisfy the conditions that $p_3(f_k) = x_k$ and $p'_3(f_k) = 1/f'_k$ for $k = n-1, n$, and it can be used to construct a root estimate $x_{n+1} = p_3(0)$ given by

$$
x_{n+1} = (\xi_{n+1}f_{n-1}^2 - 2\bar{x}_{n+1}f_n f_{n-1} + \xi_n f_n^2)/(f_n - f_{n-1})^2 \quad (4.2.12)
$$

where $\xi_{k+1} = x_k - f_k/f'_k$ is the root estimate of the Newton formula (4.2.5) and $\bar{x}_{k+1} = x_k - f_k/\hat{f}_{k-1,k}$ that of the secant rule, given by (4.2.1). This method has an order of convergence $(1 + \sqrt{3}) = 2.732$, roughly equivalent to two iterations of the secant rule. If the evaluation of the derivative $f'(x)$ involves as much computational work as $f(x)$, the efficiency of these two methods is therefore roughly comparable.

In principle, it would be beneficial to include information from further prior iterations in a scheme of inverse interpolation, but the improvement in convergence is quite limited – the order not exceeding 2 if function values alone are included, or 3 if derivative values are also used. In practice the extra complication is unlikely to be judged worth the modest increase in speed.

A comparison of the various methods discussed is given in Table 4.2, applied to the same equation as that used for Table 4.1, all with a starting-value $x_0 = 3$ on the steep-side of the root, and all with the same value of x_1 derived from one iteration of Newton-Raphson's formula. It will be seen that the use of inverse

Table 4.2

Successive root estimates of equation $\exp(x^2) = 50$ using various unbracketed methods

Iteration number	Newton-Raphson	Inverse interpolation	Secant rule	Muller's method	Hyperbolic interpolation
0	3	3	3	3	3
1	2.834362	2.834362	2.834362	2.834362	2.834362
2	2.660816	2.581738	2.734296	2.734296	2.734396
3	2.480813	2.342168	2.603929	2.682356*	2.478348
4	2.300671	2.120297	2.481646	2.576981*	2.282880
5	2.137960	1.996245	2.354853	2.486462*	2.112413
6	2.025106	1.977979	2.234303	2.397590*	2.006440
7	1.982580	1.977883	2.126547	2.297636*	1.979579
8	1.977932	,,	2.044497	2.198218*	1.977893
9	1.977883	,,	1.997238	2.095186*	1.977883
10	,,	,,	1.980610	1.986991*	,,
11	,,	,,	1.978000	1.976220	,,
12	,,	,,	1.977884	1.977891	,,
13	,,	,,	1.977883	1.977883	,,

*Stationary value estimate.

interpolation saves a couple of iterations as compared with Newton-Raphson's method, obtaining the root correct to 9 significant decimal digits after 7 evaluations of the function and its derivative. The secant rule, starting with the same values of x_0 and x_1, requires 13 function evaluations (instead of 7 each of function and derivative) to achieve the same accuracy, whilst the use of hyperbolic interpolation by the iteration formula (4.2.11) reduces the number to 9. However in this particular example, starting with the same values of x_0, x_1 and x_2 as the secant rule, Muller's method provides no increase in speed, as it repeatedly predicts complex roots; the asterisked entries in the table correspond to positions of minimum values of the interpolating quadratic, and faster convergence would be achieved by including the imaginary parts of the root estimates. Without complex arithmetic, however, hyperbolic interpolation is the most efficient method in this example (and indeed most others).

4.2.6 Termination conditions

Success in the location of a root by any of the unbracketed methods has usually to be discriminated by the condition that the value of $|f(x_n)|$ has become less than a prescribed tolerance. It may often be true that the value of $(x_{n+1} - x_n)$ derived from this value of $f(x_n)$ could be taken as an indication of the error $(\xi - x_n)$ in the value of the root-estimate x_n; but it cannot otherwise be assumed that a small value of $(x_{n+1} - x_n)$ occurring during the course of the iteration necessarily indicates that the method has converged, as it may imply that $f'(x_n)$, or $\hat{f}_{n-1, n}$ (depending on the method), is large rather than that $f(x_n)$ is small.

For instance, in applying the secant method to the location of the root of the equation $f(x) \equiv \exp(x^2) - 50 = 0$, used as an example in Tables 4.1 and 4.2, and starting with the (bracketing) values $x_0 = 3$, $x_1 = 1$, the calculation proceeds as in Table 4.3 to yield $x_5 = x_4$, despite the fact that $|f(x_4)|$ is equal to 47. Obviously the reason for this is not that the root has been located, but rather that $f(x_3)$ is extremely large, and so the increment $x_5 - x_4 = f_4/\hat{f}_{3,4}$ is very small – too small in fact to make x_5 numerically different from x_4. The calculation cannot proceed further and must be aborted as a failure. Thus we see that although it is in general necessary to test whether $|(x_{n+1} - x_n)/x_n|$ is less than some tolerance in order to avoid this kind of impasse, such a discrimination is not by itself a reliable indicator of convergence.

Table 4.3
The secant rule to the solution of $\exp(x^2) = 50$.

n	0	1	2	3	4	5
x_n	3	1	1.012	9.546	1.012	1.012
f_n	8053	−47.28	−47.22	4×10^{39}	−47.22	−47.22
$x_{n+1} - x_n$		0.012	8.534	−8.534	1×10^{-39}	?

Since convergence cannot in general be guaranteed, it is also necessary in practice to limit the number of iterations that are attempted. Necessarily this may interrupt a slowly convergent iteration. For instance, if the Newton-Raphson method is applied to the same equation, starting with $x_0 = 1$, the next iterate is $x_1 = 9.697$, and although this is the steep-side of the root and the iterations converge, they do so very slowly – just short of a hundred would be needed – and the calculation would be unlikely to be carried that far, because it might be difficult to distinguish such behaviour from a (slow) divergence.

Although global convergence may be slow and faltering, the rate of local convergence – once the root is approached – is usually so rapid for the majority of the methods we have described that little extra work is saved by stipulating a modest limit on the magnitude of $f(x_n)$. On the other hand, too small a limit on $|f(x_n)|$ below the magnitude of its numerical error can cause a lot of useless computation. If it is found that $|f(x_{n+1})| > |f(x_n)|$ at the same time as

$|f(x_{n+1})|$ and $|x_{n+1} - x_n|$ are both less than some rather coarse limits, it is a safe deduction that numerical error is dominating the function evaluation, and that the process should be terminated.

4.3 THE LOCATION OF MULTIPLE ROOTS

If $f(x)/(x - \xi)^m$ tends to a finite value as $x \to \xi$, then the equation $f(x) = 0$ is said to have a root of multiplicity m at $x = \xi$. Generally m will be integer in the examples most likely to be encountered in practice, and usually it will be equal to 2 – that is $x = \xi$ would be a double root which is both a zero and a stationary value of $f(x)$. If, in fact, m is not integer, $f(x)$ will usually be undefined on one side or other of the root; thus for example, $(x^2 - a^2)^{3/2}$ has roots of multiplicity 3/2 at $x = \pm a$ but is not defined (at least in terms of real numbers) for $|x| < |a|$.

One's first observation about multiple roots (with integer multiplicity) is that their evaluation is very badly conditioned (Fig. 4.8). If the actual error of evaluation of $f(x)$ is $M\epsilon$, then the calculated value of $f(x)$ close to the true root $x = \xi$ would be

$$\tilde{f}(x) = [f^{(m)}(\xi)(x - \xi)^m/m!] + M\epsilon \ ,$$

and the roots of $\tilde{f}(x) = 0$ can differ from those of $f(x) = 0$ by values of order $\epsilon^{1/m}$, where ϵ is the machine unit. In other words, if $m = 2$, one can usually only evaluate the position of the double-root to within half the available precision with which values of x may be expressed.

If the multiplicity is known, then of course this difficulty is overcome by finding the zero of the $(m - 1)$th derivative, which implies locating a simple root of the equation $f^{(m-1)}(x) = 0$. Where a multiple root is suspected, but its multiplicity is not known, the equation $f(x)/f'(x) = 0$ will also have a simple root at the position $x = \xi$ of the multiple root. But – assuming the function and its derivative are evaluated separately – this does not circumvent the problem of poor conditioning: the quotient $\tilde{f}(x)/\tilde{f}'(x)$ will still possibly evaluate as zero for values of $|x - \xi|$ of order $\epsilon^{1/m}$, and moreover its denominator may be zero over some generally smaller interval whose width is of order $\epsilon^{1/(m-1)}$. Thus the quotient may be calculated as divergent close to the root, and zero values possibly never even closely achieved (Fig. 4.8(b)).

Aside from the problems of loss of accuracy, there is also a considerable retardation in the local rate of convergence of the unbracketed methods of root finding, and it is these which must usually be employed, since only roots of odd multiplicity can be bracketed – obviously $f(x)$ has the same sign on either side of a root with even multiplicity. For example, on approximating $f(x)$ by $A(x - \xi)^m$ in the neighbourhood of the root $x = \xi$, where A is treated as a constant, it follows from (4.2.5) that for sufficiently small $\epsilon_n = \xi - x_n$, the Newton-Raphson iteration takes the form

$$\left. \begin{array}{l} x_{n+1} \simeq x_n - (x_n - \xi)/m \\[2mm] \epsilon_{n+1} = \xi - x_{n+1} = [(m-1)/m](\xi - x_n) = r\epsilon_n, \quad \text{say,} \end{array} \right\} \quad (4.3.1)$$

so that the local convergence is linear for $m \neq 1$, not quadratic as for a simple root. The larger the multiplicity, the closer the value of the asymptotic error ratio $r = \epsilon_{n+1}/\epsilon_n$ is to unity, and so the slower the ultimate convergence (if the iterations do in fact converge).

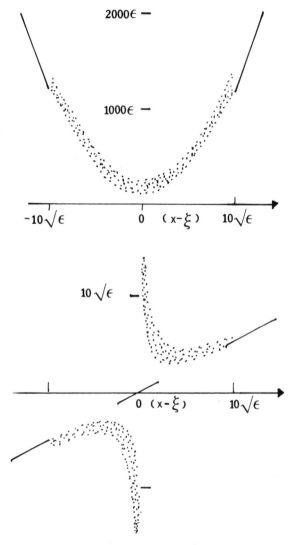

Fig. 4.8 – Computed (dotted) and precise (full-line) variations in (upper curve) value of $f(x)$, and (lower curve) value of $f(x)/f'(x)$, in neighbourhood of double zero of $f(x)$. In this example (problems 2 and 3 of section 4.3) no zero is computed.

It can be similarly shown from (4.2.1) that the secant rule likewise converges linearly in the neighbourhood of a multiple root, with the error ratio r equal to the positive root of $r^{m-1}(1 + r) = 1$, whilst iterations based on hyperbolic interpolation as in (4.2.11) converge linearly with the asymptotic value of r given by the positive root of $r^{m-1}(1 + r + r^2) = 1$. Muller's method is an important exception in so far as it maintains a super-linear rate of convergence for double roots, but converges linearly for roots of higher multiplicity with a complex error ratio — and does so by a sequence of complex root estimates.

It is possible to include a test of the rate of convergence during the iteration, and to use this measurement to assess the multiplicity of the root being located. Such a test might anyway be needed as an aid in discerning a termination condition. Thus, for example, in the Newton-Raphson method, since $u(x) = f(x)/f'(x) \simeq (x - \xi)/m$ in the neighbourhood of $x = \xi$, therefore $u'(\xi) = 1/m$, and an estimate of this derivative would be provided by $\hat{u}_{n-1,n} = (u_n - u_{n-1})/(x_n - x_{n-1})$, where $u_k = f_k/f_k'$. This becomes increasingly accurate in the later stages of the iteration, when u_n is small and x_n is close to ξ. In any event, if it is assumed that the multiplicity of the root is integer (as would in practice almost always be true) it would not matter if this is a rather rough estimate, since its reciprocal can be rounded to the nearest integer by, say,

$$m_n = \max\{\text{entier}[(1/\hat{u}_{n+1,n}) + 0.5], 1\} . \tag{4.3.2}$$

This expression also suppresses any spurious indications of non-positive multiplicity by imposing a lower limit of unity. If the convergence is rapid, then $\hat{u}_{n-1,n}$ and m_n will evaluate as 1, because $|u_n|$ would be small compared with $|u_{n-1}|$, and $(x_n - x_{n-1})$ would have been calculated as equal to $(-u_{n-1})$.

On the other hand, if it is detected that $m_n > 1$, then, in the next iteration, convergence would be enhanced by taking

$$x_{n+1} = x_n - m_n u_n = x_n - m_n f_n/f_n' . \tag{4.3.3}$$

In fact, if it is found that $m_k = m$ for all $k > n$, and if indeed the root being located *has* multiplicity m, then the modified recurrence relation (4.3.3) with $m_n = m$ restores local quadratic convergence to the iterations — at least in theory. In practice, because of the intrusion of numerical error, $u(x)$ does not necessarily behave like $(x - \xi)/m$ if $|x - \xi|$ becomes too small (Fig. 4.8(b)), and irrespective of the modification, the Newton-Raphson method will not converge in the vicinity of a multiple root if it is pressed home too far. Such a loss of local convergence, if it can be recognised, could be used to trigger termination of the process; but if not, the course of further iterations is unpredictable.

It is likely that values of $m_n > 1$ will be generated in the early stages of iteration, with x_n still distant from the root and without any implication that the root is therefore multiple. However, it can be noted that if the secant rule is applied to the equation $u(x) = 0$, then the sequence of root estimates is given from (4.2.1) by

$$x_{n+1} = x_n - u_n/\hat{u}_{n-1,n} \tag{4.3.4}$$

and it can be seen that (4.3.3) is simply a version of the same recurrence with a discretised multiple of u_n. Thus if x_n is close enough to the root that, at least, $\hat{u}_{n-1,n}$ is positive, the global convergence properties of the modified Newton-Raphson iteration applied to $f(x) = 0$ will be generally rather similar to those of the secant rule applied to $u(x) = 0$.

There appears to be no way of extending the modification to methods, like the secant rule, which do not require calculation of the function derivative. An error ratio can still be obtained, of course, and it is not too difficult to relate this to an estimate of root multiplicity. The difficulty lies in restoring super-linear convergence to the iteration. There are, however, a variety of ways of improving the rate of linear convergence which could be employed, Aitken's δ^2-process (problem 6, section 4.2.4) being the best known.

4.4 LOCATING THE REAL ZEROS OF A POLYNOMIAL

One of the most frequent applications of root-finding techniques is to algebraic equations $f(x) \equiv p_n(x) = 0$, where $p_n(x)$ is a polynomial of (genuine) degree n, and a large number of special methods have been devised which are directed towards the evaluation of their roots. Many of these methods will locate complex as well as real roots: others allude specifically to polynomials whose zeros are all real. However, these methods will not be discussed here; provided at least that only real zeros are required, then any of the battery of root-finding methods so far described can be brought to bear on this problem, as they are of course equally applicable to algebraic, as to transcendental, equations.

The particular feature of algebraic equations to be explored here lies in the possibility of calculating the number of their real roots in any arbitrary interval of the independent variable, thereby enabling a broad picture of their distribution to be obtained. Indeed, if one wanted some systematic way of doing the same thing for non-algebraic equations, then a good way would be to approximate the function by a polynomial, perhaps using Tchebyshev interpolation. Of course, whether or not the function is algebraic, the best picture of the distribution of its real zeros would no doubt be observable from a carefully constructed graphical representation. But this would be computationally a relatively expensive, and essentially interactive, exercise. What is to be described here is a purely numerical technique for performing the search, and bracketing each real root.

The first important property of polynomials which is of relevance, is that bounds on the magnitude of any root can be very easily established. If one writes the polynomial in its power form

$$p_n(x) = a_0 + a_1 x + a_2 x^2 + \ldots + a_n x^n , \qquad (4.4.1)$$

it can be assumed that both a_0 and a_n are non-zero — the former because if there were a zero root it could be removed before looking for any others, and the latter by definition of n. The quantity $(-a_0/a_n)$ is equal to the product of all the roots (real and complex), and $\mu = |a_0/a_n|^{1/n}$ is therefore the geometric mean of their magnitudes. Then (for example) the zeros of $p_n(z)$ can be shown to lie within the circle $|z| \leqslant c$ of the complex plane, and so any real zeros must lie within the interval $[-c, c]$, where

$$c/\mu = 1 + \max_{1 \leqslant j \leqslant n-1} (|a_j| \mu^j)/|a_0| . \tag{4.4.2}$$

What we shall be describing (in the next section) is a method by which the number of distinct real zeros of $p_n(x)$ within any bounded interval $[a, b]$ can be determined. Thus by applying this technique with $a < -c$ and $b > c$, the total number of real roots can be established. By generating subintervals of $[a, b]$ in some methodical way, one may locate those which include just one distinct real zero of the polynomial.

In principle, by progressively reducing the span of the subinterval including such a single zero, the same technique could be used, not merely in searching for a root, but in locating its position within any required precision. But in practice the technique does not compete with any of the bracketed methods of root-finding already described, because the computational work it entails is much heavier. It requires the evaluation of several polynomials each time an interval bound is changed. Rather it would be used only so far as is necessary to establish a sub-interval (however wide) containing a single distinct root.

However, the technique has one other important feature. It can – at least, in principle – distinguish simple from multiple roots, and can determine the multiplicity of the latter. It is therefore an effective way of preparing the ground for the application of a bracketed method, such as modified *regula falsi*, for finding all the real roots.

4.4.1 Sturm's theorem

In order to describe the method of counting real zeros of a polynomial, it is first necessary to define a **Sturm sequence**. This is a sequence of $(m + 1)$ continuous and differentiable functions $f_0(x), f_1(x), \ldots f_m(x)$ defined for $x \in [a, b]$ in such a way that :

(i) $f_m(x) \neq 0$ for all $x \in [a, b]$;

(ii) if $f_k(t) = 0$ for any $t \in [a, b]$, where $1 \leqslant k < m$, then $f_{k-1}(t)f_{k+1}(t) < 0$; and (iii) if $f_0(t) = 0$ for any $t \in (a, b)$, then the zero is simple and $f_0'(t)f_1(t) > 0$.

For instance, it can be shown that $\{(-1)^k \cos[(m - k)x]: k = 0, 1, \ldots m\}$ forms such a sequence in $[0, \pi]$, and so also does the sequence of Tchebyshev polynomials $\{T_{m-k}(x): k = 0, 1, \ldots m\}$ for any interval of x.

Further, let $V(c_1, c_2, \ldots, c_p)$ denote the number of changes of sign in the sequence $c_1, c_2, \ldots, c_p$ of real numbers. More precisely, if $\gamma_1, \gamma_2, \ldots, \gamma_q$ with

$q \leqslant p$ is the sequence $c_1, c_2, \ldots, c_p$ after deleting any zero terms of the latter, but otherwise preserving the same order, then

$$V(c_1, c_2, \ldots, c_p) = \frac{1}{2} \sum_{j=1}^{q-1} [1 - \mathrm{sgn}(\gamma_j \gamma_{j+1})] . \qquad (4.4.3)$$

The relevant property of Sturm sequences is embodied in **Sturm's theorem**: if $f_0(x)$ is a function which has only simple zeros in an interval $[\alpha, \beta]$, and $f_0(\alpha) f_0(\beta) \neq 0$, then the number of such simple zeros in $[\alpha, \beta]$ is equal to $[v(\alpha) - v(\beta)]$, where

$$v(x) = V[f_0(x), f_1(x), \ldots, f_m(x)] \qquad (4.4.4)$$

and where $\{f_k(x): k = 0, 1, \ldots, m\}$ is a Sturm sequence defined for an interval containing $[\alpha, \beta]$. Identifying $f_0(x)$ with a polynomial $p_n(x)$, it is evident that this result converts the problem of calculating the number of its real zeros in any interval, to one of constructing a Sturm sequence beginning with $p_n(x)$, and a systematic method of doing this is described in the next section.

To prove Sturm's theorem, it can be observed that a jump in the integer value of $v(x)$ can only occur at a value of $x = t$, say, in $[\alpha, \beta]$ where at least one of the functions of the Sturm sequence changes sign across a zero at $x = t$. Condition (i) clearly shows this function cannot be $f_m(x)$ since it has no zeros in $[\alpha, \beta]$. Similarly, condition (ii) shows that it cannot be $f_k(x)$ for any $1 \leqslant k < m$, because $f_{k-1}(x)$ and $f_{k+1}(x)$ are of opposite sign in some neighbour-hood of $x = t$: hence there can only be one change of sign in the sub-sequence $\{f_{k-1}(x), f_k(x), f_{k+1}(x)\}$ for any value of x in this neighbourhood, irrespective of whether $f_k(x)$ is positive, zero, or negative. Thus, such a jump in the value of $v(x)$ can only be at a zero of $f_0(x)$.

In the neighbourhood of such a zero, $f_0(x)/f_0'(x)$ has the sign of $(x - t)$, and by condition (iii), so therefore has $f_0(x)/f_1(x)$. Hence $f_0(x)$ and $f_1(x)$ have opposite signs as $x \to t$ from below, but the same signs as $x \to t$ from above. Thus the jump in $v(x)$ is such as to reduce its value by unity (associated with the disappearance of one sign change) as x increases through the value $x = t$ where $f_0(t) = 0$.

Furthermore, not only is every jump in value of $v(x)$ associated with a zero of $f_0(x)$, but every zero $x = t$ of $f_0(x)$ must cause just such a jump in the value of $v(x)$ at $x = t$ because, the zero being simple, $f_0'(t)$ is non-zero, and so by condition (iii), $f_0(x)$ and $f_1(x)$ cannot have a common zero in $[\alpha, \beta]$. Thus Sturm's theorem is established.

4.4.2 A Sturm sequence of polynomials

There are innumerable ways in which a Sturm sequence might be constructed which is defined for all real x and which begins with a particular polynomial $(p_n(x)$, say), of genuine degree n, provided that it is assumed at least that the polynomial has only simple zeros. One sets out with the intention that each successive element of the sequence shall be a polynomial of lower degree than its predecessor, and then one would anticipate that the sequence has no more than $(n+1)$ elements, terminating with $f_m(x)$ identically equal to a non-zero constant, so that condition (i) for a Sturm sequence is essentially satisfied.

With this in mind, the condition (iii) is satisfied (for instance) by taking $f_0(x) \equiv p_n(x)$ and $f_1(x) \equiv p_n'(x)$, the latter being a polynomial of genuine degree $(n-1)$. If subsequent elements are developed recursively by

$$f_{k+1}(x) = f_k(x)q_k(x) - f_{k-1}(x), \quad k = 1, 2, \ldots, (m-1) \ , \quad (4.4.5)$$

then condition (ii) is also satisfied, and letting $q_k(x)$ represent the integral quotient of $f_{k-1}(x)/f_k(x)$, it can be seen that $f_{k+1}(x)$ is the negative of the remainder polynomial, and so of lower degree than $f_k(x)$. For example, with $f_0(x) \equiv p_n(x)$ given in power form as in (4.4.1), it follows that

$$f_1(x) \equiv p_n'(x) = na_n x^{n-1} + (n-1)a_{n-1}x^{n-2} + \ldots + a_1 \ ,$$

and accordingly the integral quotient and (negative) remainder of $f_0(x)/f_1(x)$ are then the polynomials

$$q_1(x) = (na_n x + a_{n-1})/(n^2 a_n) \ ;$$

$$f_2(x) = f_1(x)q_1(x) - f_0(x),$$

$$= [1/(n^2 a_n)]\{[(n-1)a_{n-1}^2 - 2na_n a_{n-2}]x^{n-2} + \ldots$$

$$+ (a_{n-1}a_1 - n^2 a_n a_0)\} \ .$$

In general, $f_2(x)$ is a polynomial of genuine degree $(n-2)$, and similarly in general, each $q_k(x)$ is linear in x, and $f_k(x)$ will be a polynomial of degree $(n-k)$. However, exceptionally, one (or more) of the polynomials so generated may have a degree reduced by 2 (or more) compared with its predecessor, and correspondingly one (or more) of the integral quotients $q_k(x)$ may be a quadratic (or higher degree) polynomial.

It remains to be confirmed that this process leads to a non-zero constant terminator $f_m(x)$, so that condition (i) is satisfied. In fact, the process described by (4.4.5) is equivalent to the **Euclidean algorithm** for finding the highest common (polynomial) factor, or divisor, $f_m(x)$, of $f_0(x)$ and $f_1(x)$. It will be observed from (4.4.5) that such a factor (if one exists) would be common to $f_2(x)$ and all other polynomials of the sequence. Since all zeros of $p_n(x)$ have been assumed simple, $f_0(x) \equiv p_n(x)$ and $f_1(x) \equiv p_n'(x)$ can have no zeros (and so no polynomial factors) in common, and their H.C.F. is independent of x. Accordingly, $f_m(x)$ is inevitably a non-zero constant, and all the conditions for a Sturm sequence are met.

As a simple example, consider the construction of this Sturm sequence based on $f_0(x) \equiv p_4(x) \equiv 3x^4 + 4x^3 + 2$. We put $f_1(x) \equiv p_4'(x) \equiv 12x^2(x+1)$, and applying the familiar polynomial division algorithm, the hand-compution can be arranged as follows.

k	1	2	3	4	5
$q_k(x)$	1/4 1/12	12 12	$-$1/24 1/24	$-$24 $-$24	
$f_{k-1}(x)$	3 4 0 0 2	12 12 0 0	1 0 -2	-24 -24	1
	3 3 0 0	12 0 -24	1 1	-24	
	1 0 0 2	12 24 0	-1 -2	-24	
	1 1 0 0	12 0 -24	-1 -1	-24	
$-f_{k+1}(x)$	-1 0 2	24 24	-1	0	

The Sturm sequence is $\{3x^4 + 4x^3 + 2, 12x^2(x+1), x^2 - 2, -24(x+1), 1\}$ and the integral quotients are $q_1(x) = (3x+1)/12$, $q_2(x) = 12(x+1)$, $q_3(x) = (1-x)/24$ and $q_4(x) = -24\ (x+1)$. Evidently for large negative x, the Sturm sequence has signs $+-+++$ whilst for large positive x its sign sequence is $+++-+$; there is just one sign change in both sequences, and $v(-\infty) = v(\infty) = 1$. By Sturm's theorem, this polynomial therefore has no real roots.

Exceptionally, as has been pointed out, there may be fewer than $(n+1)$ polynomials in the sequence. Thus with $p_4(x) = x^4 + 4x^3 + 6x^2 - 3$ the Sturm sequence is $\{p_4(x), 4(x^2 + 3x + 3)x, 3(x+1), 4\}$, which misses out a quadratic function, and the integral quotient $q_2(x)$ is then quadratic, not linear. The sequence of signs is $+--+$ at $x = -\infty$, and is entirely positive at $x = \infty$, so that $v(-\infty) - v(\infty) = 2$; accordingly the equation has two real zeros. At $x = 0$, the sequence is $-0++$ implying that $v(0) = 1$, so therefore one zero is positive and one negative. The inequality (4.4.2) sets a bound of $|x| \leqslant 5.9$ on all zeros, so that each is now bracketed: in fact, the real zeros are at $x = -1$ and $4^{1/3} - 1$.

In computational work, it is essential to recognise these exceptions in which the degree of successive polynomials is reduced by more than one, and this is not easy because, if non-integer numbers are involved, the inevitable loss of precision in floating-point arithmetic can mean that precise cancellation (producing a computed zero leading coefficient of any of the polynomials in the sequence) is unlikely to occur. One can illustrate the possible catastrophic consequences of this by working out the Sturm sequence for the polynomial $p_4(x) = 0.333x^4 + 1.33x^3 + 2x^2 - 1$, using decimal arithmetic of floating point precision 3. To this precision, the polynomial is the same as the one just previously treated, except that all coefficients have been divided by 3. A particular implementation of the Euclidean algorithm in this arithmetic (using chopping) then gives:

$$f_q(x) \equiv p_4'(x) = 1.33x^3 + 3.99x^2 + 4.00x$$
$$f_2(x) = -0.0100x^2 + 0.992x + 1.00$$
$$f_3(x) = -13500x - 13500$$
$$f_4(x) \equiv 0$$

The fact that $f_4(x)$ is computed as zero would suggest that $f_3(x)$ is a common factor of $p_4(x)$ and $p_4'(x)$, which of course is nonsense. This error arises because the quotient $f_2(x)/f_3(x)$ has – in this precision – no significant remainder, which in turn is due to the fact that the leading coefficient (of x^2) in the expression of $f_2(x)$ is insignificant. Referring to the exact sequence previously derived, we see that $f_2(x)$ was there a linear polynomial, and consequently $f_3(x)$ was a constant.

The resolution of this difficulty depends on the introduction of **forced cancellation**: if the resulting sum or difference of two quantities has a sufficiently small magnitude compared with that of one or the other of those quantities, then the result is replaced by zero. In programming language, and using addition as an example, each such operation has to be treated as

$$c: = a + b;$$
if $\text{abs}(c) < \text{abs}(eps * a)$ **then** $c: = 0;$

and the value of eps would have been assigned equal to $N\epsilon$, say, where ϵ is the machine unit, and where N is chosen as large compared with 1. Too large a value increases the possibility of introducing, rather than correcting, generated error. What might be regarded as reasonable is a matter of judgement: there might be good reason to increase its value as the calculation proceeds and generated error gradually builds up. But in the present context, the choice is not a critical one. Note that testing instead **if** $a + c/N = a$ ... avoids the use of ϵ.

Referring to the example above, the coefficient -0.0100 of x^2 in the expression for $f_2(x)$ arises from subtracting 1 from 0.990, and the result has a magnitude of $10^{-2} = \epsilon$ compared with either operand; it would therefore be replaced by zero for any value of N larger than unity. Then with $f_2(x)$ accordingly treated as $(0.992x + 1.00)$, the value of $f_3(x)$ becomes 1.34, and terminates the sequence; the functions of the sequence are recognisably numerical approximations to those of the exact sequence (taking account of the division by a factor of 3).

Having generated a Sturm sequence, there is no great difficulty in applying the Sturm theorem to compute the number of zeros in any interval. Successive bisections of the initial interval – chosen by some such limitation as given by (4.4.2) – can then be made as part of a logarithmic search for sub-intervals containing just one (simple) root, discarding any intervals found to contain no zeros. Because of the relatively excessive computation required in evaluating $v(x)$, it would be inappropriate to take this process further than is necessary to bracket each real root. Even the bisection method of root-finding would hence-forward be easier to use, as it depends at each bisection on the sign of $p_n(x)$, and not on the sign of each polynomial of the Sturm sequence including $p_n(x)$.

4.4.3 Treatment of multiple roots by Sturm sequences

If the Euclidean algorithm (4.4.5) is applied to find the H.C.F. of $p_n(x)$ and $p_n'(x)$ when $p_n(x)$ has any multiple zeros, then the process will converge with a non-constant function $f_m(x)$, since all zeros of $p_n(x)$ of multiplicity M are also zeros of $p_n'(x)$ – and so of every function of the Sturm sequence including $f_m(x)$ – where they occur with multiplicity $(M-1)$. In practical work, in order to discriminate that $f_m(x)$ terminates the sequence (that is, that the quotient f_{m-1}/f_m has no remainder), it would generally be necessary to use forced cancellation to demonstrate that $f_{m+1}(x)$ is zero.

For example, the polynomial $p_4(x) = x^4 - x^3 - 3x^2 + x + 2$ which has a double zero at $x = -1$, and simple zeros at $x = 1$ and 2, provides the Sturm sequence $\{p_4(x), p_4'(x), 3(9x-11)(x+1)/16, 128(x+1)/81\}$, terminating with the polynomial $f_3(x)$ equal to a multiple of $(x+1)$. Performing the same calculation using truncated decimal floating-point arithmetic with precision 3, one obtains:

$$f_2(x) = 1.68x^2 - 0,375x - 2.06$$

$$f_3(x) = 1.57 + 1.58x \qquad ,$$

and $f_4(x)$ would be evaluated as $(2.06 - 2.07)$ which, although not zero, would be replaced by zero.

Of course, having supposedly found in the shape of $f_m(x)$ a polynomial which contains only, and all, the multiple zeros of $p_n(x)$, then $f_m(x)$ can itself be submitted to the same algorithm with the object of locating these zeros. This process could be envisaged as proceeding recursively so far as is necessary to reduce the multiplicity of all real zeros to unity in some derived lower-order polynomial. Thus the formation of a Sturm sequence seems a direct and effective means of dealing with what is otherwise (as we have seen) a troublesome problem. However, too sweeping a use of forced concellation imposes the same inaccuracy in discriminating multiple from closely-spaced zeros as is suffered by any other method involving arithmetic of limited precision. On the other hand, too cautious an imposition of the technique risks (in effect) treating any double zero as a complex pair with small imaginary parts, and so missing it altogether.

In the event that $p_n(x)$ has real multiple zeros within an interval $[\alpha, \beta]$, then the difference $[v(\alpha) - v(\beta)]$ in the count of sign changes in the Sturm sequence measures the number of distinct real roots in $[\alpha, \beta]$ irrespective of their multi-plicity: multiple roots count as one. In other words, the zeros which are counted are in effect those of $f_0(x)/f_m(x)$, which is a polynomial in which all zeros are simple. For instance, referring to the example of the quartic cited earlier in this section, the signs of the Sturm sequence polynomials are $+-+-$ for large negative x, and $++++$ for large positive x, implying 3 distinct real roots – known in fact to be 2 simple, and one double. However, if as a consequence of inprecise arithmetic, a non-zero constant f_4 were added to the sequence, there would appear to be either 2 real roots, or 4 real roots, depending on whether f_4 was negative or positive. The double root would therefore be misinterpreted as either a complex pair, or two (close) real zeros.

4.4.4 Polynomial deflation

The process of zero suppression is generally applicable in finding a number of zeros of any function, whether algebraic or transcendental. However, once a zero ξ of a polynomial $p_n(x)$ is determined, there arises the possibility of replacing the polynomial by the quotient $p_n(x)/(x - \xi) = p_{n-1}(x)$, say, from which this zero has been eliminated – not just suppressed. This is the process of **deflation**. The process can be repeated each time a root is found, and clearly deflation will economise on computational work.

The factorisation can be accomplished by exactly the same process of synthetic division as is used in Horner's scheme of polynomial evaluation: the recurrence

$$b_n = a_n, b_k = a_k + b_{k+1}\xi \quad (k = n-1, n-2, \ldots, 0) \, , \quad (4.4.6)$$

where the a_k's are the polynomial coefficients as in (4.4.1), is based on the relation

$$p_n(x) - p_n(\xi) = (x - \xi)(b_n x^{n-1} + b_{n-1}x^{n-2} + \ldots + b_1) \, . (4.4.7)$$

Whereas in Horner's scheme the interest centres on the value of $b_0 = p_n(\xi)$, and the coefficients $b_n, b_{n-1}, \ldots$ are just a means to this end, in deflation (with b_0 small enough for the value of $x = \xi$ to have been recognised as a root) the coefficients b_k for $k = 1, 2, \ldots, n$ are the coefficients of the deflated polynomial, $p_{n-1}(x)$.

The problem with what seems an obviously desirable simplification is that the computed value of ξ, and so also the computed coefficients of $p_n(x)$ are inevitably inprecise. Successive quotient polynomials accumulate error, and their zeros deviate more and more from those of the original polynomial. Irrespective of the accuracy required, each zero has to be evaluated as exactly as possible, and experience and analysis both show that the numerical error is then kept small if, but only if, the zeros (complex as well as real) are extracted in the order of increasing magnitude. Desirably, the zeros obtained from the deflated polynomials need to be used as initial values of an iterative process (such as Newton's method) applied to the original, undeflated polynomial, thereby eliminating inaccuracies arising from the deflation process – a safeguard called **purifying** the zeros.

To a large extent, the origin of this problem of inprecision lies in the use of the power form of representation of a polynomial (4.4.1), because (at least for equations of large degree) the successive powers of x which form the basis of the formulation are not sufficiently different from each other, and as a result cancellation between the various terms can make the polynomial value unduly sensitive to the coefficient values. Suppose, for instance, that one particular coefficient of (4.4.1), say a_j, is increased by an amount δa_j. Then in effect the equation being solved is not $p_n(x) = 0$, but rather $p_n(x) = -\delta a_j x^j$, and the root $x = \xi$ will be changed to $x = \xi + \delta\xi$, say, where

$$p_n(\xi + \delta\xi) \simeq \delta\xi p_n'(\xi) = -\delta a_j \xi^j$$

that is, $\delta\xi/\xi = -[a_j \xi^{j-1}/p_n'(\xi)](\delta a_j/a_j)$.

Particularly if the degree is high, and the root large, the relative change in the root can be huge compared with the relative error in a_j, even if $p_n'(\xi)$ is large. A well-known example is due to J. H. Wilkinson: if, in the equation

$$(x+1)(x+2)\dots(x+20) = 0 \quad \text{or} \quad x^{20} + 210x^{19} + \dots + 20! = 0$$

one changes the value of a_{19}, then for example at the root $\xi = -16$, the value of $a_{19}\xi^{18}/p_n'(\xi) = 210.16^{18}/(4!15!) = 3.16 \times 10^{10}$: adding 2^{-24} to the value of 210 – which is merely a change in the least significant bit of a mantissa of 32 binary digits – implies that $\delta\xi/\xi = 9$. In fact, several of the large roots are thereby caused to become complex, although the small real roots are virtually unaffected. The most 'sensitive' coefficients are those around $a_{12} \simeq 10^{13}$. A change in any of these causes proportionate changes up to 10^{13} times larger in the roots around $\xi = -14$. Even a mantissa of 48 binary digits (with machine unit $\epsilon = 7 \times 10^{-15}$) provides insufficient precision to avoid gross errors in the root determination. The equation is extremely badly conditioned with respect to its power-form coefficients.

On the other hand, if the same polynomial were expressed as a finite Tchebyshev series, or as a Newton form based on a well-distributed support, these difficulties are much alleviated, or avoided altogether. Deflation can still be accomplished, albeit perhaps by a more complicated process, without having to convert the polynomial to power form. The Newton form arises of course from polynomial interpolation, and if a zero is used, when found, to replace the closest support point in the interpolative scheme, the product form of the polynomial is ultimately formed (if all roots are real), which is the best conditioned of all expressions for the roots. On the other hand, if the power form is given as data, then there is little purpose in transforming to any other form, as the transformation itself embodies the ill-conditioning.

It will not escape notice that the method described for the generation of Sturm's sequence requires the power form, although (at the expense of some complication) it can be adapted for use with finite Tchebyshev series. This is yet another reason for regarding the use of Sturm's theorem as a method of bracketing, rather than precisely locating, any zero.

4.5 OPTIMISATION IN ONE DIMENSION

It quite often happens that the position of an extreme value of a function is required rather than a zero, and the subject of optimisation concerns the location of such extremes. Our concern here will be with the extremes of a function $f(x)$ of a single variable, and if it is continuous over a closed interval $[a, b]$, then evidently – unless it is a constant – $f(x)$ must reach 'extreme' values within this interval at both end-points: for example, $f(x) \equiv x$ has a maximum value of $+1$ and a minimum value of -1 in the closed interval $[-1, 1]$. However, these are obviously not the kind of extremes whose location presents any problem. They are generated merely by the limitations of the interval, and are trivial examples of **constrained** optima. Our interest is in **unconstrained** extremes which occur at an interior point of any imposed interval: often they will be stationary values of $f(x)$, but their definition can allow for other possibilities.

A function is said to have a **local minimum** at some interior point $x = \xi$, say, of its domain D, if there exists a neighbourhood N of ξ within D (say $|x - \xi| < \delta$, where $\delta > 0$), such that $f(x) \geqslant f(\xi)$ at all points $x \in N$. If the equality applies only at $x = \xi$, it is called a **strict local minimum**. In particular, if $f(x) \geqslant f(\xi)$ for all points $x \in D$, then it is called a **global minimum**, and again if $f(x) = f(\xi)$ only at $x = \xi$, it is a **strict global minimum**. Any (strict) global minimum is necessarily a (strict) local minimum; but the converse of course is not true. Similarly, replacing the inequality $\geqslant$ by $\leqslant$, one may define (strict) local and global maxima. The term 'extremum' embraces both a maximum or a minimum. In general there is no loss of generality by referring only to a minimum, since the maxima of $f(x)$ are simply the minima of $-f(x)$. Local extrema are also sometimes termed 'relative' extrema, and it is the problem of their location which is of interest here.

If a function is differentiable on an interval I and if $f(x)$ has a local extremum at an interior point $x = \xi$ of I, then $f'(\xi) = 0$. Accordingly $x = \xi$ is also a stationary value of f. The converse is not however true – for example, if $f(x) \equiv x^3$, then $f'(0) = 0$ but $f(0)$ is not an extremum. Thus if the derivative exists and is easily obtained, then almost certainly the best way of locating strict local extrema would be by finding the distinct roots of $f'(x) = 0$. One could do this using (for instance) any of the methods already described – and then eliminate any such stationary values of f which are not extrema. Thus if $f'(x)$ changes sign across $x = \xi$, or if f is twice differentiable and continuous in the neighbourhood of $x = \xi$ and $f''(\xi)$ is non-zero, it would be immediately inferred that the stationary value is indeed a strict extremum.

However, such an approach may not always be convenient, and – if $f(x)$ is not necessarily continuous or differentiable – it may not always be possible. Then a local extremum would need to be located by reference to the function values. The only reliable methods of this kind are those in which it is known that some given finite interval contains the required extremum, and furthermore that this interval in fact brackets the extremum (in a manner which will be defined).

In particular, the convergence and optimality of the methods can only be proved if the function is **unimodal** in this interval. By this it is meant − assuming that a minimum is to be located − that $f(x)$ is a strictly monotone decreasing function for all $x < \xi$, and strictly monotone increasing function for all $x > \xi$, in the interval, so that $x = \xi$ is a strict local minimum, and the only local extremum in the interval.

4.5.1 Bracketing a local minimum

Whereas a zero of $f(x)$ can be asserted to exist in any interval (a, b) if $f(a)f(b) < 0$, the existence of a minimum can only be inferred (from function values alone) by evaluating the function at some third point $x = c \in (a, b)$, and determining that $f(c)$ is not greater than either $f(a)$ or $f(b)$, nor equal to both. In such an eventuality it is said that the **minimal condition** is satisfied by $f(x)$ in (a, b) at $x = c$. For if there are no local extrema within (a, b), the sequence $f(a), f(c), f(b)$ would necessarily be strictly monotone. The minimal condition contravenes this relationship and ensures that the interval must therefore contain at least one local minimum. Of course, it does not preclude the existence of other extrema in (a, b) as well. Note, too, that although it is a sufficient condition for the existence of a local minimum in (a, b), it is not however a necessary condition − any more of course than the existence of a zero of $f(x)$ in (a, b) necessarily implies the bracketing relationship $f(a)f(b) < 0$.

Provided that a minimum can be bracketed in this way, then it is possible to reduce the bracketing interval by successive iteration. For suppose one chooses a further interior point within (a, b) distinct from $x = c$, so that there are now two interior points, say $x = p$ and $x = q$ where $a < p < q < b$, one of which coincides with $x = c$, the other being newly introduced. Then if $f(p) > f(q)$, the minimal condition is satisfied in (p, b) at $x = 1$; alternatively if $f(p) < f(q)$, it is satisfied in (a, q) at $x = p$. If it happened that $f(p) = f(q)$, then detailed consideration will confirm that the minimal condition is satisfied on at least one if not both intervals (a, q), (p, b). The effect therefore is to replace one of the end-points $x = a, b$ by an interior point of (a, b) at each iteration, thereby continually contracting the interval containing the minimum.

If it is known that $f(x)$ is unimodal in (a, b), then this same process will succeed in contracting the interval about the (unique) minimum even if neither of the interior points satisfy the minimal conditon. For then the sequence of values of the function at the points $x = a, p, q$, and b must be strictly monotone (decreasing or increasing), and the (unique) local minimum of f would be in the interval containing the abscissae of the three least function values − that is, in (p, b) if $f(p) > f(q)$, or in (a, q) if $f(p) < f(q)$. In this circumstance, if it happened that $f(p) = f(q)$, then one would know moreover that the unique strict minimum lies in (p, q). These assertions will be clearly demonstrated by consideration of Fig. 4.9. Of course if the sequence of function values is not strictly monotone, then a minimal condition is established at one of the interior points, and the process of interval contraction will apply as successfully to unimodular functions as to any other.

In other words, the prior assertion of unimodularity makes it unnecessary to prove the existence of a local minimum by establishing a minimal condition; and for this reason library implementations of a method of interval contraction are unlikely to check whether or not the minimal condition is satisfied. They will therefore still 'work' even if the function has *no* local minimum, in the sense that the interval will be progressively contracted. But the interval will then contract upon an end point of the originally imposed interval – that is, the process would locate a constrained minimum, and not the (non-existent) local minimum.

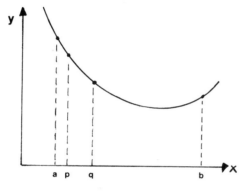

Fig. 4.9.

4.5.2 Interval reduction by golden section

The rate at which any method of interval contraction converges upon a local minimum depends upon the proportionate reduction in interval span at each iterative step. Each such step involves a selection of one of 2 alternative and overlapping sub-intervals. These sub-intervals depend upon the manner of selection of the two interior points, and there is a rule governing this selection which will make the relative contraction of the interval the same for each and every iteration, whichever of the two alternative sub-intervals is selected.

Suppose that A P B are three consecutive points on the x-axis, denoting the end-points (A, B) of the bracketing interval and the interior point (P), at which function values are available, having already been evaluated at some earlier stage of the interative scheme. These points are not necessarily ordered in the direction of increasing x; instead, it will be assumed that the length $AP = (1 - \lambda_n)l_n$ is smaller than that of $PB = \lambda_n l_n$, so that $\frac{1}{2} < \lambda_n < 1$. Evidently $\lambda_n/(1 - \lambda_n)$ is the ratio of the larger to smaller partition of the interval A B, and the subscript n has been added to denote values appropriate to the nth iteration, since both l_n and (possibly) λ_n will differ from one iteration to the next. Suppose now, as part of this nth iterative step, that the larger sub-interval is partitioned at a point Q (between P and B). The value of the function will be evaluated at the x-value corresponding to this point, and as a consequence of the relationship between this value and that of the function value already evaluated at P, the $(n+1)$th bracketing interval will consist of either the points APQ or PQB. It is required, first of all, that the lengths of AQ and PB of both these possible successor intervals shall be the same (so that the interval contraction is independent of the choice involved).

But the length of PQB is simply that of $PB = \lambda_n l_n$, so that it is required also that $AQ = \lambda_n l_n$. This means therefore that the three segments AP, PQ, QB must have lengths $(1 - \lambda_n)l_n$, $(2\lambda_n - 1)l_n$ and $(1 - \lambda_n)l_n$ respectively. In other words, the rate of interval contraction is independent of the sub-interval selected if (and only if) the points P and Q are symmetrically disposed about the mid-point of the containing interval AB. Such a method of interval contraction is therefore called *symmetric*. The length l_{n+1} of the next bracketing interval will inevitably be $\lambda_n l_n$ – that is, λ_n times that of the previous step – whether it be APQ or PQB.

However, both the intervals APQ and PQB are evidently partitioned by their interval points (P and Q) into lengths proportional to $(1 - \lambda_n)$ and $(2\lambda_n - 1)$, and the ratio of the larger of these two quantities to the total length of the new interval is simply λ_{n+1}: that is,

$$\lambda_{n+1} = \max(2\lambda_n - 1, 1 - \lambda_n)/\lambda_n . \qquad (4.5.1)$$

But if the proportionate interval contraction $l_{n+1}/l_n = \lambda_n$ is to be the same for all steps, as has been proposed, then λ_{n+1} must be the same as λ_n. Placing $\lambda_{n+1} = \lambda_n = \lambda$ in (4.5.1), we find that either $\lambda = 1$, which is excluded by the assumption that $\lambda_n < 1$, or else $\lambda^2 = 1 - \lambda$, and this quadratic equation has the (positive) solution $\lambda = (\sqrt{5} - 1)/2 = 0.618$. Note that this value λ is also the same as the ratio $(1 - \lambda)/\lambda$.

Thus if the interval APB is partitioned at P into lengths whose ratio is $PB/AP = (1 - \lambda)/\lambda = \lambda = (\sqrt{5} - 1)/2$, and Q is chosen symmetrically so that $AQ = PB$ (and so also $AP = QB$), then this same proportion is retained in the partitioning of the following contracted interval, APQ or PQB, and so in each and every interval of the iterated scheme. After n iterations, the length of the current bracketing interval will be 0.618^n times the initial interval length. This is a slower rate of interval contraction than that achieved by the bisection method of root location – about 1.44 times as many iterations are needed to obtain the same reduction in interval size – but this merely emphasises the view already expressed, that it is better to find the zeros of $f'(x)$ if the function is differentiable.

This method of interval contraction is called the **golden section method**. The name 'golden section' refers to that division of a straight line in which the ratio of the smaller sub-interval, here $AP = (1 - \lambda)l_n$, is to the larger

($PB = \lambda l_n$), as the larger is to the total length, $AB = l_n$; or in algebraic terms, $(1 - \lambda)/\lambda = \lambda$. Note that, to initiate the method, it is required that the initial interval is divided in these same proportions and that there is no guarantee, of course, that such a choice would necessarily satisfy the minimal condition at either of the appropriate internal points P, Q. Bearing in mind the comments of the previous section, it therefore must not be presumed that the method of golden section will necessarily locate a local minimum in the initial interval, even if one does exist. It will do so, however, either if the function is known to be unimodular, or if the minimal condition *is* satisfied at one of the two initial points of golden section. Otherwise it may simply converge to the lesser of the two constrained extrema at one end or the other of the initial interval.

If the method of golden section has something in common with the bisection method of root-finding, then there would be the expectation of more rapid convergence of interval contraction if function values were employed as an aid in selecting appropriate points of interval division, as in the method of *regula falsi*. However, such methods offer an improvement (at least in theory) only if the function is differentiable, as they imply that some assumed form of inter-polation can be applied to predict a zero of $f'(x)$. If this were justified, then still faster convergence would be achieved by seeking the roots of the equation $f'(x) = 0$, assuming at least that the derivative is no more difficult to evaluate than the function. Such 'improved' methods of minimisation have, therefore, only a limited applicability.

4.6 REPRISE

Many people take the rather disparaging view that the approach to the problem of solving a non-linear equation is just a collection of techniques without great depth or substance. Indeed, there is (perhaps) the appearance of contrivance in the various modifications applied to the method of *regula falsi* in order to persuade it always to perform more speedily than the entirely straightforward method of bisection. Yet these modifications, if not particularly elegant, are effective in their stated purpose, and they provide the preferred method of locating a root – assuming (as it must be hoped) that it can be bracketed. Their speed of convergence in relation to the number of function evaluations – what might be termed their *efficiency* – is surpassed only by the more complicated unbracketed methods, and then only if the latter do in fact converge.

The lack of certainty about the convergence of the unbracketed methods is a considerable disadvantage, although their simpler structure may often be preferred for hand-computation, when one can react 'intelligently' to any uncertainties. They have to be used for the evaluation of multiple roots, but frankly that is a problem which is best avoided if at all possible, as such roots can only be assessed with limited accuracy and poor efficiency.

No objections of triviality could be levelled at the accomplishment of Sturm's theorem, in enabling the bracketing of all real roots of algebraic equations. This is as near as any automatic process comes to providing a rigorous method for *searching* for real roots. The solution of any non-algebraic equation, on the other hand, requires that one must perform such a preparatory search for oneself, if success is to be guaranteed. One must either provide values bracketing a root, or — knowing its existence — observe a 'steep-sided' approach towards it, which will ensure the convergence of most unbracketed methods.

Naturally, the problems of root-finding become much more difficult if on the other hand, demands hard preparation: one must either provide values bracketing a root, or — knowing its existence — observe a 'steep-sided' approach towards it which will ensure the convergence of an unbracketed method.

Of course the problems of root-finding become much more difficult if a system of equations has to be satisfied rather than just one, and inevitably this kind of perspective reduces the significance of the processes of solving just one equation. Likewise, the problem of optimisation of a single variable pales by comparison with that involved in optimising a system. It will have been clear from our discussion that the best approach to the one-dimisional problem is usually the rather obvious one of finding the zeros of a derivative. However, one should not be disapproving if the solution of some problems is relatively straightforward!

CHAPTER 5

Numerical Quadrature: estimating the value of definite integrals

The process of interpolation can be applied either to construct a function fitting a given set of discrete data — the process of curve-drawing — or as a means of approximating a 'complicated' function by one of simpler form. In either instance, the elementary operations of the calculus could, in principle, be applied to the interpolated function to find a derivative or calculate a definite integral.

The process of differentiation of a known continuous function is 'easy' in the sense that rules are available by which the formal result can always be obtained, irrespective of how 'complicated' the function may be, and a numerical estimate of a derivative is most easily and accurately found by evaluating the differentiated function. If the function is only defined by discrete data, one can do no better than estimate its derivative from that of the interpolating curve which is fitted to the data. Obviously if a form of piecewise polynomial approximation is used for this purpose, the derivatives obtained in this way are lower-degree polynomials with degraded piecewise continuity and, as is inevitable in any polynomial representation, the high-order derivatives are zero. But however one might seek to overcome this particular deficiency, it is implicit that the process of differentiation must degrade accuracy, in the same way that the process of differencing numerical data can serve to highlight any errors or inconsistencies they may contain.

The process of integration, on the other hand, tends to enhance the quality of the data to which it is applied. It is, after all, an 'averaging' process. It is, moreover, analytically a 'difficult' process because it cannot in general be performed formally (except for certain 'standard' integrands). So a numerical estimate of an integral may often be required, whether the integrand is defined by discrete data or has a functional representation, and one starts with the confidence that such an estimate should be able to be made quite accurately. The rules of numerical integration, or **numerical quadrature** as it is more often called, are founded on the ideas of polynomial interpolation, and are mostly merely a short-cut to the obvious process of interpolating the function (or the provided data) and integrating the resulting polynomial. Such a short-cut is obviously desirable as, after all, only one number — the value of a definite integral — is required, and the detail of the interpolating function itself is largely superfluous. This would not apply if an *indefinite* integral were required, and in that event the construction of a suitable interpolate, and its formal integral, would be an obvious way of getting the answer appropriate to any value of the indefinite limit of the range of integration.

In discussing the methods of numerical quadrature, it will be assumed, as a start, that the integrand is nonsingular and continuous over the range of integration. Often it will be assumed that the integrand is infinitely differentiable (that is, analytic) over that range, though that is not always a necessary restriction. However, it is important at the outset to realise that if an integrand has discontinuities — even if only in its higher derivations — then in seeking a numerical value of the definite integral it is always good practice for the range to be segmented so that these discontinuities are at the ends of component sub-intervals, and for the integral values over each such sub-interval to be added together. Otherwise, whatever method is used, it will not perform either as efficiently or as accurately as it is capable of doing. Integrals with end-point singularities need special treatment (section 5.5).

5.1 QUADRATURE BASED ON EQUALLY-SPACED INTEGRAND EVALUATIONS

The basic intention in all forms of numerical quadrature is to approximate a definite integral by a finite sum of suitably weighted values of the integrand: that is, one seeks a relation of the form

$$\int_a^b f(x)\mathrm{d}x = \sum_{k=0}^{n} l_k f(x_k) + E_n , \qquad (5.1.1)$$

where E_n is a truncation error (supposed in some sense small or numerically insignificant) and where appropriate values of the points of evaluation x_k of the integrand, and of the coefficients l_k, are known or asserted for $k = 0, 1, \ldots, n$. The **sampling points** x_k play the same role as the support of a scheme of interpolation. If some method of linear interpolation is applied to the data pairs $(x_k, f(x_k))$ for $k = 0, 1, \ldots, n$ and one supposes that this method leads to an expression $y = y(x)$ for the interpolate, then — as for example in (3.2.3) — $y(x)$ can be expressed in terms of shape functions as $\sum_{k=0}^{n} y_k s_{nk}(x_0, x_1, \ldots, x_n; x)$, with $y_k = f(x_k)$. In so far as $y(x)$ approximates $f(x)$, the sum on the right-hand-side of (5.1.1) could accordingly be derived as the value of the integral

$$\int_a^b y(x)\mathrm{d}x = \int_a^b \sum_{k=0}^{n} y_k s_{nk}(x_0, x_1, \ldots, x_n; x)\mathrm{d}x$$

$$= \sum_{k=0}^{n} f(x_k) \int_a^b s_{nk}(x_0, x_1, \ldots, x_m; x)\mathrm{d}x . \qquad (5.1.2)$$

This immediately identifies the values of l_k in (5.1.1) with definite integrals of $s_{nk}(..)$, and leads to values of these coefficients dependent on the arrangement of the x_k's, but independent of the form or value of the integrand.

Indeed, because integration and summation are both linear operations, one might reasonably presume that the l_k ought to be independent of the integrand, even without knowing how these coefficients might be obtainable. Independence would not exist if the sum in (5.1.1) were based on an approximation to $f(x)$ obtained by a method of non-linear interpolation – by a rational function representation, for example – and consequently such methods of interpolation would not be appropriate in this context.

This method of obtaining (5.1.1) leads to formulation of the truncation error E_n as simply the definite integral $\int_a^b [f(x) - y(x)] \, dx$ of the error of the interpolate, and it is of importance to know how E_n may be influenced by the form of the integrand $f(x)$. But this is often difficult to determine. Whereas in series summation and function approximation numerical estimates – or bounding values – of error play a crucial role in determining whether or not a numerical result is acceptable, it had better be admitted at the outset that the vast majority – if not all – of the methods of numerical quadrature rely in practice on indirect and possibly insecure inference of error, rather than on strict error bounds. This is because expressions for bounds on the truncation error are either too complicated for direct calculation or else so 'loose' – so wide of the actual error – as to be misleading. For example, if $y(x)$ is the minimax polynomial approximation to $f(x)$, its derivation yields the maximum error magnitude $E_n^*(f)$, and it would then be apparent that the integral over $[a, b]$ of the actual error is certainly less than $(b - a)E_n^*(f)$ in absolute value; but this would be a wildly pessimistic estimate, because the error fluctuates in sign over the interval of approximation, and the integral of the error is accordingly much smaller. Although a tighter bound than the one quoted might be rigorously established without too much difficulty, experience shows that in general all strictly derived estimates are either too misleading, or else so complicated that their computation involves a quite disproportionate effort.

Fortunately, numerical quadrature is a 'kindly', well-conditioned process, in that apparently crude approximations can lead to excellent results: this indeed is one reason underlying the difficulty of estimating how good they are. Although estimates of error are certainly of interest, and play a role in developing a strategy for numerical quadrature, they rarely enter into the tactics of a particular algorithm, where a more pragmatic approach to error has to be taken, based on a succession of differing approximations.

In at least one important and not infrequent application they are quite irrelevant: for just in the same way as interpolation can be applied to discrete data which have no known functional representation, in order to draw a graph, so one may (in effect) want to find the area under that graph; in other words, one may want to integrate the function implied by the discrete data (usually obtained from physical experiment). In this eventuality there is no basis for judging error, and there is not even the 'eye-appeal' of a drawn curve by which to assess the answer obtained.

This will make it clear that there is interest in both the problem of quadrature with prescribed points of integrand evaluation (x_k) as well as in that of selecting the points of evaluation with the object of improving the accuracy of approximation to the integral. In both eventualities, the derivation of quadrature formulae like (5.1.1) applicable to an equi-spaced support is of special interest, and the remainder of this section relates to this particular problem.

5.1.1 The Cotes formulae

A number of the more elementary rules of quadrature are probably already well-known to the reader. The **trapezoidal rule**

$$\int_a^b f(x)\mathrm{d}x \;=\; \tfrac{1}{2}(b-a)[f(a)+f(b)] + E_T$$

in effect replaces the area under the curve $y = f(x)$ for $a < x < b$ by the trapezoidal area below the straight line joining the extreme points $(a, f(a))$, $(b, f(b))$ on the curve. It will be clear from Fig. 5.1(a) that the magnitude of

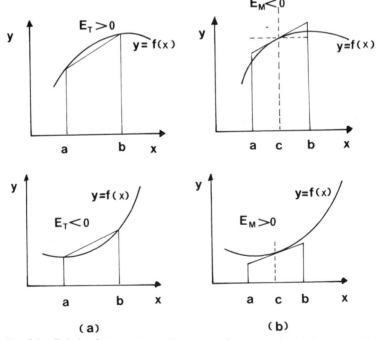

(a)   (b)

Fig. 5.1 – Relation between truncation error and curvature for (a) the trapezoidal rule, and (b) the mid-ordinate rule.

the error E_T depends on the curvature of $y = f(x)$, and that the sign of E_T changes with that of the curvature (supposing this is of one sign over the entire interval of integration). Algebraically, if $f(x)$ is twice differentiable over $[a, b]$ then it can be shown that

$$E_T = -\tfrac{1}{2} \int_a^b (b - x)(x - a) f''(x) \mathrm{d}x \ . \tag{5.1.3}$$

Using the mean value theorem for integrals, this in turn can be interpreted to imply that

$$E_T = -(b - a)^3 f''(\xi)/12 \tag{5.1.4}$$

for some $\xi \in (a, b)$, and both these expressions reveal the dependence of error on curvature.

An even simpler relationship is the **mid-ordinate rule**

$$\int_a^b f(x)\mathrm{d}x = (b - a) f\left[\tfrac{1}{2}(a + b)\right] + E_M \ . \tag{5.1.5}$$

This replaces the area under the curve by a rectangle of width $(b - a)$ and of height equal to the mid-ordinate $f(c)$ where $c = \tfrac{1}{2}(a + b)$. This seems rather crude until it is realised that this is also the area beneath any straight line through $(c, f(c))$ over the same interval $[a, b]$, and that it is in particular therefore the area beneath the tangent to the curve $y = f(x)$ at $x = c$. Here again Fig. 5.1(b) will make it clear that the error E_M depends on the curvature but with the opposite dependence on sign to that involved by the trapezoidal rule. Similarly to (5.1.3) it can be shown that

$$E_M = \tfrac{1}{2} \int_a^b (b - c - |x - c|)^2 f''(x) \mathrm{d}x \ , \tag{5.1.6}$$

or again using the mean value theorem for integrals,

$$E_M = (b - a)^3 f''(\eta)/24 \tag{5.1.7}$$

for some η in (a, b). Supposing that $f''(x)$ does not change sign over (a, b), so that the curvature is entirely convex or concave, then comparing (5.1.4) with (5.1.6) it is clear that the errors E_T and E_M must have opposite signs, and so the correct integral value must lie between the estimates provided by the trapezoidal and mid-ordinate rules. If there is no large variation in the value of $f''(x)$ over this same interval, then $|E_M|$ is only about half the magnitude of $|E_T|$, so that the mid-ordinate rule is not merely simpler than the trapezoidal rule, but twice as accurate.

In particular, if $f''(x)$ is constant over $[a, b]$, then of course the integrand $f(x)$ is a quadratic polynomial and E_M is equal exactly to $-E_T/2$. If I_T and I_M denote the integral estimates obtained respectively from the use of the trapezoidal and mid-ordinate rules, it follows in this eventuality that the weighted average $(2 I_M + I_T)/3$ has zero error. Thus the estimate given by

$$\int_a^b f(x)\mathrm{d}x = [(2I_M + I_T)/3] + E_S$$

$$= [(b - a)/6][f(a) + 4f(c) + f(b)] + E_S \tag{5.1.8}$$

with $c = \frac{1}{2}(a + b)$, is precise (that is, $E_S = 0$) if $f(x)$ is a quadratic polynomial; this approximation to the integral is of course the well-known **Simpson's rule**. The expression for the error is found to be

$$E_S = -(1/72) \int_{c-h}^{c+h} (h - |x - c|)^3 (h + 3|x - c|) f^{iv}(x) dx$$

(5.1.9)

where $(b - a)$ is here placed equal to $2h$, and this leads to

$$E_S = -(b - a)^5 f^{iv}(s)/2880$$

(5.1.10)

for some value of s in (a, b). Either of these relations shows that Simpson's rule is precise not merely if the integrand is a quadratic polynomial but also if it is any cubic polynomial (because this would imply that the fourth derivative of f, on which the error depends, is identically zero). This perhaps unexpected extension of precision arises because any cubic which passes through the three points $(x, f(x))$ for $x = a$, c and b is equal to the quadratic polynomial interpolating these points plus some multiple of the cubic $(x - a)(x - c)(x - b)$. But since $x = c$ is midway between $x = a$ and $x = b$, this cubic increment is composed of two lobes (Fig. 5.2) of equal and opposite extent on either side of the mid-point $x = c$. Whatever the factor by which it may be multiplied, it contributes nothing if it is integrated over $[a, b]$. Thus a 'good' estimate of the integral could be regarded as obtained from an apparently 'bad' representation of the integrand.

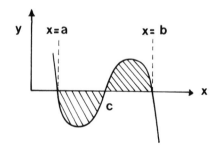

Fig. 5.2.

Generally, Simpson's rule is much more accurate than the mid-ordinate or trapezoidal rules, because the quadratic polynomial approximation to $f(x)$ it uses is almost always better than the straight line (linear polynomial) approximation implicit in the use of the simpler formulae. One would naturally anticipate that by increasing the degree of the interpolating polynomial — as would be possible if more points of integrand evaluation were included — one might be able to achieve still greater accuracy.

To examine this extension, the notation needs to be changed. Suppose, as in (5.1.1), that $x_0, x_1, \ldots, x_n$ are $(n + 1)$ points of evaluations, and define

$x_k \doteq a + k(b - a)/n$ for $k = 0, 1, \ldots, n$ and $n \geqslant 1$, so that these points are equispaced. If $p_n(x)$ is the unique nth degree polynomial interpolating $(x_k, f(x_k))$ for $k = 0, 1, \ldots, n$, then $\int_a^b p_n(x) \mathrm{d}x$ approximates $\int_a^b f(x) \mathrm{d}x$. If $n = 1$, the two points of evaluation are $x = a$ and $x = b$, and the approximation is that provided by the trapezoidal rule; if $n = 2$, the points of evaluation are augmented by the mid-point $x = \frac{1}{2}(a + b) = c$, the interpolation is by a quadratic polynomial $p_2(x)$, and the approximation is that recognised as Simpson's rule (problem 1, section 5.1). More generally, by using the shape function representation (3.2.5) of the interpolating polynomial as in (5.1.2), with $y(x) = p_n(x)$, then by placing $f(x_k) = f_k$, one obtains the approximation

$$(b - a)^{-1} \int_a^b f(x) \mathrm{d}x = \sum_{k=0}^{n} \lambda_k^{(n)} f_k + E_n , \qquad (5.1.11)$$

with $(b - a)E_n$ representing the error of the integral estimate. By dividing the value of the integral by the interval $(b - a)$, as here, one obtains an expression for the **integral mean** of the integrand $f(x)$ over $[a, b]$ where the weighting factors $\lambda_k^{(n)}$ depend on k and n, but not on the integral range $(b - a)$. Since the formulae must all be precise in the trivial eventuality that $f(x)$ is a constant, it follows that the sum of these weights $\sum_{k=0}^{n} \lambda_k^{(n)}$ must equal unity.

The quadrature rules obtained by this means are called **Cotes formulae** (or sometimes Newton-Cotes formulae), and the weighting coefficients $\lambda_k^{(n)}$ are known as **Cotes numbers**. These are given in Table 5.1, which also includes the values of the error coefficients e_n, where

$$\left. \begin{array}{c} E_n = e_n(b - a)^{2m} f^{(2m)}(\xi)/[(2m)!(2m + 1)!] , \\ m = \mathrm{entier}(n/2) + 1 , \end{array} \right\} \qquad (5.1.12)$$

and

for some ξ in (a, b). This form of error shows that the estimate is precise for any polynomial integrand of degree $\leqslant(2m - 1)$ and the value of $2m$ is called the **order** of the formula. The trapezoidal rule ($n = 1$) has an error depending

Table 5.1

Cotes numbers and error coefficients (closed type formulae).

n	$\lambda_0^{(n)}$	$\lambda_1^{(n)}$	$\lambda_2^{(n)}$	$\lambda_3^{(n)}$	$\lambda_4^{(n)}$	$\lambda_5^{(n)}$	$\lambda_6^{(n)}$	e_n	$2m$
1	1/2	1/2	—	—	—	—	—	−1	2
2	1/6	2/3	1/6	—	—	—	—	−1	4
3	1/8	3/8	3/8	1/8	—	—	—	−4/9	4
4	7/90	16/45	2/15	16/45	7/90	—	—	−15/8	6
5	19/288	25/96	25/144	25/144	25/96	19/288	—	−132/125	6
6	41/840	9/35	9/280	34/105	9/280	9/35	41/840	−28/3	8

on $f''(x)$, and so is of order 2, and is precise (that is $E_1 = 0$) for any linear polynomial integrand $f(x)$. Simpson's rule ($n = 2$), as we have already noted, is precise if $f(x)$ is any cubic polynomial (implying therefore that $f^{iv}(x) \equiv 0$) and so is of order 4. The **three-eighths rule** ($n = 3$) is likewise of order 4, being precise only for integrands which are cubics, but not for those which are quartics or higher degree polynomials. However, both quartic and also quintic integrands are precisely integrated by the sixth-order **Milne's rule** involving fourth degree interpolation ($n = 4$), the 'bonus' extension to polynomials of degree 5 arising from reasons of symmetry – in just the same way as Simpson's rule extends to cubic polynomials, although it is based on quadratic interpolation. Again, the next even-degree interpolation ($n = 6$) provides the eighth-order **Weddle's rule** which is precise for any 7th degree polynomial integrand.

This theoretical advantage of the even-degree formulae, involving an odd number of points of integrand evaluation (including the mid-point of the interval of integration), is one reason why in practice they are preferred: adding just one more point does not increase the order, and seems to produce little increase of accuracy. But one must remember that this could be to some extent illusory, because the values of the derivatives are not known, nor is the position $x = \xi$ at which they are to be evaluated in the expression (5.1.12) for the error.

These formulae are all founded on the use of equi-distant points of evaluation which include the end-points $x_0 = a$ and $x_n = a + nh = b$ of the interval, and for reasons which will in due course appear, this is a particularly convenient method of spacing. The end-points may be excluded, and if this is done one obtains what are known as the Cotes **open-type** formulae (those including the end-points being distinguished as **closed-type**). But the weights are no longer all positive – at least for most values of n – and this does not seem entirely reasonable (although the same objection would be directed at the closed-type formulae of degree $n \geqslant 8$). A more satisfactory equi-distant distribution excluding the end-points is obtained by selecting

$$x_k = a + (2k+1)(b-a)/(2n+2) \qquad (5.1.13)$$

for $n \geqslant 0$ and $k = 0, 1, \ldots, n$. This provides points of evaluation which are interleaved half-way between those used by the corresponding closed-type Cotes formula with the values of n increased by one. Thus the choice $n = 0$ leads to the mid-ordinate rule with its single point of integrand evaluation half-way between the two end-points used in the trapezoidal rule (which is the Cotes formula with $n = 1$). The mid-ordinate rule is of order 2 and exact for linear polynomial integrands, as is the two-point formula (corresponding to $n = 1$); again, the three-point open-type formula ($n = 2$) is of order 4 and exact for cubic as well as quadratic polynomial integrands, just as in Simpson's rule, and for just the same reasons.

Values of the weights and the error coefficients appropriate to this interleaved distribution of points of evaluation (5.1.13) are given in Table 5.2. This

Table 5.2

Weighting and error coefficients for interleaved (open-type) point distribution.

n	$\lambda_0^{(n)}$	$\lambda_1^{(n)}$	$\lambda_2^{(n)}$	$\lambda_3^{(n)}$	$\lambda_4^{(n)}$	$\lambda_5^{(n)}$	e_n	$2m$
0	1	–	–	–	–	–	1/2	2
1	1/2	1/2	–	–	–	–	1/8	2
2	3/8	1/4	3/8	–	–	–	7/18	4
3	13/48	11/48	11/48	13/48	–	–	103/512	4
4	275/1152	25/288	67/192	25/288	275/1152	–	669/500	6
5	247/1280	139/1280	127/640	127/640	139/1280	247/1280	5555/6912	6

shows (for example) that the 3-point formula ($n = 2$) estimates the integral $\int_0^1 f(x)\,dx$ as

$$(3/8)[f(x_0) + f(x_2)] + f(x_1)/4$$

where the interleaved points $\{x_0, x_1, x_2\}$ in this interval of integration are given by $\{1/6, 1/2, 5/6\}$; the error is

$$(7/18)f^{iv}(\xi)/(4!5!) = 7f^{iv}(\xi)/51840 \, ,$$

and this is less than half that of the Cotes 3-point closed-type formula (which is Simpson's rule). Indeed all these open-type formulae have a smaller error than that of the corresponding close-type formula employing the same number of points of evaluation (and so the same number n).

Remembering the Runge phenomenon associated with high degree polynomial interpolation on an equi-spaced support, it will come as no surprise that the extension of the Cotes formulae to higher values of n is not a reliable way of achieving greater accuracy. It will be clear from (5.1.11) that much depends on the behaviour of the high-order derivatives of the integrand, but certainly although the order of the error may increase with n, its magnitude does not necessarily decrease. For instance, to take a pathological example, if Cotes formulae with successively higher values of n (starting with $n = 1$) are applied to the estimation of $(1/10) \int_{-5}^{5}(1 + x^2)^{-1}dx$, which has the true value of $\arctan(5)/5 = 0.27 \ldots$, one obtains the sequence 0.04, 0.68, 0.21, 0.24, 0.23, 0.39, ... The formulae based on an interleaved point distribution do no better, giving the sequence $1, 0.14, 0.31, 0.21, 0.41, 0.29, \ldots$. Just as in curve-drawing, it is more effective to use piecewise-polynomial interpolation, employing a number of different low-degree polynomials, rather than a single polynomial of high degree.

5.1.2 Composite Cotes formulae

If the total interval of integration $[a, b]$ is segmented into N equal parts, and the Cotes formula of degree n is applied to each of these sub-intervals, one obtains the **composite Cotes formula** of degree n which uses $(nN + 1)$ equally-spaced points of evaluation of the integrand, including the common end-points of each interval sub-interval. Writing $h = (b - a)/(Nn)$ for the distance between the points, and putting $x_k = a + kh$, so that $x_{Nn} = b$, the N sub-intervals or **segments** are $[x_{jn}, x_{jn+n}]$ for $j = 0, 1, \ldots, (N - 1)$, and the integral mean is

$$(b - a)^{-1} \int_a^b f(x)\mathrm{d}x = (Nnh)^{-1} \sum_{j=0}^{N-1} \int_{x_{jn}}^{x_{jn}+nh} f(x)\mathrm{d}x \ .$$

Then from (5.1.11) one obtains

$$(b - a)^{-1} \int_a^b f(x)\mathrm{d}x = \sum_{j=0}^{N-1} \sum_{k=0}^{n} (\lambda_k^{(n)}/N) f(x_{jn+k}) + E_{N,n} \ , \tag{5.1.14}$$

and in particular the **composite trapezoidal rule** $(n = 1)$ is

$$(b - a)^{-1} \int_a^b f(x)\mathrm{d}x = N^{-1}(\tfrac{1}{2}f_0 + f_1 + f_2 + \ldots + f_{N-1} + \tfrac{1}{2}f_N) + E_{N,1}, \tag{5.1.15}$$

whilst the **composite Simpson's rule** $(n = 2)$ can be expressed as

$$(b - a)^{-1} \int_a^b f(x)\mathrm{d}x = (6N)^{-1}(f_0 + 4f_1 + 2f_2 + 4f_3 + \ldots + 4f_{2N-1} + f_{2N}) + E_{N,2} \ , \tag{5.1.16}$$

The error of the integral estimate is $(b - a)E_{N,n}$, and for $n = 1$ and $n = 2$ this can be determined by summing the integrals E_T and E_S given by (5.1.3) and (5.1.9) over each segment. Using the mean value theorem for integrals one obtains in general

$$E_{N,n} = (e_n/N^{2m})(b - a)^{2m} f^{(2m)}(\xi)/[(2m)!(2m + 1)!] \tag{5.1.17}$$

for some value of ξ in (a, b). Comparing this with the expression for $E_{1,n} = E_n$ given by (5.1.12), the order $2m$ is related to n as before, and it will be observed that the effect of segmentation is simply to introduce the factor N^{-2m}, although there is also no doubt some obscure effect of N on the value of ξ. However, assuming that the magnitude of $f^{(2m)}(x)$ is bounded for $x \in [a, b]$, it must follow that $E_{N,n}$ tends to zero with increase of N, so that any required accuracy may be achieved by dividing the interval into small enough segments. In other words, provided $f \in C^{(2m)}[a, b]$, increase of N causes the estimate derived from the composite formula to converge to the correct integral value; in fact, convergence is assured even if f is merely Riemann integrable (see problem 2, section 5.1.2).

There is no reason why open-type formulae could not as successfully be used to provide the integral estimate over each segment as the Cotes closed-type formula. The disadvantage is that, because the integrand is not evaluated at the common end-points of interval segments, there are more integrand values required. For instance, if the interleaved distribution of points given by (5.1.13) is used in each segment, $nN + N$ equi-spaced integrand evaluations are required (instead of $nN+1$ for the closed-type formulae). However, the **composite mid-ordinate rule:**

$$(b-a)^{-1}\int_a^b f(x)\,\mathrm{d}x = N^{-1}(f_{1/2} + f_{3/2} + \ldots + f_{N-1/2}) + E_N$$
(5.1.18)

where $f_v = f[a + v(b-a)/N]$, is more efficient than the composite trapezoidal rule. Its error $(b-a)E_N$ can be determined from (5.1.6) to be given by

$$E_N = (b-a)^2 f''(\xi)/(24N^2) ,$$

whereas from (5.1.17) the error of the composite trapezoidal rule $(b-a)E_{N,1}$ is (-2) times the same expression; since the number of required integrand evaluations is seen to be N in (5.2.18) as against $(N+1)$ in (5.1.15), the composite mid-ordinate rule is clearly the 'better' of the two.

5.1.3 Romberg's method

Let $I_T(N)$ and $I_M(N)$ denote the estimates of the integral mean obtained by the use of N equal segments of (respectively) the composite trapezoidal and mid-ordinate rule, as given by (5.1.14) and (5.1.17). Then the union of the $(N+1)$ end-points, and N mid-points, at which the integrand is evaluated in these two estimates together, make up a set of $(2N+1)$ equally-spaced points which are the end-points of $2N$ equal segments into which the interval of integration may be divided. These $(2N+1)$ points are those at which the integrand is evaluated in forming the estimate $I_T(2N)$, and it follows that

$$I_T(2N) = \frac{1}{2}[I_T(N) + I_M(N)] .$$
(5.1.19)

Again, if $I_S(N)$ denotes the integral mean value estimated by the use of N segments of the composite Simpson's rule (5.1.16), then just as in (5.1.8), it can be observed that $I_S(N)$ is a weighted average $[I_T(N) + 2I_M(N)]/3$ of the composite trapezoidal and mid-ordinate rules, or using (5.1.19),

$$I_S(N) = [4I_T(2N) - I_T(N)]/3 .$$
(5.1.20)

It is already known that this is a fourth-order estimate of the true integral mean (I, say), with error varying with N roughly as $1/N^4$; this fact could be established quite independently if it is assumed that the error of the estimate $I_T(N)$ can be expressed as

$$E_{N,1} = I - I_T(N) = a_1 N^{-2} - a_2 N^{-4} + a_3 N^{-6} - + \ldots, \quad (5.1.21)$$

where $a_1, a_2, \ldots$ are independent of N. Such an assumption is consistent with our knowledge that $I_T(N)$ is a second-order estimate of I, with error varying roughly as $1/N^2$. Consequently the error of $I_T(2N)$ will be (roughly) a quarter that of $I_T(N)$, or — more precisely — the combination $4I_T(2N) - I_T(N)$

appearing in (5.1.20) eliminates the dominant term involving a_1 in the expression for the error. In fact, from (5.1.21) in (5.1.20), the error of $I_S(N)$ is given by

$$E_{N,2} = I - I_S(N) = (a_2/4N^4) - (5a_3/16N^6) + (21a_4/64N^8) - + \dots ,$$

showing that $I_S(N)$ is a fourth-order estimate.

The significance of this result is not so much in determining the order of the error of the composite Simpson Rule — this was known already — as in indicating a process by which further estimates of still higher orders of accuracy can be obtained. Thus the error $I_S(2N)$ will be (roughly) a sixteenth of that of $I_S(N)$, and the combination $[16I_S(2N) - I_S(N)]/15$ provides an estimate of I in which the dominant term involving a_2 is eliminated from the error. In fact this particular combination is equivalent to applying Milne's Rule (the Cotes formula with $n = 4$) to N equal segments of the interval of integration: it does indeed provide an estimate of order 6, with error varying roughly as $1/N^6$. In fact, one can determine that

$$E_{N,3} = I - [16I_S(2N) - I_S(N)]/15$$
$$= (a_3/64N^6) - (21a_4/1024N^8) + - \dots ,$$

and the same process can now continue to provide estimates of order 8, 10, and so on.

A simple numerical arrangement by which these higher-order estimates can be calculated consists in setting up a triangular array of numbers I_{jk}, with $k \leqslant j$, which are formed by the recurrence relation

$$I_{j,1} = I_T(2^{j-1}), \quad I_{j,k+1} = (4^k I_{j,k} - I_{j-1,k})/(4^k - 1) , \quad (5.1.22)$$

for $k = 1, 2, \dots (j-1)$ and $j = 1, 2, \dots$. The entries in column 1 are clearly the successive estimates $I_T(1)$, $I_T(2)$, $I_T(4)$, ... obtained from the composite trapezoidal rule by successively doubling the number of (equal) segments into which the range of integration is partitioned. In column 2, the value in the jth row is that of $[4I_T(2^{j-1}) - I_T(2^{j-2})]/3$, and (5.1.20) shows that this is $I_S(2^{j-2})$; thus this column contains the estimates of the 4th-order composite Simpson's rule $I_S(1)$, $I_S(2)$, $I_S(4)$, ... Similarly, column 3 contains estimates by the 6th-order composite Milne's rule using 1, 2, 4, ... segments, and in general column k contains estimates for 1, 2, 4, ... segments based on a composite rule of order $2k$ involving $(2^{k-1} + 1)$ integrand applications in each segment. (From column 4 onwards, these implied 'rules' are not particular examples of Cotes formulae, since these latter contrive to achieve error of order $(2^{k-1} + 2)$ from $(2^{k-1} + 1)$ points of integrand evaluation, if $k > 1$). The array is usually composed row-by-row, and the jth row starts with the value of $I_{j,1} = I_T(2^{j-1})$ and proceeds through estimates of successive order 4, 6, 8, ... (based on composite $3, 5, 9, \ldots$ point 'rules') to finish with $I_{j,j}$, which represents the estimate of the integral mean obtained from a $(2^{j-1}+1)$-point rule of order $2j$ applied to the whole interval treated as a single segment.

Rather than starting with $I_T(2^{j-1})$, it is easier in fact to apply (5.1.19) and compose the first column from the recurrence

$$I_{1,1} = I_T(1), \quad I_{j,1} = \tfrac{1}{2}[I_M(2^{j-2}) + I_{j-1,1}] \ , \ldots \qquad (5.1.22a)$$

using the composite mid-ordinate rule. The addition of the jth row then involves 2^{j-2} integrand evaluations (if $j > 1$), at points not previously involved, whereas to calculate $I_T(2^{j-1})$ directly would involve more than twice that number of evaluations. It will also be noted that the array values can as easily be taken to be estimates of the integral, instead of the integral mean, since the former is merely $(b - a)$ times the latter, and of course the interval of integration $[a, b]$ is fixed.

This process is called the **Romberg method** of quadrature, and in practice the two important questions concern whether the process converges, and if so, how it may be judged to have provided an estimate of sufficient numerical accuracy. It has already been noted that increase in the number of segments ensures convergence of the composite trapezoidal and Simpson rules provided that, respectively, the second and fourth derivatives of the integrand are of bounded variation in (a, b). It is reasonable to anticipate therefore that the success of the method will depend on an extension of such conditions to higher derivatives. However, in practice, even if the integrand is infinitely differentiable at all points of (a, b), and even although the columns of values of $I_{j,1}, I_{j,2}, \ldots$ will then ultimately converge to the appropriate limit value of the integral mean, this does not necessarily imply that the estimates in later columns (having higher-order accuracy) will converge more rapidly, as presumably is the point of the exercise.

The 'difficult' example, cited in the previous section, of evaluating the integral mean $(1/10)\int_{-5}^{5}(1 + x^2)^{-1}dx$ will serve to illustrate the point. The integrand is evidently analytic over the closed interval $[a, b]$, and the array $I_{j,k}$ can be evaluated as

0.038					
0.519	0.679				
0.329	0.265	0.237			
0.278	0.262	0.262	0.262		
0.275	0.273	0.274	0.274	0.274	
0.275	0.275	0.275	0.275	0.275	0.275

The correct value is $0.27468 \ldots$, and if convergence were to be judged by values in successive columns along a row, then the process might well be terminated with the erroneous estimate 0.262 in row 4. In fact, convergence is usually judged to have occurred when successive entries in any *column* differ by less than some prescribed margin. On this basis it is evident that the trapezoidal rule converges more rapidly than any of the higher-order formulae, and the generation of the array is a redundancy.

On the other hand, evaluating the integral $\int_{-0.5}^{0.5}(1+x^2)^{-1}dx = 2\arctan(\frac{1}{2}) = 0.92730 \dots$, of the same integrand over a smaller interval, yields the $I_{j,k}$ array

0.800

0.900   0.933

0.921   0.927   0.927

0.926   0.927   0.927   0.927

Here the higher-order estimates converge (within the precision shown) more rapidly than the trapezoidal rule. Comparing successive values in column 2 or 3, one could terminate the process at a stage when two further doublings of the number of segments would be needed to observe the same convergence in the composite trapezoidal rule. In other words, the work (as judged by the number of integrand evaluations) has been reduced by a factor of about 4, and the Romberg method is obviously (in this context) extremely effective.

The fact that the integrand should be analytic over the range of integration for the method to be of benefit is a generally necessary condition, but the first example will make it clear that it is not a sufficient condition, and the second example may suggest that it is likely to be difficult to find a sufficient condition. To take this matter any further it would be necessary to look in more detail at the expression (5.1.21) for the error $E_{n,1}$ of the composite trapezoidal rule. In fact it can be shown that if $f(x) \in C^{2m}(a,b)$, then

$$
\begin{aligned}
E_{N,1} &= I - I_T(N) \\
&= \sum_{k=1}^{m-1}(-1)^k B_k (b-a)^{2k-1}[f^{(2k-1)}(b) - f^{(2k-1)}(a)]/[N^{2k}(2k)!] \\
&\quad + (-1)^m B_m (b-a)^{2m} f^{(2m)}(\xi)/[N^{2m}(2m)!] \quad (5.1.23)
\end{aligned}
$$

for some ξ in (a, b). This is the **Euler-Maclaurin summation formula**, and $B_1, B_2, \dots$ are Bernoulli numbers (see section 2.5 and section 5.1.2, problem 1). For present purposes, it suffices to say that although this formula allows certain conclusions to be reached about the convergence of the Romberg method, it is difficult to translate these into 'rules of thumb' by which one can easily recognise that use of the method is likely to be of benefit. The method, as described, cannot therefore be unreservedly recommended. Without affecting that judgement, it can nonetheless be improved by one or two modifications. The purpose of these will be better appreciated if the mechanism of Romberg quadrature is restated in a different way.

Given that $I_T(N)$ is some function of $1/N^2$ and that one wants to extrapolate its limiting value (equal to I) at $1/N^2 = 0$, the Romberg method causes polynomials in $1/N^2$ of progressively increasing degree to be fitted to the accumulating sequence of values of $I_T(N)$ for $1/N^2 = 1, 1/4, 1/16, 1/64, \ldots$, and evaluates these polynomials for $1/N^2 = 0$ in order to predict the limit. This is accomplished by using the Neville algorithm (3.2.9), and in fact the triangular array $I_{j,k}$ is simply the tabulation associated with that algorithm. Incidentally, the process of finding a limiting value of, say, $y(z)$ at $z = 0$ by fitting a polynomial to the y-values at a decreasing sequence of z's (usually reducing in geometric progression as here) is called a 'deferred approach to a limit', or **Richardson extrapolation**. The doubling of the number of segments (and the quartering of $1/N^2$) in the progression adopted by the Romberg method allows the $(2N + 1)$ points of integrand evaluation to incorporate those for the previous N segments – as implied by (5.1.22a) – so that none of the evaluations is 'wasted'. Such a sequence which generates a new set of points by augmenting the previous generation is called **common-point**.

Practical experience suggests that this progression is too rapid: each additional step of the Romberg method (adding a new row of $I_{j,k}$ values) virtually doubles the computational work. It is found generally more efficient to use two interlaced common-point sequences such as $N = 2, 3, 4, 6, 8, 12, \ldots$ so that more information is extracted from the less accurate but more quickly computed estimates. This particular sequence was originally proposed by Bulirsch and is used in **Bulirsch and Stoer's method** which also replaces the polynomial extrapolation of Romberg quadrature by continued fraction extrapolation, using the sequence of (3.5.5). Each of these modifications produces some improvement of the Romberg method in relation to what one might call 'well-behaved' integrands (Fig. 5.3). However, they do little to improve on the basic method in

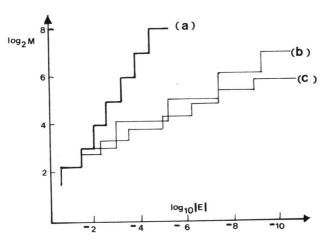

Fig. 5.3 – Dependence of discernable error E (the difference between successive estimates by the same rule) against cumulative total of integrand evaluations (M) involved in the estimation of $\int_{-1}^{+1}(1 + x^2)^{-1}\,\mathrm{d}x$ by (a) the trapezoidal rule, (b) the Romberg method, and (c) its modification by the method of Bulirsch and Stoer.

those contexts where it fails to accelerate the convergence of the trapezoidal rule estimates.

5.1.4 Integrals of periodic functions

One particular form of definite integral presents no problem in the convergence of its numerical approximation, and this is the integral of a periodic analytic function over a range equal to its period. There is no better quadrature formula for its evaluation than the composite mid-point or composite trapezoidal rule — neither being distinguishable in this context from the other. If the period of the function $f(x)$ is α (say), so that $f(x) = f(x + \alpha)$, and one places $x_k = x_0 + k\alpha/n$ and $f_k = f(x_k)$, then $f_n = f_0$ and the composite trapezoidal rule gives

$$I = (1/\alpha) \int_{x_0}^{x_0 + \alpha} f(x) \mathrm{d}x = (1/n) \sum_{k=1}^{n} f_k + \epsilon_n . \qquad (5.1.24)$$

It is clear that all the weighting coefficients are equal, and there could be no other reasonable choice. For suppose a set of n unequal coefficients $l_1, l_2, \ldots, l_n$ minimises the error in some sense, so that the integral mean I is best estimated as $\sum_{k=1}^{n} l_k f_k$. The actual integral mean would be unchanged if the upper and lower limits of integration were both shifted by an increment $j\alpha/n$, say, for some fixed but arbitrary non-zero integer j; and applying the 'optimum' rule to this integral mean would produce the estimate

$$\sum_{k=1}^{n} l_k f(x_k + j\alpha/n) = \sum_{k=1}^{n} l_k f_{k+j} = \sum_{k=1}^{n} l_{k-j} f_k$$

where we have defined $f_{k+n} = f_k$ and similarly $l_{k+n} = l_k$. Because this also minimises the error, it must equal the other estimate, and since the integrand values are in general linearly independent, it must follow that $l_{k-j} = l_k$, for arbitrary j — if indeed any 'optimum' choice really exists.

Looked at from another standpoint, the Euler-Maclaurin series (5.1.24) for the error of the composite trapezoidal rule in this context is composed entirely of zero terms, because, provided $f(x)$ is everywhere infinitely differentiable, all derivatives of $f(x)$ are equal at both ends of the range of integration. This does not, however, mean that the composite trapezoidal rule precisely evaluates the integral : obviously the number of segments ($N = n$) will affect the estimate. Rather it implies that the error decreases with n more rapidly than any positive inverse power of n, or in other words, in (5.1.24), the limit of $n^m \epsilon_n$ as $n \to \infty$ is zero for any $m > 0$. It would be a reasonable expectation that ϵ_n may decrease exponentially with some (positive) power of n, so that doubling the number of segments would result in an error ϵ_{2n} proportional to ϵ_n^r where $r > 1$. Thus an iterative process of successively approximating I by doubling the number of segments at each iteration (as in the Romberg method) would be expected to display super-linear convergence. It will be clear in this context that the composite trapezoidal rule no longer has a 'second-order' error, but rather an error of 'infinite order', and no 'higher order' approximations can exist.

For example, consider the integral of $[1 + \cos^2(x)]^{-1}$ over any range with a span equal to π. Taking say $x_0 = 0$ – the choice is not critical – and $x_k = k\pi/n$, one obtains estimates of 0.5, 0.75, 0.7083333, 0.7071078 from (5.1.24) with $n = 1, 2, 4$, and 8, whereas the correct value is $1/\sqrt{2} = 0/7071068$; accordingly the errors are (roughly) 0.2, 0.04, 0.0012, 0.0000011, perhaps suggesting quadratic convergence. An attempt to accelerate the convergence of this sequence, by (for instance) applying the Romberg method, would be a complete waste of time as the rate of convergence cannot be improved (at least by that method). In fact, the 'higher order' methods, such as Simpson's and Milne's rule, would produce less accurate estimates based on the same set of integrand evaluations. In this context, the Romberg method therefore fails because the convergence of the composite trapezoidal rule estimates is too good, rather than not good enough.

5.1.5 Gregory's formulae
It is often objected that the unequal weighting of the even and odd sub-scripted integrand values in the composite Simpson's rule (5.1.16) is somehow 'inappropriate'. This most often happens, perhaps, when these values are data obtained from some experiment, and there seems no reason why a particular subset of the data should 'count' more than the rest. In particular, if the total number of such data is even, so that the trapezoidal or three-eights rule needs to be applied to one segment, the subset of data which is more heavily weighted can be changed, and again one asks why one arrangement is better than another.

Associated with the changed weights, there will of course be a changed estimate of the integral. Although the difference can be ascribed to the ambiguity of attempting to 'define' a continuous integrand from discrete data (or if that integrand is a defined function, to the inaccuracy of the quadrature rule), this might suggest that an average of the two different estimates is equally acceptable. For instance, putting the trapezoidal rule at one end or the other, and placing h equal to the distance $(x_{k+1} - x_k)$ between successive points, the integral would be estimated as either

$$(h/2)(y_0 + y_1) + (h/3)(y_1 + 4y_2 + 2y_3 + \dots 4y_{n-1} + y_n)$$

or

$$(h/3)(y_0 + 4y_1 + 2y_2 + \dots) + (h/2)(y_{n-1} + y_n) \, ,$$

and the average would then be the composite trapezoidal rule with two 'end-corrections', $(h/12)(y_1 - y_0)$ at one end and a similar difference at the other. This, as we shall now try to show, is the basis of a good idea, and the composite trapezoidal rule of course avoids (except at the ends) the 'objectionable' unequal weighting of the integrand values.

In relation to Simpson's rule, this unequal weighting arises because, in effect, the rule is derived by interpolating the integrand in successive segments by different quadratic polynomials which have the value of the integrand at the mid- and end-points of each segment. The resulting piecewise quadratic is in general discontinuous in its first and second derivatives at every internal common end-point x_2, x_4, ... where the segments meet. The integrand is therefore rather poorly approximated in the neighbourhood of these points, but is represented by a continuous (quadratic) polynomial in the neighbourhood of the segment mid-points, where the accuracy of interpolation is accordingly better. In this sense the integrand values at the segment mid-points 'deserve' more emphasis, and the segment end-points less. Of course, this is hardly a rigorous argument, but it is undeniable that the disparity in weighting arises only as a consequence of the method of implied interpolation: it is something imposed by choice. If one wanted to refine this line of reasoning, it could be recalled that Simpson's rule remains exact for cubic pieces, and the discontinuity in slope at each segment end-point could be eliminated by suitable cubic increments, so that the implied interpolation is any one of an infinite family of piecewise cubics belonging to $C^1[a, b]$; accordingly, it is only the second-derivative which need be discontinuous at $x_2, x_4, \ldots$. Nonetheless, these points are still dealt with differently from the mid-points $x_1, x_3, \ldots$.

It is perhaps remarkable that a method of piecewise polynomial interpolation thought appropriate for the approximation of the integrand is not a method which appealed as being reasonable enough to warrant mention when the problem of curve-drawing was under consideration. There is of course no reason why what is appropriate in the one context should not be as appropriate in the other. The idea of piecewise cubic interpolation is common to both contexts, but in curve-drawing, each and every support point is taken as a break-point between the cubic pieces, and not every other point. Indeed, it does seem quite arbitrary to distinguish a particular subset of the abscissae as break-points, unless something is known about the local behaviour of the function being interpolated to suggest such a choice.

The integral of a cubic piece $\pi_k(x)$ over any sub-interval $[x_{k-1}, x_k]$ can be expressed as

$$\int_{x_{k-1}}^{x_k} \pi_k(x)\mathrm{d}x = \tfrac{1}{2}h_k\,[\pi_k(x_{k-1}) + \pi_k(x_k)] - (h_k^2/12)[\pi_k'(x_k) - \pi_k'(x_{k-1})]\,, \tag{5.1.25}$$

where $h_k = x_k - x_{k-1}$. Envisaging that it is desired to integrate a piecewise-cubic interpolate $y(x)$ as given by (3.3.2) on an equispaced support (with $h_k = h$, a constant) from $a = x_0$ to $b = x_n$, and that this interpolating function belongs to $C^1[a, b]$, so that (3.3.1) applies, then it follows from (5.1.25) that

$$\int_a^b y(x)\mathrm{d}x = \sum_{k=1}^{n} \int_{x_{k-1}}^{x_k} \pi_k(x)\mathrm{d}x = h(\tfrac{1}{2}y_0 + y_1 + \ldots + y_{n-1} + \tfrac{1}{2}y_n)$$

$$- (h^2/12)\,(y_n' - y_0')\,, \tag{5.1.26}$$

Suppose now that $y(x)$ interpolates the integrand $f(x)$, and that the derivatives y_0' and y_n' at the end-points are made equal to $f'(x_0)$ and $f'(x_n)$. Then the integral of this interpolate in (5.1.26) provides an estimate of the integral mean of $f(x)$ over $[a, b]$ which is equivalent to the composite trapezoidal rule of n equal segments, together with an end-correction involving the end-point derivatives $f'(x_0)$, $f'(x_n)$, which in fact is identical with the first term of the Euler-Maclaurin summation formula (5.1.23). Placing $N = n$ and $m = 2$ in this formula, we see that the integration of the interpolating piecewise cubic leads to the estimate

$$(b - a)^{-1} \int_a^b f(x)\mathrm{d}x = (1/n)(\tfrac{1}{2}f_0 + f_1 + \ldots + f_{n-1} + \tfrac{1}{2}f_n)$$

$$- [(b - a)/12n^2](f_n' - f_0') + (b - a)^4 f^{\mathrm{iv}}(\xi)/(720n^4)$$

$$(5.1.27)$$

for some ξ in (a, b). The end-correction changes the error of the composite trapezoidal rule from second-order to fourth-order. Putting $n = 2N$, so that the number of integrand evaluations (equal to $2N + 1$) is the same as that of the composite Simpson rule for N segments, the error of the integral mean given by (5.1.27) is seen to equal $E_{N,2}/4$ — that is, a quarter of that for the composite Simpson's rule.

One could note in passing that the addition of further terms from the expression for the error $E_{N,1}$ in the Euler-Maclaurin formula should result in still further improvements in the estimate of the integral provided by the composite trapezoidal rule. However, even though the derivatives in this expression might be readily formulated and evaluated, what is being sought is some *numerical* interpretation of the end-correction in terms of integrand values, as is achieved (in effect) by the Romberg method. In particular, of course, if the integrand is defined only by discrete data, it cannot be differentiated other than by numerical operation.

The same need to interpret the end-derivatives in terms of the data arises (as we have seen) in piecewise cubic interpolation. In Bessel cubic interpolation, for instance, it is suggested that the end-pieces, $\pi_1(x)$ and $\pi_n(x)$, might be chosen as the arcs of quadratics interpolating respectively the first and last three data: if the support is equispaced, this would imply that the values of the end-derivatives (obtained by differentiating these quadratics) are given by

$$\begin{aligned} hy_0' &= (\Delta - \tfrac{1}{2}\Delta^2)y_0 = 2y_1 - \tfrac{1}{2}(3y_0 + y_2) \\ hy_n' &= (\nabla + \tfrac{1}{2}\nabla^2)y_n = -2y_{n-1} + \tfrac{1}{2}(3y_n + y_{n-2}) \end{aligned} \Bigg\} \quad (5.1.28)$$

Noting that the Bessel interpolate belongs to the class $C^1[a, b]$, one can make these substitutions in (5.1.26). This provides a value of the integral of the Bessel piecewise-cubic interpolating the integrand at $(n + 1)$ equi-spaced points.

An expression can also be determined for the error which is involved in the use of this value to estimate the required integral. In this way, one obtains the quadrature formula:

$$(b-a)^{-1}\int_a^b f(x)\mathrm{d}x = (1/n) \{ \tfrac{1}{2}y_0 + y_1 + y_s + \ldots + y_{n-1} + \tfrac{1}{2}y_n$$

$$- (\nabla y_n - \Delta y_0)/12 - (^2y_n + \Delta^2 y_0)/24\}$$

$$- 19(b-a)^4 f^{\mathrm{iv}}(\xi)/(720 n^4)$$

$$(5.1.29)$$

for some ξ in (a, b). Although this is a fourth-order formula, the error is 19 times that given by (5.1.27) and (roughly) 5 times that of the composite Simpson rule with an equal number of segments. Nonetheless, it is founded on an interpolative scheme which would be favoured for the representation of an integrand defined only by discrete data.

When it is applied to the integration of an analytic function, clearly the deficiency of (5.1.29) must arise from the rather crude estimate of the end-derivatives. In principle, this can be remedied by estimating these from higher degree polynomials interpolating more of the points at each end of the interval of integration. Moreover, numerical estimates of the higher order derivatives in the Euler-Maclaurin expansion can also be included. By this means one obtains an estimate

$$(b-a)\int_a^b f(x)\mathrm{d}x = I_T(n) - (\nabla y_n - \Delta y_0)/(12n) - (\nabla^2 y_n + \Delta^2 y_0)/(24n)$$

$$- 19(\nabla^3 y_n - \Delta^3 y_0)/(720n)$$

$$- 3(\nabla^4 y_n + \Delta^4 y_0)/(160n)$$

$$- \ldots \qquad\qquad (5.1.30)$$

which is called **Gregory's formula**. Truncating this expansion at the fourth-differences shown, the error is of sixth order (that is, it varies as $1/n^6$). Truncated at the second-differences, we recover the formula derived on the basis of Bessel interpolation in (5.1.29).

Both these formulae meet the original objections to the unequal weighting of the generality of points which was raised at the outset of this section; what unequal weighting there is arises as an attempt to deal better with the one-sided character of the information about the integrand at the ends of the interval – which seems reasonable enough, as obviously the end-points *are* exceptional. Nonetheless, Gregory's formula is rarely, if ever, used in practical computation. If higher-order differences are included in an endeavour to increase accuracy, the error term is many times larger than that of the Cotes formula of the same order, and as with those formulae, increased numerical accuracy does not necessarily result from increased order (although it will result from increase in the number of segments, n). However, its truncated form in (5.1.29) can be recommended if the integrand is defined only by discrete data – in as much as Bessel piecewise cubic interpolation is an acceptable method of curve-drawing, judged (as it can then only be) by 'eye-appeal'; indeed perhaps the utility of (5.1.29) in this context is too often overlooked.

A piecewise cubic interpolate with still greater 'eye-appeal' is of course the cubic spline, and since this is continuously differentiable over $[a, b]$, its integral is likewise given by (5.1.26). Some computational complexity arises because the values of the end-derivatives depend upon all the integrand values as well as on the end-condition assumed for the spline. Thus although this may give the best estimate give for a data-defined integrand, the simplicity of mainly equal weighting is sacrificed. The linear system for the cubic spline can be set up directly in terms of the derivative values y'_k (see problem 3, section 3.4) and the appropriate end-correction to the composite rule, involving $(y'_n - y'_0)$ obtained from its solution. If n is sufficiently large, and quadratic end-pieces are assumed, then one finds very roughly that, in terms of shifted forward differences $\Delta y_k = y_{k+1} - y_k$,

$$hy'_0 = 1.63\Delta y_0 - 0.80\Delta y_1 + 0.22\Delta y_2 - 0.06\Delta y_3 + \dots ,$$

with a similar expression for hy'_n in terms of backward differences. Alternatively, assuming the 'not-a-knot' end-condition for the spline,

$$hy'_0 = 1.92\Delta y_0 - 1.46\Delta y_1 + 0.68\Delta y_2 - 0.18\Delta y_3 + 0.05\Delta y_4 - \dots.$$

In both instances the coefficients of further successively shifted differences decrease in roughly geometric progression by a factor of about a quarter. Consequently, the more remote differences have little effect upon the estimate of the end-derivative. These expressions can be compared with (5.1.28) for Bessel interpolation, which with the same notation gives $hy'_0 = 1.5\Delta y_0 - 0.5\Delta y_1$ and $hy'_n = 1.5\nabla y_n - 0.5\nabla y_{n-1}$.

To turn the argument back full circle, they could also be compared with the value

$$h(y'_0 - y'_n) = 2(\Delta y_0 - \nabla y_n) - 2(\Delta y_1 - \nabla y_{n-1}) + 2(\Delta y_2 - \nabla y_{n-2}) - + \dots.$$

necessary to convert (5.1.26) to the composite Simpson's rule (for even integers n). This takes all data (in effect) to be equally influential in determining the end-derivatives, rather than merely those close to the ends, and implies 'effect at a distance' which one seeks to avoid in piecewise polynomial interpolation.

5.2 QUADRATURE INVOLVING UNEQUALLY-SPACED INTEGRAND EVALUATIONS

Enough will have emerged from the comments about the quadrature rules based on sets of equispaced integrand sampling points to convey their limitations when regarded as methods of accurate and reliable estimation of integrals of prescribed functions. Of course, in relation to discrete data consisting of equi-spaced integrand values without functional representation, their use is clearly essential and unexceptionable. The situation is entirely paralleled by the differing approach to polynomial interpolation regarded as either a method of function approximation, or as a method of curve-drawing through discrete data: and an equispaced support is known to be far from ideal for interpolation if function approximation is the intent. Considerable increase in accuracy of quadrature can be anticipated if the points of integrand evaluation are able to be chosen with this aim in view.

For ease of description it will be supposed that the interval of integration is standardised as $[-1, 1]$. This involves no restriction, as any other interval may be mapped on to that by a simple change of the dummy variable of integration; thus substituting $t = a + \frac{1}{2}(x + 1)(b - a)$,

$$(b - a)^{-1} \int_a^b g(t) \mathrm{d}t = \frac{1}{2} \int_{-1}^{+1} f(x) \mathrm{d}x ,$$

$$\text{with} \quad f(x) \equiv g[a + \frac{1}{2}(x + 1)(b - a)] \tag{5.2.1}$$

In other words the integral mean of $g(t)$ over $[a, b]$ is the same as that of $f(x)$ over $[-1, 1]$, where $f(x)$ and $g(t)$ have the same values at the same fractional positions within their respective intervals of definition. Note in particular that the mth derivative of $f(x)$ at $x = \xi$ is

$$f^{(m)}(\xi) = 2^{-m}(b - a)^m g^{(m)}[a + \frac{1}{2}(\xi + 1)(b - a)] , \tag{5.2.2}$$

and a quadrature rule for the integral mean of f which has an error proportional to this term is said to be of mth order; it could also be said to be of $(m - 1)$th degree, since it would precisely evaluate the integral of any polynomial of degree $\leqslant (m - 1)$. If such a quadrature rule were applied to N equal subdivisions of $[a, b]$ then, as was seen in section 5.1.2, the error of the composite rule would be proportional to N^{-m}.

5.2.1 Gaussian quadrature rules

It is possible to choose a specified number of points of integrand evaluation so as to render the order of the error as large as possible. Let these points be $x_0, x_1, \ldots, x_n$ as before, with n supposed fixed, and if we suppose as in (5.1.11) that we express the integral mean as

$$\frac{1}{2} \int_{-1}^{+1} f(x) \mathrm{d}x = \sum_{k=0}^{n} \lambda_k f(x_k) + E_G^{(n)}, \tag{5.2.3}$$

then we require both the coefficients λ_k and the points x_k to be chosen for $k = 0, 1, \ldots, n$ in such a way that the error $E_G^{(n)}$ is zero if $f(x) \equiv x^j$ for the largest possible range of values of $j = 0, 1, \ldots$ up to $(m - 1)$. Since integration and summation are linear operations, and successive powers of x form a basis of a polynomial, this definition of m could be paraphrased as that integer value such that, if $p_k(x)$ is a polynomial of genuine degree k, then $f(x) \equiv p_{m-1}(x)$ implies zero error, whereas $f(x) \equiv p_m(x)$ implies non-zero error.

We note first of all that, as observed before, the weights must all sum to unity if $f(x) \equiv x^0 = 1$ is to be correctly evaluated. Further, if $f(x)$ is any positive odd integer power of x (that is x, x^3, x^5, ...) then its integral over $[-1, 1]$ is zero, and the value of $E_G^{(n)}$ will be zero provided the points x_0, x_1, ..., x_n and weights are symmetrically distributed about $x = 0$. More precisely, this requires that

$$\lambda_k = \lambda_{n-k}, \quad x_k = -x_{n-k}, \quad \text{for } 0 \leqslant k \leqslant n/2 , \tag{5.2.4}$$

and in particular if n is even, then $x_{n/2} = 0$ is the mid-point of the interval.

Given just one point of integrand evaluation ($n = 0$), the order of error is highest (equal to 2) for the mid-ordinate rule ($\lambda_0 = 1$, $x_0 = 0$), since no other choice of x_0 would allow $f(x) \equiv x$ to be evaluated correctly. With two points the highest achievable order of error is 4: symmetry ensures that x or x^3 would be correctly integrated provided $x_0 = -x_1$ and $\lambda_0 = \lambda_1$, and $f(x) \equiv x^0$ is correctly integrated if $\lambda_0 = \lambda_1 = 1/2$. It remains to arrange that $x_0 = -x_1$ is chosen so that $f(x) \equiv x^2$ implies zero error, and from (5.2.3) therefore it follows that

$$\tfrac{1}{2} \int_{-1}^{+1} x^2 \, dx = 1/3 = \tfrac{1}{2}(x_0^2 + x_1^2) = x_0^2 .$$

Supposing $x_0 < x_1$, therefore $x_0 = -1/\sqrt{3}$, $x_1 = 1/\sqrt{3}$.

With 3 points, a similar line of reasoning shows that 6th-order error is achieved if the mid-point $x_1 = 0$ is weighted by $\lambda_1 = 4/9$, while the other two points are selected at $x_2 = -x_0 = -\sqrt{3/5}$ and weighted by $\lambda_2 = \lambda_0 = 5/18$. In general, an $(n + 1)$ point formula can be arranged to achieve an error of order $2n + 2$ which may be shown to be expressible as

$$E_G^{(n)} = 4^{n+1} [(n + 1)!]^4 f^{(2n+2)}(\xi) / \{(2n + 3) [(2n + 2)!]^3\} \tag{5.2.5}$$

for some $|\xi| < 1$. The weights λ_k (which are all positive) are in this context called **Christoffel numbers**, and the formulae are collectively referred to as **Gauss-Legendre** quadrature rules. The association with Legendre is due to the fact that the optimal $(n + 1)$-points of evaluation can be shown to be the zeros of the polynomial $P_{n+1}(x)$, which is one of a sequence of so-called **Legendre polynomials** of ascending degree, defined recursively by

$$(n + 1)P_{n+1}(x) = (2n + 1)xP_n(x) - nP_{n-1}(x) , \tag{5.2.6}$$

given that $P_0(x) \equiv 1$ and $P_1(x) \equiv x$. However, it is not the intention for the moment to follow up the reasons for this connection, which are elucidated in a rich area of analysis relating to the properties of sequences of **orthogonal polynomials** (see problem 2 of section 5.2.2).

The zeros of the Legendre polynomials, and so the points of integrand evaluation x_0, x_1, ..., x_n, all lie within the open interval $(-1, 1)$, so that the

Gauss-Legendre formulae are of open-type. If the condition that $x_0 = -1$, $x = 1$ is imposed, and the remaining $(n-1)$-points of evaluation are then chosen to provide the largest possible order of error, one obtains a closed-type family of quadrature rules, known as the **Lobatto formulae.** The $(n+1)$-point member of this family has error order $2n$ (as if, therefore, compared with the Gaussian formulae, the end-points only count as 'half-a-point'), and the optimal points of evaluation are the $(n+1)$ zeros of $(x^2 - 1)P_n'(x)$. These $(n+1)$-points therefore interlace those of the n-point Gaussian formula having the same order of error. The 2- and 3-point Lobatto formulae are simply the trapezoidal and Simpson's rules. The 4-point formula has $x_3 = -x_0 = 1$, $x_2 = -x_1 = 1/\sqrt{5}$, and $\lambda_0 = \lambda_3 = 1/12$, $\lambda_1 = \lambda_2 = 5/12$: note the relatively low weighting given to the end-points.

It has already been remarked that the use of equally-weighted integrand evaluations is regarded as particularly appropriate where the integrand data are got from physical experiment or some other form of observation. Thus some interest attaches to **Tchebyshev's formulae** which are based on an optimal selection of $(n+1)$-points of integrand evaluation, assuming that the weights are all the same and therefore equal to $1/(n+1)$. The 1- and 2-point formulae of this family are the same as the Gaussian quadrature rules. For 3-points (for example) the appropriate selection is $x_2 = -x_0 = 1/\sqrt{2}$ and $x_1 = 0$, but the order of error is still equal to 4 (as for the 2-point formula). Other values are quoted in Table 5.3: no unique optimal solution exists for $n = 7$ or for $n > 8$ (which may seem rather curious). Where small sets of data are collected for the express purpose of providing a single integrated mean, such distributions (or something approximating to them) would commend themselves to the attention.

The major disadvantage in practical computation of any of these formulae is that they do not provide common-point sequences of points of integrand

Table 5.3
Points of integrand evaluation of Tchebyshev quadrature formulae

n	$x_k, k = 0, 1, \ldots n$	Order
0	1	2
1	$\pm 1/\sqrt{3}$	4
2	$\pm 1/\sqrt{2}, 0$	4
3	$\pm 0.7947, \pm 0.1876$	6
4	$\pm 0.8325, \pm 0.3745, 0$	6
5	$\pm 0.8662, \pm 0.4225, \pm 0.2666$	8
6	$\pm 0.8839, \pm 0.5297, \pm 0.3239, 0$	8
8	$\pm 0.9116, \pm 0.0601, \pm 0.5288, \pm 0.1679, 0$	10

evaluation. If increasing accuracy is sought in a systematic way by progressively increasing the order of error of the formula used, integrand evaluations from the lower order estimates are all 'wasted' — or at least of no use (other than perhaps the mid-, or end-, point values) for the current estimate. For instance, considering the 'difficult' example of $(1/10) \int_{-5}^{5} (1 + x^2)^{-1} dx = 0.275 \ldots$, the sequence of Gauss-Legendre estimates of increasing order 2, 4, 6, ... converges (as it does for any bounded and continuous integrand) and the successive values are

$$1, \quad 0.107, \quad 0.479, \quad 0.185, \quad 0.353, \quad 0.231, \quad \ldots$$

alternating about the correct integral; the running total of integrand evaluations forms the sequence $1, 3, 5, 9, 13, 19, 25, \ldots$. Thus after 19 evaluations the best estimate (0.231) is poor compared with that achieved by the composite trapezoidal rule (of 0.275 ...) after 17 evaluations in the Romberg scheme (p. 205). Of course, only 6 of the 19 evaluations are used to provide the 12th-order estimate of 0.231, whereas all 17 are used to give the 2nd-order estimate of 0.275; also the order of error is seen to be no guide to the actual numerical error — nor of course does it pretend to be.

As a tactical device, Gaussian quadrature is entirely successful in achieving its objective. But it is not at all clear how it can best be incorporated in a fully efficient strategy of numerical quadrature. If, for example, the number of function evaluations is doubled at each iteration, the formula employing 2^i points of integrand evaluation is reached after a total of $(2^{i+1} - 1)$ distinct evaluations, and the order of error at this stage is 2^{i+1}, or in other words, just one more than the number of evaluations. One can hope to do better than this — as will now be shown.

5.2.2 Formulae based on common-point sequences

An estimate of an integral based on an $(n + 1)$-point polynomial interpolation of the integrand would be exact for any nth degree polynomial integrand, and so its order must be at least $(n + 1)$: the success of Gaussian quadrature is that by selection of the points of interpolation — that is, of integrand evaluation — it manages to double this order of error. However, if one is looking for a common-point sequence of points of integrand evaluation, so as fully to utilise the information already obtained in some iterative process of successive approximation, then the choice of points must be restricted to those newly added at each stage of the iterative process. The remainder are all those for which integrand values are already available. Clearly, they are not open to choice.

What will now be shown is that, if there are i such fixed points to which $(j + 1)$ are added, then it is possible to select the position of the latter so that there exists a linear combination of all $(i + j + 1)$ integrand values which provides a quadrature formula of order at least $(i + 2j + 2)$. In other words, recovering the accepted use of the symbol n by writing $i + j = n$, so as to imply that the totality of points involved is $(n + 1)$, it is asserted that the newly evolved formula can have order $(2n + 2 - i)$. This is in conformity with what is already known: the Gaussian formulae have no predisposed points $(i = 0)$ and have error order $(2n + 2)$; the Lobatto formulae all utilise the two end-points $(i = 2)$ and have error order $2n$.

In order to find such a quadrature formula of order $(2n + 2 - i)$, it can be observed that it should provide a precise value of the integral over $[-1, 1]$ of any polynomial of degree $(2n + 1 - i)$: but if the $(n + 1)$-points of evaluation are taken as $x_0, x_1, \ldots, x_n$, as before, then such a polynomial could be expressed as

$$p_{2n+1-i}(x) = [\prod_{k=0}^{n} (x - x_k)] (\sum_{r=0}^{n-i} a_r x^r) + (\sum_{s=0}^{n} b_s x^s) . \quad (5.2.7)$$

For here the right-hand-side involves a linear combination of $(2n + 2 - i)$ arbitrary coefficients $a_0, \ldots, a_{n-i}, b_0, \ldots, b_n$; they multiply the same number of linearly independent polynomials of degree $\leqslant 2n + 1 - i$, which therefore form a basis for $p_{2n+1-i}(x)$. To solve the problem, suppose that the $(n - i + 1)$ points of evaluation which are open to choice can be selected so that

$$\int_{-1}^{+1} x^r \prod_{k=0}^{n} (x - x_k) dx = 0: \quad r = 0, 1, \ldots, (n - i). \quad (5.2.8)$$

Then, with all the $(n + 1)$ points $x_0, x_1, \ldots, x_n$ known, the $(n + 1)$ weighting coefficients of the required quadrature formula can be evaluated as in (5.1.2), by supposing that the integrand is replaced by its nth degree interpolating polynomial $y(x)$ which has these $(n + 1)$ points as support.

In the event that the integrand is the arbitrary polynomial $p_{2n+1-i}(x)$, this interpolating polynomial $y(x)$ would simply be the second term (involving the coefficients b_s) on the right-hand-side of (5.2.7), because this has the same value as $p_{2n+1-i}(x)$ at $x_0, x_1, \ldots, x_n$. On the other hand, the first term on the right-hand-side makes no contribution to the integral of $p_{2n+1-i}(x)$, by virtue of (5.2.8). Thus the latter integral and that of $y(x)$ are equal. In other words, the formula correctly integrates any arbitrary polynomial of degree $(2n + 1 - i)$, and therefore has an error of order $\geqslant (2n + 2 - i)$.

Of course it is by no means obvious that (5.2.8) *can* be satisfied. To take a trivial example, if the number i of fixed points is odd and symmetrically distributed in $[-1, 1]$, there can be no optimal addition of an odd number of points since symmetry could not be preserved. Relying on intuition, however, it would seem plausible that an optimal common-point sequence could be built up on the supposition that all points are interior to the interval, and that the added points will *interlace* the fixed points. This would require that the number of added points $j = n + 1 - i$ is one more than the number i of fixed points, so that $n = 2i$. If $i_0, i_1, i_2, i_3, \ldots$ denote the number of fixed points at each stage of the process, then the total number $(n_j + 1) = (2i_j + 1)$ of points at the jth stage are the i_{j+1} fixed points of the next $(j + 1)$th generation, so that

$$i_{j+1} = 2i_j + 1 \ ,$$

that is, $i_j = 2^j(i_0 + 1) - 1 \quad$. $\qquad\qquad$ (5.2.9)

Starting with no fixed points ($i_0 = 0$), one generates in this way a common-point sequence of $1, 3, 7, 15, \ldots$ points, by adding $1, 2, 4, 8, \ldots$ points to the existing $0, 1, 3, 7, \ldots$ fixed points.

The initial optimum point must by symmetry be $x = 0$, and the optimal addition of 2 points must lead to the Gauss-Legendre 3-point formula. Each $(n + 1)$-point formula in the iteration correctly integrates polynomials of degree $2n_j + 1 - i_j = 3i_j + 1$, which implies the sequence $1, 4, 10, 22, \ldots$. Because the formulae are symmetric in the interval $[-1, 1]$, odd functions are always integrated exactly, and the error order is therefore two more than the degree and so equal to 3.2^j for $j > 0$, implying that the order increases in the sequence $2, 6, 12, 24, \ldots$. Thus it is $(3n/2) + 3$ for the $(n + 1)$-point formula (if $n > 0$), reached after $(n + 1)$ function evaluations.

An optimal common point sequence generated by interlacing in this way is called an open-type **Patterson family**, and by including the end-points of the interval there are corresponding closed-type families. The derived sequence of quadrature formulae constitute the **Patterson method** of quadrature. Evidently, the difficulty of the process lies in determining the position of the points and the values of the weighting coefficients. If the positions of the newly added points at each stage are expressed as the zeros of a polynomial $G_{i+1}(x) = c_0 + c_1 x + \ldots + c_{i+1} x^{i+1}$, say, with $c_{i+1} = 1$, whilst the known fixed points are at the zeros of the polynomial $F_i(x)$, say, then (5.2.8) becomes a system of $(i + 1)$ linear equations for the unknown coefficients $c_0, c_1, \ldots, c_i$: namely

$$\sum_{k=0}^{i+1} [\textstyle\int_{-1}^{+1} F(x) x^{j+k} \, dx] c_k = 0: \quad j = 0, 1, \ldots, i \ .$$

These can be solved and the zeros of $G_{i+1}(x)$ determined, but the computational work would clearly be very heavy. Library implementations of the method would simply rely on stored values of the weights and abscissae obtained from prior calculations.

There is room for a simpler method which, although without the strictly conceived theoretical advantages of Patterson's method, allow weights and abscissae to be more readily found. Best known in this respect is **Clenshaw-Curtis quadrature**, which assumes a common-point sequence for the sampling integrand evaluation, given at the jth generation by

$$x_k = \cos(k\pi/2^j): \quad k = 0, 1, \ldots 2^j \ ,$$

the even-subscripted values being common with the points used in the previous generation. The weighting coefficients are found by assuming the integrand is replaced by the polynomial interpolating the function at these points: this polynomial is most readily obtained as a truncated Tchebyshev series, and the integral is then related to the coefficients of the series. Moreover, these coefficients can be used to construct some estimate of the error at each step, so that it is unnecessary to proceed to a higher order formula to check the accuracy of the result; this halves the amount of computation otherwise needed. If one wanted some justification for this choice of points for integrand evaluation, one might note that they interlace the abscissae of Tchebyshev interpolation. But the proof of this particular pudding is in the eating – in practice it provides accurate results almost as quickly as Patterson's method, and sometimes in fact more quickly, discounting in either method the cost of computing the weights and abscissae.

Both Patterson's method and that of Clenshaw and Curtis require merely that the integrand is continuous to be assured of convergence. However, in practice they are only fully efficient if the integrand is analytic over the range of integration. In the context of what we have come to call 'well-behaved' integrands, for which the convergence acceleration technique of the Romberg scheme is efficient, these methods (based on unequally spaced points of integrand evaluation) are highly efficient – generally much more so than the Romberg scheme. But for 'badly-behaved' integrands, they require more or less comparable numbers of integrand evaluations as the composite trapezoidal rule, and it is arguable whether they offer any real advantage.

5.2.3 Transformation of the integrand

One completely different approach to numerical quadrature, which banishes the customary distinctions between the behaviour of various integrands, is afforded by the **transformation method**. The idea here is to substitute a new variable of integration (t, say) which is chosen in such a way that all the derivatives of the transformed integrand are caused to vanish at both ends of the range of integration. Such a device places the definite integral in the same category as that of a periodic function integrated over its complete period. The terms of the Euler-Maclaurin summation formula (5.1.23) are all zero, and the composite trapezoidal rule (employing equispaced points of integral evaluation) generally displays super-linear convergence if the number of segments is successively doubled. However, although the points are equispaced in terms of t, they are unequally spaced in terms of the original variable of integration.

At first it might seem likely to be difficult to find such a substitution: but this is not so – there are in fact many possibilities, and one example will show the general principle involved. Consider the use of the substitution:

$$x = \phi(t) \equiv \tanh[2ct/(1-t^2)], \quad |t| \leqslant 1, \qquad (5.2.10)$$

where $c > 0$ is an arbitrary parameter. The range of integration assumed to be for x between -1 and $+1$ remains unaltered in terms of the new variable t, and

$$\int_{-1}^{+1} f(x)\,dx = \int_{-1}^{+1} g(t)\,dt, \quad \text{say},$$

where $\quad g(t) \equiv f[\phi(t)]\phi'(t),$

$$= 2c(1 + t^2)f[\phi(t)]/\{(1 - t^2)\cosh[2ct/(1 - t^2)]\}^2 \qquad (5.2.11)$$

$$= 2c(1 + t^2)(1 - t^2)^{-2}[1 - \phi^2(t)]f[\phi(t)].$$

It is now required to be shown that the new integrand $g(t)$ and all its derivatives vanish at the ends of the range of integration $t = \pm 1$: it will suffice to show this in relation to just one end, say $t = 1$, corresponding to $x = 1$.

Since $f(x)$ is by implication integrable, let us assume that as $x \to 1$, the value of $f(x)$ behaves like $A(1 - x)^{v-1}$, where A is non-zero and finite and v is positive. This is not quite a necessary consequence of integrability, but almost so: it overlooks some forms of integrable logarithmic singularity in the behaviour of $f(x)$ at $x = 1$. However, we have chiefly in mind for the present that $f(x)$ is non-singular, so that not merely is v positive but it is integer – generally its value would be unity. Now as x tends to unity, so does the corresponding value of t, and noting the identity $[1 - \tanh(s)] = 2/[\exp(2s) - 1]$, (5.2.10) shows that the value of $(1 - x)$ will tend to zero like $2\exp[-c/(1 - t)]$. Using this information, the behaviour of the new integrand $g(t)$ as $t \to 1$ is given from (5.2.10) and (5.2.11) as

$$g(t) \sim 2^{1+v}Ac(1 - t)^{-2}\exp[-vc/(1 - t)],$$

and it is not difficult to verify that $g(1) = g'(1) = g''(1) = \dots = 0$, or in other words $g(t)$ tends to zero as $t \to 1$ faster than any positive power of $(1 - t)$.

This result (as anticipated) prompts the use of the composite trapezoidal rule as a means of estimating the value of $\int_{-1}^{+1} g(t)\,dt$, so that evaluating the integrand at $t_j = (2j - n)/n$ for $j = 0, 1, \dots, n$, the integral of $f(x)$ is expressed as a sum of values of $g(t_j)$, or using (5.2.10) and (5.2.11) as

$$\tfrac{1}{2}\int_{-1}^{+1} f(x)\,dx = (1/n)\sum_{j=1}^{n-1}\phi'(t_j)f(x_j) + \epsilon_n$$

where

$$\phi'(t_j) = 2c(1 + t_j^2)(1 - t_j^2)^{-2}(1 - x_j^2) \qquad (5.2.12)$$

and

$$x_j = \tanh[2ct_j/(1 - t_j^2)]$$

If a method of successive approximation is adopted by which n is doubled at each iteration, then the t_j-values, and so also the corresponding x_j-values, form a common point family, and all previous function values *and* in this context also all previous weights $\phi'(t_j)$ are used. In other words, if $I(n)$ denotes the estimate using n segments,

$$I(2n) = \tfrac{1}{2}[I(n) + (1/n)\sum_{j=1}^{n}\phi'(t_{2j-1})f(x_{2j-1})] \qquad (5.2.13)$$

Further, as was also anticipated, the error ϵ_{2n} would be expected to be proportional to some power of ϵ_n (perhaps ϵ_n^2) and the convergence would be rapid, at least for large enough values of n.

Evidently the transformation (5.2.10) we have chosen depends on the parameter c, and if c is large compared with unity the distribution of x-values corresponding to equispaced values of t would show a strong concentration of points close to the ends of the interval (Fig. 5.4). Indeed, depending on the precision of machine arithmetic, many of the points close to the ends may be indistinguishable from the end-points themselves: for instance, if floating point numbers are represented with binary precision p, then this applies to the set of values for which $1 - |t_j|$ is less than approximately $c/(p\ln 2)$. The weighting coefficients $\phi'(t_j)/n$ associated with these 'lost' points are all, however, insignificant compared with those over the middle of the range (the maximum being $\phi'(0)/n = 2c/n$). This produces complications if values of $f(x)$ close to $x = \pm 1$ are very much larger than elsewhere – which would certainly happen if $f(x)$ were infinite at the end-points of its range of integration – but otherwise the coalesced points can be ignored.

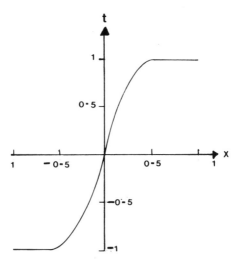

Fig. 5.4 – From equation (5.2.10) with $c = 2$.

The accuracy and rate of convergence of the formula (5.2.12) are quite sensitive to the choice of the parameter c; the best way of demonstrating this is to use it to integrate $f(x) \equiv 1$, for although the order is 'infinite', the formula is inexact for any polynomial, even one of zero degree. The estimate of the integral is in this example simply the mean value of the weights and the results are shown in Fig. 5.5: evidently the value of c giving least error increases with n. Although the accuracies achieved may not seem particularly impressive in regard to such a

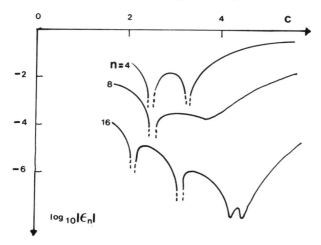

Fig. 5.5 – The truncation error ϵ_n of the transformation method in estimating the integral of unity, as affected by the parameter c and the number of points $(n + 1)$. The broken curves denote a change in sign of error.

trivial integral, it is important to realise that similar convergence is obtained for non-trivial integrands. The variation of error with n for three different integrals is as follows (using the parameter value $c = 2$ in each example):

n	$\int_{-0.5}^{0.5}(1 + x^2)^{-1}\mathrm{d}x$	$\frac{1}{2}\int_{-1}^{+1}(1 - x^2)^{-\frac{1}{2}}\mathrm{d}x$	$(1/10)\int_{-5}^{+5}(1 + x^2)^{-1}\mathrm{d}x$
4	1.41×10^{-1}	4.39×10^{-2}	7.29×10^{-1}
8	3.17×10^{-3}	-2.32×10^{-3}	2.54×10^{-1}
16	1.78×10^{-6}	1.13×10^{-5}	5.71×10^{-2}
32	1.04×10^{-10}	9.58×10^{-8}	4.38×10^{-3}
64	$-$	$-$	3.01×10^{-5}

Convergence is apparent, even in the second example – despite the singular integrand (which would defeat any of the methods previously described). On the other hand, the performance with the 'badly behaved' integrand is very poor, and certainly worse than for the composite trapezoidal rule applied to the untransformed integrand.

As we have already mentioned, there are a number of possible transformations which can be used to produce formulae similar to (5.2.12). Extensive investigations have been conducted using the rather more complicated substitution (originally proposed by Iri and Japanese co-workers):

$$x = \phi(t) \equiv (1/k)\int_0^t \exp[-p^2 t^2/(1 - t^2)]\mathrm{d}t \qquad (5.2.14)$$

where $p > 0$ is again an arbitrary parameter, and k is chosen so that $\phi(1) = 1$, and so also $\phi(-1) = -1$ (since the indefinite integral is clearly an odd function of t). Experience suggests that p is best chosen as equal to about 4, and with this value one obtains rather better results than by the substitution (5.2.10). For instance, errors for both the following integrals

n	$\int_{-1}^{+1} dx$	$\frac{1}{2} \int_{-1}^{+1} (1 - x^2)^{-\frac{1}{2}} dx$
4	1.89×10^{-1}	
8	1.74×10^{-4}	1.07×10^{-4}
16	6.12×10^{-8}	5.74×10^{-7}
32	–	5.48×10^{-10}

are clearly smaller, but the integral $\int_{-5}^{+5} (1 + x^2)^{-1} dx$ is no better dealt with than before.

In fact, the bulk of the contribution to the value of this last-named integral derives from near the centre of the range, whereas both the transformation methods, as well as the optimal formulae discussed in the previous section, concentrate the work of integrand evaluation on the two ends of the range of integration. It is time to turn attention to the ways in which this kind of problem can be overcome. Obviously it would be better in this particular example to observe the symmetry of the integrand and evaluate $2 \int_{0}^{5} (1 + x^2)^{-1} dx$. But of course this symmetry would not usually exist, and in any event it does not seem reasonable that the success of numerical quadrature should depend on such insight and intervention.

5.3 ADAPTIVE QUADRATURE

It has been characteristic of all the quadrature methods so far described that the points of integrand evaluation are unaffected by the form of the integrand; or perhaps rather more precisely, they are chosen (except for the transformation method) to enable a precise integration of any polynomial integrand (of some specified degree) and therefore a more or less accurate integration of an integrand which can be approximated by such a polynomial. They can all be characterised as *fixed-point* methods. All perform well – judged by the computational work required – for certain integrands (which presumably are able to be accurately represented by a polynomial), but not for others. These specially appropriate integrands we have typified as 'well-behaved', and the others as 'badly-behaved', but except for the understanding that the distinction lies in the behaviour of high-order derivatives, no clear means of identification has been asserted.

In the **adaptive methods** now to be described, the computation proceeds by partitioning the range of integration and testing the estimated error (involving a numerical estimate of a relevant derivative of the integrand): if the error in any segment exceeds a specified tolerance, then that segment is partitioned further until at length the estimated error becomes tolerable. This produces a non-uniform and generally unpredictable set of points of integrand evaluation, concentrated in regions where the greatest error is judged to arise, and therefore distributed in a manner related to numerical estimates of the integrand derivative.

The technique involved is recursive, and can be illustrated by considering the adaptive use of the trapezoidal rule. At some stage in the process, the range of integration is (say) the interval $[\alpha, \beta]$, and let us assume that the integrand values $f_\alpha = f(\alpha), f_\beta = f(\beta)$ have already been calculated; further let ϵ be an error tolerance. What is required is an estimate of the integral $\int_\alpha^\beta f(x) \mathrm{d}x$ with error magnitude estimated to be less than ϵ: let the estimate be $I(\alpha, h, f_\alpha, f_\beta, \epsilon)$ where, to simplify the notation, h is supposed to be the interval range $(\beta - \alpha)$. Using the trapezoidal rule, one estimate of the required integral is clearly given by $(f_\alpha + f_\beta)h/2$. However, if the value of the mid-ordinate $f_\gamma = f(\alpha + h/2)$ is also calculated, there are further estimates available: either the mid-ordinate rule, giving simply $f_\gamma h$, or the composite (2-segment) trapezoidal rule $(f_\alpha + 2f_\gamma + f_\beta)h/4$, or Simpson's rule $(f_\alpha + 4f_\gamma + f_\beta)h/6$. These three 'new' estimates are all interrelated; and of course we know that Simpson's rule has the highest (fourth-) order error. Thus by subtracting the Simpson's rule estimate from that derived from the two-segment composite trapezoidal rule, one can obtain a reasonable estimate of the error of the latter – not entirely a reliable one, admittedly, because the Simpson rule estimate itself will be in error, but an estimate which will be a good approximation if h is 'small enough'. This difference is $(2f_\gamma - f_\alpha - f_\beta)h/12$, and should its magnitude be less than ϵ, the error in $(f_\alpha + 2f_\gamma + f_\beta)h/4$ is regarded as 'tolerable'. In this event one might as well regard the value of $I(\alpha, h, f_\alpha, f_\beta, \epsilon)$ as given by Simpson's rule $(f_\alpha + 4f_\gamma + f_\beta)h/6$, which should have an even smaller error, so that (as it were) one is doubly insured against exceeding the imposed tolerance. Alternatively, if the error estimate found in this way has a larger magnitude than ϵ, then the interval is bisected, and the required integral calculated as

$$I(\alpha, h, f_\alpha, f_\beta, \epsilon) = I(\alpha, h/2, f_\alpha, f_\beta, \epsilon/2) + I(\gamma, h/2, f_\gamma, f_\beta, \epsilon/2) . \qquad (5.3.1)$$

This implies that $I(..)$ is a recursively defined function.

Note that the error tolerance is halved in the next recursive level, or 'depth'. this ensures that each half-segment contributes an error of smaller magnitude than $\epsilon/2$, and consequently the total magnitude at the current level of recursion cannot exceed ϵ, as stipulated. Since the trapezoidal rule provides an integral mean with error roughly varying as h^2, the error of each of the two component estimates of the integral over $[\alpha, \gamma]$ and $[\gamma, \beta]$ can be expected to be reduced in proportion to the cube of the range, and each error should be roughly one-eighth of that over $[\alpha, \beta]$; thus the ratio of error to tolerance should be quartered by moving to the next deeper level of recursion. Ultimately, then, provided the integrand is not discontinuous, or 'noisy' owing to random generated error in its computation, the error will become 'tolerable' at some level of recursion, and an increment to the integral will be accepted. However, as each new depth of recursion produces two calls on $I(..)$, a binary tree of evaluations of $I(..)$ may remain to be resolved at any one time, and their evaluation may of course sprout further branches of calls to take the recursion back once more to deeper levels.

The process is admirably suited to any algorithmic language which allows recursive generation of functions: what has been described might, for example, be written as

real procedure $I(a, h, fa, fb, tol)$; **value** a, h, fa, fb, tol; **real** a, h, fa, fb, tol;
    begin real fc;
        $fc := f(a + h/2)$;
        $I :=$ **if** $\text{abs}(2 * fc - fa - fb) \leqslant tol$ **then**
                $(fa + fb + 4 * fc) * h/6$
           else $I(a, h/2, fa, fc, tol) + I(a + h/2, h/2, fc, fb, tol)$
    end; .

In practice, it would also be necessary to impose some limit on the depth of recursion, and to flag the run-time activation of such a limit, as otherwise indefinite recursion may result from a discontinuity in the integrand, or from the 'noise' of numerical error in its evaluation. In the quoted instruction sequence, the value of tol would be taken as $(12\epsilon/h)$. Since h and ϵ are both halved together as the depth of recursion is increased, this remains a constant independent of depth; it would therefore be initialised as 12 times the accuracy required for the integral mean.

This use of tol in place of ϵ shows that what is achieved is a partitioning of the initial interval $(b - a)$ into segments of variable semi-span $s_d = (b - a)/2^d$, where d is the least depth for which

$$tol \geqslant f(\gamma + s_d) + f(\gamma - s_d) - 2(\gamma) \simeq f''(\gamma)s_d^2 . \tag{5.3.2}$$

Here the approximation on the right-hand-side, of the second-difference of f by a second-derivative, follows simply by substituting the appropriate Taylor series expansions for $f(\gamma \pm s_d)$ about γ, and neglecting terms of degree 4 (or more) in s_d. Since the density of function evaluations depends on $(1/s_d)$, one can restate the effect of the tolerance requirement by saying that this density is proportional (roughly) to $[f''(x)/tol]^{1/2}$. Consequently, this particular use of adaptive quadrature will result in a large number of evaluations covering regions of high curvature, but relatively few close to any points of inflection (where the second derivative is zero).

This can be restated in another way. It follows by placing m equal to 1 in the Euler-Maclaurin formula (5.1.23) that the error of the integral mean over $[\gamma - s_d, \gamma + s_d]$ derived from the 2-segment composite trapezoidal rule is equal to $-f''(\xi)s_d^2/12$, for some value of ξ in $(\gamma - s_d, \gamma + s_d)$. Assuming that there is no large variation in the second derivative over this interval, (5.3.2) shows that this error is forced to be no greater than $tol/12 = \epsilon/(b - a)$ in every 'tolerated' estimate of a component integral mean. The error of the aggregated integral over

$[a, b]$ should therefore be less than ϵ. In other words, the process estimates the derivative value (in this instance, the second derivative) from the integrand evaluations available, and adapts the size of the partitioning to suit.

Applied to the 'well-behaved' integral $\int_{-0.5}^{+0.5} (1 + x^2)^{-1}dx$ with an error tolerance of $\epsilon = 0.002$, the following 13 points of integrand evaluation would be invoked by the algorithm described above.

level	0	$-1/2$							1/2
1					0				
2			$-1/4$				1/4		
3		$-3/8$		$-1/8$		1/8		3/8	
4			$-3/16$	$-1/16$	1/16	3/16			

Evidently the central half-range $[-1/4, 1/4]$, where the value of $|f''(x)| = 2|3x^2 - 1|/(1 + x^2)^2$ is larger, is covered at twice the density of the rest of the range. The error of the integral estimate obtained from these 13 integrand values is 1.65×10^{-6}, obviously very much better than the prescribed tolerance; on the other hand, the Romberg scheme (p. 206) provides a less accurate answer, but nevertheless one discernible to be correct within 3 decimal places as required, after only 9 integrand evaluations. More nearly comparable numbers of evaluations would be determined for the 'difficult' integral $\int_{-5}^{+5}(1 + x^2)^{-1}dx$, but clearly the trouble with the algorithm as presented is that it is doing a lot of unnecessary work, merely because its error estimate is much too conservative. In other words, Simpson's rule (which in fact provides the integral estimates) has generally a much smaller error than the trapezoidal rule (which provides the error estimates).

However, the idea of adaptive partitioning of the range of integration can just as easily be applied to other quadrature formulae. An adaptive use of Simpson's rule would compare one application of the rule to the current segment with the composite Simpson's rule applied to two equal component segments. Richardson extrapolation applied to the elimination of the fourth-order error of these two different estimates of the integral mean leads to Milne's formula applied to the undivided segment (as has been observed before in discussing the Romberg scheme). This result provides a yardstick whereby the error of the composite Simpson's rule is assessed. It works out as a fourth-order difference of the integrand values (proportional roughly to the mid-point value of $f^{iv}(x)$). Accepting the Milne formula estimate when this fourth difference is less in magnitude than some prescribed tolerance for the integral mean now produces a distribution of points of evaluation roughly proportional to $[f^{iv}(x)/tol]^{1/4}$.

The process still involves a generally conservative error estimate, but the error of the sixth-order Milne formula is less markedly different from that of the fourth-order Simpson's rule, in that the order is only increased by half, rather than doubled – as in the comparison of errors derived from the use of Simpson's rule and the second-order trapezoidal rule. Evaluating $\int_{-0.5}^{+0.5}(1+x^2)^{-1}dx$ with $\epsilon = 0.002$ by this method provides the required estimate at the first level of recursion (after just 5 integrand evaluations corresponding to 2 applications of Simpson's rule), and the actual error is 2.36×10^{-4}. Applied to the 'difficult' example $\int_{-5}^{+5}(1+x^2)^{-1}dx$, any prescribed error value between 2×10^{-4} and 3×10^{-5} for the integral mean provides an estimate with actual error 3.6×10^{-6} by a total of 33 integrand evaluations, 25 of them in the half-range $[-2.5, 2.5]$ and 17 in the central quarter $[-1.25, 1.25]$. This can be compared with the result of applying the Romberg method to the same integral: after the same number of evaluations – in this context, of course, equispaced over the entire range of integration – the actual error is 2.4×10^{-5}, almost 10 times as large, and the discerned error (based on a comparison of successive estimates) is 7×10^{-5}.

The benefit of using any of the Cotes formulae in an adaptive scheme is that it is possible to produce a common-point sequence by bisections of each partition. Each function evaluation, even if initiated for an integral estimate which is rejected, is subsequently used in an accepted incremental estimate. The same benefit can accrue from the adaptive use of open-type quadrature formulae for the equispaced interleaved distributions of points of integrand evaluation (Table 5.2) by trisecting the interval. As a simple example, an adaptive form of the mid-ordinate rule would compare the integrand at the mid-point of any segment with the mean of the 3 values at the mid-points of 3 equal component segments: the three-point open-type formula replaces Simpson's rule as a fourth-order estimate by which the error can be assessed, and if this error is too large, the process is repeated in turn for each of the three component segments, and the integral value accumulated as the sum of the 3 component integrals. This is evidently entirely similar to the adaptive use of the trapezoidal rule, except that the error in the integral mean should reduce by a factor of 9, instead of 4, at each level of recursion; the depth of recursion may therefore be less, but, on the other hand, since there are no common end-points shared between contiguous segments, the number of function evaluations may not be so very different.

Similarly, the three-point open-type formulae can be used as the basic quadrature relation of an adaptive scheme. Moreover, another variation is now possible, since there is no need for the three component sub-intervals of each partition to be of equal width – though they should of course be symmetrically arranged. Thus, supposing the current segment is mapped onto $[-1, 1]$, an adaptive use of the mid-ordinate rule could involve partitioning this segment into subintervals with end-points at $-1, 1 - 2\sqrt{(3/5)}, 2\sqrt{(3/5)} - 1, 1$ and mid-points at $-\sqrt{(3/5)}, 0, +\sqrt{(3/5)}$ which would allow the sixth-order (3-point) Gauss-Legendre formula to be employed in place of the 3-point equispaced formula from Table 5.2. In **Robinson's method**, this same unequal partitioning is used, but the 3-point Gauss formula is used to estimate the integral over the unpartitioned interval, and also over each of the three component segments; as the integrand has already been evaluated at the mid-points of each component segment, 6 further evaluations are produced at each level of recursion. Richardson extrapolation can still be applied (on the supposition that $f^{vi}(x)$ is roughly constant over the interval) to provide an eighth-order estimate of the integral over the entire segment; but as there are 9 integrand values available in all, a tenth-order estimate is in fact possible.

It was at one time the fashion to prefer low-order adaptive schemes, and the adaptive Simpson's rule was particularly favoured; but more recently attention has turned to the adaptive use of such higher order formulae. The danger of going too far in this direction is that the error reduction at each recursion becomes so very large that superfluous precision is difficult to avoid; thus even if the error is not grossly overestimated, the process may nonetheless still lead to a lot of superfluous calculation.

Nearly all the schemes can be made to produce a wrong estimate if one knows their inner structure. The algorithm for the adaptive trapezoidal rule which was set out earlier in this section will be found (for instance) to assign a zero value to the essentially positive integral $\int_{-\pi}^{+\pi} \sin^2(x)\,dx$, irrespective of error tolerance; the adaptive Simpson rule would likewise fail to evaluate $\sin^2(2x)$ over the same range. This kind of difficulty could be overcome by randomly dividing the total range into two segments before the work begins; or it might be agreed that, provided a minimum number of function evaluations has been performed, such chance misinterpretations can be safely ignored. Always one has to weigh the *robustness* of any quadrature algorithm — its ability to deal with all eventualities — against its *aptitude* in dealing with the most likely applications.

5.4 AREAS AND INTEGRALS DEFINED BY IRREGULAR DISCRETE DATA

Sometimes it is necessary to provide numerical estimates of the value of an integral

$$I = \int_\gamma y(s)x'(s)\,ds\ ,\tag{5.4.1}$$

where s is the distance along a rectifiable curve γ in the (x, y)-plane (Fig. 5.6(a)).

(a)

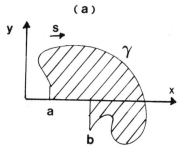

Fig. 5.6 – (a) The curvilinear integral I

If y is a single valued function of x, this expression can be reinterpreted as implying simply that $I = \int_a^b y(x)\,\mathrm{d}x$, where a and b are the end abscissae of γ, so it is particularly in view that $y(x)$ may not be single-valued, and likewise x may not be expressible as a single-valued function of y. It is this generality which causes the expression on the right-hand-side of (5.4.1) to be termed a **curvilinear integral**. In the particular instance that γ is a closed curve, then obviously neither x nor y can be expressed as a single-valued function of the other; the integral (5.4.1) then describes the area within the closed curve, and it is referred to as a **circuit-integral**.

Of course, if both $x(s)$ and $y(s)$ are known functions, then I can be interpreted as $\int_0^l f(s)\,\mathrm{d}s$, where $f(s)$ is a known single-valued function of s, and l is the length of γ. Its evaluation is the familiar quadrature problem. Or it may be possible to express the position coordinates in terms of some other parameter than s, with the same outcome. The difficulty of evaluating I therefore arises only where γ is defined merely by discrete data — as the curve interpolated (in some way) between a given sequence of (say) $n + 1$ Cartesian coordinates (x_k, y_k) for $k = 0, 1, \ldots, n$, ordered in respect of their position along γ, so that γ extends between (x_0, y_0) and (x_n, y_n). The easiest approach to this problem seems to be afforded by considering the data to be interpolated by a piecewise continuous function, leading to a composite quadrature rule of some kind. But because the sequence of x-values is not necessarily monotonic — since y is not necessarily a single-valued function of x — the customary option of expressing y as a piecewise polynomial in x is not usually available.

However, this restriction does not apply if successive data coordinates are interpolated by straight lines, since y can then in general be expressed as a piecewise-linear polynomial in x, and (for instance) over the first segment the contribution to I would be simply

$$\int_{x_0}^{x_1} y\,\mathrm{d}x = \tfrac{1}{2}(y_0 + y_1)(x_1 - x_0) = \tfrac{1}{2}(y_0 x_1 - y_1 x_0) + \tfrac{1}{2}(y_1 x_1 - y_0 x_0) \ ,$$

whilst adding a second segment gives

$$\int_{x_0}^{x_2} y\,\mathrm{d}x = \tfrac{1}{2}(y_1 x_2 - y_2 x_1) + \tfrac{1}{2}(y_0 x_1 - y_1 x_0) + \tfrac{1}{2}(y_2 x_2 - y_0 x_0) \ .$$

Proceeding in this way, one is led to a general expression by which the required integral is approximated as

$$I \simeq \tfrac{1}{2} \sum_{k=1}^{n} (y_{k-1} x_k - y_k x_{k-1}) + \tfrac{1}{2}(x_n y_n - x_0 y_0) \ . \tag{5.4.2}$$

In particular, if the coordinates (x_0, y_0) and (x_n, y_n) are coincident, implying that the curve γ is closed, then the last term on the right-hand-side vanishes, and the sum is evidently an expression for the area within an n-sided polygon whose vertices are the n distinct data coordinates (x_k, y_k) for $k = 1, 2, \ldots, n$.

This may not seem a particularly pleasing geometric form of approximation to the 'curve' γ, but at least the representation of the integral is succinct, and one is led to enquire whether there may not be more realistic estimates which have a similar simplicity; this suggests one looks for formulae of the general form:

$$J \equiv I - \tfrac{1}{2}(x_n y_n - x_0 y_0) \simeq \sum_{j=0}^{n} \sum_{k=0}^{n} c_{jk} y_j x_k = \mathbf{y}^T \mathbf{C} \mathbf{x} \ . \quad (5.4.3)$$

Here the right-hand-side approximation is called a **bilinear form** of the given data. The quadrature weighting coefficients c_{jk} are supposed to be numbers independent of the data, except of course that they may depend on the value of n. As indicated in (5.4.3), if $\mathbf{x} = (x_0 \ x_1 \dots x_n)^T$ and $\mathbf{y} = (y_0 \ y_1 \dots y_n)^T$ are the $(n+1)$-vectors of data abscissae and ordinates, then any coefficient c_{jk} can be identified as an element (in row j, column k) of a square matrix $\mathbf{C}$ of order $(n+1)$. Reinterpreting the result of (5.4.2) in this light, the appropriate representation for $\mathbf{C}$ can be identified as the skew-symmetric tri-diagonal band matrix

$$\mathbf{C} = \tfrac{1}{2}
\begin{bmatrix}
0 & 1 & 0 & \dots & 0 & 0 \\
-1 & 0 & 1 & \dots & 0 & 0 \\
0 & -1 & 0 & \dots & 0 & 0 \\
 & \dots & & \dots & & \dots \\
0 & 0 & 0 & \dots & 0 & 1 \\
0 & 0 & 0 & \dots & -1 & 0
\end{bmatrix}
\quad (5.4.4)$$

This skew-symmetry is inherent in all formulations of $\mathbf{C}$, and reflects the fact that if the x- and y-values are interchanged, the sign of J is reversed. This is a consequence of the identity

$$J = \int_{\gamma} y(s) x'(s) \mathrm{d}s - \tfrac{1}{2}(x_n y_n - y_0 x_0)$$
$$= -\int_{\gamma} x(s) y'(s) \mathrm{d}s + \tfrac{1}{2}(y_n x_n - x_0 y_0) \ .$$

In fact, the integral J measures the area between the curve γ and the two radial lines joining its end-points to the origin (Fig. 5.6(b)). In terms of radial

(b)

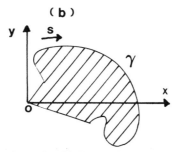

Fig. 5.6 – (b) The curvilinear integral J.

polar coordinates (r, θ), it could also be expressed as

$$J = -\frac{1}{2} \int_{\theta_0}^{\theta_n} r^2(\theta) \, d\theta \ ,$$

where θ_0 and θ_n are the polar coordinates of the end-points (x_0, y_0), (x_n, y_n) of the curve γ. This assumes that the polar equation of γ is $r = r(\theta)$, where $r(\theta)$ is a single-valued function of θ: if it is single-valued, it needs to be kept in mind that the value of J (and therefore of I) might more readily be obtainable by applying such a transformation than by using Cartesian coordinates and the bilinear form (5.4.3).

In general, a composite quadrature rule based on the piecewise interpolation of (say) N successive and contiguous sets of m consecutive data coordinates, where $n = (m-1)N$, provides a pattern for elements of the square matrix $\mathbf{C}$, of order $(n+1)$, most easily pictured as a superposition of N identical diagonal skew-symmetric sub-matrices $\mathbf{C}^{(m)}$ of order m, overlapped at their top- and bottom-corner diagonal (zero) elements:

$$\begin{pmatrix} \begin{pmatrix} 0 & \mathbf{C}^{(m)} \\ & \ddots \begin{pmatrix} 0 \end{pmatrix} & \mathbf{C}^{(m)} \\ & & \ddots \begin{pmatrix} 0 \end{pmatrix} & \mathbf{C}^{(m)} \\ & & & \cdots \cdots \cdots \\ & & & & \cdots \cdots \cdots \cdots \\ & 0 & & & \begin{pmatrix} & \ddots \\ & & 0 \end{pmatrix} \end{pmatrix} & 0 \\ \end{pmatrix}$$

Clearly, this implies that $\mathbf{C}$ is a skew-symmetric band matrix with generally $(2m-1)$ non-vanishing diagonals. For instance, with this understanding, the matrix of (5.4.4) can be more concisely represented by the elementary 2×2 diagonal sub-matrix

$$\mathbf{C}^{(2)} = \begin{bmatrix} 0 & 1/2 \\ -1/2 & 0 \end{bmatrix} \tag{5.4.5}$$

and, like (5.4.4), this is a skew-symmetric tri-diagonal matrix.

It is indeed possible to find more appealing approximations to the required integral than this form of generalised trapezoidal rule: for example, on interpolating γ by a succession of piecewise continuous parabolae through contiguous sets of 3 consecutive data ordinates, the result (see problem 3) is embodied in the sub-matrix

$$\mathbf{C}^{(3)} = (1/6) \begin{bmatrix} 0 & 4 & -1 \\ -4 & 0 & 4 \\ 1 & -4 & 0 \end{bmatrix} \tag{5.4.6}$$

which builds a composite skew-symmetric 5-diagonal band matrix $\mathbf{C}$, of order $2N + 1$. Applied to just 3 data, and supposing $x_0 < x_1 < x_2$, this is consistent with the formula:

$$\int_{x_0}^{x_2} y(x)\,dx = [(x_2 - x_0)/6](y_0 + 4y_1 + y_2) + [(y_2 - y_0)/3](x_0 - 2x_1 + x_2)$$

$$= \tfrac{1}{2}(x_2 - x_0)(y_0 + y_2) - [(x_2 - x_0)\delta^2 y_1 - (y_2 - y_0)\delta^2 x_1]/3$$

$$\dots (5.4.7)$$

which is **Brun's rule** for numerical quadrature based on three unequally-spaced integrand evaluations. It specialises to Simpson's rule in the event that $x_1 - x_0 = x_2 - x_1$, as might be anticipated, since the piecewise quadratic polynomial interpolation on which Simpson's rule is based is also equivalent to interpolation by a parabolic arc. However, (5.4.7) remains valid even if $(x_1 - x_0)$ has the opposite sign to $(x_2 - x_1)$, provided that the integral on the right-hand-side is interpreted as a curvilinear integral. (Readers aware of the concept of Stieltj integration will realise that this provides an alternative and equivalent interpretation, in this context, of the Riemann integral (5.4.7).)

It is difficult to find a meaningful way of expressing the error involved by quadrature formulae like (5.4.3), and of course if the integrand *is* merely defined by discrete data, the error is in any event strictly undefined. What can be said is that if a *known* continuous contour of integration γ is represented by $(n + 1)$-points, and it is supposed that n is allowed to increase, then (5.4.2) has an error which is proportional to n^{-2} as $n \to \infty$, provided that the points are everywhere dense (meaning that the number of points in any finite segement of γ also tends to infinity as $n \to \infty$). Thus the composite trapezoidal rule can in this sense be said to be of second order. The composite Brun's rule, obtained by forming $\mathbf{C}$ in (5.4.3) from a superposition of the sub-matrices $\mathbf{C}^{(3)}$ in (5.4.6), is in this sense of fourth order (like the composite Simpson rule), provided that the slope and curvature of γ are continuous. Indeed, it is possible to derive (5.4.6) by applying Richardson extrapolation to the result (5.4.2) – comparing the results of using the trapezoidal rule on set of data (x_k, y_k) for $k = 0, 1, \dots, n$, and on the subset (x_{2k}, y_{2k}) for $k = 0, 1, \dots, (n/2)$. This would suggest that a matrix $\mathbf{C}^{(5)}$ could be developed from $\mathbf{C}^{(3)}$ by similar extrapolation of the composite Brun's rule (in an analogous way to that by which Milne's rule can be developed from Simpson's).

The easiest approach to the generation of bilinear forms for quadrature (5.4.3) lies in the use of cardinal parametric interpolation (section 3.6), which hypothesises the existence of a parametric representation $x = x(t)$, $y = y(t)$ for the curve γ, by which the data values x_k, y_k correspond to parameters t_k which are equispaced. Then the sub-matrix forms (5.4.5) and (5.4.6) are readily derived by assuming that $x(t)$ and $y(t)$ are respectively piecewise linear or quadratic polynomials in t. However, this approach also shows that there are no unique forms of $\mathbf{C}^{(m)}$ for $m > 3$, but rather an infinity of possible forms, all possessing the same order of error. Thus two possible forms for the sub-matrix $\mathbf{C}^{(4)}$ are

$$
\mathbf{C}_1^{(4)} = (1/16) \begin{bmatrix} 0 & 9 & 0 & -1 \\ -9 & 0 & 9 & 0 \\ 0 & -9 & 0 & 9 \\ 1 & 0 & -9 & 0 \end{bmatrix} ; \quad \mathbf{C}_2^{(4)} = (1/8) \begin{bmatrix} 0 & 5 & -1 & 0 \\ -5 & 0 & 6 & -1 \\ 1 & -6 & 0 & 5 \\ 0 & 1 & -5 & 0 \end{bmatrix}
$$

and both have fourth-order accuracy, as does any linear combination, $\lambda \mathbf{C}_1^{(4)} + (1 - \lambda) \mathbf{C}_2^{(4)}$. The form of $\mathbf{C}_1^{(4)}$ was originally proposed by Selmer.

Both forms specialise to the three-eighths rule if all x-values are equispaced, and one such sub-matrix $\mathbf{C}^{(4)}$ can be used (like one segment of the three-eighths rule) along with several sub-matrices $\mathbf{C}^{(3)}$ to provide a form of $\mathbf{C}$ with fourth-order error, if the number of data is even rather than odd as would be required for the composite Brun's rule. These higher-order forms are rendered unique by requiring that the extent of the non-vanishing diagonals of the band matrix $\mathbf{C}$ is minimised. In this event the extent equals one more than the order 2 entier$[(m + 1)/2]$ of each formula – as for instance is true of the particular matrix $\mathbf{C}_2^{(4)}$. This, like $\mathbf{C}^{(3)}$, is a 5-diagonal band matrix. However, such a criterion seems of no more than pictorial significance.

Bearing in mind that our concern is with discrete data, perhaps the most useful of the bilinear forms having fourth-order error is obtained by constructing the cardinal parametric representation of $x(t)$ and $y(t)$ by Bessel (piecewise cubic) interpolation, which provides continuity of slope to the constructed curve of integration, γ. The coefficient matrix is then

$$
\mathbf{C} = (1/12) \begin{bmatrix} 0 & 15/2 & -3/2 & 0 & 0 & 0 & \cdots \\ -15/2 & 0 & 17/2 & -1 & 0 & 0 & \cdots \\ 3/2 & -17/2 & 0 & 8 & -1 & 0 & \cdots \\ 0 & 1 & -8 & 0 & 8 & -1 & \cdots \\ 0 & 0 & 1 & -8 & 0 & 8 & \cdots \\ \cdots & \cdots & \cdots & \cdots & \cdots & \cdots & \cdots \\ & & & & 0 & 17/2 & -3/2 \\ & & & & -17/3 & 0 & 15/2 \\ & & & & 3/2 & -15/2 & 0 \end{bmatrix}
$$

$$(5.4.8)$$

This is a skew-symmetric 5-diagonal band matrix, and its diagonal elements are constant, except over the first and last three rows and columns. Observe that, these extremes apart, this matrix has elements $\pm 2/3$ next to the principal diagonal, like the composite Brun's rule derived from (5.4.6), and values of $\pm 1/12$ on the outside diagonals of the band, where Brun's rule assigns alternate values of $\pm 1/6$ and 0. If all the x-values are equispaced (so that it can be assumed that $x(t) \equiv t$), the use of this form of $\mathbf{C}$ in (5.4.3) reduces the expression for the integral to the truncated Gregory formula (5.1.29), also resulting from Bessel interpolation of the integrand. This again is no mere coincidence because, with $x(t) \equiv t$, $y(t)$ is of course the same function of x as of t.

These quadrature formulae all become considerably simplified if γ is a closed curve (Fig. 5.7), in which event the curvilinear (circuit) integral measures the

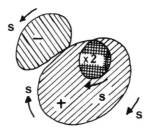

Fig. 5.7 – Circuit integral (I or J) about a closed contour with multiple points, showing relation between sign of area and direction of integration.

area enclosed by γ (positive if the data are ordered along γ in the clockwise sense). Then, defining the points (x_0, y_0) and (x_n, y_n) to be coincident, and noting that there is nothing to distinguish the particular coordinate on the closed circuit which is enumerated as the 'first' datum (x_0, y_0), it must follow that the coefficient matrix $\mathbf{C} = (c_{ij})$ of the bilinear form $\sum_{j=1}^{n} \sum_{k=1}^{n} c_{jk} y_j x_k$ is circulant. Accordingly, if one defines $x_{k+n} = x_k$, the circuit integral must be estimated by

$$\oint_\gamma y(t) x'(t)\, dt \simeq \sum_{j=1}^{n} y_j \left[\sum_{k=1}^{m} b_k^{(m)} (x_{j+k} - x_{j-k}) \right] , \qquad (5.4.9)$$

where $2m \leqslant n$ is the number of non-zero diagonals of $\mathbf{C}$, and where $2m$ is also an upper limit to the order of the corresponding formula. For that class of formulae of order $2m$ which involve the least number $(2m)$ of non-zero columns (and so of non-zero coefficients $b_k^{(m)}$), there exists the simple relation

$$\left. \begin{aligned} k b_k^{(m)} &= (-1)^{k+1} (m+1-k)_k / (m+1)_k \\ &= (-1)^{k+1} (m!)^2 / [(m+k)!(m-k)!] \end{aligned} \right\} \qquad (5.4.10)$$

for $k = 1, 2, \ldots, m$, and some values are tabulated in Table 5.4. The formula for $m = 1$ corresponds to the composite trapezoidal rule, derived of course by interpolating the data by piecewise straight lines, and that for $m = 2$ corresponds to

(5.4.8) without the end-corrections, derived from the use of Bessel interpolation to construct the periodic parametric expressions $x(t)$ and $y(t)$. The
higher-order formulae correspond to the use of higher degree piecewise polynomial interpolation satisfying successively higher order continuity conditions –
of class C^m – and with the end-conditions determined from the periodicity of
the functions $x(t)$ and $y(t)$.

Table 5.4
Quadrature coefficients for circuit integrals

m	$b_1^{(m)}$	$b_2^{(m)}$	$b_3^{(m)}$	$b_4^{(m)}$
1	1/2	–	–	–
2	2/3	−1/12	–	–
3	3/4	−3/20	1/60	–
4	4/5	−1/5	4/105	−1/260

In relation to circuit integrals for which the contour has a known continuous
(periodic) parametric representation, if the data coordinates are computed corresponding to equal increments of that parameter (for example, points equally
spaced round a circle), then the formulae given by (5.4.9) and (5.4.10) may well
display an error reducing with the order, m. But these integrals are usually more
efficiently evaluated by applying a quadrature rule to the Riemann integral of
$f(t) \equiv y(t)x'(t)$, since $x'(t)$ is as easily calculable as $x(t)$. Rather, the formulae
(5.4.9) are intended for use with discrete data derived from experiment, without
any information about derivatives. The quality of those data is unlikely to justify
the sophistication of high-degree interpolation involved by the higher-order
approximations, particularly as their spacing may not conform with the basic
hypothesis of a 'regular choice' in the positioning of coordinates, implicit in the
use of cardinal parametric interpolation. Whilst the representation of γ by an
n-sided polygon involved by the second-order formula (with $m = 1$) may certainly
seem too crude, in practice the fourth- or sixth-order estimates are likely to be
as 'accurate' as any others, and the difference between them would give some
feeling for possible 'error'.

It needs to be emphasised that, since the formulae with $m > 1$ are based on
interpolation of the closed curve by a scheme involving continuity of the first
$(m - 1)$ derivatives, they are clearly unsuitable where γ is known (for example)
to have a slope discontinuity – a corner – at one or more points. In such an
instance it would be better to segment γ into one or more component arcs
between discontinuities, and to estimate the component curvilinear integrals
by one of the formulae previously described. Similar segmentation of the circuit
integral should also be applied if a different criterion of choice is known to
apply to the spacing of the coordinates on different segments of the curve γ.

It should also be noted that the formulae remain valid, though of course not necessarily accurate, even if γ is not a simple curve – that is, if it crosses itself at multiple points – whether or not it is a closed curve. If any such multiple points are known among the data, they should appear as many times as their multiplicity demands and in the appropriate sequence. In any case, multiple points could also inadvertently be implied to exist by the interpolative scheme, and pass unnoticed.

5.5 SINGULAR INTEGRANDS AND IMPROPER INTEGRALS

Turning attention back to the problem of the estimation of the definite integrals of known functions, it will have been obvious that the rapidity of convergence of all methods depends on the property of 'sufficient differentiability' of the integrand over the closed interval of integration: generally it is desirable that the integrand be infinitely differentiable. The sole exception to this general rule is provided by the transformation method, which allows the integrand to be singular at either or both of the end-points of the range of integration. Most of the methods will be convergent merely if $f(x)$ is continuous, but not any of its derivatives. However, a lot of unnecessary computation may be caused if one seeks to take advantage of the slow process, which is all that then would be implied.

There are analytical techniques which can be applied to eliminate, or at least change, the singularities in the integrand or its derivatives, and as a general principle these should always be applied before submitting the integral to numerical estimation. Where the singularities exist at more than one point in the range, it is always possible to partition the range into sub-intervals each containing not more than one singularity, which allows different techniques to be applied to each in turn. Moreover, as another general rule, it is usually possible to isolate the singularity within a small interval and apply a series development to the integrand which can be integrated term-by-term. The numerical problem then becomes a 'hybrid' one of series summation and numerical quadrature. However, care needs to be exercised in the discrimination of an appropriate size for the interval of isolation so that the computational work associated with one or the other of the two component tasks is not disproportionately large.

However, it is almost always possible to transform the singular integrand (by the substitution of a new variable of integration) into a function which is either infinitely differentiable, or else 'infinitely flat' – that is, which has a singularity, of the form $\exp[-k/(x-a)]$ with $k > 0$ at the lower end of the range of integration $[a, b]$, or $\exp[-k/(b-x)]$ at the upper end. For instance, if $f(x)$ is supposed infinitely differentiable, then an integral of the form $\int_0^b x^{m/n} f(x)\,\mathrm{d}x$, where $n \geqslant 2$ and $m > -n$ are integers, is rendered innocuous by the substitution $x = t^n$, the transformed integrand $t^{n+m-1} f(t^n)$ being infinitely differentiable. On the other hand, $\int_0^b x^\alpha f(x)\,\mathrm{d}x$, where $\alpha > -1$ is an irrational number, would have to be dealt with by a substitution such as $x = \exp(-1/t)$, and the same technique is generally needed if the integrand has a logarithmic singularity. However, $\int_0^b \ln^m(x) f(x) \mathrm{d}x$ is amenable to integration by parts if m is a positive integer, as in

$$\int_0^b \ln(x) f(x) \mathrm{d}x = -\int_0^b x^{-1}[\int_0^x f(t)\mathrm{d}t] \mathrm{d}x \ .$$

Here the right-hand-side has an infinitely differentiable integrand, although care may be needed in its evaluation close to $x = 0$. Likewise, the Cauchy principal part of an improper integral, such as $\int_a^b f(x)(x - \xi)^{-1}\mathrm{d}x$ where $a < \xi < b$, can be found by expressing its value as

$$f(\xi)\ln[(b - \xi)/(\xi - a)] + \int_a^b [f(x) - f(\xi)] (x - \xi)^{-1}\mathrm{d}x \ .$$

Improper integrals with an infinite upper limit to the range of integration can of course be transformed into integrals over a finite range by substitutions such as $x = \ln(1/t)$, or $x = 1/t$.

It is arguable that in all instances where the singularity can only be treated by transforming it into an 'infinitely flat' singularity, one might as well use the transformation method – which of course has precisely this same effect. The only drawback of using the numerical transformation, rather than performing it analytically beforehand, is that the untransformed integrand has to be evaluated very close to the singularity, and considerable care is necessary to avoid inaccuracy or even, perhaps, overflow. The loss in accuracy is usually avoided if the interval of integration is adjusted, so that the singularity (at one end of the interval) is placed at $x = 0$, thereby allowing very small values of x to be represented with full precision.

There are well-known quadrature formulae of the Gaussian type, by which the points of integrand evaluation are optimally selected to maximise the order of the error in the numerical estimation of an integral $\int_a^b w(x)f(x)\mathrm{d}x$, where $w(x)$ has a prescribed form with particular singularities at one or both end-points of the range of integration, and where $f(x)$ is assumed analytic over $[a, b]$. In particular, those in which $a = -1$, $b = 1$ and $w(x) = (1 + x)^\alpha(1 - x)^\beta$ are known as **Gauss-Jacobi quadrature** formulae, and the special case $w(x) = (1 - x^2)^{-\frac{1}{2}}$ yields the simple result that the points of integrand evaluation are optimally chosen as the Tchebyshev abscissae (of appropriate degree), and the weights are all equal. In general, following the same line of argument that led to the formulation of the equations (5.2.8), the points of integrand evaluation are optimally chosen as the $(n + 1)$ zeros of the polynomial $p_{n+1}(x)$ whose coefficients satisfy the $(n + 1)$ homogeneous linear equations determined by

$$\int_a^b w(x)p_{n+1}(x)x^r\mathrm{d}x = 0: \quad r = 0, 1, \ldots, n \ . \tag{5.5.1}$$

These polynomials belong to the sequence which is orthogonal on $[a, b]$ with weight $w(x)$; those orthogonal on $[-1, 1]$ with weight $(1 - x^2)^{-\frac{1}{2}}$ are the Tchebyshev polynomials $\{T_n(x): n = 0, 1, 2, \ldots\}$, which is the reason why the choice of the Tchebyshev abscissae is optimal for integrals of the form $\int_{-1}^{+1}(1-x^2)^{-\frac{1}{2}}f(x)dx$. The $(n+1)$ weights of the quadrature formula can be found by the method of undetermined coefficients – that is, by assuming that the formula is exact for $f(x)\equiv x^k$ and $k=0,1,2,\ldots,n$, which yields a set of $(n+1)$ linear equations for the weights. In fact, the integral will be exactly evaluated if $f(x)$ is any polynomial of degree $(2n-1)$ – and so will have error of order $2n$ – just as in the context of the Gauss-Legendre formulae.

Applied to integrals

$$\int_0^\infty x^{-\alpha}\exp(-x)f(x)dx, \quad \int_{-\infty}^{+\infty}\exp(-x^2)f(x)dx \;,$$

this approach produces what are known respectively as the **Gauss-Laguerre** and **Gauss-Hermite** quadrature formulae. The latter are simply a particular example of the former with $\alpha=\frac{1}{2}$, since on substituting $x=t^2$

$$\int_0^\infty x^{-\frac{1}{2}}\exp(-x)f(x)dx = 2\int_0^\infty\exp(-t^2)f(t^2)dt = \int_{-\infty}^{+\infty}\exp(-t^2)\phi(t)dt$$

if $\phi(t)=\phi(-t)=f(t^2)$. In using either of the formulae, one must question whether the behaviour of $f(x)$ can indeed be well approximated by a polynomial for all $x\geq 0$, as is assumed in the derivation of the formulae. The error in the n-point formulae depends on $f^{(2n)}(\xi)$ for some $\xi>0$. So it is only to that restricted class of functions (including of course polynomials) whose higher derivatives are bounded in magnitude for all $x>0$ that this form of quadrature estimate can confidently be applied. The transformation method is still available here as an alternative approach (problem 4, section 5.2.3).

Definite integrals of the form $\int_0^{m\alpha}\omega(x)f(x)dx$ where $\omega(x)$ is a periodic function of period α, and where m is an integer, are 'difficult' to evaluate if m is large, even if $f(x)$ is not singular, and the problem of the evaluation of the improper integral resulting from taking $m\alpha$ to be infinite is especially complicated. Clearly they can be rewritten as $\int_0^\alpha\omega(x)S_m(x)dx$, where $S_m(x)=\sum_{k=0}^{m-1}f(x+k\alpha)$, and this becomes an infinite series if the range is infinite. It is in just such a context that the economy resulting from the use of higher-order quadrature formulae would be a clear advantage. Equally it is also a context where conversion of the problem to one of series evaluation, by use of formal term-by-term integration, might provide a more direct and economic approach.

5.6 REPRISE

It is tempting to conclude by trying to recommend particular methods of numerical quadrature for particular types of problem integrals. A choice of different methods may be available in a computational library, and it is important to have a feeling for which is the more appropriate. In relation to integrands defined by discrete data the choice is fairly clear. If the data are at equispaced intervals, then the composite Simpson-rule, or the truncated Gregory formula (5.1.29), or (better still) a formula based on cubic spline interpolation, would be appropriate. Two different methods will give an indication of the 'accuracy' of the process (which of course is, to a greater or less extent, purely specualtive). If the data are not equispaced, then spline interpolation may still be applicable. Also there are analogues of the other formulae derived from the use of cardinal parametric interpolation (section 5.4) which, used sensitively, provide realistic answers; these are especially easy to compute if the data define a smooth closed curve, in which event of course the integral will evaluate its enclosed area.

More firm advice is possible if the integrand is not everywhere infinitely differentiable (section 5.5): then contrive that it becomes so, or else use the transformation method (section 5.2.3) with the singularities at one or both ends of the range of integration. An integrand which is a periodic function is another exception where the advice is unambiguous (section 5.1.4): use the composite trapezoidal rule, as it cannot be bettered.

In other instances advice becomes more difficult to provide: a distinction has to be drawn between 'badly-behaved' and 'well-behaved' integrands, and the fact that this has to do with values of the higher derivatives is no great help in practice. A study of functions of a complex variable (which is beyond the scope of this book) would indicate that this 'bad behaviour' is also related in some manner to the position of the singularities of the integrand in its analytic extension to the complex domain. Thus $\int_{-5}^{+5}(1 + x^2)^{-1}dx$ is 'badly-behaved' because it has poles at $x = \pm\sqrt{-1}$ which are a small fraction of the range $(b - a) = 10$ away from the real-axis in the complex plane; whereas for the same integral with a range from -0.5 to 0.5, that fraction is much larger. The integral $\int_0^1 \sin(x)dx$ is 'well-behaved' because $\sin(z)$ has no singularities in the finite complex plane; but then neither has $\int_0^1 \exp(10x)dx$, and this is not 'well-behaved', as can be judged by the fact that its higher derivatives – the nth is equal to $10^n \exp(10x)$ – are large compared with unity.

Usually adaptive quadrative (section 5.3) is to be recommended if the integrand *is* 'badly-behaved', and in so far as one may not know whether an integrand deserves this description or not, this may therefore seem a 'safe' method for general use. More sophisticated adaptive routines may employ the Romberg method (section 5.1.3) if the omens are judged to be favourable, and in any event the Romberg method also usually performs well in practice – even though (in the worst extremes) its extrapolative content may be of little use, so that its estimate relies essentially on the composite trapezoidal rule.

The amount of preparation needed in order to implement the high-order methods (section 5.2.2) will most likely mean that these are only considered if they are in a computational library to which one has access: they are especially valuable if heavy computation is needed to evaluate the integrand. But their economy is only manifest if the integrand is 'well-behaved', which brings us back to the question as to how that may be judged. Having set out originally to show that numerical computation is not so straightforward as is sometimes thought, that note of uncertainty seems no bad place at which to stop.

Bibliography

It is the intention in what follows to provide a short list of books which may be found useful for further reading. No attempt is made, however, to refer all the material covered in this book back to its original publication in research papers, unless this is the only source of more information (as happens in one or two instances). Several of the books listed themselves contain comprehensive bibliographies.

NUMERICAL ANALYSIS – GENERAL

There are so many excellent texts on numerical analysis that it seems invidious to attempt a selection. Of the four mentioned in the preface which have had special appeal to this author, the one which demands the least mathematical preparation from the reader is that by Fröberg (1969). That book was written in the expectation that the student had access to a calculator (at the time of its publication, an electromechanical one) or a set of mathematical tables. Geared much more to the use of a current digital computer, and available in a student's paperback edition, is Ralston & Rabinowitz (1978). This contains a wealth of problems (solutions to which are available separately) and detailed bibliographies. Another excellent text with a still wider scope, covering most of the subject matter treated here (and much more besides) is that of Dahlquist & Björck (1974). A similarly comprehensive treatment, but requiring a higher preparatory attainment in mathematical skills, is the more recent book of Stoer & Bulirsch (1980), now also available revised (1983) and in a student's edition.

At the other end of the mathematical spectrum, the simple introductory treatment of McCracken & Dorn (1964) has stood the test of time very well. This provides listings of FORTRAN programs alongside the mathematics, as also does the rather more advanced text of Conte & de Boor (1980), now deservedly in its third edition. Referring again to books with a somewhat lighter touch, an excellent one designed particularly for the non-mathematical specialist is Hamming (1973).

A few introductory mathematical treatments have provided a particularly distinctive approach which has given them a special value. The elementary treatment of Henrici (1964), surviving to an old-age, was one of these. Certainly this could be said of the book by Hildebrand (1974) along with its separate Instructor's Manual, and of the more recent text by Atkinson (1978).

Books containing collected papers by various authors are usually of most appeal to experts, but a few set out to provide surveys which are of value to a wider audience. The volume edited by Evans (1974) is of this kind, describing and comparing a variety of available numerical software. The earlier volumes of Ralston & Wilf (1960, 1967) contain articles on a variety of topics together with flow charts and program text of the algorithms described.

It may happen that the reader has access to a library of computer software, but if programs need to be devised for any purpose, it would be as well first to consult a compilation such as the *Collected Algorithms from CACM*, which is being continually updated on an annual basis by its publishers (the Association of Computing Machinery). The programs in the 'Handbook' series of the journal *Numerische Mathematik* are also of particularly high quality. If working on a microcomputer in BASIC, the software listed in Ruckdeschal (1981) may be useful; the many Pascal users will benefit from the excellent integrated study of language and numerical method in Atkinson & Harley (1983). Those with a pocket calculator will find Henrici (1982) of interest.

All these books of a 'general' category refer to most of the material treated by the author in this book. The following notes relate to more specialised publications which have particular reference to the topics discussed in each chapter. Rice (1983) gives an up-to-date and highly practical account of computational methods in linear algebra, including least-squares approximation and linear regression – major topics which have been excluded from our treatment.

COMPUTATIONAL ARITHMETIC AND EVALUATION TECHNIQUES

Most books on numerical analysis contain an introductory chapter on arithmetic error. The definitive work in this field is by Wilkinson (1963); much of its emphasis is directed towards the elucidation of error in the processes of numerical linear algebra. An extensive discussion of error analysis is contained in Knuth (1981), along with a treatment of the development and analysis of computer algorithms. This book also contains a discussion of computer arithmetic. A more complete discussion of this topic is to be found in Sterbenz (1974).

Series are widely used in analysis as well as computation, and most analysis texts – at least, those above the most elementary level – treat them to some extent. Among the mathematical texts which deal exclusively with infinite series are the 'classics' of Bromwich (1926) and Knopp (1928), as well as the more recent and distinguished book by Hirschman (1978). Various methods of series summation by computer are given in the general textbook of Hamming (1973), to which we referred earlier.

There are also a number of books describing methods of deriving asymptotic series, including de Bruijn (1982) and Copson (1965). Both these are theoretically oriented; a more practical approach is given by Murray (1974) and Oliver (1974). The latter also includes a discussion of methods of convergence acceleration (exemplified at the simplest level by the method of repeated averaging). These techniques are widely used in all areas of successive approximation, and Wimp (1981) provides a scholarly treatment of the theory.

The easiest introduction to continued fractions is that of Khovanskii (1963). The general theory, including that of S-fractions, is discussed fully by Wall (1948).

The problem posed by the algebraic solutions of difference equations is extensively covered by Goldberg (1958). Practical demonstrations of numerical instability will be found in Henrici (1982), cited earlier.

A remarkable collection of series which can be summed precisely, or in terms of special functions, is due to Hansen (1975). It is also frequently of assistance to have access to a compilation of formulae relating to the better-known special functions, such as the one edited by Abramowitz & Stegun (1965), or the tables of integrals, series, and products of Gradshteyn & Ryzhik (1980).

INTERPOLATION

The older 'classical' approach, motivated by the problems of tabular interpolation, is well expounded in Steffensen (1950). A more modern treatment is the highly respected book by Davis (1975), and there is a particularly effective and detailed treatment in the general textbook of Hildebrand (1974), referred to earlier.

Hildebrand also gives a full description of interpolation by rational functions. The original work in this area is described in the recently reissued treatise of Milne-Thompson (1980) which contains much else of relevance to interpolation, including the difference-operator notation. The latter is concisely summarised in Fröberg's (1969) textbook. There is also a detailed description of rational function interpolation in the general text of Stoer & Bulirsch (1980).

There is an extensive literature on spline functions. The first expositions in book form were due to Ahlberg et al. (1967) and the collection of papers edited by Greville (1969). More recent work in this actively researched area is in the excellent 'practical guide' of de Boor (1979). Applications to computational geometry of the kind employed in computer-aided design are described in Faux & Pratt (1978).

The definitive work on Tchebyshev interpolation, as well as on Fourier series, is Lanczos (1956), notable as much for its penetrating insights into practical methods as for its mathematical exposition. More recent books on

Tchebyshev polynomials are by Fox & Parker (1968) and Rivlin (1974) which also contain accounts of nearly-minimax approximation. The coefficients of Tchebyshev series representations of various functions are quoted in Luke (1969) as well as in the handbook of Abramowitz & Stegun (1965), to which reference was made earlier.

Approximation theory has become an extremely popular branch of mathematics, and it is difficult to make a selection of the published work. The book by Davis has already been mentioned. Among many others Achieser (1956), Cheney (1966), and Lorentz (1966) are all highly respected works, all now (unhappily) out-of-print. Approximation by rational functions is especially well treated in Meinardus (1967) and in the collection of lectures, many with a practical orientation, edited by Hanscomb (1966). It is also well covered in the authoritative and exhaustive two-volume treatise on approximation by Rice (1969).

A compilation of computer approximations to special functions is contained in Hart *et al.* (1978), and in Lyusternik *et al.* (1965).

SOLVING EQUATIONS

The most complete treatment of functional iteration is in Traub (1981), which contains a lot of material to be found nowhere else. Other excellent and comprehensive books on the solution of single non-linear equations are by Householder (1970) and Ostrowski (1973), the latter containing a particularly good discussion of the convergence properties of the standard methods, and comments on convergence acceleration techniques. A detailed treatment of methods accessible to the reader without specialised mathematical background is given by Balfour & McTernan (1967). The recent work on modified *regula falsi* originated in a paper of Anderson & Björck (1973).

The solution of polynomial equations is a topic with a very long history. All the specialised texts quoted above deal with the problems it poses, and the general textbook of Ralston & Rabinowitz (1978) provides a particularly good summary of methods of solution, as also does Wilf's paper in the collection of Ralston & Wilf (1960). The report of a conference exclusively given over to the problem, edited by Dejon & Henrici (1969), also forms an excellent introduction as well as listing algorithms. The QD-algorithm is also relevant to this problem – see in particular the textbook of Henrici (1964) and Wall's (1948) treatise on continued functions. A simple but detailed discussion of Sturm's theorem is in Turnbull (1952) and algorithms for polynomials are described in Knuth (1981). The numerical accuracy of solutions is explored in Wilkinson's (1963) treatise on error cited earlier (as well as in subsequent research papers by the same author).

The subject of optimisation is again richly endowed in publications and very diversified in extent; we shall be content to refer to the simple introduction to non-linear optimisation by Dixon (1972) and the more advanced comprehensive treatment of practical problems by Gill *et al.* (1981).

NUMERICAL QUADRATURE

The acknowledged authoritative survey is by Davis & Rabinowitz (1975) which contains computer programs, and selected methods for special forms of integral. Another comprehensive, though now somewhat dated, treatment is by Krylov (1962). The error term in the Cotes formulae is considered in a number of books, including those on interpolation by Steffensen (1950) and Hildebrand (1974). The ideas of extrapolation to the limit, implicit in the Romberg method, are discussed and extended by Oliver (1947) and Wimp (1981).

The Gaussian abscissae and weights can be found in an extensive collection of tables by Stroud & Secrest (1966), as well as in the handbook of Abramowitz & Stegun (1965). Patterson's (1973) method is algorithm 468 (in FORTRAN) of the *Collected Algorithm from CACM*, and is described in the excellent chapter on quadrature in Atkinson's (1978) textbook, cited earlier, which also makes (brief) reference to the transformation method. Automatic numerical integration, adaptive or non-adaptive, is an active area of research, and a comparative survey of software is in the collection of papers edited by Evans (1974). The work on curvilinear integrals originates from Nonweiler (1977).

Before using numerical quadrature, it is of course always advisable to ensure that the integral in question cannot be formally evaluated, even if only in terms of special functions. The Abramowitz handbook, or one of the standard compilations of integrals, such as that of Gradshteyn & Ryzhik (1980), will be found invaluable for this purpose.

REFERENCES

Abramowitz, M. & Stegun, J. A. (eds.): (1965) *Handbook of mathematical functions*. Gannon.

Achieser, N. I.: (1956) *Theory of approximation*. Frederick Ungar Publ. Co.†

Ahlberg, J. H. E. *et al*: (1967) *The theory of splines and their applications*. Academic Press Inc.

Anderson, N. & Björck, Å.: (1973) A new high-order method of *regula falsi* type for computing the root of an equation. *BIT* 13 253–264.

Atkinson, K. E.: (1978) *An introduction to numerical analysis*. John Wiley & Sons.

Atkinson, L. V. & Harley, P. J.: (1983) *An introduction to numerical methods with Pascal*. Addison-Wesley Publ. Co.

Balfour, A. & McTernan, A. J.: (1967) *The numerical solution of equations*. Heinemann.†

Bromwich, T. J. I'a: (1926) *Theory of infinite series*. 2nd edn. Macmillan.†

Cheney, E.W.: (1966) *Introduction to approximation theory*. McGraw-Hill.†

Conte, S. D. & de Boor, C.: (1980) *Elementary numerical analysis, an algorithmic approach*. 3rd edn. McGraw-Hill Kogakusha.

† Denotes volume out-of-print, 1983.

Copson, E. T.: (1965) *Asymptotic expansions*. Cambridge University Press.

Dahlquist, G. & Björck, Å.: (1974) *Numerical methods*. Prentice-Hall Inc.

Davis, P. J.: (1975) *Interpolation and approximation*. Dover.

— & Rabinowitz, P.: (1975) *Methods of numerical integration*. 2nd edn. Academic Press.

de Boor, C.: (1979) *A practical guide to splines*. Springer Verlag.

de Bruijn, N. G.: (1982) *Asymptotic methods in analysis*. Dover.

Dejon, B. & Henrici, P. (eds.): (1969) *Constructive aspects of the fundamental theorem of algebra*. Wiley-Interscience.†

Dixon, L. C. W.: (1972) *Non-linear optimisation*. English Universities Press.

Evans, D. (ed.): (1974) *Software for numerical mathematics*. Academic Press.

Faux, I. D. & Pratt, M. J.: (1978) *Computational geometry for design and manufacture*. Ellis Horwood.

Fox. L. & Parker, I. P.: (1968) *Chebyshev polynomials in numerical analysis*. Oxford University Press.†

Fröberg, C.: (1969) *Introduction to numerical analysis*. 2nd edn. Addison-Wesley Publ. Co.†

Gill, P. E., Murray, W. & Wright, M. H.: (1981) *Practical optimisation*. Academic Press.

Goldberg, S.: (1958) *Introduction to difference equations*. John Wiley & Sons.†

Gradshteyn, I. S. & Ryzhik, I. M.: (1980) *Table of integrals, series & products*. 2nd edn. Academic Press.

Greville, T. N. E. (ed.): (1969) *Theory and application of spline functions*. Academic Press.†

Hamming, R. W.: (1973) *Numerical methods for scientists and engineers*. 2nd edn. McGraw-Hill.

Hanscomb, D. C. (ed.): (1966) *Methods of numerical approximation*. Pergamon.†

Hansen, E. R.: (1975) *A table of series and products*. Prentice-Hall Inc.

Hart, J. F. *et al.*: (1978) *Computer approximations*. Kreiger.

Henrici, P. K.: (1964) *Elements of numerical analysis*. John Wiley & Sons.

— : (1982) *The essentials of numerical analysis with pocket calculator demonstrations*. John Wiley & Sons.

Hildebrand, F. B.: (1974) *Introduction to numerical analysis*. 2nd edn. McGraw-Hill.

Hirschman, I. I.: (1978) *Infinite series*. Greenwood.

Householder, A. S.: (1970) *The numerical treatment of a single non-linear equation*. McGraw-Hill.

Khovanskii, A. N.: (1963) *The application of continued fractions and their generalisations to problems in approximation theory*. P. Noordhoff.†

Knopp, K.: (1928) *Theory and application of infinite series*. Blackie & Sons.†

Knuth, D. E.: (1981) *The art of computer programming*. Vol. 2: Semi-numerical algorithms. 2nd edn. Addison-Wesley Publ. Co.

† Denotes volume out-of-print, 1983.

Krylov, V. I.: (1962) *Approximate calculation of integrals.* Macmillan.†

Lanczos, C.: (1956) *Applied analysis.* Prentice-Hall Inc.†

Lorentz, G. G.: (1966) *Approximation of functions.* Holt, Rinehart & Winston.†

Luke, Y.: (1969) *The special functions and their applications.* Vols. 1 and 2, Academic Press.

Lyusternik, L. A. *et al.*: (1965) *Handbook for computing elementary functions.* Pergamon.†

McCracken, D. D. & Dorn, W. S.: (1964) *Numerical methods and FORTRAN programming.* John Wiley & Sons.†

Meinardus, G.: (1967) *Approximation of functions: theory and numerical methods.* Springer Verlag.

Milne-Thompson, L. M.: (1980) *Calculus of finite differences.* 2nd edn. Chelsea Publ.

Moore, R. E.: (1979) *Methods and applications of interval analysis.* Soc. Indus-Appl. Math.

Murray, J. D.: (1974) *Asymptotic analysis.* Oxford University Press.

Nonweiler, T.: (1977) Integrals and areas expressed as bilinear forms of discrete defining data. *J. Inst. Maths. Applics.,* **19** 159–168.

Oliver, F. W. J.: (1974) *Asymptotic and special functions.* Academic Press.†

Ostrowski, A.: (1973) *Solution of equations in Euclidean and Banach spaces.* 3rd edn. Academic Press.†

Patterson. T.: (1973) Algorithm 468: Algorithm for numerical integration over a finite interval. *Comm. A.C.M.,* **16** 694–699.

Ralston, A. & Rabinowitz, P.: (1978) *A first course in numerical analysis.* 2nd edn. McGraw-Hill Kogakusha.

— & Wilf, H. S. (eds.): (1960, 1967) *Mathematical methods for digital computers.* Vols 1 and 2. John Wiley & Sons.

Rice, J.: (1964, 1969) *The approximation of linear functions.* Vols 1 and 2. Addison-Wesley.

—: (1983) *Matrix computation and mathematical software.* McGraw-Hill International.

Rivlin, T.: (1974) *The Chebyshev Polynomials.* John Wiley & Sons.

Ruckdeschal, F. R.: (1981) *BASIC scientific subroutines.* Vols 1 and 2. McGraw-Hill.

Steffensen, J. F.: (1950) *Interpolation.* Chelsea Publ.

Sterbenz, P.: (1974) *Floating-point computation.* Prentice-Hall.†

Stoer, J. & Bulirsch, R.: (1980) *Introduction to numerical analysis.* Springer Verlag.

Stroud, A. & Secrest, D.: (1966) *Gaussian quadrature formulas.* Prentice-Hall.†

Troub, J. F.: (1981) *Iterative methods for the solution of equations.* 2nd edn. Chelsea Publ.

† Denotes volume out-of-print, 1983.

Turnbull, H.W.: (1952) *Theory of equations*. 5th edn. Oliver & Boyd.†
Wall, H. S.: (1948) *The analytic theory of continued fractions*. Chelsea Publ.
Wilkinson, J. H.: (1963) *Rounding-errors in algebraic processes*. Prentice-Hall.†
Wimp, J.: (1981) *Sequence transformations and their applications*. Academic Press.

† Denotes volume out-of-print, 1983.

Appendix : some numerical subroutines in basic

A library of about 50 subroutines written in a subset of BASIC has been assembled. It implements many of the numerical methods referred to in this book, and a photographic copy of the listing appears at the end of this Appendix.

The BASIC language is not a convenient medium for communicating the structure of numerical algorithms, and explanatory comments have been edited out of the listing. However, in most instances the algorithms used are explained elsewhere in the text. References are given in the descriptions of each subroutine to where these explanations can be found.

All subroutines require a driver program to be supplied. They have been designed to allow the user to observe, and interact with, many of the processes involved. It will also be found that there are groupings of some of the subroutines together which perform computations of considerable complexity (such as minimax approximation).

A.1 THE SUBSET OF BASIC USED

In order to improve the portability of this library, a restricted subset of BASIC is used. Certain features (like the absence of multiple-statement lines) will be immediately apparent. Less obvious restrictions and assumptions are listed below.

1) The only statement types included are arithmetic assignments, FOR. . . NEXT loops, GOTO, GOSUB, IF. . .THEN, PRINT, REM, and RETURN statements.

2) The only operands used are string literals and arithmetic real-type simple, and one- or two-dimensional array, variables. The lower bound of array subscripts is assumed to be zero. The value of any unused variable is taken as undefined.

3) The only operators used are relational and arithmetic. Exponentiation is not used. Multiplication and division are assumed to have equal priority in determining the order of operations. Addition and subtraction are taken to

have lower equal priority. The order of operations is rendered otherwise unimportant by the use of brackets.

4) No assumption is made about the precision of the floating point arithmetic. Its range is supposed to extend between at least 10^{20} and 10^{-20}, or thereabouts.

5) The functions ABS, SGN, EXP, LOG, SQR and INT are used, the last three applied only to non-negative arguments.

6) All FOR statements are guarded to prevent entry into a null-loop. All NEXT statements are tagged with the loop variable. Exit from a FOR-loop is only through the appropriate NEXT statement (unless an irrecoverable failure condition is discerned). Within a loop, there is no assignment to the loop variable, nor to any variables appearing in the TO and STEP expressions. The value of a loop variable is supposed undefined after exhausation of the loop.

7) GOSUB statements are envisaged as implemented by stacking the return address. Some routines use 1 or 2 levels of internal subroutine calls.

8) The THEN-clause of an IF-statement is allowed to be a statement (including another IF-statement).

9) The only separator used in a PRINT-list is a semi-colon, which is supposed to have the effect of concatenation.

10) In normal operation, every subroutine terminates by executing a RETURN statement.

11) All variable names used consist of two alpha-numeric characters.

A.2 HOW THE SUBROUTINES ARE SPECIFIED

Each subroutine is described (in the next section) by a conventionalised specification which has the form

⟨ title ⟩   ⟨ entry spec ⟩   ⟨ entry spec ⟩ . . . .

The ⟨ title ⟩ is a single line of the form:

SUBdddd: ⟨ title ⟩   ⟨ includes ⟩   ⟨ entries ⟩

The ⟨ includes ⟩ statement gives the first and last line numbers of the subroutine in the annexed (photocopy) listing; all line numbers between and including these limits that appear in this listing form the subroutine instructions. The ⟨ entries ⟩ statement gives the numbers of one or more lines in the subroutine to which control may be transferred by a GOSUB statement, for the purpose of performing a defined calculation. For each such line number there is a following ⟨ entry spec ⟩, beginning '*ENTRYdddd*', where *dddd* is that line number.

In each ⟨ entry spec ⟩ the descriptions following the headings *Purpose* and *Method* are self-explanatory. There then follow two lists of variable names and associated descriptions, under the headings *Reads* and *Writes*.

Any simple variable whose name appears under the heading *Reads* must have been assigned a pertinent value (as indicated by the associated description) prior to entering the subroutine at the relevant ⟨ entry spec ⟩ line number. Any one- or two-dimensional array variables under this heading appear respectively as

$$A1 \ (l1 : s1) \text{ or } A2 \ (l1 : s1, l2 : s2)$$

where $l1 : u1$ and $l2 : u2$ are subscript bound-pairs denoting the range of values of the relevant subscript for which array elements must have defined values (as indicated by the description). The expressions $l1, s1, l2, s2$ contain either constants, or the name of a simple variable (appearing elsewhere under the heading *Reads*), implying in the latter instance the value assigned to that variable on entering the subroutine (at the relevant ⟨ entry spec ⟩ line number). In addition $l2$ and $u2$ may contain the dummy variable '*sub1*' to denote the value of the first subscript. Thus for example A2 $(1:N, \emptyset: sub1)$ implies the range of values of A2 (j,k) for $0 \leqslant k \leqslant j$ and $1 \leqslant j \leqslant n$.

Variable names under the heading *Reads* which begin with initial letter I through N refer to, and must be assigned, (small) integer values: failure to observe this restriction will have undefined consequences.

Any simple variable whose name appears under the heading *Writes* is assigned a value (as indicated by the associated description) within the subroutine, before the execution of the RETURN statement which terminates the relevant entry. Array variables appearing under this heading are designated with subscript bound-pair lists (as shown above), in which $l1 = l2 = 0$ and $u1, u2$ are either constants or expressions involving the name of a variable, to denote the range of subscript values of elements to which values *may* be assigned before RETURN is executed. Where $u1$ or $u2$ contain a variable name, the value implied is that assigned to this variable on entry to the subroutine at the relevant ⟨ entry spec ⟩ line number (and this name will also therefore appear under the heading *Reads*). A suitable DIMension statement for each array under the heading *Writes* must be executed *in the calling sequence* prior to entering the subroutine.

Variable names under the heading *Writes* which begin with initial letter I through N refer to, and will only be assigned, (small) integer values. Assignments are made within the subroutine *only* to simple or array variables whose names appear under the heading *Writes*. If the same name appears under both headings *Reads* and *Writes*, it is to be inferred that the variable's entry value(s) will be overwritten; otherwise, however, variables under the heading *Reads* have their values preserved.

Descriptions of both simple and array variables under the headings *Reads* and *Writes* usually take the form of simple equivalencing to symbols introduced and defined under the heading *Purpose* in the ⟨ entry spec ⟩. The convention '=*' is used to refer to a description in the *Programming Notes* appended to the ⟨ entry spec ⟩. Alternative (conditional) descriptions are grouped within brackets [. .]. Where elements of one array variable refer to different vectors (or matrices) defined by different symbols, these descriptions are grouped within braces $\{..\}$. The notation of subscript bound-pair lists is extended to such multiple descriptions in an obvious way. Thus an entry AA $(\emptyset : N , \emptyset : N) \equiv \{$AA $(\emptyset , \emptyset) = d$, AA $(1 : N , \emptyset) \equiv x$, AA $(1 : N , 1 : sub1) \equiv l$, AA $(1 : N , 1 + sub1 : N) \equiv u\}$ under the heading *Writes* would imply a partitioning of AA $(..)$ into a scalar d, a vector x_j, and lower and upper diagonal sub-matrices identified by l_{jk} and u_{jk}, the remaining elements AA $(\emptyset, 1 : N)$ being undefined.

Simple variables with names Nd or Zd where d is a digit 0 through 6, which appear within broken brackets $\langle\rangle$ under the heading *Writes*, are given no description. Likewise certain other simple and array variables, or portions of arrays (as in the example of AA above), may have no description. These variables (or elements) are used as working storage by the subroutine, and their values are undefined on execution of RETURN.

Certain array variables whose names appear under the heading *Reads* are given the description 'preserved'. This happens only if a subroutine is to be entered iteratively, and implies that *all elements of that array must have been preserved* since the call on the associated subroutine which initialises their values.

Following the lists of variables and descriptions, each $\langle$ entry spec $\rangle$ will include, if relevant, the heading *Uses* followed by either SUB500 or GOTO 999, to denote the presence in the body of the subroutine of a GOSUB or GOTO statement which transfers control to the stated (non-local) line number. SUB500 is a subroutine to be supplied by the user. Its specification is given in parenthesis in similar form to $\langle$ entry spec $\rangle$. Line 999 is used as an abnormal failure exit. If a condition is discerned which prevents further execution, then an explanatory message is PRINTed, and the GOTO999 statement executed with the variable N$\emptyset$ set equal to the line number at which the subroutine was entered. The first RETURN statement executed after this GOTO wll transfer control to the calling sequence, as in normal termination. However, the variables under the heading *Writes* will then generally have undefined values, and FOR-loops initiated from within the failing subroutine may still be active unless, and until, de-activated by RETURN. It will be appreciated that a wide range of possible error conditions, including those which are the consequence of non-conformity to specification, cannot be trapped in this manner.

A.3 SUBROUTINE SPECIFICATIONS

A.3.1 Machine arithmetic

SUB6800 : PRECISION. Include 6800–6814; entry 6800

ENTRY 6800

Purpose:	to find the base (b), precision (p) and inverse machine unit ($1/\epsilon = b^{p-1}$) of the real-type number representation of the machine in use.
Method:	See problems 4 and 6, § 1.2.

Reads: (nothing).
Writes: IB = b; IP = p; BP = $1/\epsilon$; $\langle$ N$\emptyset$; N1 $\rangle$.

SUB6900: *ROUNDING METHOD. Include 6900–6970; entries 6900, 6920,*
                                6940, 6960

ENTRY 6900
Purpose: to find the method of rounding used in adding two positive operands.
               The subroutine (in effect) arranges such an operation whose precise
               result would be $(b^{p-1} + f)$, where f is a given positive (non-integer)
               value $\ll b^{p-1} = 1/\epsilon$. The computed result of the operation will be
               $(b^{p-1} + l)$ where l is an integer. The mapping of $f \to l$ and the rela-
               tive error $\delta = (f - l)/(b^{p-1} + l)$ can be explored.
Method: See problem 6, § 1.3.
Reads: IB = b; BP = $1/\epsilon$; FP = f.
Writes: FL = l; ER = δ; $\langle$ Z$\emptyset$; Z1; Z2 $\rangle$.
• *Programming*
notes: IB, BP are composed by SUB6800. FP is interpreted as ABS(FP).

ENTRY 6920
Purpose: to find (similarly) the method of rounding used in differencing two
               positive operands.
Method &c: as for ENTRY 6900.

ENTRY 6940
Purpose: to find (similarly) the method of rounding used in multiplying two
               operands.
Method: See problem 4, § 1.7.
Reads &c: as for ENTRY 6900.

ENTRY 6960
Purpose: to find (similarly) the method of rounding used in forming the
               quotient of two operands.
Method: see problem 5, § 1.7.
Reads &c: as for ENTRY 6900.

A.3.2 Evaluation techniques

SUB7000: *INFINITE SERIES. Include 7000–7028; entries 7000, 7020*

ENTRY 7000
Purpose: To initialise an iterative procedure for forming the sum of an
               indefinite number of terms of the series $\sum\limits_{k=0}^{\infty} t_k$, and to supply t_0.
Method: Initialises workspace for Kahn's device.
Reads: TM = t_0.
Writes: SM = t_0; NS = $\emptyset$; W1($\emptyset$: 3).

ENTRY 7020

Purpose: to be entered iteratively after an initialising call GOSUB7$\emptyset\emptyset\emptyset$, so that by supplying t_m on the mth consecutive entry, the partial sum

$$s_m = \sum_{k=0}^{m} t_k \text{ is computed. The occurrence is noted of } n \text{ successive}$$

entries which have left the computed sum unchanged (i.e. $s_m = s_{m-1} = \ldots = s_{m-n}$).

Method: s_m accumulated using Kahn's device.

Reads: TM = t_m; W1($\emptyset$: 3) *preserved.*

Writes: SM = s_m; NS = n; W1($\emptyset$: 3).

Programming

notes: By way of example, the instruction sequence

2$\emptyset$ TM = 1 : GOSUB7$\emptyset\emptyset\emptyset$

3$\emptyset$ FOR M = 1 TO 99: TM = X∗TM/M: GOSUB7$\emptyset$2$\emptyset$: IF NS = $\emptyset$ THEN NEXT

4$\emptyset$ PRINT SM, NS

will print the saturated sum of exponential series (if NS = 1), or failing this, the partial sum of the first 100 terms (if NS = $\emptyset$).

SUB7050:   *INFINITE CONTD. FRACTION. Include 7000-7090; entries 7050, 7070*

ENTRY 7050

Purpose: To initialise an iterative procedure for evaluating an indefinite number of successive convergents of the continued fraction $\sum_{j=1}^{\infty} p_j \downarrow q_j$, and to supply p_1, q_1.

Method: initialisation of workspace.

Reads: PN = p_1; PD = q_1.

Writes: NS = [if q_1 = 0 then −1 else 0] ; SM = [if NS = 0 then p_1/q_1] ; W1($\emptyset$: 5).

ENTRY 7070

Purpose: To be entered iteratively after an initialising call on GOSUB7$\emptyset$5$\emptyset$, so that on the $(m-1)$th successive entry, the mth convergent $c_m = \underset{j=1}{\overset{m}{K}} \; p_j \downarrow q_j$ is formed from supplied values of the mth partial numerator and denominator (p_m, q_m). The occurrence is noted of n successive entries which have left the computed convergents unchanged (i.e. $c_m = c_{m-1} = \ldots = c_{m-n}$).

Method: see § 2.8, in particular (2.8.1) and problem 9. The increments between successive convergents are summed by Kahn's device.

Reads: PN = p_m; PD = q_m; W1($\emptyset$: 5) *preserved.*

Writes: NS = n; SM = [if NS $\geqslant$ 0 then c_m] ; W1($\emptyset$: 5).

Uses: GOTO 999

Programming

notes: Calling sequence similar to that of SUB7020 (q.v.). NS may be given a negative value (to imply that SM is undefined). Delete the word **PARTIAL** from the error message text at line 7087.

SUB7100: *POLYNOMIAL POWER FORM. Include 7100-7105; ENTRY 7100*

ENTRY 7100

Purpose: to evaluate $y = \sum\limits_{k=0}^{n} a_k x^k$.

Method: Horner's scheme.

Reads: NI = n; AP($\emptyset$: NI) $\equiv a$; XV = x.

Writes: YV = y; $\langle$ N$\emptyset$ $\rangle$.

SUB 7150: *FINITE TCHEBYSHEV EXPN. Include 7150-7160; entry 7150*

ENTRY 7150

Purpose: To evaluate $y = (b_0/2) + \sum\limits_{k=1}^{n} b_k T_k(x)$

Method: Clenshaw's technique.

Reads: NI = n; TC($\emptyset$: NI) $\equiv b$; XV = x.

Writes: YV = y; $\langle$ N$\emptyset$; Z$\emptyset$; Z1; Z2 $\rangle$.

SUB7200: *FINITE CONTD. FRACTION. Include 7200-7257; entries 7200, 7220*

ENTRY 7200

Purpose: To evaluate $y = \mathop{K}\limits_{j=m}^{n} p_j \downarrow q_j$.

Method: reverse recurrence (2.8.2).

Reads: MP = m; NP = n; PN(MP: NP) $\equiv p$; PD(MP: NP) $\equiv q$.

Writes: YV = y; $\langle$ N$\emptyset$; N1; Z$\emptyset$; Z1 $\rangle$.

Uses: GOTO 999.

ENTRY 7220

Purpose: To evaluate $y = r_0 \mathop{K'}\limits_{j=0}^{n} r_j x \downarrow 1$.

Method: reverse recurrence (2.8.2).

Reads: NS = n; RS($\emptyset$: NS) = r; XV = x.

Writes &

Uses: as for **ENTRY 7220**.

SUB7300: *INTERPOLATING CONTD. FRAC. Include 7250-7332; entries 7300, 7320*

ENTRY 7300

Purpose: To evaluate the continued fraction $y = q_0 + \mathop{K}\limits_{j=1}^{n} (x - \xi_{j-1}) \downarrow q_j$ interpolating the data pairs (x_k, y_k) for $k = 0, 1, \ldots n$, where $\xi_j = x_{p_j}$ for $j = 0, 1, \ldots n$ (Normally, $p_j = j$).

Method: reverse recurrence (2.8.2).

Reads: $NI = n$; $XI(\emptyset: NI) \equiv x$; $YI(\emptyset: NI) \equiv y$; $QI(\emptyset: NI) \equiv q$; $NQ(\emptyset: NI) \equiv p$;
$XV = x$.

Writes: $YV = y$; $\langle N\emptyset; NI; N2; Z\emptyset; Z1 \rangle$.

Programming
notes: $QI(..)$ and $NQ(..)$ can be composed by SUB8300.

ENTRY 7320

Purpose: similarly to evaluate the continued fraction $y = \overset{n}{\underset{j=0}{K'}} (x - \xi_{j-1}) \downarrow q_j$.

Method &c: as for ENTRY 7300.

SUB7400: *RATIONAL POWER FORM*. *Include 7400-7419; entry 7400*

ENTRY 7400

Purpose: to evaluate $y = (\overset{m}{\underset{k=0}{\Sigma}} a_{0k} x^k)/(\overset{n}{\underset{k=0}{\Sigma}} a_{1k} x^k)$.

Method: Horner's scheme.

Reads: $MR = m$; $NR = n$; $AP(\emptyset: 1, \emptyset: MR + (NR - MR)*sub1) \equiv a$.

Writes: $YV = y$; $\langle N\emptyset; N1; N2; Z\emptyset \rangle$.

Uses: GOTO 999.

SUB7500: *REVERSED NEWTON FORM*. *Include 7500-7505; entry 7500*

ENTRY 7500

Purpose: to evaluate (3.2.16), namely $y = \overset{n}{\underset{j=0}{\Sigma}} c_j^{(n)} \overset{n}{\underset{k=j+1}{\Pi}} (x - x_k)$.

Method: nested multiplication

Reads: $NI = n$; $CN(\emptyset: NI) \equiv c^{(n)}$; $XI(\emptyset: NI) \equiv x$; $XV = x$.

Writes: $YV = y$; $\langle N\emptyset \rangle$.

Programming
notes: $CN(..)$ may be composed by SUB8200.

SUB7600: *PIECEWISE HERMITIAN INTERP*. *Include 7600-7668; entry 7600*

ENTRY 7600

Purpose: to evaluate the value of $y = y(x)$ as given by (3.3.2), where $\pi_k(x)$
for $k = 1, 2, \ldots n$ is the cubic interpolating the data triads (x_j, y_j, y_j') for $j = k - 1, k$, and where $x_0, x_1, \ldots x_n$ is a strictly monotonic sequence, by identifying $y = \pi_p(x)$ as appropriate.

Method: locates interval by SUB7650 (q.v.).

Reads: $NI = n$; $XI(\emptyset: NI) \equiv x$; $YI(\emptyset: NI) \equiv y$; $DI(\emptyset: NI) \equiv y'$; $XV = x$.

Writes: $YV = y$; $IP = p$; $\langle N\emptyset; N1; N2; N3; N4; Z\emptyset; Z1; Z2; Z3 \rangle$.

Uses: GOTO 999.

SUB7650: *LOCATE INTERVAL*. *Include 7650-7668; entry 7650*

ENTRY 7650

Purpose: given x and a strictly monotonic sequence $x_0, x_1, \ldots x_n$, to find p
such that $x \in (x_{p-1}, x_p)$, the values of x_0 and x_n being taken as
infinite.

Method: see problem 1, § 3.3.

Reads: $NI = n; XI(\emptyset: NI) \equiv x; XV = x.$
Writes: $IP = p; \langle N\emptyset; N1; N2; N3; N4 \rangle.$
Uses: GOTO 999.
Programming
notes: Does not check ordering of XI(. .).

A.3.3 Conversion techniques

<u>SUB7700</u>: *POLYNOMIAL TO TCHEB.EXPN. Include 7700-7734; entries 7700, 7720*

ENTRY 7700
Purpose: To convert $\sum\limits_{j=0}^{n} a_j(x - \xi)^j$ into the form $(b_0/2) + \sum\limits_{j=1}^{n} b_j T_j(x).$

Method: (not described in text).

Reads: $NI = n; AP(\emptyset: NI) \equiv a; XS = \xi.$
Writes: $TC(\emptyset: NI) \equiv b; \langle N\emptyset; N1; N2; Z\emptyset; Z1; Z2 \rangle.$
Programming
notes: May be used (possibly in conjunction with SUB7800) for power series economisation.

ENTRY 7720
Purpose: to convert $\sum\limits_{j=0}^{n} [c_j^{(n)} \prod\limits_{k=j+1}^{n} (x - x_k)]$ into the form $(b_0/2) + \sum\limits_{j=1}^{n} b_j T_j(x).$

Method: not described in text.
Reads: $NI = n; CN(\emptyset: NI) \equiv c^{(n)}; XI(\emptyset: N) \equiv x.$
Writes: $TC(\emptyset: NI) \equiv b; \langle N\emptyset; N1: N2; Z\emptyset; Z1; Z2 \rangle.$

<u>SUB7800</u>: *TCHEB.EXPN. TO POWER FORM. Include 7800-7811; entry 7800*
ENTRY 7800
Purpose: To convert $(b_0/2) + \sum\limits_{j=1}^{n} b_j T_j(x)$ into the form $\sum\limits_{j=0}^{n} a_j x^j.$

Method: not described in text.
Reads: $NI = n; TC(\emptyset: NI) \equiv b.$
Writes: $AP(\emptyset: NI) \equiv a; \langle N\emptyset; N1; N2; Z\emptyset; Z1 \rangle.$

<u>SUB7900</u>: *REVERSED NEWTON TO POWER FORM. Include 7900-7908; entry 7900*

ENTRY 7900
Purpose: to convert $\sum\limits_{j=0}^{n} [c_j^{(n)} \prod\limits_{k=j+1}^{n} (x - x_k)]$ into the form $\sum\limits_{j=0}^{n} a_j(x - \xi)^j.$
Method: similar to problem 1, § 3.2.2.
Reads: $NI = n; CN(\emptyset: NI) \equiv c^{(n)}; XI(\emptyset: NI) \equiv x; XS = \xi.$
Writes: $AP(\emptyset: NI) \equiv a; \langle N\emptyset; N1; Z\emptyset \rangle.$

SUB8000: POWER SERIES TO S-FRAC. Include 8000–8057; entries 8000, 8020

ENTRY 8000

Purpose:	To generate the first $(n + 1)$ coefficients of the S-fraction represen-tation $r_0 \overset{\infty}{\underset{j=0}{K'}} r_j x \downarrow 1$, given the first $(n + 1)$ coefficients of its power series representation $\overset{\infty}{\underset{j=0}{\Sigma}} a_j x^j \ (a_j \neq 0 \ \forall \ j)$.
Method:	the QD-algorithm.
Reads:	NI $= n$; AP($\emptyset$: NI) $\equiv a$.
Writes:	RS($\emptyset$: NI) $\equiv r$; NS $= *$; TR($\emptyset$: NI); $\langle$ N$\emptyset$; N1; Z$\emptyset$; Z1; Z2; Z3 $\rangle$.
Uses:	GOTO 999.
Programming notes:	If NS $= m < n$ then $r_{m+1} = r_{m+2} = \ldots = r_n$ and the S-fraction (apparently) terminates.

ENTRY 8020

Purpose:	the same, except that $\rho_0, \rho_1, \ldots, \rho_n$ are given (instead of the a's) where $\rho_0 = a_0$, and $\rho_j = - a_j/a_{j-1}$ for $j \geqslant 1$.
Method:	the QD-algorithm.
Reads:	NI $= n$; TR($\emptyset$: NI) $\equiv \rho$.
Writes:	as for ENTRY 8000.
Uses:	GOTO 999.
Programming notes:	The preferred data presentation. Note that TR (. .) will be over-written.

SUB8100: CONTD. FRACT. TO RATIONAL. Include 8100–8162; entries 8100, 8120

ENTRY 8100

Purpose:	To convert $q_0 + \overset{n}{\underset{j=1}{K}} (x - \xi_{j-1}) \downarrow q_j$, where $\xi_j = x_{p_j}$, into the form $(\overset{m}{\underset{k=0}{\Sigma}} a_{0k} x^k)/(\overset{n}{\underset{k=0}{\Sigma}} a_{1k} x^k)$.
Method:	adapted from problem 11, § 2.8.
Reads:	NI $= n$; XI($\emptyset$: NI) $\equiv x$; QI($\emptyset$: NI) $\equiv q$; NQ($\emptyset$: NI) $\equiv p$.
Writes:	MR $= m$; NR $= n$; AR($\emptyset$: 1, $\emptyset$; INT((NI + 1)/2)) $\equiv \{$AR($\emptyset$: 1, MR + (NR − MR) $*$ sub1$) \equiv a \}$; $\langle$N$\emptyset$; N1; N2; N3; Z$\emptyset$; Z1; Z2 $\rangle$.
Programming notes:	QI(. .) and NQ(. .) are composed by SUB8300.

ENTRY 8120

Purpose:	similarly to convert $\overset{n}{\underset{j=0}{K'}} (x - \xi_{j-1}) \downarrow q_j$ into rational power form.
Method &c:	as for ENTRY 8100.

SUB8150: FINITE S-FRAC TO RATIONAL. *Includes 8150-8180; entry 8150*

ENTRY 8150

Purpose: To convert $r_0 \overset{n'}{\underset{j=0}{K}} r_j x \downarrow 1$ into the form $(\overset{m}{\underset{k=0}{\Sigma}} a_{0k} x^k)/(\overset{n}{\underset{k=0}{\Sigma}} a_{1k} x^k)$.

Method: see problem 11, § 2.8.

Reads: NS $= n$; RS($\emptyset$: NS) $\equiv r$.

Writes: MR $= m$; NR $= n$;
              AR($\emptyset$: 1, $\emptyset$: INT((NS + 1)/2)) $\equiv \{$AR($\emptyset$: 1, MR + (NR $-$ MR) $*$
              $sub1) \equiv a\}$; $\langle$ N$\emptyset$; N1; Z $\emptyset$ $\rangle$.

A.3.4 Interpolation techniques

SUB8200: REVERSED NEWTON FORM. *Include 8200-8290: entries 8200,*
                    8220, 8240, 8260, 8280

ENTRY 8200

Purpose: to compose $p_n(x) = \overset{n}{\underset{j=0}{\Sigma}} [c_j^{(n)} \overset{n}{\underset{k=j+1}{\Pi}} (x - x_k)]$, as in equation
              (3.2.16), interpolating (x_k, y_k) for $k = 0, 1, \ldots n$, using the con-
              vention that successive identical values of $x_k = x_{k+1} = \ldots = x_{k+p}$
              imply that $y_k, y_{k+1}, \ldots y_{k+p}$ respectively equal $y_k, y_k', \ldots y_k^{(p)}$.
              (See § 3.2.4).

Method: algorithm (3.2.15) for divided difference table, extended as in
              (3.2.22) to include derivative values.

Reads: NI $= n$; XI($\emptyset$: NI) $\equiv x$; YI($\emptyset$: NI) $\equiv y$.

Writes: CN($\emptyset$: NI) $\equiv c^{(n)}$; $\langle$ N$\emptyset$; N1; N2; Z$\emptyset$; Z1; Z2 $\rangle$.

Uses: GOTO 999.

Programming

notes: Identical support points must be consecutive.

ENTRY 8220

Purpose: to modify the vector of support points, and to change the co-
              efficients $c_j^{(n)}$ to $c_j^{(n+1)}$ where $n \geqslant -1$, by interpolating an extra
              data pair (ξ, η).

Method: as for ENTRY 8200.

Reads: NI $= n$; XI($\emptyset$: NI) $\equiv x$; CN($\emptyset$: NI) $\equiv c^{(n)}$; XI $= \xi$; YI $= \eta$.

Writes: XI($\emptyset$: NI + 1) $\equiv x$; CN($\emptyset$: NI + 1) $\equiv c^{(n+1)}$; $<$ N$\emptyset$; N1; Z$\emptyset$; Z1 $\rangle$.

Uses: GOTO 999.

Programming

notes: as for ENTRY 8200. Note that YI(NI + 1) is not assigned.

ENTRY 8240

Purpose: To modify the vector of support points, and to change the co-
              efficients $c_j^{(n)}$ to $c_j^{(n+1)}$ for $j = 0, 1, \ldots (n + 1)$ where $n \geqslant 0$, by
              adding an extra support point $x = \xi$ without changing the inter-
              polating polynomial (that is, the ordinate at $x = \xi$ is $p_n(\xi)$). The
              support at $x = x_{n-\nu}$ is (as a consequence) withdrawn, where $\nu \leqslant n$

Method: as for ENTRY 8200.

Reads: NI = n; XI($\emptyset$: NI) $\equiv x$; CN($\emptyset$: NI) $\equiv c^{(n)}$; XI = ξ.

Writes: XI($\emptyset$: NI + 1) $\equiv x$; CN($\emptyset$: NI + 1) $\equiv c^{(n+1)}$; YI = $p_n(\xi)$; $\langle$ N$\emptyset$; Z$\emptyset$ $\rangle$.

Programming

notes: If ξ is already a support point, the effect is to increase the multi-
             plicity of contact at that point by 1.

ENTRY 8260

Purpose: To modify the vector of support points, and to change the co-
             efficients $c_j^{(n)}$ to $c_j^{(n-1)}$ for $j = 0, 1, \ldots (n-1)$, by removing $x = x_i$
             from the interpolating scheme, where $0 \leqslant i \leqslant n$.

Method: see problem 9, § 3.2.2.

Reads: NI = n; XI($\emptyset$: NI) $\equiv x$; CN($\emptyset$: NI) $\equiv c^{(n)}$; IR = i.

Writes: XI($\emptyset$: NI − 1) $\equiv x$; CN($\emptyset$: NI − 1) $\equiv c^{(n-1)}$; $\langle$ N$\emptyset$; Z$\emptyset$ $\rangle$.

Programming

notes: The element x_i is removed from XI(. .) but the order is otherwise
             unchanged. If $x = x_i$ is a multiple point of interpolation, the multi-
             plicity of contact is reduced by one.

ENTRY 8280

Purpose: As for ENTRY 826$\emptyset$, but to modify also the vector of ordinate
             values $\{y_k: k = 0, 1, \ldots n\}$ assuming the convention for repeated
             support points as in entry 8200.

Method: If $x = x_i$ is a multiple point, the highest derivative value is removed.

Reads: NI = n; XI($\emptyset$: NI) $\equiv x$; YI($\emptyset$: NI) $\equiv y$; CN($\emptyset$: NI) $\equiv c^{(n)}$; IR = i.

Writes: XI($\emptyset$: NI − 1) $\equiv x$; YI($\emptyset$: NI − 1) $\equiv y$; CN($\emptyset$: NI − 1) $\equiv c^{(n-1)}$;
             $\langle$ N$\emptyset$; Z$\emptyset$ $\rangle$.

<u>SUB8300:</u> *CONTD. FRACTION INTERPOLATION. Include 8300–8348:*
              entries 8300, 8320

ENTRY 8300

Purpose: To compose the continued fraction $q_0 + \overset{m}{\underset{j=1}{K}} (x - \xi_{j-1}) \downarrow q_j$, with
             $m \leqslant n$, interpolating the distinct data (x_k, y_k) for $k = 0, 1, \ldots n$,
             where $\xi_j = x_{p_j}$ (and $p_j = j$ if possible).

Method: Composes inverted difference table (3.5.4), reordering data if
             necessary.

Reads: NI = n; XI($\emptyset$: NI) $\equiv x$; YI($\emptyset$: NI) $\equiv y$.

Writes: NQ = m; QI($\emptyset$: NI) $\equiv \{$QI($\emptyset$: NQ) $\equiv q\}$; NQ($\emptyset$: NI) $\equiv p$; $\langle$ N$\emptyset$; N1;
             N2; Z$\emptyset$; Z1; Z2 $\rangle$.

Programming

notes: If NQ = $m < n$, then $q_{m+1} = \infty$, and the interpolating rational
             function is reducible.

ENTRY 8320

Purpose: similarly to compose the interpolating continued fraction
             $\overset{n}{\underset{j=0}{K'}} (x - \xi_{j-1}) \downarrow q_j$.

Method &c: as for ENTRY 8300.

SUB8400: CUBIC SPLINE INTERPOLATION. Include 8400–8570; entries
                              8400, 8420, 8440
ENTRY 8400

Purpose: to compose the $(m + 1)$-vectors of distinct and ordered support points x_j, ordinates y_j and derivative values y'_j, corresponding to the cubic spline satisfying the 'not-a-knot' end-condition and interpolating the $(n + 1)$ data pairs (x_k, y_k) for $k = 0, 1, \ldots n$ with $n \geqslant m$, where $x_0, x_1, \ldots x_n$ is a monotonic sequence containing $(n - m)$ identical pairs of values. If $x_k = x_{k+1}$ is such a pair, then y_k is taken as prescribed ordinate, and $y'_k = y_{k+1}$ as prescribed derivative at $x = x_k$.

Method: Constructs and solves the tridiagonal system as in problem 3, § 3.4, modified to include the appropriate end-condition and the presence of prescribed derivative values (if any).

Reads: NI = n; XI($\emptyset$: NI) $\equiv x$, YI($\emptyset$: NI) $\equiv y$.

Writes: NR = m; XI($\emptyset$: NI) $\equiv \{$XI($\emptyset$: NR) $\equiv x\}$; YI($\emptyset$: NI) $\equiv \{$YI($\emptyset$: NR) $\equiv y\}$; DI($\emptyset$: NI) $\equiv \{$DI($\emptyset$: NR) $\equiv y'\}$; AA($\emptyset$: 2, $\emptyset$: NI); $\langle$ N$\emptyset$, N1, N2, Z$\emptyset$, Z1, Z2, Z3 $\rangle$.

Uses: GOTO 999.

Programming
notes: If $m < n$, then the repeated support points are removed from XI(. .) and the prescribed derivative values removed from YI(. .), leaving XI(..) a strictly monotonic sequence of distinct values. Note that if y'_0 and y'_m are prescribed then the end-conditions are redundant. SUB7600 may be used for evaluation of interpolated values.

ENTRY 8420
Purpose: as in ENTRY 8400 except that the cubic spline has quadratic end-pieces.

Method &c: as for ENTRY 8400.

ENTRY 8440
Purpose: as in ENTRY 8400 except that a natural cubic spline is used (satisfying the free-end conditions).

Method &c: as for ENTRY 8400.

SUB8500: BESSEL INTERPOLATION. Include 8500–8524; entry 8500
ENTRY 8500

Purpose: To compose the derivative values y'_j for $j = 0, 1, \ldots n$ at the support points appropriate to the piecewise cubic Bessel interpolation of the $(n + 1)$ data pairs (x_k, y_k) for $k = 0, 1, \ldots n$, the sequence $x_0, x_1, \ldots x_n$ being assumed strictly monotonic.

Method: uses equation (3.3.3).

Reads: NI = n; XI($\emptyset$: NI) $\equiv x$; YI($\emptyset$: NI) $\equiv y$.

Writes: DI($\emptyset$: NI) $\equiv y'$; $\langle$ N$\emptyset$; N1; Z$\emptyset$; Z1; Z2; Z3 $\rangle$.

Uses: GOTO 999.

Programming
notes: If a derivative value is prescribed at any support point, then the appropriate element of DI(. .) should be overwritten *after* the RETURN to the calling sequence. SUB7600 may be used for evaluation of interpolated values.

A.3.5 Solving equations

SUB8550: TRIDIAGONAL SYSTEM. Include 8550–8568; entry 8550

ENTRY 8550

Purpose: To solve the linear system C x = y for x, when x and y are n-vectors and C is an $n \times n$ tridiagonal matrix with unit diagonal elements and terms l_j below, and u_j above, the diagonal in row j. (The subscript range is taken as 0 to $(n - 1)$, inclusive.)

Method: Gaussian elimination and back substitution. (Not described in text.)

Reads: NA = $n - 1$; AA($\emptyset$: 2, $\emptyset$: NA) $\equiv$ {AA($\emptyset$, 1: NA) $\equiv l$, AA(1, $\emptyset$: NA − 1) $\equiv u$, AA(2, $\emptyset$: NA) $\equiv y$}.

Writes: AA(1: 2, $\emptyset$: NA) $\equiv$ {AA(2, $\emptyset$: NA) $\equiv x$}; ⟨ N$\emptyset$, N1, N2, Z$\emptyset$ ⟩.

Uses: GOTO 999.

Programming
notes: AA(1, $\emptyset$: NA) is overwritten and undefined.

SUB8600: LINEAR SYSTEM. Include 8600–8669; entries 8600, 8650

ENTRY 8600

Purpose: To perform a triangular decomposition of the $n \times n$ (non-singular) matrix A = (a_{jk}), finding d = det A, and to solve the linear system Ax = y for x, where x and y are n-vectors. (The subscript range is taken as 1 to n, inclusive.)

Method: Crout's method, using partial pivoting with implicit equilibration, and accumulating inner products using Kahn's device. (Not described in text.)

Reads: NA = n; AA(1: NA, 1: NA + 1) $\equiv$ {AA(1: NA, 1: NA) $\equiv a$; AA(1: NA, NA + 1) $\equiv y$}.

Writes: AA($\emptyset$: NA, $\emptyset$: NA + 1) $\equiv$ {AA($\emptyset$. $\emptyset$) = d; AA(1: NA, $\emptyset$) $\equiv x$; AA($\emptyset$: NA, 1: NA) = *}.

Uses: GOTO 999.

Programming
notes: AA($\emptyset$: NA, 1: NA) contains the decomposed matrix, and AA(1: NA, NA + 1) is overwritten (and undefined).

ENTRY 8650

Purpose: To solve a linear system with the same coefficient matrix A as has already been decomposed by GOSUB 8600, but with a different given vector y.

Method: as for ENTRY 8600.

Reads: NA = n; AA($\emptyset$: NA, 1: NA + 1) $\equiv$ {AA($\emptyset$: NA, 1: NA) *preserved;* AA(1: NA, NA + 1) $\equiv y$}.

Writes: as for ENTRY 8600.

SUB8700: TWO-TERM FUNCTION ITERATION. Include 8700–8766; entries 8700, 8750

ENTRY 8700

Purpose: To initialise any one of a choice (denoted by code m) of two-term iterative procedures for the solution of $f(x) = 0$, given two distinct starting values x_0, x_1, and to compute the first iterated root estimate, x_2.

Method: $m \leqslant 0$ for secant method (§ 4.2.1), $m = 1$ for interval bisection (§ 4.1.1), $m = 2$ for *regula falsi* (§ 4.1.2), and $m \geqslant 3$ for various modified methods of *regula falsi* (§ 4.1.3), viz. $m = 3$ for Illinois, and $m = 4$ for Pegasus method, $m = 5$ for hyperbolic, and $m \geqslant 6$ for parabolic, interpolative modification.

Reads: $MD = m$; $X\emptyset = x_0$; $F\emptyset = f(x_0)$; $X1 = x_1$; $F1 = f(x_1)$.

Writes: $XN = x_2$; $DX = x_2 - x_1$; $IB = *$; $SX = |x_2 - x_1|$; $WQ(\emptyset: 7)$; $\langle N\emptyset \rangle$.

Uses: GOTO 999.

Programming

notes: IB is set equal to -1 if root is bracketed (i.e. $F\emptyset * FI < \emptyset$), else zero.

ENTRY 8750

Purpose: To be entered iteratively after an initialising call GOSUB 8700, so that on the $(n - 1)$th successive entry to line 8750, the value of $f(x_n)$ is read-in, and the values of x_{n+1} and the bracketing span s_{n+1} are returned, as calculated by the chosen method. (Except that, if a bracketed method is chosen, but the root is not as yet bracketed, the secant method will be substituted).

Method: as initialised; see ENTRY 8700.

Reads: $FX = f(x_n)$; $W2(\emptyset: 7)$ *preserved*.

Writes: $XN = x_{n+1}$; $DX = x_{n+1} - x_n$; $IB = *$; $SX =$ [if $IB < 0$ and $m > 0$ then s_{n+1} else $|x_n - x_{n-1}|$] ; $W2(\emptyset: 7)$; $\langle N\emptyset \rangle$.

Uses: GOTO 999.

Programming

notes: IB is set equal to -1 if root is bracketed, else zero. A root, once bracketed, remains so, except in the secant method ($m = 0$).

SUB8800: POLYNOMIAL STURM SEQUENCE. Include 8800–8832, entry 8800

ENTRY 8800

Purpose: To generate a polynomial Sturm sequence $\{f_j(x): j = 0, 1, \ldots m\}$ in power form as $f_j(x) = \sum_{k=0}^{n} b_{jk}x^k$, given n and $b_{0k} = a_k$ for $k = 0, 1, \ldots n$, taking $f_1(x) = f_0'(x)$, and finding the genuine degree v_j of $f_j(x)$ for $j \geqslant 1$.

Method: Euclidean algorithm (§ 4.4.2), using forced cancellation where necessary.

Reads: $\quad$ NP $= n$; AP($\emptyset$: NP) $\equiv a$.

Writes: $\quad$ NS $= m$; SS($\emptyset$: NP, $\emptyset$: NP) $\equiv \{$ SS($\emptyset$, $\emptyset$: NP) $\equiv a$; SS(1: NS, 0: v_{sub1}) $\equiv b$; SS(1: NS, NP) $\equiv v \}$; $\langle$ N$\emptyset$; N1; N2; N3; N4; Z$\emptyset$; Z1; Z2; Z3 $\rangle$.

<u>*SUB8900:*</u> *BRACKET POLYNOMIAL ZEROS. Include 8900–8949; entry 8900*

ENTRY 8900

Purpose: $\quad$ To number and bound the discrete real zeros $\{ x_k \in (a_k, b_k)$: $k =$ 1, 2, $\ldots r \}$ of the polynomial of degree n which is the base of a given Sturm sequence of $(m + 1)$ elements.

Method: $\quad$ uses Sturm's theorem, and initial root-bounds developed as in problem 3, § 4.4.

Reads: $\quad$ NP $= n$; NS $= m$; SS($\emptyset$: NS, $\emptyset$: NP) $\equiv *$.

Writes: $\quad$ NR $= r$; RB($\emptyset$: 1, $\emptyset$: NP) $\equiv \{$ RB($\emptyset$, 1: NR) $\equiv a$; RB(1, 1: NR) $\equiv b \}$; $\langle$ N$\emptyset$; N1; N2; N3; N4; N5; Z$\emptyset$; Z1; Z2 $\rangle$.

Programming
notes: $\quad$ SS(. .) is as composed by GOSUB 8800. The particular polynomial $(ax^{2p} - a)$ is not correctly dealt with.

A.3.6 Numerical quadrature

<u>*SUB9000:*</u> *ROMBERG METHOD. Include 9000–9075; entries 9000, 9050*

ENTRY 9000

Purpose: $\quad$ To initialise workspace for the Romberg tableau $I_{j,k}$ of estimates of the integral mean $(b - a)^{-1} \int_a^b f(x)\,dx$, and to calculate $I_{1,1}$ by the composite trapezoidal rule using n equal segments ($n \geqslant 1$).

Method: $\quad$ see § 5.1.3.

Reads: $\quad$ XA $= a$; XB $= b$; NI $= n$.

Writes: $\quad$ QR($\emptyset$: 1) $\equiv \{$ QR($\emptyset$) $= 1$, QR(1) $= I_{1,1} \}$; NQ $= 1$; W$\emptyset$($\emptyset$; 6); XQ.

Uses: $\quad$ GOSUB5$\emptyset\emptyset$ (*SUB5$\emptyset\emptyset$: Reads* XQ $= x$. *Writes* FQ $= f(x)$).

ENTRY 9050

Purpose: $\quad$ To be entered iteratively after an initialising call GOSUB 9000, so that on the $(j - 1)$th entry, the jth row $i_k = I_{j,k}$ of the Romberg table is formed for $1 \leqslant k \leqslant \min(j, m)$, and the increments e_k recorded, where $e_1 = |I_{j,1} - I_{j-1,1}|$, $e_k = |I_{j,k} - I_{j-1,k-1}|$ for $k = 2, 3, \ldots \min(m, j)$, and e_0 the subscript value of the smallest value of e_k.

Method: $\quad$ see § 5.1.3.

Reads: $\quad$ MQ $= m$; QR($\emptyset$: MQ) preserved; W$\emptyset$($\emptyset$: 6) preserved.

Writes: $\quad$ NQ $= \min(j, m)$; QR($\emptyset$: MQ) $\equiv \{$ QR($\emptyset$) $= j$; QR(1: NQ) $\equiv i \}$; ER($\emptyset$: MQ) $\equiv \{$ ER($\emptyset$: NQ) $\equiv e \}$; W$\emptyset$($\emptyset$: 6); XQ.

Uses: $\quad$ GOSUB5$\emptyset\emptyset$ (*SUB5$\emptyset\emptyset$: Reads* XQ $= x$. *Writes* FQ $= f(x)$).

<u>*SUB9100:*</u> *ADAPTIVE SIMPSON. Include 9100–9147; entry 9100*

ENTRY 9100

Purpose:	to evaluate the integral mean $I = (b - a) \int_a^b f(x)\,dx$ with error judged less than $|\epsilon|$ in magnitude, by comparing Milne's rule with Simpson's on successively reducing bisected partitions of the interval of integration (subject to an over-riding limit of n such bisections), and to estimate the error magnitude (δ) of the mean.
Method:	adaptive Simpson's rule (§ 5.3).
Reads:	XA $= a$; XB $= b$; ER $= \epsilon$; MS $= 2n + 13$.
Writes:	AS $= I$; ES $= \delta$; LM $= *$; ST($\emptyset$: MS); XQ.
Uses:	GOSUB5$\emptyset\emptyset$ (*SUB5$\emptyset\emptyset$: Reads* XQ $= x$. *Writes* FQ $= f(x)$.).
Programming notes:	LM $= 1$ if the limit on stacking level (n) has been imposed, else 0.

SUB9200: TRANSFORMATION METHOD. Include 9200–9253; entries 9200, 9250

ENTRY 9200

Purpose:	to estimate the integral $I = \int_0^1 x^{a-1}(1-x)^{b-1}f(x)\,dx$ by a transformation of the variable of integration (with parameter value m) using the composite trapezoidal rule with n equal segments applied to the transformed integrand.
Method:	problem 3, § 5.2.3.
Reads:	AI $= a$; BI $= b$; NI $= n$; PM $= m$.
Writes:	TI $= I$; W$\emptyset$($\emptyset$: 9); XQ; $\langle$ N$\emptyset$; N1; Z$\emptyset$; Z1; Z2 $\rangle$.
Uses:	SUB5$\emptyset\emptyset$ (*SUB5$\emptyset\emptyset$: Reads* XQ $= x$. *Writes* FQ $= f(x)$.).
Programming notes:	The constant (88) assigned to W$\emptyset$(8) on line 92$\emptyset\emptyset$ is machine dependent. The value required is (nearly) the largest positive number p (say) such that EXP($-p$) is neither zero, nor under-flows.

ENTRY 9250

Purpose:	to be entered iteratively after an initialising call GOSUB9200, so that on the jth successive entry, the integral I is estimated using the trapezoidal rule with $(2^j n)$ equal segments applied to the transformed integrand.
Method:	as for ENTRY 9200.
Reads:	W$\emptyset$($\emptyset$: 9) preserved.
Writes:	TI $= I$; NI $= 2^j n$; W$\emptyset$($\emptyset$: 9); XQ; $\langle$ N$\emptyset$; N1; Z$\emptyset$; Z1; Z2 $\rangle$.
Uses:	SUB5$\emptyset\emptyset$ (*SUB5$\emptyset\emptyset$: Reads* XQ $= x$. *Writes* FQ $= f(x)$.).

A.4 SUBROUTINE LISTING

All subroutines have been tested, but no check has been made of the accuracy of this following (machine-generated) listing.

For machine-readable copy, please contact the publishers.

```
6799 REM *** BEGIN SUB6800 ***          7053 W1(4) = 1
6800 IP = 0                             7054 IF PD = 0 THEN GOTO 7058
6801 BP = 1                             7055 SM = PN * W1(4) / PD
6802 N1 = 2                             7056 W1(4) = PD
6803 IF 1.1 - 1 = 0.1 THEN N1 = 10      7057 GOTO 7074
6804 IP = IP + 1                        7058 W1(4) = PN * W1(4)
6805 BP = BP * N1                        7059 GOTO 7077
6806 IF (BP + 1) - BP = 1 THEN GOTO 6804 7070 IF W1(3) < 0 THEN GOTO 7079
6807 N0 = 0                             7071 W1(4) = PN / W1(4) + PD
6808 IB = 1                             7072 IF W1(4) / 16 + PD = PD THEN GOTO 7076
6809 N0 = N0 + 1                        7073 SM = (PD / W1(4) - 1) * W1(5)
6810 IB = IB * N1                       7074 W1(5) = SM
6811 IF (BP + IB) - BP <> IB THEN GOTO 6809 7075 GOTO 7021
6812 IP = IP / N0                       7076 W1(4) = PD * W1(5)
6813 BP = BP / IB                       7077 NS = - 2
6814 RETURN                             7078 GOTO 7027
6815 REM ***** END SUB6800 ***          7079 IF W1(3) = - 1 THEN GOTO 7054
6899 REM *** BEGIN SUB6900 ***          7080 IF PN = 0 THEN GOTO 7085
6900 Z0 = BP + ABS (FP)                 7081 W1(4) = W1(4) / PN
6901 GOTO 6922                          7082 SM = PD * W1(4)
6920 GOSUB 6963                         7083 NS = - 1
6921 Z0 = (BP + Z1) - (Z1 - ABS (FP))   7084 GOTO 7022
6922 Z2 = BP                            7085 PRINT
6923 FL = Z0 - Z2                       7086 PRINT "SUB7070 FAILS"
6924 ER = ( ABS (FP) - FL) / Z0         7087 PRINT "PN=0 MAKES PARTIAL ";
6925 RETURN                             7088 PRINT "DENOMINATOR ZERO"
6940 GOSUB 6967                         7089 N0 = 7070
6941 Z0 = (BP + Z1) * Z2 / BP           7090 GOTO 999
6942 Z2 = Z1 + Z2                       7091 REM ***** END SUB7050 ***
6943 GOTO 6923                          7099 REM *** BEGIN SUB7100 ***
6960 GOSUB 6967                         7100 YV = AP(NI)
6961 Z0 = BP * (Z2 - Z1) / (BP - Z1)    7101 IF NI <= 0 THEN RETURN
6962 GOTO 6923                          7102 FOR N0 = NI - 1 TO 0 STEP - 1
6963 Z1 = 1                             7103 YV = YV * XV + AP(N0)
6964 IF ABS (FP) <= Z1 THEN RETURN      7104 NEXT N0
6965 Z1 = IB * Z1                       7105 RETURN
6966 GOTO 6964                          7106 REM ***** END SUB7100 ***
6967 GOSUB 6963                         7149 REM *** BEGIN SUB7150 ***
6968 Z1 = IB * Z1                       7150 Z0 = 0
6969 Z2 = (BP / Z1) * ABS (FP) + BP     7151 Z1 = 0
6970 RETURN                             7152 IF NI <= 0 THEN GOTO 7159
6971 REM ***** END SUB6900 ***          7153 Z2 = 2 * XV
6998 REM *** BEGIN SUB7000 ***          7154 FOR N0 = NI TO 1 STEP - 1
6999 REM *** BEGIN SUB7050 ***          7155 YV = Z1 * Z2 - Z0 + TC(N0)
7000 SM = TM                            7156 Z0 = Z1
7001 NS = 0                             7157 Z1 = YV
7002 W1(0) = SM                         7158 NEXT N0
7003 W1(1) = 0                          7159 YV = TC(0) / 2 + XV * Z1 - Z0
7004 GOTO 7026                          7160 RETURN
7020 SM = TM                            7161 REM ***** END SUB7150 ***
7021 NS = 0                             7199 REM *** BEGIN SUB7200 ***
7022 W1(0) = ((W1(1) - W1(2)) + W1(0)) + SM 7200 YV = 0
7023 W1(1) = W1(2)                      7201 IF NP < MP THEN RETURN
7024 SM = W1(2) + W1(0)                 7202 N0 = - 7200
7025 IF SM = W1(1) THEN NS = W1(3) + 1  7203 FOR N1 = NP TO MP STEP - 1
7026 W1(2) = SM                         7204 Z0 = PN(N1)
7027 W1(3) = NS                         7205 Z1 = PD(N1)
7028 RETURN                             7206 GOTO 7227
7029 REM ***** END SUB7000 ***          7207 PRINT
7050 SM = 0                             7208 PRINT "SUB";N0;" FAILS"
7051 NS = - 1                           7209 PRINT "0/0 AT ZERO PARTIAL ";
7052 GOSUB 7002                         7210 PRINT "NUMERATOR ";N1
```

```
7211 GOTO 999
7220 YV = RS(NS)
7221 IF NS <= 0 THEN RETURN
7222 N0 = - 7220
7223 Z1 = 1
7224 FOR N1 = NS - 1 TO 0 STEP - 1
7225 YV = YV * XV
7226 Z0 = RS(N1)
7227 GOSUB 7250
7228 IF N0 > 0 THEN GOTO 7207
7229 NEXT N1
7230 RETURN
7249 REM *** BEGIN SUB7300 ***
7250 YV = YV + Z1
7251 IF YV = 0 THEN GOSUB 7254
7252 YV = Z0 / YV
7253 RETURN
7254 YV = 5E - 20
7255 IF Z1 <> 0 THEN YV = YV * Z1
7256 IF Z0 = 0 THEN N0 = - N0
7257 RETURN
7258 REM ***** END SUB7200 ***
7300 N0 = 2
7301 GOSUB 7321
7302 YV = YV + QI(0)
7303 RETURN
7320 N0 = 1
7321 YV = 0
7322 N2 = N0 - 2
7323 IF NI <= N2 THEN RETURN
7324 FOR N1 = NI - 1 TO N2 STEP - 1
7325 IF N0 < 0 THEN GOTO 7331
7326 Z0 = 1
7327 IF N1 >= 0 THEN Z0 = XV - XI(NQ(N1))
7328 Z1 = QI(N1 + 1)
7329 GOSUB 7250
7330 IF N0 < 0 THEN YV = YI(NQ(N1))
7331 NEXT N1
7332 RETURN
7333 REM ***** END SUB7300 ***
7399 REM *** BEGIN SUB7400 ***
7400 N0 = MR
7401 FOR N1 = 0 TO 1
7402 YV = AR(N1,N0)
7403 IF N0 <= 0 THEN GOTO 7407
7404 FOR N2 = N0 - 1 TO 0 STEP - 1
7405 YV = YV * XV + AR(N1,N2)
7406 NEXT N2
7407 N0 = NR
7408 IF N1 = 0 THEN Z0 = YV
7409 NEXT N1
7410 IF YV = 0 THEN GOSUB 7413
7411 YV = Z0 / YV
7412 RETURN
7413 YV = 5E - 20
7414 IF Z0 <> 0 THEN RETURN
7415 PRINT
7416 PRINT "SUB7400 FAILS"
7417 PRINT "VALUE 0/0"
7418 N0 = 7400
7419 GOTO 999
7420 REM ***** END SUB7400 ***
```

```
7499 REM *** BEGIN SUB7500 ***
7500 YV = CN(0)
7501 IF NI = 0 THEN RETURN
7502 FOR N0 = 1 TO NI
7503 YV = (XV - XI(N0)) * YV + CN(N0)
7504 NEXT N0
7505 RETURN
7506 REM ***** END SUB7500 ***
7599 REM *** BEGIN SUB7600 ***
7600 N0 = 7600
7601 GOSUB 7651
7602 IF N0 < 0 THEN GOTO 7663
7603 Z0 = XI(IP) - XI(IP - 1)
7604 IF Z0 = 0 THEN GOTO 7663
7605 Z1 = (XV - XI(IP - 1)) / Z0
7606 Z2 = (XI(IP) - XV) / Z0
7607 YV = 0
7608 FOR N0 = IP - 1 TO IP
7609 YV = (YI(N0) * (2 - Z2 + Z1) + DI(N0) *
     Z0 * Z1) * Z2 * Z2 + YV
7610 Z3 = Z2
7611 Z2 = Z1
7612 Z1 = Z3
7613 Z0 = - Z0
7614 NEXT N0
7615 RETURN
7649 REM *** BEGIN SUB7650 ***
7650 N0 = 7650
7651 IP = 1
7652 N3 = SGN (XI(NI) - XI(0))
7653 IF N3 = 0 THEN GOTO 7661
7654 N4 = NI
7655 IF N4 <= IP THEN RETURN
7656 N1 = INT ((N4 + IP) / 2)
7657 N2 = SGN (XI(N1) - XV) - N3
7658 IF N2 = 0 THEN N4 = N1
7659 IF N2 <> 0 THEN IP = N1 + 1
7660 GOTO 7655
7661 N0 = - N0
7662 IF N0 <> - 7650 THEN RETURN
7663 N0 = ABS (N0)
7664 PRINT
7665 PRINT "SUB";N0;" FAILS"
7666 PRINT "X-VALUES NOT STRICTLY MONOTONIC"
7667 PRINT "VALUE ";XI(IP - 1);" DUPLICATED"
7668 GOTO 999
7669 REM ***** END SUB7650 ***
7670 REM ***** END SUB7600 ***
7699 REM *** BEGIN SUB7700 ***
7700 N2 = 1
7701 Z2 = XS
7702 GOTO 7721
7720 N2 = 0
7721 FOR N0 = 0 TO NI
7722 Z0 = 0
7723 Z1 = 0
7724 IF N2 = 0 THEN Z2 = XI(N0)
7725 IF N0 = 0 THEN GOTO 7731
7726 FOR N1 = N0 - 1 TO 0 STEP - 1
7727 TC(N1 + 1) = (TC(N1) + Z1) / 2 - Z2 * Z0
7728 Z1 = Z0
7729 Z0 = TC(N1)
```

```
7730 NEXT N1
7731 IF N2 = 0 THEN TC(0) = 2 * CN(N0) - Z0 *
     Z2 + Z1
7732 IF N2 <> 0 THEN TC(0) = 2 * AP(NI - N0)
     - Z0 * Z2 + Z1
7733 NEXT N0
7734 RETURN
7735 REM ***** END SUB7700 ***
7799 REM *** BEGIN SUB7800 ***
7800 Z0 = .5
7801 FOR N0 = 0 TO NI
7802 Z1 = Z0 * TC(N0)
7803 AP(N0) = Z1
7804 IF N0 <= 1 THEN GOTO 7809
7805 FOR N1 = N0 - 2 TO 0 STEP - 2
7806 Z1 = (N1 + 1) * (N1 + 2) * Z1 / ((N1 -
     N0) * (N1 + N0))
7807 AP(N1) = AP(N1) + Z1
7808 NEXT N1
7809 Z0 = Z0 + Z0
7810 NEXT N0
7811 RETURN
7812 REM ***** END SUB7800 ***
7899 REM *** BEGIN SUB7900 ***
7900 FOR N0 = 0 TO NI
7901 AP(NI - N0) = CN(N0)
7902 IF N0 = 0 THEN GOTO 7907
7903 Z0 = XI(N0) - XS
7904 FOR N1 = NI - N0 TO NI - 1
7905 AP(N1) = AP(N1) - AP(N1 + 1) * Z0
7906 NEXT N1
7907 NEXT N0
7908 RETURN
7909 REM ***** END SUB7900 ***
7999 REM *** BEGIN SUB8000 ***
8000 N0 = 8000
8001 TR(0) = AP(0)
8002 IF NI <= 0 THEN GOTO 8021
8003 FOR N1 = 0 TO NI - 1
8004 IF AP(N1) = 0 THEN GOTO 8009
8005 TR(N1 + 1) = - AP(N1 + 1) / AP(N1)
8006 NEXT N1
8007 GOTO 8021
8008 N1 = 1
8009 PRINT
8010 PRINT "SUB";N0;" FAILS"
8011 PRINT "POWER SERIES COEF.";N1;" ZERO"
8012 GOTO 999
8013 PRINT
8014 PRINT "SUB";N0;" FAILS"
8015 PRINT "ZERO DIVIDE: ROW ";N1;" COL.";N2
8016 PRINT "OF QD-ALGORITHM"
8017 GOTO 999
8020 N0 = 8020
8021 RS(0) = TR(0)
8022 NS = NI
8023 IF NI <= 0 THEN RETURN
8024 IF RS(0) = 0 THEN GOTO 8008
8025 Z0 = 0
8026 FOR N1 = 1 TO NI
8027 RS(N1) = 0
8028 IF ABS (TR(N1)) > Z0 THEN Z0 = ABS
```

```
     (TR(N1))
8029 NEXT N1
8030 IF NI <= 1 THEN GOTO 8055
8031 Z0 = Z0 * 6
8032 FOR N2 = 2 TO NI
8033 Z1 = 0
8034 IF NS < 0 THEN GOTO 8047
8035 IF NS <> NI THEN GOTO 8054
8036 FOR N1 = NI TO N2 STEP - 1
8037 Z2 = RS(N1 - 1) - TR(N1 - 1) + TR(N1)
8038 IF Z0 + Z2 = Z0 THEN Z2 = 0
8039 Z3 = Z1
8040 IF ABS (Z2) > Z1 THEN Z1 = ABS (Z2)
8041 RS(N1) = Z2
8042 NEXT N1
8043 IF Z3 <> 0 THEN GOTO 8052
8044 NS = N2
8045 IF Z1 = 0 THEN NS = NS - 1
8046 GOTO 8054
8047 FOR N1 = NI TO N2 STEP - 1
8048 IF RS(N1 - 1) = 0 THEN GOTO 8013
8049 TR(N1) = TR(N1 - 1) * RS(N1) / RS(N1 - 1)
8050 IF ABS (TR(N1)) > Z1 THEN Z1 = ABS
     (TR(N1))
8051 NEXT N1
8052 Z0 = 6 * Z1 + Z0
8053 NS = - NS
8054 NEXT N2
8055 FOR N1 = 1 TO NS STEP 2
8056 RS(N1) = TR(N1)
8057 NEXT N1
8058 RETURN
8059 REM ***** END SUB8000 ***
8099 REM *** BEGIN SUB8100 ***
8100 N2 = 1
8101 GOTO 8121
8120 N2 = 0
8121 N3 = 1 - N2
8122 AR(N2,0) = 1
8123 AR(N3,0) = QI(NI)
8124 NR = INT ((NI + 1) / 2)
8125 IF NR <= 0 THEN GOTO 8151
8126 GOSUB 8158
8127 FOR N0 = NI - 1 TO 0 STEP - 1
8128 Z0 = QI(N0)
8129 Z1 = XI(NQ(N0))
8130 FOR N1 = INT ((NI + 1 - N0) / 2) TO 0
     STEP - 1
8131 Z2 = AR(N3,N1) * Z0 - AR(N2,N1) * Z1
8132 AR(N2,N1) = AR(N3,N1)
8133 IF N1 <> 0 THEN AR(N3,N1) = AR(N2,N1 -
     1) + Z2
8134 NEXT N1
8135 AR(N3,0) = Z2
8136 NEXT N0
8137 GOTO 8151
8149 REM *** BEGIN SUB8150 ***
8150 GOSUB 8164
8151 MR = NR + 1
8152 MR = MR - 1
8153 IF MR > 0 THEN IF AR(0,MR) = 0 THEN GOTO
     8152
```

```
8154 IF NR <= 0 THEN RETURN
8155 IF AR(1,NR) <> 0 THEN RETURN
8156 NR = NR - 1
8157 GOTO 8154
8158 FOR N0 = 1 TO NR
8159 AR(0,N0) = 0
8160 AR(1,N0) = 0
8161 NEXT N0
8162 RETURN
8163 REM ***** END SUB8100 ***
8164 AR(0,0) = 1
8165 AR(1,0) = 1
8166 NR = INT ((NS + 1) / 2)
8167 IF NR <= 0 THEN GOTO 8176
8168 GOSUB 8158
8169 FOR N1 = NS TO 1 STEP - 1
8170 Z0 = RS(N1)
8171 FOR N0 = INT ((NS - N1 + 2) / 2) TO 1
     STEP - 1
8172 AR(0,N0) = AR(1,N0)
8173 AR(1,N0) = AR(0,N0 - 1) * Z0 + AR(1,N0)
8174 NEXT N0
8175 NEXT N1
8176 Z0 = RS(0)
8177 FOR N0 = 0 TO NR
8178 AR(0,N0) = Z0 * AR(0,N0)
8179 NEXT N0
8180 RETURN
8181 REM ***** END SUB8150 ***
8199 REM *** BEGIN SUB8200 ***
8200 N0 = - 8200
8201 N2 = NI
8202 FOR NI = - 1 TO N2 - 1
8203 Z0 = XI(NI + 1)
8204 Z1 = YI(NI + 1)
8205 GOSUB 8224
8206 IF N0 < 0 THEN NEXT NI
8207 NI = N2
8208 IF N0 > 0 THEN GOTO 8212
8209 RETURN
8210 N0 = - N0
8211 IF N0 <> 8220 THEN RETURN
8212 PRINT
8213 PRINT "SUB";N0;" FAILS"
8214 PRINT "REPEATED SUPPORT AT ";Z0
8215 PRINT "NOT CONSECUTIVE"
8216 GOTO 999
8220 N0 = - 8220
8221 Z0 = XI
8222 Z1 = YI
8223 XI(NI + 1) = Z0
8224 CN(NI + 1) = Z1
8225 IF NI < 0 THEN RETURN
8226 FOR N1 = NI TO 0 STEP - 1
8227 IF Z0 = XI(N1) THEN GOTO 8231
8228 CN(N1) = (CN(N1 + 1) - CN(N1)) / (Z0 -
     XI(N1))
8229 NEXT N1
8230 RETURN
8231 IF Z0 <> XI(N1 + 1) THEN GOTO 8210
8232 Z1 = CN(N1)
8233 CN(N1) = CN(N1 + 1) / (NI - N1 + 1)
```

```
8234 CN(N1 + 1) = Z1
8235 GOTO 8229
8240 YI = 0
8241 IF NI < 0 THEN GOTO 8247
8242 FOR N0 = 0 TO NI
8243 Z0 = (XI - XI(N0)) * YI + CN(N0)
8244 CN(N0) = YI
8245 YI = Z0
8246 NEXT N0
8247 CN(NI + 1) = YI
8248 XI(NI + 1) = XI
8249 RETURN
8260 IF IR * (NI - IR) < 0 THEN RETURN
8261 IF IR = 0 THEN GOTO 8266
8262 Z0 = XI(IR)
8263 FOR N0 = 0 TO IR - 1
8264 CN(N0) = (XI(N0) - Z0) * CN(N0) + CN(N0
     + 1)
8265 NEXT N0
8266 IF IR = NI THEN RETURN
8267 FOR N0 = IR + 1 TO NI
8268 CN(N0 - 1) = CN(N0)
8269 XI(N0 - 1) = XI(N0)
8270 NEXT N0
8271 RETURN
8280 GOSUB 8260
8281 IF IR < 0 THEN RETURN
8282 N0 = IR
8283 IF N0 >= NI THEN RETURN
8284 IF Z0 = XI(N0) THEN GOTO 8289
8285 FOR N0 = N0 + 1 TO NI
8286 YI(N0 - 1) = YI(N0)
8287 NEXT N0
8288 RETURN
8289 N0 = N0 + 1
8290 GOTO 8283
8291 REM ***** END SUB8200 ***
8299 REM *** BEGIN SUB8300 ***
8300 N0 = 0
8301 GOTO 8321
8302 QI(N2) = 0
8303 GOTO 8346
8304 QI(N0) = Z0
8305 QI(N1) = Z1
8306 N2 = NQ(N0)
8307 NQ(N0) = NQ(N1)
8308 NQ(N1) = N2
8309 N0 = N1
8310 RETURN
8320 N0 = 1
8321 Z0 = 1.8E19
8322 FOR N2 = 0 TO NI
8323 NQ(N2) = N2
8324 Z3 = YI(N2)
8325 IF N0 = 0 THEN GOTO 8328
8326 IF Z3 <> 0 THEN Z3 = 1 / Z3
8327 IF Z3 = 0 THEN Z3 = Z0
8328 QI(N2) = Z3
8329 NEXT N2
8330 NQ = - 1
8331 N0 = NQ
8332 FOR N1 = 0 TO NI
```

```
8333 IF N0 = NI THEN GOTO 8347
8334 N0 = N0 + 1
8335 IF QI(N0) = Z0 THEN GOTO 8333
8336 NQ = N1
8337 IF N1 = NI THEN GOTO 8347
8338 Z1 = QI(N0)
8339 Z2 = XI(NQ(N0))
8340 IF N1 <> N0 THEN GOSUB 8304
8341 FOR N2 = N1 + 1 TO NI
8342 IF QI(N2) = Z0 THEN GOTO 8302
8343 Z3 = QI(N2) - Z1
8344 IF Z3 = 0 THEN QI(N2) = Z0
8345 IF Z3 <> 0 THEN QI(N2) = (XI(NQ(N2)) -
     Z2) / Z3
8346 NEXT N2
8347 NEXT N1
8348 RETURN
8349 REM ***** END SUB8300 ***
8399 REM *** BEGIN SUB8400 ***
8400 N0 = - 8400
8401 NR = NI
8402 FOR N1 = 0 TO NR - 1
8403 IF XI(N1 + 1) = XI(N1) THEN GOSUB 8422
8404 NEXT N1
8405 GOSUB 8501
8406 IF N0 > 0 THEN GOTO 8431
8407 IF NI <= SGN (8440 + N0) + 1 THEN GOTO
     8447
8408 FOR N1 = 1 TO NI - 1
8409 Z0 = (XI(N1 + 1) - XI(N1 - 1)) * 2
8410 AA(1,N1) = (XI(N1) - XI(N1 - 1)) / Z0
8411 AA(0,N1) = (XI(N1 + 1) - XI(N1)) / Z0
8412 AA(2,N1) = 1.5 * DI(N1)
8413 NEXT N1
8414 IF N0 <> - 8400 THEN GOTO 8442
8415 AA(1,0) = 0.5 / AA(0,1)
8416 AA(0,NI) = 0.5 / AA(1,NI - 1)
8417 AA(2,0) = DI(1) * AA(1,0) + DI(0)
8418 AA(2,NI) = DI(NI - 1) * AA(0,NI) + DI(NI)
8419 GOTO 8447
8420 N0 = - 8420
8421 GOTO 8401
8422 Z0 = YI(N1 + 1)
8423 FOR N2 = N1 + 2 TO NI
8424 XI(N2 - 1) = XI(N2)
8425 YI(N2 - 1) = YI(N2)
8426 NEXT N2
8427 XI(NI) = N1
8428 YI(NI) = Z0
8429 NI = NI - 1
8430 RETURN
8431 GOSUB 8462
8432 GOTO 8520
8433 GOSUB 8462
8434 GOTO 8565
8440 N0 = - 8440
8441 GOTO 8401
8442 Z0 = (8460 + N0) / 40
8443 AA(1,0) = Z0
8444 AA(0,NI) = Z0
8445 AA(2,0) = (1 + Z0) * (DI(0) + DI(1)) / 2
8446 AA(2,NI) = (1 + Z0) * (DI(NI - 1) +
```

```
     DI(NI)) / 2
8447 IF NI = NR THEN GOTO 8455
8448 FOR N1 = NI + 1 TO NR
8449 N2 = XI(N1)
8450 AA(0,N2) = 0
8451 AA(1,N2) = 0
8452 DI(N2) = YI(N1)
8453 AA(2,N2) = DI(N2)
8454 NEXT N1
8455 IF NI <= SGN (8440 + N0) + 1 THEN GOTO
     8462
8456 N1 = NI
8457 GOSUB 8553
8458 IF N0 > 0 THEN GOTO 8433
8459 FOR N2 = 0 TO NI
8460 DI(N2) = AA(2,N2)
8461 NEXT N2
8462 N1 = NI
8463 NI = NR
8464 NR = N1
8465 RETURN
8499 REM *** BEGIN SUB8500 ***
8500 N0 = - 8500
8501 N2 = 1'
8502 Z0 = XI(1) - XI(0)
8503 IF Z0 = 0 THEN GOTO 8518
8504 Z1 = (YI(1) - YI(0)) / Z0
8505 DI(0) = Z1
8506 IF NI <= 1 THEN GOTO 8515
8507 FOR N2 = 2 TO NI
8508 Z2 = XI(N2) - XI(N2 - 1)
8509 IF Z2 * Z0 <= 0 THEN GOTO 8518
8510 Z3 = (YI(N2) - YI(N2 - 1)) / Z2
8511 DI(N2 - 1) = (Z1 * Z2 + Z3 * Z0) / (Z0 +
     Z2)
8512 Z0 = Z2
8513 Z1 = Z3
8514 NEXT N2
8515 DI(NI) = Z1 + Z1 - DI(NI - 1)
8516 DI(0) = 2 * DI(0) - DI(1)
8517 RETURN
8518 N0 = - N0
8519 IF N0 <> 8500 THEN RETURN
8520 PRINT
8521 PRINT "SUB";N0;" FAILS"
8522 PRINT "XI(";N2;") NOT STRICTLY MONOTONE"
8523 GOTO 999
8524 REM ***** END SUB8500 ***
8549 REM *** BEGIN SUB8550 ***
8550 N0 = - 8550
8551 IF NA <= 0 THEN RETURN
8552 N1 = NA
8553 FOR N2 = 1 TO N1
8554 Z0 = 1 - AA(1,N2 - 1) * AA(0,N2)
8555 IF 1 + Z0 / (N2 + N2) = 1 THEN GOTO 8563
8556 IF N2 <> N1 THEN AA(1,N2) = AA(1,N2) / Z0
8557 AA(2,N2) = (AA(2,N2) - AA(2,N2 - 1) *
     AA(0,N2)) / Z0
8558 NEXT N2
8559 FOR N2 = N1 - 1 TO 0 STEP - 1
8560 AA(2,N2) = AA(2,N2) - AA(1,N2) * AA(2,N2
     + 1)
```

```
8561 NEXT N2                              8648 RETURN
8562 RETURN                               8650 N1 = NA + 1
8563 N0 = - N0                            8651 N2 = 1
8564 IF N0 <> 8550 THEN RETURN            8652 FOR N3 = 0 TO NA - 1
8565 PRINT                                8653 N0 = AA(0,N3 + 1)
8566 PRINT "SUB";N0;" FAILS"              8654 GOSUB 8636
8567 PRINT "SYSTEM SINGULAR"              8655 AA(N0,N1) = Z0 / AA(N0,N3 + 1)
8568 GOTO 999                             8656 NEXT N3
8569 REM ***** END SUB8550 ***            8657 N3 = NA
8570 REM ***** END SUB8400 ***            8658 FOR N2 = NA + 1 TO 2 STEP - 1
8599 REM *** BEGIN SUB8600 ***            8659 N0 = AA(0,N2 - 1)
8600 FOR N1 = 1 TO NA                     8660 GOSUB 8636
8601 AA(0,N1) = N1                         8661 AA(N0,N1) = Z0
8602 Z0 = 0                                8662 AA(N2 - 1,0) = Z0
8603 FOR N2 = 1 TO NA                      8663 NEXT N2
8604 IF ABS (AA(N1,N2)) > Z0 THEN Z0 = ABS 8664 RETURN
     (AA(N1,N2))                          8665 PRINT
8605 NEXT N2                              8666 PRINT "SUB8600 FAILS"
8606 IF Z0 = 0 THEN GOTO 8665             8667 PRINT "MATRIX SINGULAR (ROW ";N1;")"
8607 AA(N1,0) = Z0                         8668 N0 = 8600
8608 NEXT N1                              8669 GOTO 999
8609 Z4 = 1                               8670 REM ***** END SUB8600 ***
8610 N2 = 1                               8699 REM *** BEGIN SUB8700 ***
8611 FOR N3 = 0 TO NA - 1                 8700 IB = 2
8612 Z3 = 0                               8701 IF ABS (F1) > ABS (F0) THEN IB = 4
8613 N1 = N3 + 1                          8702 W2(IB) = X0
8614 FOR N5 = N1 TO NA                    8703 W2(IB + 1) = F0
8615 N0 = AA(0,N5)                        8704 W2(6 - IB) = X1
8616 GOSUB 8636                           8705 W2(7 - IB) = F1
8617 AA(N0,N1) = Z0                       8706 W2(6) = W2(4)
8618 IF ABS (Z0) > Z3 * AA(N0,0) THEN GOSUB 8707 W2(0) = INT (MD)
     8646                                 8708 IF MD * (6 - MD) < 0 THEN W2(0) = ( SGN
8619 NEXT N5                                   (MD) + 1) * 3
8620 IF N1 + Z3 = N1 THEN GOTO 8665       8709 W2(1) = 0
8621 N0 = AA(0,N6)                        8710 IF F0 * F1 < 0 THEN GOTO 8763
8622 Z3 = AA(N0,N1)                       8711 N0 = 8700
8623 IF N6 = N1 THEN GOTO 8627            8712 GOTO 8724
8624 AA(0,N6) = AA(0,N1)                  8713 XN = 0
8625 AA(0,N1) = N0                        8714 IF W2(3) = 0 THEN GOTO 8726
8626 Z4 = - Z4                            8715 PRINT
8627 Z4 = Z3 * Z4                         8716 PRINT "SUB";N0;" FAILS"
8628 FOR N1 = N1 + 1 TO NA + 1            8717 PRINT "ZERODIVIDE IN ROOT ESTIMATE"
8629 GOSUB 8636                           8718 GOTO 999
8630 AA(N0,N1) = Z0 / Z3                  8719 N0 = 8750
8631 NEXT N1                              8720 W2(2) = W2(4)
8632 NEXT N3                              8721 W2(3) = W2(5)
8633 AA(0,0) = Z4                         8722 W2(4) = W2(6)
8634 N1 = NA + 1                          8723 W2(5) = FX
8635 GOTO 8657                            8724 IF W2(3) = W2(5) THEN GOTO 8713
8636 Z0 = AA(N0,N1)                       8725 XN = W2(5) / (W2(3) - W2(5))
8637 IF N3 < N2 THEN RETURN               8726 SX = W2(4) - W2(2)
8638 Z1 = 0                               8727 XN = W2(4) + SX * XN
8639 Z2 = Z0                              8728 DX = XN - W2(6)
8640 FOR N4 = N2 TO N3                    8729 W2(6) = XN
8641 Z2 = ((Z1 - Z0) + Z2) - AA(N0,N4) *  8730 IB = 0
     AA(AA(0,N4),N1)                      8731 IF W2(3) * W2(5) <= 0 THEN IB = - 1
8642 Z1 = Z0                              8732 W2(7) = W2(0) * IB
8643 Z0 = Z0 + Z2                         8733 RETURN
8644 NEXT N4                              8734 XN = FX / W2(4 - IB)
8645 RETURN                              8735 SX = (SX + XN) / 2
8646 N6 = N5                              8736 SX = SQR (SX * SX - XN) + SX
8647 Z3 = ABS (Z0 / AA(N0,0))             8737 RETURN
```

```
8750 IF W2(7) * FX = 0 THEN GOTO 8719
8751 IB = SGN (FX * W2(5))
8752 IF W2(0) <= 2 THEN GOTO 8761
8753 IF W2(1) = IB THEN GOTO 8761
8754 SX = 1 - FX / W2(IB + 4)
8755 N0 = W2(0)
8756 IF N0 = 6 THEN GOSUB 8734
8757 IF N0 = 5 THEN IF SX <= 0 THEN SX = .0625
8758 IF N0 = 4 THEN SX = 1 / (2 - SX)
8759 IF N0 = 3 THEN SX = .5
8760 W2(4 - IB) = W2(4 - IB) * SX
8761 W2(IB + 4) = FX
8762 W2(IB + 3) = W2(6)
8763 W2(1) = - IB
8764 IF W2(0) <> 1 THEN GOTO 8725
8765 XN = - .5
8766 GOTO 8726
8767 REM ***** END SUB8700 ***
8799 REM *** BEGIN SUB8800 ***
8800 NS = 0
8801 SS(0,0) = AP(0)
8802 IF NP <= 0 THEN RETURN
8803 FOR N0 = 1 TO NP
8804 SS(0,N0) = AP(N0)
8805 SS(1,N0 - 1) = N0 * AP(N0)
8806 NEXT N0
8807 N2 = 1
8808 N1 = NP
8809 N0 = N1
8810 N1 = N1 - 1
8811 IF SS(NS + 1,N1) = 0 THEN GOTO 8831
8812 NS = NS + 1
8813 SS(NS,NP) = N1
8814 IF N1 = 0 THEN RETURN
8815 Z0 = SS(NS,N1)
8816 FOR N3 = 0 TO N0
8817 SS(NS + 1,N3) = - SS(NS - 1,N3)
8818 NEXT N3
8819 FOR N3 = N0 TO N1 STEP - 1
8820 Z1 = SS(NS + 1,N3) / Z0
8821 IF Z1 = 0 THEN GOTO 8829
8822 N2 = N2 + 1
8823 FOR N4 = N3 - N1 TO N3 - 1
8824 Z2 = SS(NS,N4 + N1 - N3) * Z1
8825 Z3 = SS(NS + 1,N4) - Z2
8826 IF Z2 + Z3 / N2 = Z2 THEN Z3 = 0
8827 SS(NS + 1,N4) = Z3
8828 NEXT N4
8829 NEXT N3
8830 GOTO 8809
8831 IF N1 <> 0 THEN GOTO 8810
8832 RETURN
8833 REM ***** END SUB8800 ***
8899 REM *** BEGIN SUB8900 ***
8900 NR = 0
8901 IF NP * NS <= 0 THEN RETURN
8902 N0 = SS(1,NP)
8903 Z0 = 0
8904 FOR N1 = 0 TO N0
8905 Z0 = ABS (SS(0,N1)) + Z0
8906 NEXT N1
8907 Z0 = Z0 / ABS (SS(0,N0 + 1))
```

```
8908 IF Z0 < 1 THEN Z0 = 1
8909 GOSUB 8937
8910 N5 = N0
8911 Z0 = - Z0
8912 GOSUB 8937
8913 NR = N0 - N5
8914 N4 = N0
8915 IF NR <= 0 THEN RETURN
8916 FOR N0 = 1 TO NR
8917 RB(0,N0) = + Z0
8918 RB(1,N0) = - Z0
8919 NEXT N0
8920 IF NR = 1 THEN RETURN
8921 FOR N5 = 1 TO NR - 1
8922 IF RB(1,N5) <= RB(0,N5 + 1) THEN GOTO
     8935
8923 Z0 = (RB(0,N5) + RB(1,N5)) / 2
8924 GOSUB 8937
8925 N0 = N4 - N0
8926 IF N0 < N5 THEN GOTO 8932
8927 FOR N1 = N5 TO N0
8928 RB(1,N1) = Z0
8929 NEXT N1
8930 IF N0 >= NR THEN GOTO 8922
8931 IF RB(0,N0 + 1) >= Z0 THEN GOTO 8922
8932 N0 = N0 + 1
8933 RB(0,N0) = Z0
8934 GOTO 8930
8935 NEXT N5
8936 RETURN
8937 Z1 = 0
8938 N0 = 0
8939 N1 = SS(1,NP) + 1
8940 FOR N2 = 0 TO NS
8941 IF N2 <> 0 THEN N1 = SS(N2,NP)
8942 Z2 = 0
8943 FOR N3 = N1 TO 0 STEP - 1
8944 Z2 = Z2 * Z0 + SS(N2,N3)
8945 NEXT N3
8946 IF Z2 * Z1 < 0 THEN N0 = N0 + 1
8947 IF Z2 <> 0 THEN Z1 = SGN (Z2)
8948 NEXT N2
8949 RETURN
8950 REM ***** END SUB8900 ***
8999 REM *** BEGIN SUB9000 ***
9000 W0(2) = INT (NI)
9001 IF NI < 1 THEN W0(2) = 1
9002 W0(0) = XB
9003 W0(1) = (XB - XA) / W0(2)
9004 W0(4) = XA
9005 W0(5) = - 1
9006 W0(5) = W0(5) + 1
9007 XQ = W0(5) / W0(2)
9008 XQ = W0(0) * XQ + W0(4) * (1 - XQ)
9009 GOSUB 500
9010 IF W0(5) = 0 THEN W0(3) = - FQ / 2
9011 W0(3) = W0(3) + FQ
9012 IF W0(5) <> W0(2) THEN GOTO 9006
9013 QR(1) = (W0(3) - FQ / 2) / W0(2)
9014 W0(3) = 1
9015 QR(0) = 1
9016 GOTO 9074
```

```
9017 W0(6) = ER(NQ)
9018 ER(0) = NQ
9019 RETURN
9050 W0(1) = W0(1) / 2
9051 W0(2) = W0(2) * 2
9052 IF W0(3) < MQ THEN W0(3) = W0(3) + 1
9053 W0(4) = 0
9054 W0(5) = W0(2) - 1
9055 XQ = W0(0) - W0(5) * W0(1)
9056 GOSUB 500
9057 W0(4) = (FQ - QR(1)) + W0(4)
9058 W0(5) = W0(5) - 2
9059 IF W0(5) > 0 THEN GOTO 9055
9060 W0(4) = W0(4) / W0(2)
9061 W0(5) = QR(1)
9062 W0(6) = ABS (W0(4)) * 2
9063 XQ = 1
9064 FOR NQ = 1 TO W0(3)
9065 ER(NQ) = ABS (XQ * W0(4))
9066 IF ER(NQ) < W0(6) THEN GOSUB 9017
9067 W0(5) = W0(5) + W0(4)
9068 IF NQ = W0(3) THEN GOTO 9071
9069 XQ = 4 * XQ
9070 W0(4) = (W0(5) - QR(NQ)) / (XQ - 1)
9071 QR(NQ) = W0(5)
9072 NEXT NQ
9073 QR(0) = QR(0) + 1
9074 NQ = W0(3)
9075 RETURN
9076 REM ***** END SUB9000 ***
9099 REM *** BEGIN SUB9100 ***
9100 ST(4) = INT (MS) - 3
9101 ST(5) = XA
9102 ST(6) = XB - XA
9103 ST(7) = 180 * ABS (ER)
9104 ST(1) = XB
9105 ST(0) = 3
9106 ST(0) = ST(0) - 1
9107 XQ = ((2 - ST(0)) * ST(1) + ST(0) *
     ST(5)) / 2
9108 GOSUB 500
9109 ST(ST(0) + 10) = FQ
9110 IF ST(0) <> 0 THEN GOTO 9106
9111 ST(0) = 11
9112 ST(1) = 0
9113 ST(2) = 1
9114 ST(3) = 0
9115 ST(8) = 0
9116 ST(9) = 0
9117 ST(1) = 2 * ST(1)
9118 ST(2) = 2 * ST(2)
9119 XQ = .5
9120 GOSUB 9146
9121 ST(ST(0) + 2) = FQ
9122 XQ = 1.5
9123 GOSUB 9146
9124 LM = ST(0)
9125 XQ = 4 * (2 * ST(LM) - FQ - ST(LM + 2))
     - (2 * ST(LM) - ST(LM - 1) - ST(LM +
     1))
9126 IF ABS (XQ) <= ST(7) THEN GOTO 9134
9127 IF LM >= ST(4) THEN GOTO 9133
9128 ST(LM + 3) = ST(LM + 1)
9129 ST(LM + 1) = ST(LM)
9130 ST(LM) = FQ
9131 ST(0) = ST(0) + 2
9132 GOTO 9117
9133 ST(3) = 1
9134 ST(9) = XQ / ST(2) + ST(9)
9135 ST(8) = (2 * (ST(LM + 2) + FQ) - ST(LM))
     / ST(2) + ST(8)
9136 ST(1) = INT (ST(1) / 2)
9137 ST(2) = ST(2) / 2
9138 IF ST(1) <> INT (ST(1) / 2) * 2 THEN
     GOTO 9136
9139 ST(1) = ST(1) + 1
9140 ST(0) = ST(0) - 2
9141 IF ST(0) > 9 THEN GOTO 9117
9142 AS = (30 * ST(8) + 7 * ST(9)) / 45
9143 ES = ABS (ST(9) / 90)
9144 LM = ST(3)
9145 RETURN
9146 XQ = (ST(1) + XQ) * ST(6) / ST(2) + ST(5)
9147 GOTO 500
9148 REM ***** END SUB9100 ***
9199 REM *** BEGIN SUB9200 ***
9200 W0(8) = 88
9201 W0(0) = - AI
9202 W0(1) = AI + BI
9203 W0(2) = BI
9204 W0(3) = 0
9205 W0(4) = INT (NI)
9206 IF NI < 2 THEN W0(4) = 2
9207 W0(5) = 2 * PM / W0(4)
9208 W0(6) = 1
9209 W0(7) = W0(4) - 1
9210 N0 = W0(4)
9211 N1 = N0 - 2 * W0(7)
9212 Z0 = (N0 - N1) * (N0 + N1) / (N0 * N0)
9213 Z1 = N1 * W0(5) / SQR (Z0)
9214 Z2 = 0
9215 IF ABS (Z1) < W0(8) THEN Z2 = EXP ( -
     ABS (Z1))
9216 XQ = 1 / (1 + Z2)
9217 W0(9) = 1.5 * LOG (Z0) - W0(1) * LOG
     (XQ) + W0( SGN (N1) + 1) * Z1
9218 IF W0(9) > W0(8) THEN GOTO 9222
9219 IF N1 < 0 THEN XQ = Z2 * XQ
9220 GOSUB 500
9221 W0(3) = EXP ( - W0(9)) * FQ + W0(3)
9222 W0(7) = W0(7) - W0(6)
9223 IF W0(7) > 0 THEN GOTO 9210
9224 IF W0(6) = 2 THEN NI = W0(4)
9225 TI = 2 * W0(5) * W0(3)
9226 RETURN
9250 W0(4) = 2 * W0(4)
9251 W0(5) = W0(5) / 2
9252 W0(6) = 2
9253 GOTO 9209
9254 REM ***** END SUB9200 ***
```

Problems

Section 1.2

1. Write down the representation of the following numbers with decimal base and precision $p = 2$, using each of the normalisations (1.2.2), (1.2.3), and (1.2.4):
   (a) 1, (b) 1/20, (c) 10^{-6}.

2. (a) A floating point number is written with binary base ($b = 2$) and precision $p = 4$: write down in binary, and in decimal, the largest and smallest values of the mantissa according to the normalisation (1.2.4).

   (b) The *most significant digit* of a mantissa is the leading digit (written on the left), and the *least significant digit* is the trailing digit (written on the right). Show that the most significant (binary) digit of the mantissa of a normalised floating point number with binary base is always 1, irrespective of the precision. (Hence this digit can be 'understood', and does not need explicit representation by the machine.)

3. (a) Using the result $(1 - x)^{-1} = 1 + x + x^2 + \dots$, sum the geometric series $1 + (1/16) + (1/16)^2 + \dots$, and hence determine the rational decimal fraction represented by zero-point-one-recurring in hexadecimal, i.e., by $(0.\dot{1})_{16}$.

   (b) Using result (a) represent the decimal numbers 1/5, 1/3, 4/5 as hexadecimal recurring fractions.

   (c) A hexadecimal integer like $(123)_{16}$ is evaluated as $1 \times 16^2 + 2 \times 16 + 3$; using result (b) find the floating point number with hexadecimal base and precision 3, normalised in conformity with (1.2.4), which approximates the rational fraction 1/5. What rational decimal number is *precisely* represented by this hexadecimal number?

4. Find out whether the base of a calculator/computer available to you is decimal, or some integer power of 2, by the following method.

Let $x_0 = 1$, and $x_k = x_{k-1}/10$ for $k = 1, 2 \ldots$ (so that $x_k = 10^{-k}$). For each k, calculate $y_k = (x_k + 1) - 1$: stop when $y_k \neq x_k$. If this least value of k (for which $y_k \neq x_k$) is greater than 1, the base is decimal and k is its precision; but if $k = 1$, the base is an integer power of 2.

5. Let k be a positive integer and $k < b$. What sequence of digits represents each of the following integers when written with base b, and which of them can be precisely represented as a floating point number of base b and precision p?

(a) b^p, (b) $b^p + k$, (c) $b^p/2 + k$, (d) $b^p + b$.

Assume that p is some fixed integer > 1, and that b is exactly divisible by 2. (Hint: The component digits are the same irrespective of the base b, so it may be easier first to consider the special case $b = 10$.)

6. If the base of your computer is some integer power of 2 (say, $b = 2^l$), its base and precision (p) can be evaluated by the following algorithm, which is based on the answers to problem 5.

> **real** x; **integer** lp, l, b;
> **comment** first find $(l \times p) = lp$, say, as the least integer for which
> $\qquad (x + 1)$ is not precise, where $x = 2 \uparrow lp$;
> $lp := 0; x := 1.0$;
>
> FINDLP: $lp := lp + 1; x := x + x$; **comment** $lp = 1, 2, \ldots$ and $x = 2 \uparrow lp$;
> $\qquad$ **if** $(x + 1) - x = 1$ **then goto** FINDLP;
>
> $\qquad\qquad\qquad$ **comment** repeat if $(x + 1)$ is precise;
> $\qquad$ **comment** lp is now $(l \times p)$ and x is $2 \uparrow (l \times p) = b \uparrow p$. Next find l by
> $\qquad\qquad$ discriminating the base $b = 2 \uparrow l$ as the least integer b for
> $\qquad\qquad$ which $(x + b)$ is precise:
>
> $\qquad l := 0; b := 1$;
> FINDB: $\quad l := l + 1; b := b + b$; $\quad$ **comment** $l = 1, 2, \ldots$ and $b = 2 \uparrow l$;
> $\qquad$ **if** $(x + b) - x \neq b$ **then goto** FINDB;
>
> $\qquad\qquad\qquad$ **comment** repeat if $(x + b)$ is not precise;
> $\qquad$ output (b); output (lp/l); **comment** $lp/l =$ precision p;

(Note: In this and following examples where an ALGOL program appears, the operator $\uparrow$ signifies 'to the power of'; thus $2 \uparrow lp$ is 2^{lp}.)

Section 1.3

1. (a) Show that *entier* $(x) = -ceil\ (-x)$ for any real number x.
   (b) If rounding is achieved by defining $|m| = ceil\ (|y| - 0.5)$, for what values of $|y|$, if any, does this rule produce a different value of $|m|$ from that given by $(1.3.3)$?

2. In general, for what range of the exact mantissa $|y|$, as bounded by (1.3.1), will the machine exponent e equal $i + 1$, if (a) correct rounding or (b) boosting is used to float the value of x?

3. Would the bounds (1.3.8) be altered if $RE(x)$ were defined by $(x - n)/n$ instead of as in (1.3.5)?

4. What are the (approximate) decimal values of the machine units of the following machines/systems?

    (a) Burroughs 6000 series: $b = 8$, $p = 13$.
    (b) Cyber : $b = 2$, $p = 48$.
    (c) DEC 11   : $b = 2$, $p = 24$.
    (d) HP 67   : $b = 10$, $p = 10$.
    (e) IBM 370 : $b = 16$, $p = 6$.
    (f) IBM 7094 : $b = 2$, $p = 27$.

5. (a) If $c = b^j$, where j is an integer, and b is the base of the floating-point representation, show that $fl(cx) = c\,fl(x)$ and $fl(-cx) = -c\,fl(x)$ for any real x.

    (b) Is there any other value of c for which these relations are true for any x?

6. Using a machine of known base (b) and precision (p), find the method of floating used in addition and subtraction, by the following algorithm.

real m, $frac$; **integer** b, p, k ;
$input$ (b); $input$ (p); $m := 1$;   **comment** read base & precision;
for $k := 1$ **step** 1 **until** $p - 1$ **do** $m := b * m$;
                                comment form $m = b \uparrow (p - 1)$;
$m := m + 1$;   **comment** $m = b \uparrow (p - 1) + 1$ precisely ;

comment now input various fractional parts $(frac)$ bètween 0 and 1 which are to be added to or subtracted from the integer value of the mantissa (m), either by operation, or by sign of operand. The change in the machine representation of this incremented mantissa is then output;

$MORE$: $input$ $(frac)$;   **comment** $0 < frac < 1$;
$REPEAT$: $output$ $(frac)$; $output$ $((m + frac) - m)$;
             $frac := -frac$; $output$ $((m - frac) - m)$; $newline$;
             if $frac < 0$ **then goto** $REPEAT$;
                            comment repeat with negative $frac$;
             if $frac \neq 0$ **then goto** $MORE$;
                            comment to end, let $frac = 0$;

(Note that, because of the double negative, $(m - frac)$ in the second output has the same exact mantissa as $(m + frac)$ in the first, but the operator is changed.)

Section 1.4

1. Use the following program to chop the value of randomly generated 'exact' mantissas y, and so find the sample mean ($abar$) of alpha values and the unbiased estimator (var) of their variance, using simulated arithmetic of arbitrary base (b) and precision (p). The value of the 'simulated' machine unit ($1/b^{p-1}$) should be large compared with the actual machine unit (ϵ, say) of the computer/ calculator in use; say $b^{p-1} \approx \epsilon^{-1/2}$. The simulated base b does not need to be the same as that of the machine being used. The size of the sample (n) should not be less than 30, and preferably much larger. The program uses the function $random(0)$ which is supposed to provide the value of a pseudo-random number having a uniform distribution in (0, 1).

```
integer b, p, n, k; real lb, ub, sum, sumsqs, y, m, alpha, abar, var;
input (b); input (p); input (n);
output (b); output (p); output (n); lb: = 1;
for k: = 1 step 1 until p − 1 do lb: = b * lb;
                                          comment lb = b ↑ (p − 1);
ub: = lb * b;                             comment ub = b ↑ p;
sum: = 0; sumsqs: = 0;                    comment initialise accumulators;
for k: = 1 step 1 until n do begin       comment find an alpha value;
NEWY: y: = random(0) * ub;               comment y random in (0, ub);
    if y < lb then goto NEWY;            comment discard y if in (0, lb);
    comment y is now 'exact' mantissa in [b ↑ (p − 1), b ↑ p). Next form
            simulated rounded mantissa (m) of precision p;
    m: = entier (y);                     comment chopping;
    alpha: = (y − m) * lb/y;             comment rel. error * b ↑ (p − 1);
    sum: = alpha + sum;                  comment accumulate alpha;
    sumsqs: = alpha * alpha + sumsqs;    comment and alpha ↑ 2;
    end;                                 comment repeat n times;
abar: = sum/n;                           comment sample mean;
var: = (sumsqs − abar * sum)/(n − 1);   comment variance = (S.D.) ↑ 2;
output (abar); output (var); output (sqrt(var));
```

2. What assignment to m in the above program would simulate correct rounding instead of chopping? Rerun the program incorporating this change.

3. Rearrange the above program to provide a frequency histogram of alpha values.

4. It is a consequence of the central limit theorem that, provided the sample size n is large, sample means ($\bar{a}$, say) of alpha values (drawn from a distribution with true mean $\bar{\alpha}$ and standard deviation σ_α) are *normally distributed* about $\bar{\alpha}$ with standard deviation $\sigma_\alpha/n^{1/2}$. Using a table of values of the normal distribution, find the probability that $|\bar{a} - \bar{\alpha}|n^{1/2}/\sigma_\alpha$ should equal or exceed values of this difference fround from the above program. (Use values of $\bar{\alpha}$ and σ_α in Table 1.1.)

5. Supposing that $\epsilon/(b+1)$ gives a more realistic estimate of 'average error' than ϵ, and that a machine mantissa consists of 24 bits (i.e., binary digits), compare the (decimal) values of $\epsilon/(b+1)$ corresponding to the use of those 24 bits to accommodate (a) binary, (b) octal, or (c) hexadecimal digits.

6. Calculate values of $\epsilon/(b+1)$ for the machines listed in problem 4 of section 1.3.

Section 1.5

1. Find expressions for the condition number of the following functions, and determine the values of x for which they are (a) ill-conditioned, (b) well-conditioned.

(i) $\exp(x)$,   (ii) $\ln(|x|)$,   (iii) $\sin(x)$,   (iv) $\cos(x)$,   (v) $\tan(x)$,
(vi) $(x^2+1)^{1/2}-|x|$,   (vii) $|x|-(x^2-1)^{1/2}$ with $|x| \geqslant 1$.

2. If the x-values quoted below are correctly rounded in decimal to precision 3, find the bounds upon the transmitted relative error in evaluating the given functions:

(a) $\sin(x)$ at $x = 3.01$; (b) $(x^2-10)^{1/2}$ for $x = 3.26$; (c) x^x for $x = 6.67$.

3. (a) Show that the condition number of $y = f(x)$ can be written as $d(\ln y)/d(\ln x)$.
   (b) If $\phi(x)$ is the condition number of $y = f(x)$ show that the condition number of the inverse function $x = f^{-1}(y)$ is $1/\phi[f^{-1}(y)]$.

4. Using the result of problem 3(b), evaluate the condition number of the following pairs of inverse functions:
   (a) x^n, $y^{1/n}$; (b) $\exp(x)$, $\ln(|y|)$; (c) $\tan(x)$, $\arctan(y)$.

5. (a) If $f(x)$ is the product of k functions $f_1(x) f_2(x) \ldots f_k(x)$ then use the result of problem 3(a) to show that the condition number of $f(x)$ is the *sum* of the condition numbers of each factor $f_j(x)$; i.e.,
$$\text{cond}\{f(x)\} = \sum_{j=1}^{k} \text{cond}\{f_k(x)\}.$$
   (b) It is required to calculate $f(x) = (x+1)(x+2)\ldots(x+20)$ at $x = -20.15$. What is the condition number of this calculation?

6. (a) A number of dependent variables $x_2, x_3, \ldots$ each depend on a single independent variable, x_1 say. Therefore each variable can, in general, be related to any other, which we express by defining $x_j = f_{jk}(x_k)$. Show that $f'_{ik}(x_k) = dx_i/dx_k = f'_{ij}(x_j) f'_{jk}(x_k)$ and hence verify that the condition number of $f_{ik}(x_k)$ is the *product* of the condition numbers of $f_{ij}(x_j)$ and of $f_{jk}(x_k)$.
   (b) If a sequential calculation is performed as $x_2 = f_{21}(x_1)$, $x_3 = f_{32}(x_2)$, ... and we denote the condition number of $f_{jk}(x_k)$ by ϕ_{jk}, then prove by induction that the condition number ϕ_{j1} of the function $f_{j1}(x_1)$ which

expresses x_j as a function of x_1, is the *product* of all the condition numbers of the sequential calculation leading to x_j. That is, $\phi_{j1} =$
$$\phi_{j(j-1)}\phi_{(j-1)(j-2)} \cdots \phi_{21} = \prod_{l=1}^{j-1} \phi_{(l+1)l}, \text{ for } j = 2, 3, \ldots.$$

Section 1.6

1. Verify that (1.6.2) is consistent with the recurrence relation for β_j as given by (1.6.1).

2. (a) The following expressions have to be computed for small values of $|x| \leqslant 1$, but as written this would involve cancellation error in the final stage of the calculation: (i) $(1 + 2x)^{-1} - (1 - x)(1 + x)^{-1}$; (ii) $(1 + x^2)^{1/2} - (1 - x^2)^{1/2}$; (iii) $1 - \cos x$; (iv) $\sin(1 + x) - \sin(1)$. Show how these expressions may be rearranged before computation to avoid this problem.

   (b) Using a value of $x = 0.0123$ and decimal arithmetic of precision 3, evaluate the above expressions (i) to (iv) from left to right precisely as written, employing correct rounding. Compare the results with the same process applied to the rearranged expressions, and with the answer obtained from using higher precision but correctly rounded to 3 significant figures.

3. The value of $(x - \sin x)$ is to be evaluated at $x = 0.0123$ using decimal arithmetic of precision 3. Show how the calculation can be performed without loss of significance.

4. (a) Write down the power series in z for $\ln(1 + z)$ and $(1 + z)^{-1}$. Then by substituting $x = (1 + z)$, express $f(x) = (2 \ln x - x + x^{-1})$ as a series in $z = (x - 1)$. This being also the Taylor series expansion for $f(x)$ about $x = 1$, what do you infer to be the general expression for $f^{(k)}(1)$, i.e. for the kth derivative of $f(x)$ at $x = 1$?

   (b) Show that the calculation of $f(x)$ is ill-conditioned close to $x = 1$. Use the series to evaluate $f(1.010)$ correct to 3 significant figures.

   (c) If the input value of $x = 1.010$ is correctly rounded to 4 significant figures, what are the bounds of the inherent error in evaluating $f(1.010)$?

5. A calculator has a special function key to evaluate π and $\arctan(x)$. The handbook indicates that to evaluate $\arccos(x)$ the identity
$$\arccos(x) = \pi/2 - \arctan(x/\text{sqrt}(1 - x^2))$$
should be used. For what parts of the defined range $|x| \leqslant 1$ is the evaluation of this principal value of $\arccos(x)$ ill-conditioned? For what part(s) of this range is the above method of evaluation likely to be unnecessarily inaccurate? Can you suggest a better method of calculation? Is this 'improved' formulation better for *all* $|x| \leqslant 1$ than the one of the handbook?

6. Let $N = b^i$, where b is the base of your calculator/computer, and i is a positive integer equal (about) to $p/2$, p being the machine's precision. For arbitrary (random) values of x in $(0, 1)$ compute the values of $y = (x/N) + 1$, $z_1 = (y \uparrow N) \uparrow (1/N) - y$, and $z_2 = [y \uparrow (1/N)] \uparrow N - y$, where the operator $\uparrow$ denotes exponentiation. Evidently z_1 and z_2 should both be zero. You may find that z_1 is sometimes zero, but $|z_2|$ is rarely zero and is almost always very much larger than $|z_1|$. Why is this?

7. (a) Confirm (for instance by Taylor series expansion) that, if $f(x)$ is thrice differentiable, then $f'(x) - [\delta f(x)/h] \to Ch^2$ as $h \to 0$, where C is (in the limit) independent of h, and where $\delta f(x) = f(x + \frac{1}{2}h) - f(x - \frac{1}{2}h)$. What is the value of C? (This result shows that $\delta f(x)/h$ is an approximation to the derivative $f'(x)$, provided h is 'sufficiently small'.)

   (b) If there is no numerical error in forming the values of x, h or $(x \pm \frac{1}{2}h)$, and the only generated error in computing $\tilde{f}(..)$ is rounding error, show that $f'(x) - [\delta\tilde{f}(x)/h]$ is approximately given by $Ch^2 + \alpha f(x)\epsilon h^{-1}$, where α is random within a fixed interval $[-A, A]$. What is the value of A if $\tilde{f}(..)$ is correctly rounded? (This result shows why the computed approximation $\delta\tilde{f}(x)/h$ to $f'(x)$ may be poor if h is 'too small'.)

   (c) The difference between $\delta\tilde{f}(x)/h$ and $f'(x)$ is largest when $|\alpha| = A$, and $\alpha f(x)$ has the same sign as C. What choice of h would minimise the magnitude of the difference in such an eventuality? Assuming the function $f(x)$ and its derivatives have values not very different from unity, and that the computation is carried out with base 2 and precision 24, what (very roughly) would be the appropriate choice of h?

Section 1.7

1. (a) According to (1.7.1), the commutative law applies to machine addition and multiplication. However, the associative law does not. Give examples in decimal floating point arithmetic of precision 3. (The associative law for addition states that $(a + b) + c = a + (b + c)$.)

   (b) Show that neither does the distributive law apply (which states that $a(b + c) = ab + ac$).

2. (a) Verify the formulae of (1.7.2) for addition and division.

   (b) Deduce the recursion formula for summation, (1.7.5).

3. The machine multiplication of two floating point numbers $n_1 \tilde{*} n_2$, each of precision p, is the only arithmetic operation whose *precise* result has a necessarily limited precision (setting aside bounds imposed by the range of exponent values permitted). What is the maximum precision required to represent $n_1 * n_2$ precisely? Give an example.

4. Suppose b is the base of machine arithmetic, p its precision, and i is an integer such that $p > i > p/2$. Let $n_1 = b^i + k_1$, $n_2 = b^i + k_2$, and $n_3 = b^{2i} + (k_1 + k_2)b^i$,

where k_1 and k_2 are small integers and $(k_1 + k_2) > 0$. Then n_1, n_2 and n_3 are all exactly representable by the machine, and the subtraction in $n_0 = n_1 \tilde{*} n_2 - n_3$ has no rounding error. Determine how your machine adjusts the result of multiplication to limited precision p, by comparing n_0 with the precise value $(n_1 * n_2 - n_3) = k_1 k_2$, for various integer pairs k_1, k_2. (N.B. You may find that the precision effectively used in multiplication is less than that of the representation of numbers by the machine.)

5. With $n_1 = (b^i + k_1) * b^i$, $n_2 = 1 - k_2/b^i$, but n_3 as in question 4, compare $n_0 = n_1/n_2 - n_3$ with the approximate value $(n_1/n_2 - n_3) \simeq k_2(k_1 + k_2)$ and so determine how your machine floats the result of division. Take $k_1 + k_2 > 0$ as before. (N.B. See note to problem 4, which may also apply here.)

6. The accumulated error in forming a sum of n positive terms is detected by the following algorithm. Use it to observe the rate of growth of relative error with n (Fig. 1.3). It sums terms generated by random(0) which is supposed to return a pseudo-random number from a uniform distribution in (0, 1). *Make sure the least significant digits of* random(0) *are not zero*. If they are zero, then multiply it by π say (an upper limit of 1 for the terms is unimportant), or alternatively assign *term* as equal to (random(0) + *eps* * random(0)), where *eps* is small (say, equal to $b^{-p/2}$).

real *sum, toterror, term, oldsum, whatadded*; **integer** k, n;
input (n); *sum*: $= 0$; *toterror*: $= 0$;   **comment** initialise;
for $k: = 1$ **step** 1 **until** n **do begin**   **comment** set-up loop;
        term: $= random(0)$;   **comment** *term* > 0 is random;
        oldsum: $= sum$;   **comment** copy what *sum* was;
        sum: $= term + sum$;   **comment** increment *sum*;
        whatadded: $= sum - oldsum$;   **comment** *error* $= term - whatadded$;
        toterror: $= (term\text{-}whatadded) + toterror$;
        newline; *output* (k);   **comment** output counter;
        output $(toterror/sum)$;   **comment** and relative-error of sum;
        end *of loop*;

7. (a) Compare the generated errors in the floating-point implementation of the two algebraically equivalent calculations: (i) $x_0 = x_1 * x_1 - x_2 * x_2$, and (ii) $x_0 = (x_1 + x_2) * (x_1 - x_2)$. Show that the former calculation is likely to have a much larger generated error if $|x_1/x_2|$ is close (but unequal) to unity.

  (b) If values of α_k are independent random variables from a distribution with mean $\bar{\alpha}$ and variance σ_α^2, then values of $A\alpha_1 + B\alpha_2$ are randomly distributed with mean $(A + B)\bar{\alpha}$ and variance $(A^2 + B^2)\sigma_\alpha^2$. What, therefore, are the mean generated errors and their variances for the calculations (i) and (ii) of part (a)?

(c) Show that, if correct rounding is used, calculation by (ii) can be expected to have a greater generated error than (i) if $(x_1/x_2)^2$ or $(x_2/x_1)^2$ exceeds $2 + \sqrt{3}$.

8. By successively squaring x, we form $x^2, x^4, x^8, x^{16}, \ldots$ in turn, and any positive integer power of x can be composed of a product of such squares (e.g. $x^{53} = x^1 * x^4 * x^{16} * x^{32}$). Program your calculator/computer to perform this operation (of **integer-exponentiation**). What are the greatest, and least, number of multiplications involved in calculating x^k by this method if $2^{i+1} > k \geqslant 2^i$, where i is a positive integer?

9. If $x_0 = x_1 \uparrow x_2$ for any real x_1, x_2 with $x_1 > 0$, is defined as $\exp(x_2 * \ln(x_1))$, find an (approximate) expression for the rounding error $\beta_0 \epsilon$ of the machine representation of this operation, $n_0 = n_1 \widetilde{\uparrow} n_2$, where n_1 and n_2 are both floating-point numbers. Show that the generated relative error is proportional to $\ln(x_0)$. (Large magnitude results of *integer* exponentiation are therefore more accurately evaluated by repeated multiplication, as in problem 8.)

10. If $f(x)$ is correctly evaluated in floating-point arithmetic of base b and precision p, for what range of values of $f(x)$ would the computed difference $[f(x) - 1]$ be likewise correct?

11.(a) Let $a_1, a_2, a_3, \ldots$ denote successive values of some variable. Suppose $\bar{a}_n = \sum\limits_{k=1}^{n} a_k/n$ is the arithmetic mean of the first n values, and $\sigma_n^2 = \sum\limits_{k=1}^{n} (a_k - \bar{a}_n)^2/(n-1)$ is their variance. Confirm that

$$\sigma_n^2 = [\sum\limits_{k=1}^{n} a_k^2 - n(\bar{a}_n)^2]/(n-1) .$$

(b) If $\epsilon_k = a_k - \alpha$, where α is any constant (say, an estimate of the mean value of the a_k's), then verify that

$$\bar{a}_n = \alpha + (\sum\limits_{k=1}^{n} \epsilon_k/n) = \alpha + \bar{\epsilon}_n, \quad \text{say},$$
and
$$\bar{\sigma}_n^2 = [\sum\limits_{k=1}^{n} \epsilon_k^2 - n(\bar{\epsilon}_n)^2]/(n-1) .$$

(c) If $\delta_k = a_k - \bar{a}_{k-1}$, and $\bar{a}_0$ and σ_1 are defined as zero, then prove (by induction) that **running** values of mean and variance are formed as

$$\bar{a}_n = \sum\limits_{k=1}^{n} (\delta_k/k), \quad \sigma_n^2 = \sum\limits_{k=2}^{n} [(\delta_k^2/k) - \sigma_{k-1}^2/(k-1)] .$$

(Note that this avoids accumulator saturation, without the need for prior estimation of the mean.)

Section 1.8

1. Rearrange the following expressions (subject, if necessary, to certain stated conditions) so as to avoid unnecessary exceptional conditions:

(a) $((x_1)\uparrow N + (x_2)\uparrow N)\uparrow(1/N)$ ($N>0$ integer, $x_1 \geqslant x_2 > 0$);
(b) $\exp(-x)*(\text{abs}(x)\uparrow N)$ ($N>0$ and integer);
(c) $\arctan(\text{sqrt}((1-x)/(1+x)))$;
(d) $x*\sin(1/x)$.

2. If a computer represents an overflowed value (positive or negative) by a symbolic value (VR, say) and an underflowed value by 0, then $x/0$ could reasonably be represented by VR if $x \neq 0$. Define suitable values of the results of the four arithmetic operations involving $x, 0$, or VR as operands, where $x \neq 0, VR$. Note any operations which defy definition (like $0/0$).

3. If the range of a floating point number is 'symmetric', then the *product* of the largest and smallest magnitudes which can be represented without overflow or underflow is (almost) unity. If the quotient of these magnitudes is b^{2R} say, what would be the greatest and least exponent values, assuming that p is even and standardisation is effected by (1.2.4)?

4. How you would program the evaluation of

(a) $\arctan[2(1-\sin x)\tan^2 x]$
(b) $(x/2) + \ln|[1-\exp(-x)]/x|$

so as to preserve maximum precision and avoid exceptional conditions? It may be assumed that all trigonometric and hyperbolic functions, and their inverses, are available and accurate.

5. (a) The **Gudermanian function** is defined by $\text{gd}(x) = \int_0^x \text{sech}(t)dt$, and it is bounded and real for all real x. It can be expressed in terms of trigonometric and hyperbolic functions in numerous ways: e.g.
  (i) $\text{gd}(x) = \arctan(\sinh x) + C_1$,
  (ii) $\text{gd}(x) = \arcsin(\tanh x) + C_2$,
  (iii) $\text{gd}(x) = 2\arctan(\exp x) + C_3$,
  (iv) $\text{gd}(x) = \arccos(\text{sech} x) + C_4$,
  where the C's are constants. What are their values?

  (b) Supposing that evaluation of any of the hyperbolic or inverse trigonometric functions in the above expressions presents no problem, and that it is desired accurately to determine $\text{gd}(x)$ on a calculator or computer for *any* value of x, large or small, positive or negative, which of the formulae should be chosen, and why?

PROBLEMS: CHAPTER 2

Section 2.2

1. The **binomial theorem** states that, if $|z| < 1$ then

$$(1-z)^{-c} = \sum_{k=0}^{\infty} (c)_k z^k / k!$$

where $(c)_0 = 1$ and $(c)_k = c(c+1)(c+2)\dots(c+k-1)$ for positive integer k is **Pochhammer's symbol**.

  (a) Show that the series terminates if $c = -m$ is a negative integer, and express the coefficients in terms of factorials.

  (b) Verify that the series is the geometric series if $c = 1$, and express the coefficients in terms of factorials if $c = m$ is a positive integer.

2. Find the series expansions of:

  (a) $(1-z)^{-c} + (1+z)^{-c}$; and (b) $(1+z)(1-z)^{-c}$.

3. (a) Noting the identity $(1-z)^{-c} = 2^c[1-(2z-1)]^{-c}$, expand $(1-z)^{-c}$ as a series in ascending powers of $(z-\frac{1}{2})$.

  (b) If $\alpha \neq 1$, expand $(1-z)^{-c}$ as a series in ascending powers of $(z-\alpha)$.

4. (a) Given that $(1-2z\cos\theta + z^2)^{-1} = \sum_{k=0}^{\infty}[\sin(k+1)\theta]z^k/\sin\theta$ for $|z|<1$, write down the first few terms of the series corresponding to $\theta = \pi/3$, $\pi/2$, $2\pi/3$ together with their respective sums.

  (b) Deduce the limit of $[\sin(k+1)\theta]/\sin\theta$ for $\theta \to 0$, and repeat part (a) using this limit of θ. Confirm that the result is consistent with the binomial theorem (see problem 1(b)).

Section 2.3

1. (a) If $\Gamma(z+1) = z\Gamma(z)$, then prove by induction that for any positive integer m, $\Gamma(z+m) = (z)_m\Gamma(z)$ where $(z)_m = z(z+1)\dots(z+m-1)$.

  (b) If, conversely, $(z)_m$ is defined for any integer m as equal to $\Gamma(z+m)/\Gamma(z)$, show that $(z)_0 = 1$, and that $(z)_{(-m)} = 1/(z-m)_m$.

  (c) Express $(z)_{(-m)}$ for positive integers m as a rational function of z. (Use the results of parts (a) and (b).)

  (d) Verify that $(1/2)_m = [(2m)!/(4^m m!)] = \Gamma(2m)/[2^{2m-1}\Gamma(m)]$.

2. (a) Show that $(z+k)_{(m+1)} - (z+k-1)_{(m+1)} = (m+1)(z+k)_m$ and hence verify the following expression for the sum of the **factorial series**:

$$\sum_{k=1}^{N}(z+k)_m = (m+1)^{-1}[(z+N)_{(m+1)} - (z)_{(m+1)}] .$$

  (b) Assuming that n is a positive integer, and $x > 0$, show that the *infinite* factorial series:

$$\sum_{k=0}^{\infty}[(x+k)(x+k+1)\dots(x+k+n)]^{-1}$$
$$= [nx(x+1)\dots(x+n-1)]^{-1} = [n(x)_n]^{-1} .$$

  (c) Prove that

$$1 + (1/3) + (1/6) + (1/10) + (1/15) + \dots = 2\sum_{k=1}^{\infty}[k(k+1)]^{-1} .$$

3. Explain the following paradox:

$$1-(1/2)+(1/3)-(1/4)+(1/5)-(1/6)+\ldots$$
$$= [1+(1/2)+(1/3)+(1/4)+(1/5)+(1/6)\ldots] - 2[(1/2)+(1/4)+(1/6)+\ldots]$$
$$= [1+(1/2)+(1/3)+\ldots] - [1+(1/2)+(1/3)+\ldots] = 0.$$

4. Confirm the identities

(a) $\quad \sum_{k=0}^{\infty}(2k+1)^{-2} = (3/4)\sum_{k=0}^{\infty}(k+1)^{-2}$

(b) $\quad 1-2\sum_{k=1}^{\infty}(-1)^k(36k^2-1)^{-1} = (4/3)\sum_{k=0}^{\infty}(-1)^k(2k+1)^{-1}$.

(Hint: the series on the left in (b) is $1+(1/5)-(1/7)-(1/11)+(1/13)+\ldots$
and simply omits reciprocal multiples of 3 from that on the right.)

Section 2.3.1

1. (a) Show that on a machine using correctly rounded floating-point arith-
metic, the condition $\tilde{S}_{n+1} = \tilde{S}_n$ of (2.3.9) implies that $|\tilde{t}_{n+1}/\tilde{S}_n| \leqslant \epsilon/2$,
where $\epsilon = b^{1-p}$ is the machine-unit.

(b) What limits on $(\tilde{t}_{n+1}/\tilde{S}_n)$ are implied if chopping is used?

2. Show that on a machine using correctly rounded arithmetic of base b
and precision p, the condition $|\tilde{t}_{n+1}| \leqslant (b^{-p}/2)|\tilde{S}_n|$ implies in general that
$\tilde{S}_{n+1} = \tilde{S}_n$. (This is a generally *sufficient* condition that $\tilde{S}_{n+1} = \tilde{S}_n$, but problem
1(a) shows that it is not a *necessary* condition.)

3. If summation of an alternating series of terms with decreasing magnitude
is terminated when $|t_{n+1}| \leqslant e|S_n|$, then show that the precisely evaluated
estimate $S_e = S_n + (t_{n+1}/2)$ has a relative error not greater in magnitude than
approximately $|e/2|$.

4. (a) What is the term-ratio (t_k/t_{k-1}) in the binomial series for $(1+z)^{-c}$?
(b) Program a summing loop for the binomial series which employs iterative
development of t_k using the term-ratio, and which terminates either by
summing to saturation or when $|t_{n+1}| \leqslant e|S_n|$.
(c) Sum the alternating series for $(1+z)^{-c}$ with $z = 7/9$ and $c = 3/2$ and
compare with the precise result ($= 27/64$), using this program.

5. Write down in a column the successive partial sums $S_0, S_1, \ldots, S_8$ of the
binomial series $(1+z)^{-c}$ with $z = 7/9$ and $c = 3/2$, correct (say) to 4 decimal
places. Call these (respectively) the values $S_{j0} = S_j$ for $j = 0, 1, \ldots$. Then form
the triangular array of numbers S_{jk}:

S_{00}

$S_{10} \quad S_{11}$

$S_{20} \quad S_{21} \quad S_{22}$

$S_{30} \quad S_{31} \quad S_{32} \quad S_{23}$

$\cdots \qquad\qquad \cdots$

which are generated for each column $k = 1, 2, \ldots, 8$ in turn by the recurrence relation

$$S_{jk} = \tfrac{1}{2} \left(S_{j,k-1} + S_{j-1,k-1} \right) \quad (j = k, k+1, \ldots, 8) \ .$$

It will be observed that each column consists of values which (ultimately) alternately over-estimate or under-estimate the true sum $(= 27/64)$, and that these estimates improve as each new column is generated. This is called summation by **repeated averaging**, and is a particular case of **Euler's transformation** of an alternating series. It provides a means of accelerating the convergence of a slowly-converging alternating series.

6. (a) The second column of the array of repeated averages (defined in the last question) is made up of estimates of the sum $S_{j1} = S_j'$, say. Show that
$S_{j+1}' = S_j' + t_{j+1}'$ (i.e. $S_n' = \sum\limits_{k=0}^{n} t_k'$) where the terms

$$t_{j+1}' = (t_{j+1} + t_j)/2 \quad (j = 0, 1, 2, \ldots) \ .$$

What is the value of S_0'?

(b) Show that if the magnitude of the terms of the alternating series $\sum\limits_{k=0}^{\infty} t_k$ decreases monotonically to zero for all $k > K$, then the terms t_j' of the averaged series are also of alternating sign for $j > K + 1$.

Section 2.3.3

1. (a) If the term ratio $r_k = t_k/t_{k-1}$ of a series of positive terms is less than 1 and decreases monotonically with k for all $k > K$ (say), show that the sum of the series tail $\tau_n = \sum\limits_{k=n+1}^{\infty} t_k$ is less than $t_{n+1}/(1 - r_{n+2})$, provided that $n + 2 > K$.

(b) If $t_k = [(z/2)^k/k!]^2$, provide an upper bound on τ_n/t_{n+1} for sufficiently large n.

2. (a) Find a *lower* bound on the tail of $\sum\limits_{k=1}^{\infty} k^{-m-1}$, where m is any positive integer, by comparison with the infinite factorial series (see problem 2(b) of section 2.3).

(b) Show similarly that an *upper* bound on the tail τ_n of the series $\sum\limits_{k=1}^{\infty} k^{-3}$ is given by $[2n(n+1)]^{-1}$.

(c) Use the *lower* bound found in part (a) to provide an improved estimate S_e of $S_\infty = \sum\limits_{k=1}^{\infty} k^{-3}$, based on a correction added to the partial sum S_n, and find an upper bound on the truncation error of S_e.

3. Show by comparison with the infinite factorial series that an *upper* bound on the tail τ_n of the series $\sum\limits_{k=1}^{\infty} k^{-m-1}$ is given by

$$\tau_n < [m(n+1 - \tfrac{1}{2}m)_m]^{-1} \ .$$

4. (a) Show that $[(a+1)/a]^k > (a+k)/a$, for all $a > 0$, and all integers $k > 1$.
   (b) Find upper and lower bounds on the tail of the series $\sum_{k=0}^{\infty} x^k/(k+\frac{1}{2})$
   for $0 < x < 1$, by comparison with a suitable geometric series.
   (c) Use the lower bound together with S_n to provide an improved estimate
   of the sum of the series in part (b), and find an upper bound on its
   truncation error.

Section 2.3.4

1. (a) Confirm that the bounds of (2.3.16) apply to the tail τ_n of the series
   $\sum_{k=1}^{\infty} k^{-m-1}$, where $m > 0$, and show that
   $$1 + m/[2(n+1)] < m(n+1)^m \tau_n < [1 + (2n+1)^{-1}]^m .$$
   (b) The bounds on the tail of the series $\sum_{k=1}^{\infty} k^{-3}$ given by the comparison
   series of problem 2(c) of section 2.3.3 are $(n+2)^{-1} < 2(n+1)\tau_n < n^{-1}$.
   Are these larger or smaller than the bounds derived from (2.3.16)?

2. (a) The series $\sum_{k=1}^{\infty} [k^{-1} - \ln(1 + k^{-1})]$ sums to **Euler's constant**
   $(C = 0.5772 \ldots)$. Find the value of $\int_v^{\infty} t(u)du$, where $t(u)$ is the term
   magnitude function of this series.
   (b) The terms of this series, and this integral of $t(u)$, involve cancellation
   error in their evaluation. Suggest how this can be overcome, and evaluate
   t_n, and the value of the integral with $v = n + \frac{1}{2}$, using the value $n = 100$
   and correctly rounded decimal arithmetic of precision 4.

3. Sum the series $\sum_{k=0}^{\infty} (k^3 + 1)^{-1/2}$ with a truncation error less than 10^{-3}.

4. (a) Writing $x = t^2$, the series $\sum_{k=0}^{\infty} x^k/(k+1/2)$ represents for $1 > x \geqslant 0$ the
   value of $t^{-1}\ln[(1+t)/(1-t)]$. Verify by differentiation of the term
   magnitude function that the bounds of (2.3.16) apply to the tail of this
   series. (Hint: substitute $\exp(-p)$, say, for x.)
   (b) Show that the estimate of (2.3.17) is $S_e = S_n + x^{-1/2}E_1[(n+1)p]$,
   where
   $$E_1(z) = \int_z^{\infty} e^{-s}s^{-1}ds$$
   is known as the **exponential integral**.

5. (a) Transform the series $\sum_{k=0}^{\infty} (-1)^k(2k+1)^{-1}$ into an averaged alternating
   series. How many terms of this averaged series are required to be summed
   if the truncation error is to be less than 10^{-4} in magnitude?
   (b) Convert the series into a Leibnitz sum of positive terms. If the truncation
   error of the estimate (2.3.17) is assumed to be correctly given by (2.3.18),
   how many terms of this series of positive terms must be summed in order
   that this estimate has a truncation error less than 10^{-4} in magnitude?

Section 2.4

1. Using correctly rounded decimal arithmetic of precision 2 in developing the terms and performing the additions, sum the series $\sum_{k=1}^{24} [1/(k * k)]$.

    (a) in natural order,

    (b) in reverse order,

    (c) in natural order but adding the terms using a double-precision accumulator, and

    (d) in natural order either using a second (spill) single-length accumulator, or employing Kahan's device.

2. Repeat the calculations of problem 1 using chopping instead of correct rounding in developing the terms and performing additions.

3. The **Bessel function** $J_0(x)$ is represented by the power series expansion $\sum_{k=0}^{\infty} (-x^2/4)^k/(k!)^2$. For values of $x \simeq 20$ it has a magnitude of about 0.1. If it is to be evaluated correctly to 4 significant decimal digits, estimate the required precision with which the series must be summed.

4. (a) Justify the statement that: ignoring the error in forming the terms, the generated numerical relative error in accumulating the sum of a large number (N, say) of terms of a convergent series, in the forward direction, is likely to be equal roughly to about $N\bar{\alpha}\epsilon$, where $\bar{\alpha}\epsilon$ is an estimate of the mean relative error of addition. (Refer to equation (1.7.5).)

    (b) If the series were alternating and the machine arithmetic used chopping in adding the positive numbers, but boosting in subtracting them, what would be a suitable estimate of $\bar{\alpha}$ in this context?

    (c) Supposing that the computation of successive terms of a convergent series of positive decreasing terms involves a generated relative error which increases in proportion to N, justify the assertion that this source of error is likely to be insignificant compared with $N\bar{\alpha}\epsilon$ if chopping is used for addition, and N is large. (Hint: for such a series, $\sum_{k=1}^{n} jt_j/n \rightarrow 0$ as $n \rightarrow \infty$).

    (d) Would a generated error increasing in proportion to the number of terms evaluated be likely to be insignificant if the series were summed in reverse order?

Section 2.5

1. Using integration by parts, establish that the exponential integral has an asymptotic series representation for $x > 0$ given by:

$$E_1(x) = \int_x^{\infty} (e^{-t}/t)\,dt = (e^{-x}/x)\left(\sum_{k=0}^{n} t_k + E_n\right), \quad \text{with } t_k = k!(-x)^{-k},$$

and confirm that the truncation error E_n has the same sign as, but a smaller magnitude than, the first omitted term of the series t_{n+1}. Is the infinite series $\sum_{k=0}^{\infty} k!(-x)^{-k}$ convergent or divergent?

2. (a) If $m = \text{entier}(x)$, show that for any given $x > 0$, the least magnitude of the upper bound $|t_{n+1}|$ to the truncation error E_n of the asymptotic series for $E_1(x)$ in problem 1 is provided by selecting $n = m - 1$, and show that $|t_m| < m!/m^m$.

   (b) Verify that $|t_m| < (2\pi m)^{\frac{1}{2}} \exp(-m)$.

3. What is the least value of x for which the asymptotic series for $E_1(x)$ may be employed on a machine which uses correctly rounded arithmetic of base $b = 2$ and precision $p = 24$, if the truncation error is to be compatible with the possible rounding error? How many terms of the series will be required?

4. Find the value of $\lim_{N \to \infty} \{[1 + (a/N)]^{bN+c}\}$ and hence using (2.5.7) establish for large values of N, that $N!/(N + a)^{N+c}$ is approximately equal to $(2\pi)^{1/2} N^{1/2 - c} \exp(-N - a)$.

5. (a) Find approximately the values of z for which successive terms (t_j and t_{j+1}) of the asymptotic series in Stirling's formula (2.5.4) are equal to each other (for $j = 1, 2, \ldots, 6$), and hence establish the range of z-values for which the upper bound $|t_{n+1}|$ to the truncation error $|E_n|$ is rendered least by choosing n equal (in turn) to 1, 2, 3, 4 or 5. What is the upper bound to the truncation error at the change-over points?

   (b) Show, using (2.5.6), that as $j \to \infty$, $t_{j+1} = t_j$ at $\pi z = j - 1/4$. What is the approximate value of the least upper bound to the truncation error (of Stirling's formula) at this value of z? Compare these approximate values with those obtained in part (a), taking $j = 6$.

6. (a) The **Euler-Maclaurin summation formula** states that the sum S_∞ of a series of positive terms t_k can be represented in terms of its partial sum S_n and the term magnitude frunction $t(..)$ by

$$S_\infty - [S_n + (t_{n+1}/2) + \int_{n+1}^{\infty} t(u)du] = \sum_{j=1}^{m+1} b_{nj} + E_{nm}$$

   where $b_{nj} = [(-1)^j B_j/(2j)!] t^{(2j-1)}(n + 1)$, and the magnitude $|E_{nm}|$ truncation error is less than that of the last term $|b_{n(m+1)}|$, provided $t^{(2m+2)}(u) > 0$ for all $u > (n + 1)$. This represents a powerful extension of the method of integral approximation to the tail of a series. Show that if $t(u) = u^{-2}$, the terms $b_{nj} = (-1)^{j+1} B_j(n + 1)^{-2j-1}$, and the series $\sum_{j=1}^{\infty} b_{nj}$ is semi-convergent.

   (b) Taking $n = 2$, find the value of m which renders the upper bound on $|E_{2m}|$ least, and estimate the value of this least bound, if $t(u) = u^{-2}$.

   (c) Taking $n = m = 5$, estimate the sum of the series $\sum_{k=1}^{\infty} k^{-2}$ and the bounds on the truncation error.

7. The **modified Euler-Maclaurin formula** is

$$S_\infty - [S_n + \int_{n+1/2}^\infty t(u)du] = -\sum_{j=1}^{m+1} b'_{nj} + E'_{nm}$$

where $b'_{nj} = [(-1)^j(1-2^{1-2j})B_j/(2j)!]t^{(2j-1)}(n+\frac{1}{2})$, and the magnitude of the truncation error $|E'_{nm}|$ is less than $|b'_{n(m+1)}|$, provided that $t^{(2m+2)}(u) > 0$ for all $u > n + \frac{1}{2}$. Confirm that the inequalities (2.3.16) follow as a special case of these two Euler-Maclaurin summation formulae with $m = 0$.

8. **Riemann's zeta function** $\zeta(x)$ is defined for $x > 1$ as the sum $\sum_{k=1}^\infty k^{-x}$, and values of $\zeta(2m)$, where m is integer, are related to the Bernoulli numbers B_m by (2.5.5). Applying the modified Euler-Maclaurin formula to this series it follows that

$$\zeta(x) = \sum_{k=1}^n k^{-x} + v^{1-x}\sum_{j=0}^{m+1} a_j(x)_{(2j-1)}/(-v^2)^j + E'_{nm}$$

where $v = n + 1/2$, $a_j = (1-2^{1-2j})B_j/(2j)!$ with a_0 defined as unity, and $(x)_{(2j-1)}$ is Pochhammer's symbol (see problem 1 of section 2.2). Taking $n = m = 3$, estimate the range of values in this $\zeta(1.01)$ must lie. About how many terms of the series $\sum_{k=1}^\infty k^{-1.01}$ would be needed to be directly summed to achieve this accuracy?

9. For a series of alternating terms, application of the Euler-Maclaurin formulae to the individual sums of positive and negative terms shows that

$$S_\infty - S_n - (t_{n+1}/2) = \text{sgn}(t_{n+1})\sum_{j=1}^{m+1}(4^j - 1)b_{nj} + E_{nm}$$

where b_{nj} is defined as in problem 6, and the magnitude of the truncation error is less than $(4^{m+1} - 1)|b_{(m+1)j}|$.

Apply this formula with $n = m = 5$ to evaluate the sum of the series $\sum_{k=1}^\infty (-1)^{k+1}/k$, and estimate the bounds of the truncation error.

Section 2.6

1. (a) Write a sequence of machine instructions which serves to assign the value of $\Gamma(z)$ to the variable *gamma*. You may assume that a reference to $strlg(x)$ will provide a correct value of $\Gamma(x)$ provided that the argument value $x \geqslant N$. Suitable values may be assumed to have been assigned to non-local variables N, z and $pi(=\pi)$. An attempt to evaluate $\Gamma(z)$ at a pole of this function should be allowed to fail by 'zero-divide'. (Use the functional equations: $\Gamma(z + 1) = z\Gamma(z)$ and $\Gamma(z)\Gamma(1 - z) = \pi/\sin(\pi z)$. If the programming language you use allows the definition of recursive functions, then *gamma* (..) may be assumed to be such a function.)

   (b) Taking $N = 2$, check that your program gives the values $\Gamma(1.5) = 2s/3$, $\Gamma(0.5) = 4s/3$ and $\Gamma(-0.5) = -3\pi/2s$ where s is the value of $strlg(2.5)$. What will be the value of $\Gamma(-1.5)$ in terms of s? And what will be the value of s?

2. (a) A function $f(z)$ obeys the functional equations:

$$f(z) + f(1-z) = A(z) \tag{A}$$

$$f(z) + f(1/z) = B(z) \tag{B}$$

where $A(z)$, $B(z)$ are known functions of z. Verify that $A(z) = A(1-z)$, $B(z) = B(1/z)$ and that these functions are related by the consistency equation:

$$A(z) + A(1/z) + A[z/(z-1)] = B(z) + B(1-z) + B[(z-1)/z] .$$

(b) The value of $f(z)$ may be related linearly to each of $f(1-z), f(1/z)$, $f[1/(1-z)], f[(z-1)/z]$, and $f[z/(z-1)]$. Equations (A) and (B) above are two such relations: establish from these the other three.

(c) Whatever the value of z, these five relations enable the value of $f(z)$ to be expressed in terms of $f(x)$ where $0 \leqslant x \leqslant \frac{1}{2}$, and x is some function of z. Establish which relation should be used for this purpose according to the value of z.

3. (a) The definite integral $L_2(z) = -\int_0^z t^{-1} \ln|1-t| dt$ is called **Euler's dilogarithm**. It obeys the functional equations:

$$L_2(z) + L_2(1-z) = 2c - \ln(|1-z|)\ln(|z|) \tag{A}$$

$$L_2(z) + L_2(1/z) = [3\,\mathrm{sgn}(z) + 1]\,c - (\ln|z|)^2/2 \tag{B}$$

where $c = \pi^2/12$. What are the values of $L_2(z)$ at $z = -1, 1/2, 1$, and 2? Verify that $z = 2$ is the only stationary value of $L_2(z)$, and that $L_2(z) = 0$ for $z = 0$ and some $z > 2$.

(b) From problem 2 it will be apparent that $L_2(..)$ only needs to be directly evaluated for arguments which lie in the interval $[0, 1/2]$. Two expansions are available:

$$L_2(x) = \sum_{k=1}^{\infty} x^k/k^2 \quad (|x| \leqslant 1)$$
$$= -w + (w^2/4) + \sum_{k=1}^{\infty} (-1)^k B_k w^{2k+1}/(2k+1)! \quad (|w| \leqslant 2\pi)$$

where $w = \ln(1-x)$, and the B_k are the Bernoulli numbers. What considerations would determine the choice of expansion on a machine using correctly rounded arithmetic of base 2 and precision 24?

Section 2.7

1. Prove, using induction, that (A_k/B_k) as given by (2.7.5) is the kth convergent of the continued fraction $\underset{j=1}{\overset{\infty}{\mathrm{K}}} p_j \downarrow q_j.$. (Hint: the $(k+1)$th-convergent is found from the kth by replacing q_k by $q_k + (p_{k+1}/q_{k+1}).$)

2. (a) Show that $\underset{j=0}{\overset{\infty}{\mathrm{K}'}} r_j z^{-2} \downarrow 1 = z \underset{j=0}{\overset{\infty}{\mathrm{K}'}} r_j \downarrow z.$

(b) If the partial numerators of the continued fraction $\displaystyle\mathop{K}_{j=1}^{\infty} p_j \downarrow q_j$ are made unity by the similarity transform (2.7.1) with $c_0 = 1$, then verify that $c_0 = 1$, $c_1 = 1/p_1$ and $c_j = p_{j-1}c_{j-2}/p_j$ for $j > 1$. Consequently:

$$c_{2j} = (p_1 p_3 \ldots p_{2j-1})/(p_2 p_4 \ldots p_{2j}),$$

$$c_{2j+1} = (p_2 p_4 \ldots p_{2j})/(p_1 p_3 \ldots p_{2j+1})$$

for $j = 1, 2, \ldots$.

(c) Express the S-fraction $\displaystyle\mathop{K'}_{j=0}^{\infty} (j/2)z^{-2} \downarrow 1$ in the form $\displaystyle z \mathop{K}_{j=1}^{\infty} 1 \downarrow s_j z$, and identify the values of s_j.

3. (a) Deduce an expression for the 4th convergent of the S-fraction $\displaystyle\mathop{K'}_{j=0}^{\infty} r_j z \downarrow 1$.

(b) Necessary conditions which must be satisfied for a function $f(z)$ to be representable as an S-fraction in terms of z are that $f(0) = 1$, and further that each of its derivatives $f^{(k)}(0)$ at $z = 0$ must be finite and non-zero. (These conditions are far from sufficient.) Show that $(1 + az)/(1 + z^2)$ is expressible as the terminating S-fraction $\displaystyle\mathop{K'}_{j=0}^{3} r_j z \downarrow 1$, if $a \neq 0$, and identify the values of r_1, r_2 and r_3.

(c) What necessary condition is not met by this particular (rational) function which prevents its representation as an S-fraction in terms of z, if $a = 0$?

4. (a) Show that $\displaystyle\mathop{K}_{j=1}^{\infty} p_j \downarrow q_j = p_1/(q_1 + \mathop{K}_{j=2}^{\infty} p_j \downarrow q_j)$.

(b) If $p_j = p$, $q_j = q$ are constant for all j show that, if the continued fraction $\displaystyle\mathop{K}_{j=1}^{\infty} p \downarrow q$ converges, it will equal one of the two values $\pm [(q^2/4) + p]^{\frac{1}{2}} - (q/2)$.

(c) Show from the similarity transform that $\displaystyle\mathop{K}_{j=1}^{\infty} p \downarrow q$ must change sign if q changes sign, and hence that $\displaystyle\mathop{K}_{j=1}^{\infty} p \downarrow q = (q/2) \{ \text{sqrt}[1 + (4p/q^2)] - 1 \}$ where sqrt(..) denotes the positive square-root.

5. (a) The particular S-fraction representation

$$\mathop{K'}_{j=0}^{\infty} z \downarrow 1 = \frac{1}{1+} \frac{z}{1+} \frac{z}{1+} \ldots = 2/[1 + \text{sqrt}(4z + 1)]$$

is valid for $z \geq -1/4$. Verify that the recurrence relations (2.7.6) imply that $B_k = A_{k+1}$ and that if $z = 2$, then they are satisfied by $A_k = [2^k - (-1)^k]/3$. Hence show that, for large k, $A_k/B_k \simeq (1/2)[1 + 3/(-2)^{k+1}]$ if $z = 2$.

(b) If $z = -1/4$, verify that $A_k = k/2^{k-1}$ and find the corresponding expression for (A_k/B_k).

6. (a) If $p_j = b_j$, $q_j = 1$ for all $j = 1, 2, \ldots$ and A_j, B_j satisfy the recurrence (2.7.5) so that $\displaystyle\mathop{K}_{j=1}^{n} b_j \downarrow 1 = A_n/B_n$, show that

$$A_{k+2} = (1 + b_{k+1} + b_{k+2})A_k - b_k b_{k+1} A_{k-2}$$

for all $k = 1, 2, \ldots$ (with a similar expression for B_{k+2} in terms of B_k and B_{k-2}).

(b) If n is any positive integer, prove that $\overset{2n}{\underset{j=1}{K}} b_j \downarrow 1 = \overset{n}{\underset{j=1}{K}} p_j \downarrow q_j$ where

$$\left. \begin{array}{l} p_1 = b_1, \; p_j = -b_{2j-2}b_{2j-1} , \\[4pt] q_1 = 1 + b_2, \; a_j = 1 + b_{2j-1} + b_{2j} \end{array} \right\} \quad \forall j = 2, 3, \ldots .$$

(c) If $\overset{\infty}{\underset{j=0}{K'}} r_j z^{-1} \downarrow 1$ is convergent, show that it is equal (but not equivalent) to the continuied fraction $z \overset{\infty}{\underset{j=0}{K'}} (-r_{2j-1}r_{2j}) \downarrow (z + r_{2j} + r_{2j+1})$, with r_0 defined as zero. (Continued fractions like this, with partial denominators linear in z, and partial numerators independent of z, are called **J-fractions** in z.)

7. A continued fraction $q_0 + \overset{n}{\underset{j=1}{K}} p_j \downarrow q_j$ can be described as **augmented** by the addition of the term q_0, which is taken as its 'zero-th' convergent. If its kth convergent is represented for any $k \leq n$ by A'_k / B'_k, say, verify that A'_k, B'_k are given by the same recurrences as A_k, B_k respectively in (2.7.5), except that $A'_0 = q_0$ replaces $A_0 = 0$.

8. (a) Prove that $\overset{2n+1}{\underset{j=1}{K}} \alpha_j \downarrow 1 = \alpha_1 + \overset{n}{\underset{j=1}{K}} (-\alpha_{2j-1}\alpha_{2j}) \downarrow (1 + \alpha_{2j} + \alpha_{2j+1})$. (Hint: use the results of problems 6(a) and 7.)

(b) If the augmented continued fraction $\alpha_1 + \overset{2n}{\underset{j=1}{K}} b_j \downarrow 1$ with $\alpha_1 \neq 0$ is to be expressed for any $n \geq 0$ as $\overset{2n+1}{\underset{j=1}{K}} \alpha_j \downarrow 1$, confirm from the result of problem 6(b) that successive α-values are calculable as

$$\alpha_2 = -b_1/\alpha_1, \quad \alpha_3 = b_2 - \alpha_2 , \alpha_{2j} = b_{2j-2}b_{2j-1}/\alpha_{2j-1},$$

$$\alpha_{2j+1} = b_{2j-1} + b_{2j} - \alpha_{2j}, \quad \forall j = 2, 3, \ldots, n.$$

9. Prove by induction (on the value of n) that $\overset{n}{\underset{j=0}{K'}} r_j z \downarrow 1$ is a rational function of order $m/(n - m)$, where $m = \text{entier}(n/2)$.

Section 2.8

1. Verify that the recursion formula (2.8.1) is consistent with (2.7.5).

2. (a) Verify that if $q_j = p_1 = 1$ and $p_{j+1} = r_j z$ for all $j = 1, 2, \ldots$, then

$$t_{k+1} = (-z)^k \overset{k}{\underset{j=1}{\Pi}} r_j \beta_j \beta_{j+1} .$$

(b) Prove that the nth and $(n+1)$th convergents of the S-fraction $\overset{\infty}{\underset{j=0}{K'}} r_j z \downarrow 1$ have the same values at $z = 0$, and the same values of the kth derivative at $z = 0$, for all $k = 1, 2, \ldots (n-1)$. (In other words, their Taylor expansions about $z = 0$ have the same first n terms.)

3. At $x = 9.65$, the first 9 terms of the semi-convergent asymptotic series for the exponential integral $E_1(x)$ are needed to provide its value with a relative error of less than 5×10^{-4}. The S-fraction representation of this function is given by:

$$E_1(x) = \frac{(e^{-x}/x)}{1+} \frac{x^{-1}}{1+} \frac{x^{-1}}{1+} \frac{2x^{-1}}{1+} \frac{2x^{-1}}{1+} \frac{3x^{-1}}{1+} \cdots$$

$$= (e^{-x}/x) \underset{j=1}{\overset{\infty}{K'}} entier\,(j/2)\,x^{-1} \downarrow 1 \;.$$

Calculate the 9th convergent of this continued fraction at $x = 9.65$, and provide a bound on its truncation error. Is the relative error larger or smaller than that of the semi-convergent series?

4. The **theorem of Van Vleck** states that a continued fraction $\underset{j=1}{\overset{\infty}{K}} 1 \downarrow q_j$ with unit partial numerators and positive partial denominators (i.e. $q_j > 0$ for $j = 1, 2, \ldots$) converges if, and only if, the infinite series $\sum_{j=1}^{\infty} q_j$ diverges. Convert the S-fraction $\underset{j=1}{\overset{\infty}{K'}} entier\,(j/2)\,x^{-1} \downarrow 1$ into such a form (assuming $x > 0$), and hence show that it converges. (Hint: use similarity transform given in problem 2 of section 2.7.)

5. Calculate values (correct to 5 decimal places) of the terminating S-fraction $\underset{j=0}{\overset{n}{K'}} (j/2) \downarrow 1$ for $n = 4, 5, \ldots, 10$, and apply repeated averaging to estimate the value of the infinite S-fraction correct to 4 decimal places.

6. (a) If $q_j = 1$ and $0 < p_j \to p$ as $j \to \infty$, where p is a (positive) constant, then verify in (2.8.1) that, $\beta_j \to 2/[1 + \sqrt{(1 + 4p)}]$ as $j \to \infty$, and find the value of the asymptotic term-ratio, $\lim_{j \to \infty} (t_{j+1}/t_j)$.

   (b) The inverse tangent has the following power series and continued fraction representations:

   $$(1/x)\arctan(x) = \sum_{j=0}^{\infty} (-x^2)^j/(2j+1) = \underset{j=0}{\overset{\infty}{K'}} j^2 x^2 \downarrow (2j+1)$$

   the power series being convergent for $|x| \leqslant 1$. By expressing the continued fraction as the sum of an alternating series, and comparing its asymptotic term-ratio with that of the (alternating) power series, show that it converges much more rapidly than the power series.

   (c) Show that, already, the second convergent of the continued fraction has less error than the second partial sum.

7. (a) Placing $a_j = p_j/(q_j q_{j-1})$ and $b_j = q_j \beta_j$, confirm that (2.8.1) becomes

   $$b_1 = 1, \quad b_j = (1 + a_j b_{j-1})^{-1}$$

   $$t_1 = p_1/q_1, \quad t_j = (b_j - 1)t_{j-1} = -b_j b_{j-1} a_j t_{j-1} \;.$$

   (b) If $t_j \sim (-1)^{j+1} c j^{-\alpha}$ as $j \to \infty$, where c and α are positive constants, then show that $p_j/(q_j q_{j-1}) \sim j^2/\alpha^2$ as $j \to \infty$.

8. (a) Let $S_{m,n} = \sum\limits_{j=m}^{n} t_j$ and $C_{m,n} = \underset{j=m}{\overset{n}{K}} (-r_j) \downarrow (1 + r_j)$, where $r_j = t_j/t_{j-1}$ and $m \leq n$. Then prove that $C_{m,n} = -S_{m,n}/S_{m-1,n}$. (Hint: use induction, decreasing the value of m.)

   (b) Show that $\sum\limits_{j=1}^{n} t_j = t_1 \underset{j=1}{\overset{n}{K'}}(-r_j)/(1 + r_j)$ with r_1 defined as zero. (The expression on the right-hand-side is the nth convergent of **Euler's continued fraction**.)

   (c) Show that, already, the second convergent of the continued fraction has $\underset{j=1}{\overset{n}{K}} p_j \downarrow q_j$ to convert it to the form $t_1 \underset{j=1}{\overset{n}{K'}}(-r_j) \downarrow (1 + r_j)$, with $r_1 = 0$, and hence establish (2.8.1).

9. It could happen in evaluating (2.8.1) that some value of β_j, say β_m, becomes infinite because $q_m/p_m = -\beta_{m-1}$. Verify that in this exceptional circumstance, although t_m and t_{m+1} are both infinite,

$$t_{m+1} + t_m = (q_{m+1}q_m/p_{m+1})t_{m-1} \,,$$

$$t_{m+2} = (p_{m+2}q_m/p_{m+1}q_{m+2})t_{m-1} \,,$$

$$\beta_{m+2} = 1/q_{m+2} \,.$$

(Occurrence of large values of β_j can lead to loss of significance in computational work, due to cancellation error in adding t_{j+1} to t_j.)

10.(a) Evaluate the continued fraction $\underset{j=1}{\overset{3}{K'}} p_j \downarrow 1$, with $p_2 = -1.01$ and $p_3 = 0.0123$, using correctly rounded decimal arithmetic of precision 3, by (i) the forward recurrence (2.7.5), (ii) the series given by (2.8.1), and (iii) the backward recurrence (2.8.2). Note the differing effects of rounding.

   (b) Repeat part (a) but this time place $p_2 = 0.0123$ and $p_3 = -1.01$.

11. (a) Another form of the backward recurrence relation (2.8.2) is obtained by placing $C_k = p_k Z_k/Z_{k-1}$. Confirm then that the Z's are given by

$$Z_N = 1, \; Z_{N-1} = q_N, \; Z_{k-1} = q_k Z_k + p_{k+1}Z_{k+1}: \; 0 < k < N.$$

Express the value of the augmented continued fraction $q_0 + C_1$ in terms of the Z's.

   (b) The conversion of the terminated S-fraction $\underset{j=1}{\overset{N}{K'}} r_{j-1}z \downarrow 1$ into the form of a rational function $(\sum\limits_{j=0}^{M} a_j z^j)/(\sum\limits_{j=0}^{M} b_j z^j)$, where $M = \text{entier}(N/2)$, is most easily contrived using the recurrence of part (a), the numerator and denominator polynomials after completion of the $(N-k)$th iteration being Z_k and Z_{k-1} respectively, for $k = N, N-1, \ldots, 1$. Devise an algorithm which iteratively assigns values to the polynomial coefficient vectors, a and b, by this method. (Note that $a_0 = b_0 = 1$ throughout all iterations, and that the degree of the polynomial Z_j is entier $[(N-j)/2]$ for any j. It is easiest to assign a_j and b_j in decreasing order of subscript j, for each value of k.)

Section 2.9

1. Verify that the first 4 terms of the series development (in inverse powers of x) of the terminated S-fraction $\overset{4}{\underset{j=1}{K'}}\,entier\,(j/2)\,x^{-1} \downarrow 1$, are the same as the first 4 terms of the asymptotic series for $x\,e^x\,E_1(x)$ as given in problem 1 of section 2.5. (This S-fraction is the 4th convergent of the infinite S-fraction representation of the same function, as in problem 3 of section 2.8.)

2. (a) The asymptotic series $\sum\limits_{j=0}^{\infty}(-1)^j\,(4^j-2)\,B_j\,z^{-2j}$ is semi-convergent, and represents $(z/2)$ times the sum of the slowly converging series $\sum\limits_{k=0}^{\infty}[k+(z+1)/2]^{-2}$, as $z \to \infty$. This semi-convergent series is a particular example of that given in problem 8 of section 2.5 with $v = z/2$ and $x = 2$, resulting from the Euler-Maclaurin formula being applied to the summation of the tail of the series $\tfrac{1}{2}\sum\limits_{k=0}^{\infty}(k+1)^{-2}$. Its first few terms are

$$1 - 1/(3z^2) + 7/(15z^4) - 31/(21z^6) + 127/(15z^8) - + \dots$$

Calculate the values of r_1, r_2, r_3, and r_4 in its S-fraction representation $\overset{\infty}{\underset{j=0}{K'}}\,r_j\,z^{-2} \downarrow 1$, using rational arithmetic.

   (b) Can you discern the rule by which r_j appears to be expressible in terms of j? If so, determine the range of values of z for which the S-fraction converges.

   (c) Assuming $z = 1$, calculate the 4th and 5th convergents of the S-fraction and compare the result with the precise value $\tfrac{1}{2}\sum\limits_{k=0}^{\infty}(k+1)^{-2} = \pi^2/12$.

3. (a) Suppose that the diagonal entries $r_{m+k,\,k}$ for $k = 1, 2, \dots, n$ of the QD-algorithm provide the partial numerators of an S-fraction $a_m\,\overset{n}{\underset{k=0}{K'}}\,r_{m+k,\,k}\,z \downarrow 1 = C_{m,\,m+n}(z)$, say, for any $n \geqslant 0$, which has a Taylor expansion about $z = 0$ given by

$$C_{m,\,m+n}(z) = \sum_{j=m}^{m+n} a_j z^{j-m} + O(z^{n+1}) \quad (n \geqslant 0) .$$

Verify that $C_{m,\,m+2l}(z)$ and the augmented S-fraction $[a_m + zC_{m+1,\,m+2l}(z)]$ are both rational functions of order l/l. (Provided that there is only one such rational function which has the expansion $\sum\limits_{j=m}^{m+2l} a_j z^{j-m} + O(z^{2l+1})$ for prescribed values of a_j, they are therefore equal.)

(b) Use the result of problem 8(b) of section 2.7 to show that this (presumed) equality involving $C_{m,m+2l}(z)$ and $C_{m+1,m+2l}(z)$ implies the recurrence relations (2.9.1) and (2.9.2) governing the construction of the QD-algorithm.

(c) Placing $S_n(x) = \sum_{j=0}^{n} a_j x^j$, and $R_{m,n}(x) = S_{m-1}(x) + x^m C_{m,m+n}(x)$, verify that $S_\infty(x) = R_{m,n}(x) + O(x^{n+m+1})$. What is the order of the rational function $R_{m,n}(x)$? (Ordered arrangements of such rational function approximations to $S_\infty(x)$ constitute a **Padé Table**.)

4. (a) The QD-algorithm can be applied in reverse to construct the power series expansion $\sum_{j=0}^{\infty} a_j z^j$ of a given S-fraction representation $\underset{j=0}{\overset{\infty}{K'}} r_j z \downarrow 1$ for a function $f(z)$. What are the approximate recurrence relations governing the construction of the tableau of $r_{j,k}$ values in this context?

(b) Find the first few terms of the power series expansion of $\underset{j=1}{\overset{\infty}{K'}}(z/2) \downarrow 1$.

(c) If $\underset{j=0}{\overset{\infty}{K'}} p_j z \downarrow 1$ is a known S-fraction representation of $\phi(z)$, show how the power series expansion $\sum_{j=0}^{\infty} a_j z^j$ of $\underset{j=0}{\overset{\infty}{K'}} r_j z \downarrow 1$, where $r_j = p_{j+1}$ for $j = 1, 2, \ldots$, leads to a power series expansion of $1/\phi(z)$.

5. (a) Any element in an even-numbered column ($k = 2l$, say) of the array of the QD-algorithm may be small (or zero) and lose significance in computation due to cancellation error. By considering that such an element is $a_{j,2l} = \epsilon$, say, show that this loss of significance extends to all elements of row $j + 2$ in columns $2l + 2$ onwards, and so to the values of $r_{j+2}, r_{j+3}, \ldots$

(b) Since an S-fraction expansion of a function may not exist, this may be indicated by the appearance of a zero element of the array. Alternatively, the presence of zero elements can indicate that the function in question is rational (and has a terminated S-fraction representation). Compare the first few rows of the QD-algorithm applied to the two functions:

$$(1+z)/(1+5z) = 1 - 4z + 20z^2 - 100z^3 + 500z^4 - 2500z^5 + \ldots$$

$$(1 + 6z + 10z^2)/(1 + 4z)^{5/2}$$
$$= 1 - 4z + 20z^2 - 100z^3 + 490z^4 - 2352z^5 + \ldots$$

6. In the notation most often used in describing the QD-algorithm, one writes

$$r_{j,2l} = e_l^{j-2l}, \quad r_{j,2l-1} = q_l^{j-2l+1}$$

where l is an integer. What form do the recurrence relations then take?

7. (a) The great majority of continued fractions which have a simple form and which represent familiar transcendental functions are examples of **Gauss's continued fraction** $c \underset{j=0}{\overset{\infty}{K'}} G_j z \downarrow (j + c)$, where for given values of a, b, and c,

$$G_{2j+1} = -(j+a)(j+c-b) \quad (j = 0, 1, 2, \ldots)$$

$$G_{2j} = -(j+b)(j+c-a) \quad (j = 1, 2, \ldots) \ .$$

It converges, in general, for all $z < 1$ and is equal in value to the quotient of two so-called **hypergeometric series** $F(a, b+1; c+1; z)/F(a, b; c; z)$. These series are convergent for $|z| < 1$ and are given by

$$F(a, b; c; z) = \sum_{m=0}^{\infty} (a)_m (b)_m z^m / [(c)_m m!]$$

in terms of Pochhammer's symbol $(a)_m = \Gamma(a+m)/\Gamma(a)$. They specialise to many well-known functions for particular values of the parameters a, b, and c. Show, for example, that $F(a, c; c; z)$ is the binomial series for $(1-z)^{-a}$, irrespective of the value of c, as also is $F(c, a; c; z)$.

(b) Given that $x^{-1} \arcsin(x) = F(1/2, 1/2; 3/2; x^2)$ find the first few terms of the continued fraction representation of $\arcsin(x)/[x(1-x^2)^{1/2}]$.

(c) The **incomplete beta function** $B_x(p, q) = \int_0^x t^{p-1}(1-t)^{q-1} dt$ with $0 \leqslant x \leqslant 1$ occurs in mathematical statistics, and can be expanded as the series $(x^p/p) F(p, 1-q; p+1; x)$, if p and q are positive. Find the first few terms of the continued fraction representation of

$$B_x(p, q)/[x^p(1-x)^q] \ .$$

8. (a) Find a continued fraction representation of the hypergeometric series

$$F(a, 1; c+1; z) = \sum_{m=0}^{\infty} (a)_m z^m / (c+1)_m \ .$$

(b) Identify the values of a, c, and z such that this continued fraction represents (i) $x^{-1} \ln(1+x)$, (ii) $x^{-1} \operatorname{argtanh}(x) = \sum_{k=0}^{\infty} x^{2k}/(2k+1)$, (iii) $x^{-1} \arctan(x) = \sum_{k=0}^{\infty} (-x^2)^k/(2k+1)$, and (iv) $[(1+x)^\alpha - 1]/(\alpha x)$; and hence determine continued fraction representations of these functions.

9. (a) The value of $\lim_{a \to \infty} F(a, b; c; x/a) = \sum_{m=1}^{\infty} (b)_m x^m / [(c)_m m!] = \Phi(b, c; x)$ is known as a **confluent hypergeometric series**. It converges for all values of x. Verify from Gauss's result that

$$\Phi(\alpha+1, \beta+1; x)/\Phi(\alpha, \beta; x) = \beta \overset{\infty}{\underset{j=0}{K'}} U_j x \downarrow (j+\beta)$$

$$\Phi(\alpha, \beta+1; x)/\Phi(\alpha, \beta; x) = \beta \overset{\infty}{\underset{j=0}{K'}} V_j x \downarrow (j+\beta)$$

where $U_{2j+1} = V_{2j} = -(j+\beta-\alpha)$, and $U_{2j} = V_{2j+1} = j+\alpha$.

(b) Find a continued fraction representation of $[\exp(x) - 1]/x$.

(c) Identify the series expansions (2.6.2) and (2.6.3) for $\operatorname{erf}(x)$ as examples of confluent hypergeometric series, and show that both lead to the same continued fraction representation of $\operatorname{erf}(x)$.

10.(a) Recast Gauss's continued fraction as an S-fraction.

(b) Writing $I_\alpha(x) = \int_x^\infty t^{-\alpha}\exp(-t)dt$, confirm by repeated integration by parts that:

$$I_\alpha(x) = x^{-\alpha}\exp(-x) \sum_{m=0}^{n-1} (\alpha)_m(-x)^{-m} + (-1)^n(\alpha)_n I_{\alpha+n}(x) \ .$$

Verify that, formally, the semi-convergent series $\sum_{m=0}^{\infty} (\alpha)_m(-x)^{-m}$ is equal to $\lim_{c\to\infty} F(\alpha, 1; c+1; -c/x)$, and hence derive an S-fraction representation of the series which converges for all $x>0$. (The S-fractions for $E_1(x)$ and erfc(x) are particular examples.)

11. (a) The Bessel function $J_\nu(z)$ is related to the confluent hypergeometric series by:

$$J_\nu(z) = [(z/2)^\nu/\Gamma(\nu+1)] \lim_{\alpha\to\infty} \Phi(\alpha, \nu+1; -z^2/4\alpha)$$

$$= \sum_{m=0}^{\infty} (-1)^m(z/2)^{\nu+2m}/[m!\Gamma(\nu+m+1)] \ .$$

Find a continued fraction representation of $J_\nu(z)/J_{\nu-1}(z)$.

(b) Since $J_{-1/2}(z) = (2/\pi z)^{1/2}\cos z$ and $J_{1/2}(z) = (2/\pi z)^{1/2}\sin z$, a continued fraction representation for $z^{-1}\tan(z)$ is obtainable, or changing z to iz, for $z^{-1}\tanh(z)$. Find a continued fraction representing $[z\coth(z) - 1]$.

Section 2.10

1. (a) Using a computer/calculator available to you, calculate successive convergents (A_k/B_k) of the recurrence relation (2.7.5) leading to the value of the continued fraction (2.9.3) for tan(x) at $x = \pi/2$. Is this value 'reasonable' in terms of the explanation given in the text?

(b) If the successive convergents (A_k/B_k), as recorded in the numerical experiment of part (a) of this problem, appear to have reached an asymptotic value at $k = N$, use the reverse recurrence relation (2.8.2) to generate a value of the same continued fraction. Observe how the 'infinite' value of tan$(\pi/2)$ is obtained.

(c) Repeat part (b) but with $x = \pi$, and observe how the 'zero' value of tan(π) is generated. Can you conjecture how the continued fraction represents the 'value' of tan$(n\pi/2)$, where n is any positive integer?

2. Verify that (2.10.2), with the values of λ_1 and λ_2 given by (2.10.3), is a solution of the recurrence relation $Z_k = qZ_{k-1} + pZ_{k-2}$, where p and q are constants.

3. (a) Let $Z_k = X_k$ and $Z_k = Y_k$ for $k = -1, 0, 1, 2, \ldots$ be any two linearly-independent solutions of the recurrence relation (2.10.1), such that $X_0Y_1 - X_1Y_0 = p_1 \neq 0$. Noting that A_k and B_k as defined by (2.7.5) are also solutions of the same recurrence relation, show that

$$A_k = X_0Y_k - Y_0X_k, \quad B_k = Y_{-1}X_k - X_{-1}Y_k \ .$$

(b) If $Z_k = X_k$ is the decrescent solution of (2.10.1) in the sense that $\lim\limits_{m \to \infty} (X_m/Y_m) = 0$, then show that the continued fraction $\overset{\infty}{\underset{j=1}{K}} p_j \downarrow q_j$ converges to the value $-(X_0/X_{-1})$ provided that no partial numerator p_j is zero. (The equation (2.10.8) is a straightforward generalisation of this result, obtainable by incrementing all subscript values by n.)

(c) Conversely, if the continued fraction $\overset{\infty}{\underset{j=1}{K}} p_j \downarrow q_j$ converges and none of the partial numerators is zero, show that there exists a decrescent solution $Z_k = X_k'$ of the recurrence equation such that $\lim\limits_{n \to \infty} (X_n'/Y_n) = 0$. (In fact, the more rapidly the continued fraction converges, the more rapidly the magnitudes of X_n' and Y_n diverge as $n \to \infty$, so that if there exists bad numerical instability in forming X_n', then (2.10.8) is a rapidly convergent continued fraction; and *vice versa*.)

4. (a) If $J_\nu(x)/J_{\nu-1}(x) = x \overset{\infty}{\underset{j=0}{K'}} (-x^2) \downarrow 2(j+\nu)$ where $J_\nu(x)$ is the Bessel function of the first kind and of order ν, find a decrescent solution of the recurrence relation

$$Z_k = 2(k+\nu)Z_{k-1} - x^2 Z_{k-2}, \quad \text{for} \quad k = 0, 1, 2, \ldots,$$

for given x, such that $Z_{-1} = J_\nu(x)$.

(b) Given that $Z_{-1} = J_0(\pi/4) = 0.85163191$, and taking $Z_0 = 0.28524706$, calculate the values of Z_k for $\nu = 0$, $x = \pi/4$, and $k = 1, 2, \ldots, 9$, by forward recurrence, rounded either to 8 significant decimal digits or to the (lesser) precision of your calculator/computer.

(c) Use backward recursion, with $\nu = 0$ and $x = \pi/4$, to find the values Z_k for $k = 0, 1, \ldots, 9$, corresponding to the decrescent solution with $Z_{-1} = J_0(\pi/4)$ as given above, to either 8 significant decimal digits or the (lesser) precision of your calculator/computer. (Compare the value of Z_0 with that given in part (b) – it should be the same – and so observe the effects of numerical instability in the calculation of part (b).)

5. (a) A recurrence relation such as $Z_n - cZ_{n-1} = f(n)$ is said to be *non-homogeneous* because of the presence of the term $f(n)$ independent of the Z's. Its *general* solution (i.e., the algebraic solution valid for any starting value) is simply any *particular* solution of the equation added to the general solution of the corresponding homogeneous equation – in other words, added to the algebraic solution of the equation without the term $f(n)$. Verify that if b and c are constants, and $c \neq 1$, then the general solution of $Z_n - cZ_{n-1} = b$ is given by $Z_n = Ac^n - b/(c-1)$, where A is an arbitrary constant determined by a (single) starting value.

(b) Numerical instability can arise in solving even two-term non-homogeneous equations if the magnitude of the required particular solution diverges less rapidly than that of the general solution of the corresponding homogeneous equation. Thus $Z_n = 12Z_{n-1} - 6$ has the particular solution

$Z_0 = Z_1 = Z_2 = \ldots = 6/11$; however, using correctly rounded arithmetic with decimal precision 3, and starting with the correctly rounded value of Z_0 appropriate to this solution, show that Z_1, Z_2, Z_3 and Z_4 as developed numerically by this recurrence formula, are not all equal. Explain the source of the error.

(c) If $Z_n = 12Z_{n-1} - 6$, express Z_n in terms of Z_0 and find the condition number of Z_4 with respect to Z_0 if $Z_0 \simeq 6/11$.

6. (a) If $I_n = \int_0^{1/3} x^n (x+1)^{-1} dx$, then show that $I_n + I_{n-1} = 1/(3^n n)$. Starting with the value of I_0, use this recurrence relation to estimate I_5. (The precision used is not important – decimal precision 3 will suffice.)

(b) Evaluate I_5 to the same precision as in part (a) by expanding $(x+1)^{-1}$ within the integrand as a series, integrating term-by-term, and summing. If this results in a different answer to that of part (a), which do you believe and why?

(c) Show that the problem posed in parts (a) and (b) is equivalent to summing the tail τ_5 of the infinite series $S_\infty = \sum_{k=1}^{\infty} (-1)^{k+1}/(3^k k)$, and explain the two numerical results of parts (a) and (b) from this standpoint.

PROBLEMS: CHAPTER 3

Section 3.2

1. (a) If the quadratic curve $y = p_2(x)$ interpolating $(x_0, y_0), (x_1, y_1)$ and (x_2, x_2) is expressed in the form:

$$p_2(x) = a_0 + a_1(x - x_1) + a_2(x - x_1)^2$$

identify the values of $a_0, a_1,$ and a_2 in terms of the data.

(b) Placing $h_1 = x_1 - x_0$ and $h_2 = x_2 - x_1$, express the values of $p_2'(x_1)$ and $p_2''(x_1)$ in terms of h_1, h_2 and the data ordinates.

(c) Specialise the results of part (b) to the particular condition in which $h_1 = h_2 = h$, say. What condition must be satisfied by the data ordinates y_0, y_1 and y_2 if $p_2(x)$ is to have a stationary (maximum or minimum) value in the interval $|x - x_1| \leqslant h$?

2. (a) Write down the shape functions for the interpolation of the 4 data pairs (x_k, y_k) for $k = 0, 1, 2, 3$ by a cubic polynomial.

(b) Assuming that $x_k = kh$ where h is a constant (so that the data support is equi-spaced), locate the position of stationary values of s_{30} and s_{31}, and roughly sketch these shape functions. (Hint: place $\xi = x/h$.)

3. (a) If $\Phi_n(x) = \prod_{j=0}^{n} (x - x_j)$, verify that the shape function defined by (3.2.5) can be written as:

$$s_{nk}(x_0, x_1, \ldots, x_n; x) = \Phi_n(x)/[(x - x_k)\Phi_n'(x_k)] \quad (k = 0, 1, \ldots, n)$$

where $\Phi_n'(\xi)$ is the derivative of $\Phi_n(x)$ at $x = \xi$.

(b) Show that the derivative of the shape function $s_{nk}(x_0, x_1, \ldots, x_n; x)$ with respect to x at $x = x_i$ is equal to $\Phi'_n(x_i)/[(x_i - x_k)\Phi'_n(x_k)]$ if $i \neq k$. What is the derivative at $x = x_k$?

(c) If $x_j = jh$ for $j = 0, 1, \ldots, n$, where h is a constant (so that the support is equi-spaced), verify that $\Phi'_n(x_j) = (-1)^{n-j}(n - j)! \, j! \, h^n$.

(d) If $x_j = j/n$ for $j = 0, 1, \ldots, n$ so that the data points are equi-spaced over the interval $[0, 1]$, find the value of the derivative of the shape function s_{nk} with respect to x at $x = 0$, corresponding to $n = 12, k = 6$. (Note: the derivative is large compared with 1.)

4. Prove the identities:

$$\sum_{k=0}^{n} s_{nk}(x_0, x_1, \ldots, x_n; x) = 1$$

$$\sum_{k=0}^{n} x_k^m s_{nk}(x, x_1, \ldots, x_n; x) = x^m \quad (1 \leqslant m \leqslant n)$$

$$\sum_{k=0}^{n} x_k^{n+1} s_{nk}(x_0, x_1, \ldots, x_n, x) = x^{n+1} - \Phi_n(x)$$

where $\Phi_n(x) = \prod_{j=0}^{n} (x - x_j)$, and $s_{nk}(..)$ is defined by (3.2.5). (Hint: appeal to uniqueness of polynomial interpolation.)

5. (a) Another (less satisfactory) way of finding the polynomial $p_n(x)$ which interpolates the data (x_k, y_k) for $k = 0, 1, \ldots, n$ is to place $p_n(x) = a_0 + a_1 x + \ldots + a_n x^n$ and to solve the system of linear equations for $a_0, a_1, \ldots, a_n$ which result from writing $p_n(x_j) = y_j$ for $j = 0, 1, \ldots, n$: that is, the system of $(n + 1)$ equations

$$a_0 + a_1 x_j + \ldots + a_n x_j^n = y_j \quad (j = 0, 1, \ldots, n) \ .$$

In matrix notation, defining the $(n + 1)$-vectors $\mathbf{a} = (a_0, a_1, \ldots, a_n)^{\mathrm{T}}$, $\mathbf{y} = (y_0, y_1, \ldots, y_n)^{\mathrm{T}}$, this system can be written as $\mathbf{V_n a} = \mathbf{y}$ where $\mathbf{V_n}$ is a square matrix of order $(n + 1)$ known as the **Vandermonde matrix**. Identify the element v_{jk} of $\mathbf{V_n}$ in the jth row and kth column for $j, k = 1, 2, \ldots, (n + 1)$.

(b) Assuming $n = 2$ and $x_1 = 0$, show that the solution of this system of equations is consistent with that obtained in problem 1(a).

(c) Expand the last row of the determinant $\det(\mathbf{V_n})$ of the Vandermonde matrix by its minors so as to show that it is a polynomial in x_n. Find the coefficient of x_n^n and so obtain the formula

$$\det(\mathbf{V}_n) = (x_n - x_0)(x_n - x_1) \ldots (x_n - x_{n-1}) \det(\mathbf{V}_{n-1}) \quad (n \geqslant 1).$$

(d) Using induction on n, show that

$$\det(\mathbf{V}_n) = \prod_{j=1}^{n} \prod_{k=0}^{j-1} (x_j - x_k)$$

and hence show that $\mathbf{V_n a} = \mathbf{y}$ has a unique solution provided that the x-data are all distinct.

Section 3.2.1

1. The cubic polynomial shape function $s_{3,1}(0,1,2,3;x)$, defined in conformity with (3.2.5), has a maximum value at $x = 6/(5 + \sqrt{7}) = 0.7847 \ldots$. Find this maximum value using (a) Aitken's and (b) Neville's method.

2. (a) The array generated by Aitken's method can also as easily be constructed by columns, assigning a_{j0} for all j, then calculating a_{j1} for $j = 1, 2, \ldots, n$, followed by a_{j2} for $j = 2, \ldots, n$, and so on. Write an algorithm which assigns each column to the same vector d_j, partially overwriting the previous column, and finishing with the vector equal to the diagonal elements.

   (b) In the arrangement of Neville's algorithm for machine computation, suppose that the calculation proceeds by rows but each element is assigned to a vector element c_{j-k} (say), rather than r_{n-j+k}. Rewrite the instruction sequence of the text with this modification. What vector element finally contains the value of $p_n(x)$?

   (c) If the triangular array of Neville's algorithm is calculated by diagonals, starting at a_{no}, proceding to $a_{n-1,0}$, $a_{n,1}$, then to $a_{n-2,0}$, $a_{n-1,1}$, $a_{n,2}$, and so on, finishing with a_{00}, $a_{11}, \ldots, a_{nn}$, only the last-formed diagonal needs to be preserved for future reference. Write an algorithm which assigns these intermediate diagonals to a vector d, so that its elements are finally equal to those on the diagonal (a_{ll}), just as results from the algorithm for Aitken's method.

3. (a) By substituting the expression (3.2.8) for a_{jk} in Aitken's recursion (3.2.6), identify this with the recursion (3.2.7), and so determine the intersection set $\mathbf{S} = \mathbf{S}_j \cap \mathbf{S}_k$ used in forming $a_{j,k+1}$ by (3.2.6).

   (b) Determine the intersection set $\mathbf{S}$ as defined for (3.2.7) which is used in forming $a_{j,k+1}$ by Neville's recursion (3.2.9).

4. (a) If the data points are close to each other, the numerical evaluation of the formulae (3.2.6) and (3.2.9) is liable to suffer from loss of significance due to cancellation. An effective way of modifying Aitken's method to overcome this problem is to define

   $$a_{jk} = a'_{jk} + \sum_{i=0}^{k-1} a'_{ii}$$

   and to generate the triangular array of the quantities a'_{jk} rather than a_{jk}. Verify that the values of a'_{jk} obey the recursion formula:

   $$a'_{j,0} = y_j, \quad a'_{j,k+1} = (a'_{jk} - a'_{kk})(x - x_k)/(x_j - x_k)$$

   for $k = 0, 1, \ldots, (j-1)$ and $j = 0, 1, \ldots, n$. (The value of the polynomial given before by a_{nn} is now the *sum* of the diagonal elements $\sum_{i=0}^{n} a'_{ii}$.)

(b) The recursion formula for $a'_{j,k+1}$ shows that the elements in column $k+1$ for $k = 0, 1, \ldots, n-1$ are all multiplied by $(x - x_k)$; it economises on computational effort, and reduces numerical error, if these common factors are extracted. This is accomplished by defining

$$a'_{jk} = \hat{a}_{jk} \prod_{i=0}^{k-1} (x - x_i) \quad \text{for} \quad k = 1, 2, \ldots, n \;,$$

and calculating the triangular array of the quantities $\hat{a}_{jk}$ rather than a'_{jk} (or a_{jk}). Verify that $\hat{a}_{jk}$ obeys the recursion:

$$\hat{a}_{j,0} = y_j, \quad \hat{a}_{j,k+1} = (\hat{a}_{jk} - \hat{a}_{kk})/(x_j - x_k)$$

and that $p_n(x) = \hat{a}_{00} + \hat{a}_{11}(x - x_0) + \hat{a}_{22}(x - x_0)(x - x_1)$

$$+ \ldots + \hat{a}_{nn} \prod_{i=0}^{n-1} (x - x_i) \;.$$

This is a Newton's form of the interpolating polynomial (discussed in the next section), and the values of $\hat{a}_{jk}$ are called **divided differences**.

Section 3.2.2

1. (a) Define $q_0(x) \equiv d_n$, and $q_j(x) \equiv d_{n-j} + (x - x_{n-j})q_{j-1}(x)$ for $j = 1, 2, \ldots, n$, so that $q_n(x)$ is identical with the Newton form $p_n(x)$ of (3.2.11). Suppose that $q_j(x) = \sum_{k=0}^{j} q_{j,k} x^k$. Then confirm that the triangular array $q_{j,k}$ can be generated by placing $q_{00} = d_n$ and using the recurrence

$$q_{j,0} = d_{n-j} - x_{n-j} q_{j-1,0} \;,$$

$$q_{j,k} = q_{j-1,k-1} - x_{n-j} q_{j-1,k} \;, \quad \text{for} \quad k = 1, 2, \ldots, (j-1), \text{ and}$$

$$q_{j,j} = q_{j-1,j-1}$$

for $j = 1, 2, \ldots, n$. (The elements of the last row $q_{n,k} = a_k$, say, are then the coefficients of the power form of $p_n(x) \equiv \sum_{j=0}^{n} a_k x^k$.)

(b) Devise an algorithm which converts the Newton form of an interpolating polynomial into its equivalent power form $a_0 + a_1 x + \ldots + a_n x^n$, assuming vectors d_k and x_k are available as data, and assigning a_k, for $k = 0, 1, \ldots, n$. (Hint: assign values of $q_{j,k}$ to a_{n-j+k} for $k = 0, 1, \ldots, j$ and $j = 0, 1, \ldots, n$.)

(c) Use your algorithm to find the vector a_k corresponding to $x_k = k$, $d_k = k + 1$, and $n = 3$.

(d) What change in the algorithm would be necessary if the series were to be expressed in powers of $(x - \xi)$, rather than x, where ξ is any given value? (This is called a **shifted power form**.)

(e) If the polynomial evaluation is ill-conditioned, one needs to be careful how it is manipulated. Evaluate the cubic Newton form $(x - 99\pi)$

$(x - 100\pi)(x - 101\pi)$ at $x = 314.15$. Then use the methods of parts
(a) and (d) to convert it into a power form in x, and a shifted power
form in $(x - 100\pi)$. Evaluate these power forms and compare answers.

2. Show by induction that if $p_m(x)$ for $0 \leqslant m \leqslant n$ is the polynomial defined
by replacing n by m in (3.2.11), but leaving the values of d_k and x_k unaltered
for all k, then $p_m(x) = y_{0,1,\ldots,m}$.

3. Interpret the recursion (3.2.15) as a relation between suitably subscripted
values of $\hat{y}$.

4. (a) Verify that if $\Phi_n(x) = \prod\limits_{i=0}^{n} (x - x_i)$, and $s_{nk}(x_0, x_1, \ldots, x_n; x)$ is
the polynomial shape function defined by (3.2.5) for $n \geqslant 0$, then
$\hat{s}_{nk}(x_0, x_1, \ldots, x_n; x) = 1/\Phi_n'(x_k)$.

(b) Express $\hat{y}_{0,1,\ldots,m}$ terms of a linear combination of the $(m + 1)$ data
ordinates y_k for $k = 0, 1, \ldots, m$, and hence show that $\hat{y}_{0,1,\ldots,m}$ is a
symmetric function of the data.

(c) Generalise the results of parts (a) and (b) by providing an expression for
$s_{m-1,k}(x_1, x_2, \ldots, x_m; x)$ for $m \geqslant 1$, and express $\hat{y}_{1,2,\ldots,m}$ in terms of
a linear combination of the m data ordinates y_k for $k = 1, 2, \ldots, m$.

(d) Verify that $(x_k - x_m)\hat{s}_{mk}(x_0, x_1, \ldots, x_m; x) = \hat{s}_{m-1,k}(x_0, x_1, \ldots, x_{m-1}; x)$
and similarly interpret $\hat{s}_{m-1,k}(x_1, x_2, \ldots, x_m; x)$ in terms of $\hat{s}_{mk}$. Hence
prove the identity

$$\hat{y}_{1,2,\ldots,m} - \hat{y}_{0,1,\ldots,m-1} = (x_m - x_0)\hat{y}_{0,1,\ldots,m} .$$

5. Find the Newton form of the cubic interpolating the points $(-1, 1), (0, -2)$,
$(1, 3)$, and $(1\frac{1}{2}, 1)$, and calculate its value at $x = 2$.

6. By applying the iterated instruction sequence of the text for the generation
of d_{n+1} in terms of $d_0, d_1, \ldots, d_n$ and $x_0, x_1, \ldots, x_n$, write down an algebraic
expression by which d_3 is calculated, and verify algebraically that it is correct.

7. (a) Let $p_2(x)$ be a quadratic through three data pairs $(x_0, y_0), (x_1, y_1)$,
(x_2, y_2). Show that $\hat{y}_{02}$ is equal to the value of $p_2'(x)$ at the point
$x = (x_0 + x_2)/2$, midway between $x = x_0$ and $x = x_2$. More generally,
if $\{i, j, k\}$ is some permutation of $\{0, 1, 2\}$, show that $\hat{y}_{ij}$ is the value of
$p_2'(x)$ at $x = (x_i + x_j)/2$.

(b) Verify that, if $h_1 = x_1 - x_0$ and $h_2 = x_2 - x_1$, then

$$p_2'(x_1) = (\hat{y}_{01} h_2 + \hat{y}_{12} h_1)/(h_1 + h_2) .$$

8. (a) If the mth and $(m + 1)$th data pairs are interchanged, then verify that
only the coefficient d_m of Newton's interpolating polynomial is altered,
and that (in terms of the original ordering of the data) its value becomes
$\hat{y}_{0,1,\ldots,(m-1),(m+1)}$.

(b) Use (3.2.14) to show how the new value of $d_m = d'_m$ (say) resulting from this interchange can be calculated from the coefficients of the Newton interpolating polynomial (applying to the original ordering of the data).

(c) The effect of omitting the nth data pair on the form of Newton's interpolating polynomial (3.2.11) is simply to delete the term involving d_n (i.e. to make d_n zero). What would be the effect on the coefficients of this polynomial of omitting the mth data pair where $0 \leqslant m < n$?

9. Following a line of reasoning like that of the last problem, determine the effect of omitting the mth data pair on the coefficients of the reverse Newton form (3.2.16)? (Hint: start by considering an interchange of the mth and $(m-1)$th data pairs.)

Section 3.2.3

1. (a) Set up a difference table for the following tabulated values of $\sin(x)$, where x is measured in degrees

x	$\sin(x)$	x	$\sin(x)$
28	0.46947 15628	34	0.55919 29035
30	0.50000 00000	36	0.58778 52523
32	0.52991 92642	38	0.61566 14753

(b) Ordering the support values of x as follows: $x_0 = 32$, $x_1 = 34$, $x_2 = 30$, $x_3 = 36$, $x_4 = 28$, $x_5 = 38$, and putting $y(x_k) = \sin(x_k) = y_k$, identify the entries of this difference table in terms of appropriately scaled and subscripted values of $\hat{y}$; and hence determine the coefficients of the Newton form of the quintic interpolating the above data, $d_0 + d_1(x - x_0) + d_2(x - x_0)(x - x_1) + \dots$, appropriate to this ordering.

(c) Find the value at $x = 32.12345$ interpolated by the mth degree polynomial $y_{0,1,\dots,m}$ through the first $(m + 1)$ data pairs taken in the order given in part (b), for $m = 1, 2, \dots, 5$; and hence confirm that, taken in this order, successive data pairs have a decreasing effect upon the interpolated value. (Hint: evaluate each term of the Newton polynomial separately at $x = 32.12345$ and then form progressive sums.)

2. (a) If $y(x)$ is a polynomial of degree n, then show that $\Delta^n y_j$ is a constant, independent of j, and so $\Delta^k y_0 = 0$ for all integer $k > n$.

(b) The following ordinates are those of a polynomial at equi-spaced abscissae: $-5, 1, 1, 1, 7, 25$. What is its degree?

(c) Using the operator identity $E \equiv \Delta + 1$, show that if $y(x)$ is a polynomial of degree n, and v is arbitrary (and not necessarily integer) then

$$y_v = y(x_0 + vh) = \sum_{k=0}^{n} (v - k + 1)_k \Delta^k y_0 / k!$$

which is called the Newton **forward difference formula**, first discovered by a Scots contemporary of Newton named James Gregory. Verify that it is identical with the Newton form of the polynomial interpolating $(x_0 + kh, y_k)$ for $k = 0, 1, \ldots, n$.

3. (a) Consider the effect of an error in a data ordinate by constructing a difference table for a function which is zero at all points except one, where it is ϵ (say). If each and every data ordinate has an error equal to ϵ in magnitude, show that the worst possible combination of errors occurs if the errors of successive ordinates alternate in sign, and that then the associated differences also alternate in sign, and double as each new difference is introduced.

(b) If the errors of the data ordinates are regarded as random and independent and drawn from a single distribution with mean μ and variance σ^2, show that the inherent error of any difference is a random variable drawn from a distribution with zero mean. Further verify (for the first few integer values of k) that the variance of the kth differences is equal to $\sigma^2(2k)!/(k!)^2$. Use this last result to show (for large k) that the standard deviation of the kth differences is approximately $2^k \sigma/(\pi k)^{1/4}$.

4. (a) Construct a graph by hand (preferably on transparent paper) interpolating the following (x, y) data pairs over the range of $-3 \leqslant x \leqslant 3$:

$$(-2.5, -3), (-1.5, -3.33), (-0.5, -1.78), (0.5, 1.78),$$
$$(1.5, 3.33), (2.5, 3).$$

(b) Find the Newton polynomial through the above data, convert it to power form, and compare its graph over $x \in (-3, 3)$ with that constructed by hand. (No conclusions can be drawn, but close resemblance is unlikely. Keep these curves for future reference.)

(c) List the values of the first and second derivatives of the interpolating polynomial found in part (b) at the support points; and find the maximum value of the polynomial in $(-2.5, 2.5)$ together with its position.

Section 3.2.4

1. (a) Suppose $y(x) = y_{i, i+1, \ldots, i+p}$ is the polynomial of degree $\leqslant p$ which interpolates the $(p + 1)$ data pairs (x_{i+k}, y_{i+k}) for $k = 0, 1, \ldots, p$. Show that if $y(x)$ is written in shifted power form as $a_0 + a_1(x - x_i) + \ldots + a_n(x - x_i)^n$ then, irrespective of the data values, $a_k = y^{(k)}(x_i)/k!$ and verify that $\hat{y}_{i, i+1, \ldots, i+p} = a_p$.

(b) If, consistent with the convention of incorporating derivative data $x_{i-1} < x_i$ and $x_i = x_{i+1} = \ldots = x_{i+p}$, then confirm that the interpolating polynomial $y(x) = y_{i, i+1, \ldots, i+p}$ will be of the required form if $a_k = y_{i+k}/k!$.

2. (a) Interpret the conventionalised meaning of the sequence of data pairs: $(-1, 1), (0, 3), (0, 5), (0, 14), (1, 29), (1, 78)$.

   (b) How many permutations of these six data pairs are possible if the conventionalised meaning is not to be impaired?

   (c) Find the Newton form, and the reversed Newton form, interpolating the data ordered as in part (a).

3. The value of $E_1(x) = \int_x^\infty t^{-1} \exp(-t) dt$ is equal to 2.9591 at $x = 0.03$ and 2.6813 at $x = 0.04$. Interpolate a value at $x = 0.0357$ by an osculating cubic polynomial.

4. (a) Construct an algorithm which assigns the coefficients $c_0, c_1, \ldots, c_n$ of the reversed Newton form (3.2.16) interpolating $(n+1)$ data pairs including (possibly) derivative values recognisable by the usual convention.

   (b) According to the convention, any repeated data abscissa must be consecutive. Does the algorithm you have constructed check that this convention is followed? If not, arrange for it to do so by jumping to non-local label *FAIL* in the event that the condition is not met.

5. It is possible to express the interpolating Hermitian polynomial as

$$y(x) = \sum_{k=0}^{n} [y_k u_{nk}(x) + y'_k v_{nk}(x)]$$

where $u_{nk}(x)$ and $v_{nk}(x)$ are polynomials in x whose coefficients depend on the support data $x_0, x_1, \ldots, x_n$. What conditions must be met by these polynomials (supposing that y_k and y'_k are given for $k = 0, 1, \ldots, n$)? Determine expressions for these polynomials in terms of x and the support data.

Section 3.3

1. (a) A basic computational problem in piecewise polynomial interpolation lies in the identification of the interval $I_k = [x_{k-1}, x_k)$ in which a particular value of $x = \xi$, say, lies. Some convention is needed to deal with the possibility of extrapolation as well as interpolation. So suppose that the nodes are ordered in increasing x-value, and that I_k is defined as above for $k = 2, 3, \ldots, (n-1)$, but the first interval is $I_1 = (-\infty, x_1)$, and the nth is $I_n = [x_{n-1}, \infty)$. Write instructions which will determine the value of k such that $\xi \in I_k$ if the support points are equally spaced, given ξ, x_0, n, and $h = x_j - x_{j-1}$.

   (b) Write instructions which similarly determine k if the support points are not equally spaced, given $n > 0$ and the vector $x_0, x_1, \ldots, x_n$, which can be assumed ordered so that $x_0 < x_1 < \ldots < x_n$.

2. (a) In piecewise Hermitian interpolation one can express

$$h_k \pi_k(x) = y_k(x - x_{k-1}) + y_{k-1}(x_k - x)$$
$$+ (x - x_{k-1})(x_k - x)[A(x - x_{k-1}) + B(x_k - x)]$$

for $x \in [x_{k-1}, x_k]$, where $h_k = x_k - x_{k-1}$. Find expressions for the constants A and B in terms of the data x_j, y_j, and y_j' for $j = k-1, k$.

(b) The same form of expression may be used for the pieces of the Bessel cubic interpolation formula: verify that it is appropriate then to place

$$A = -\hat{y}_{k-1, k, k+1}, \quad B = -\hat{y}_{k-2, k-1, k}$$

where these are divided differences, i.e. $\hat{y}_{i,j,k} = (\hat{y}_{i,j} - \hat{y}_{i,k})/(x_j - x_k)$.

(c) Prove that if the data points lie on a curve $y = p_2(x)$, where $p_2(x)$ is a quadratic polynomial, then the piecewise cubic Bessel interpolate coincides everywhere precisely with this curve.

3. (a) If $y_j = \delta_{jk}$ for some positive $k < n$ and all $j = 0, 1, \ldots, n$, what are the assumed derivative values at the nodes $x = x_j$ for $j = 1, 2, \ldots, (n-1)$ in cubic Bessel interpolation?

(b) Verify that if the support is equi-spaced, so that $h_j = x_j - x_{j-1} = h$ for all $j = 1, 2, \ldots, n$, then if $y_j = \delta_{jk}$, the assumed derivative values are all zero except for $y'(x_{k-1}) = -y'(x_{k+1}) = 1/(2h)$. Hence deduce an expression for the shape functions $s_k(..)$ given in (3.3.5) for an equi-spaced support in terms of $\xi = (x - x_k)/h$.

(c) Find the minimum value of $s_k(..)$ as determined in part (b).

4. (a) Construct and graph over $x \in (-3, 3)$ the cubic Bessel interpolate with quadratic end-pieces for the data given in problem 4 of section 3.2.3. (Compare with the hand-drawn curve and the graph of the Lagrangian polynomial interpolating the same data, and keep for later reference.)

(b) List the values of the first derivative of this cubic Bessel interpolate at the nodes; and find the maximum value of this interpolate in $(-2.5, 2.5)$ together with its position.

Section 3.4

1. (a) Verify that (3.4.1) and (3.4.2), together with the condition that $y'(x_0) = y_0'$, lead to the recurrence (3.4.3).

(b) Placing $y'(x_j) = y_j'$, show that (3.4.3) may be rewritten as the recurrence relation $y_{j-1}' + y_j' = 2\hat{y}_{j-1, j}$ for $j = 1, 2, \ldots, n$; and that this is equivalent to the set of equations

$$\int_{x_{j-1}}^{x_j} y'(x) dx = y_j - y_{j-1} \quad (j = 1, 2, \ldots, n)$$

in which the integral is evaluated by the trapezoidal rule.

(c) Confirm that (3.4.4) is the solution of this recurrence relation (3.4.3).

(d) Assuming that $y_0' = 0$ and that the nodes are equi-spaced, so that $h_k = h$, a constant, for $k = 1, 2, \ldots, n$, find the shape function $s_{nm}(..)$ associated with the ordinate y_m at any internal node $m = 1, 2, \ldots, (n-1)$ for the piecewise quadratic interpolation defined by (3.3.2), (3.4.2), and (3.4.4). (Hint: place $y_k = \delta_{km}$ and then $y(x)$ is the required shape function.)

2. (a) Regarding y_0' as a variable, then (3.4.4) shows that μ_k is a linear function of y_0' such that $d\mu_k/dy_0' = (-1)^k$, for $k = 1, 2, \ldots, n$. Hence deduce that $\sum_{k=1}^{n} \mu_k^2/h_k$ has a minimum value with respect to y_0' if $\sum_{k=1}^{n} (-1)^k \mu_k/h_k = 0$.

   (b) Assuming that the nodes are equi-spaced, so that $h_k = h$, a constant, express the value of y_0', which minimises $\sum_{k=1}^{n} \mu_k^2$, in terms of the data ordinates $y_0, y_1, \ldots, y_n$ for values of $n = 2, 3$, and 4.

3. (a) Show that the piecewise cubic interpolation defined by (3.3.2) and (3.4.6) implies that $y''(x)$ is continuous and piecewise linear in $[a, b]$.
   (b) Verify that (3.4.5) and (3.4.6), together with the conditions that $y'(x_0) = y_0'$ and $y'(x_n) = y_n'$, lead to the system of equations (3.4.7).
   (c) Show that, in (3.4.6), M_{k-1} and M_k are connected to the values of $\pi_k'(x_{k-1}) = y_{k-1}'$ and $\pi_k'(x_k) = y_k'$ by

$$M_{k-1} = 2(3\hat{y}_{k-1,k} - 2y_{k-1}' - y_k')/h_k$$
$$M_k = 2(y_{k-1}' + 2y_k' - 3\hat{y}_{k-1,k})/h_k .$$

   Hence verify that $y_1', y_2', \ldots, y_{n-1}'$ satisfy the system of $(n-1)$ linear equations:

$$(1 - \lambda_j)y_{j-1}' + 2y_j' + \lambda_j y_{j+1}' = 3[(1 - \lambda_j)\hat{y}_{j-1,j} + \lambda_j \hat{y}_{j,j+1}]$$

   for $j = 1, 2, \ldots, (n-1)$, the values of y_0' and y_n' being given as data. Here $\lambda_j = h_j/(h_j + h_{j+1})$, and the right-hand-side is three times the value of the tangent to the interpolating quadratic $y_{j-1,j,j+1}$ at $x = x_j$, as given in (3.3.3).
   (d) Show that, if the nodes are equi-spaced so that $h_k = h$, a constant, then the linear system for $y_1', y_2', \ldots, y_{n-1}'$ is equivalent to the set of equations

$$\int_{x_{j-1}}^{x_{j+1}} y'(x)dx = y_{j+1} - y_{j-1}, \quad j = 1, 2, \ldots, (n-1)$$

   in which the integral is evaluated by Simpson's rule.

4. (a) Suppose $y(x) \in C^2[x_0, x_n]$ and $\eta(x) \in C^1[x_0, x_n]$ but $\eta''(x)$ is possibly discontinuous only at $x = x_1, x_2, \ldots, x_{n-1}$; then show that if $\eta'(x_0) = \eta'(x_n) = 0$,

$$\sum_{k=1}^{n} \int_{x_{k-1}}^{x_k} y''(x)\eta''(x)dx = -\sum_{k=1}^{n} \int_{x_{k-1}}^{x_k} y'''(x)\eta'(x)dx .$$

   (b) Supposing further that $y'''(x)$ is piecewise constant with discontinuities at $x = x_1, x_2, \ldots, x_{n-1}$, verify that if $\eta(x_0) = \eta(x_1) = \ldots = \eta(x_n)$ then $\int_{x_{k-1}}^{x_k} y'''(x)\eta'(x)dx$ is zero for $k = 1, 2, \ldots, n$.
   (c) If $y = y(x)$ is now identified with the cubic spline interpolating the data (x_k, y_k) for $k = 0, 1, \ldots, n$, having prescribed tangents y_0' and y_n'

at $x = x_0, x_n$ respectively, and if $y = y(x) + \eta(x)$ is taken to be any other function interpolating the same data and having the same end tangents, where $\eta(x) \in C^1[x_0, x_n]$ and $\eta''(x)$ is possibly discontinuous only at $x = x_1, x_2, \ldots, x_{n-1}$, then

$$\sum_{k=1}^{n} \int_{x_{k-1}}^{x_k} y''(x)\eta''(x)dx = 0 \ .$$

(d) With $y(x)$ and $\eta(x)$ defined as in part (c), show that

$$\sum_{k=1}^{n} \int_{x_{k-1}}^{x_k} [y''(x) + \eta''(x)]^2 dx \geqslant \int_{x_0}^{x_n} [y''(x)]^2 dx$$

with the equality being achieved if and only if $\eta(x)$ is identically zero. (This establishes the variational principle underlying the definition of a cubic spline.)

(e) If the tangents at $x = x_0, x_n$ are not prescribed, verify that the result of part (d) is still true provided that the cubic spline satisfies the free-end conditions $y''(x_0) = y''(x_n) = 0$.

5. (a) A convenient set of basis functions for a cubic spline defined on a given set of nodes $x_0, x_1, \ldots, x_n$ is formed by what are called **B-splines**. The B-spline centred on the node $x = x_k$ is that (unique) function which belongs to $C^2[a, b]$, which vanishes identically outside the interval (x_{k-2}, x_{k+2}), which has a non-zero value (say, unity) at $x = x_k$, and which reduces to a cubic polynomial in each interval between nodes. If the nodes are equi-spaced, find the expression for the B-spline $B_k(x)$ centred on $x = x_k$. (Hint: it is a function symmetric about $x = x_k$: express it in terms of $\xi = (x - x_k)/h$.)

(b) If the nodes are equi-spaced, find the values of $B_k(x)$, and $B_k'(x)$ and $B_k''(x)$ at the nodes $x = x_j$, for $j = k - 2, k - 1, \ldots, k + 2$.

(c) If $y = y(x)$ interpolates the $(n + 1)$ equi-spaced data (x_k, y_k) where $x_k = x_0 + kh$ for $k = 0, 1, \ldots, n$, and $y(x)$ is expressed for $x \in [x_0, x_n]$ as a linear combination of the $(n + 3)$ B-splines $B_k(x)$ centred at $x = x_k$ for $k = -1, 0, \ldots, (n + 1)$, namely as

$$y(x) = c_{-1}B_{-1}(x) + c_0 B_0(x) \ldots + c_{n+1}B_{n+1}(x) \ ,$$

find the $(n + 1)$ equations connecting the $(n + 3)$ coefficients $c_{-1}, c_0, \ldots, c_{n+1}$ which ensure that $y = y(x)$ interpolates the data. Two extra end-conditions are of course necessary fully to define the coefficients. Set down the two extra equations corresponding in turn to:

(i) given end-tangents, $y'(x_0) = y_0', y'(x_n) = y_n'$;
(ii) free-end conditions, $y''(x_0) = y''(x_n) = 0$;
(iii) quadratic end-pieces, $y'''(x) = 0$ for $x \in [x_0, x_1]$ and $x \in [x_{n-1}, x_n]$; and
(iv) the 'not-a-knot' conditions, that $y'''(x)$ is continuous at $x = x_1, x_{n-1}$.

(d) Assuming the existence and uniqueness of the cubic spline interpolate on the given set of nodes, show that $y = y(x)$ as defined by (c) is identical over $[x_0, x_n]$ with the cubic spline interpolating the given data and satisfying the given end-conditions. (In developing computational algorithms, this expression of the cubic spline by a linear combination of B-splines, generalised to allow for unequally spaced nodes, is usually the preferred approach; but there are complications in describing the suitable extrapolate outside $[x_0, x_n]$.)

6. (a) Construct and graph over $x \in (-3, 3)$ the cubic spline satisfying the 'not-a-knot' end-conditions which interpolates the data given in problem 4 of section 3.2.3. (Compare with the hand-drawn curve and the graphs of the Lagrangian and the cubic Bessel interpolates, and keep the graph.)

   (b) List the values of the first and second derivatives of this spline at the nodes; and find the maximum value of $y(x)$ in $(-2.5, 2.5)$ together with its position.

   (c) Find the coefficients $c_{-1}, c_0, \ldots, c_6$ in the expression of this cubic spline as a linear combination of B-splines, $y(x) = \sum_{j=-1}^{6} c_j B_j(x)$, as given in problem 5. (Hint: in this, and the other parts of the question, use the fact that $y(x)$ will be an odd function of x.)

7. Repeat problems 6(a) and (b) for the cubic spline with quadratic end-pieces. (Note its closer resemblance to the Bessel interpolate of problem 4, section 3.3, which also has quadratic end-pieces.)

8. A thin, uniformly flexible designer's spline is pinned at the three points $(-1, 1)$, $(0, 0.425)$, $(1, 1)$. If its shape is given by $y = y(x)$, and its radius of curvature by $r(x)$, particular values of $y(x)$ and $r(x)$ are given below:

$\lvert x\rvert$	0	0.147	0.261	0.363	0.458	0.546	0.630	0.709	0.858	1
$y(x)$	0.425	0.438	0.467	0.505	0.552	0.605	0.663	0.726	0.860	1
$r(x)$	0.787	0.877	0.974	1.096	1.253	1.462	1.754	2.193	4.385	∞

Its free-ends (for $\lvert x \rvert \geqslant 1$) extend along the straight lines $y = \lvert x \rvert$, at $45°$ to the y-axis. Determine the natural cubic spline interpolating the same three points, and calculate its y-ordinate and its radius of curvature at the values of $\lvert x \rvert$ listed above. What are the inclinations of its end-tangents to the y-axis? (The **radius of curvature** of a curve $y = y(x)$ is given by $[1 + y'(x)^2]^{3/2}/y''(x)$. Note that the mathematical spline is an approximation to the physical spline.)

Section 3.5

1. Show that not all the values of b_k for $k = 0, 1, \ldots, \nu$ and $\nu \geqslant 0$, as defined by the non-trivial solution of (3.5.2), can be zero. (Hint: how many zeros would $A_\mu(x)$ then have?)

2. (a) Defining $Q_k(x) = q_k + \overset{\mu+\nu}{\underset{i=k+1}{K}} (x - x_{i-1}) \downarrow q_i$ for all $k = 0, 1, \ldots, (\mu + \nu)$,
   show that $Q_k(x_k) = q_k$ and that $Q_{k+1}(x) = (x - x_k)/[Q_k(x) - q_k]$.
   (b) Given that the values of q_{jk} are equal to $Q_k(x_j)$, confirm that the
   relations of (3.5.4) are valid, and that $q_{kk} = q_k$ for $k = 0, 1, \ldots, (\mu + \nu)$.
   (c) The $Q_k(x)$ are rational functions of x: what is their order in terms of μ
   and ν for values of $k = 0, 1, \ldots, (\mu + \nu)$?

3. (a) Use rational arithmetic to find the continued fractions which are the
   rational functions (i) of order $2/1$, and (ii) of order $1/2$, interpolating the
   4 data $(0, 1), (1, 3), (2, 6), (3, 10)$.
   (b) Convert the continued fractions of part (a) each to rational functions
   of order $2/2$, by adding another data point $(4, 15)$, and show that they
   represent identical interpolating functions.

4. Find the rational function of order $2/1$ interpolating the four data $(0, 2)$,
   $(1, 2), (3, 3.5), (4, 4.4)$.

5. (a) If a rational function of order $1/1$ interpolates three distinct data, verify
   that the rank of the system (3.5.2) is never less than 2, and equals 2 if
   and only if the three data ordinates are all equal.
   (b) In the event that the rank is 2, confirm that the partial denominators
   q_1 of (3.5.3) and (3.5.5) are necessarily infinite (and so $q_2 = 0$). At
   what value of x (if any) is the value of this continued fraction undefined?

6. (a) Using correctly rounded floating-point decimal arithmetic of precision 3,
   the value of $y = x/(4.56 + x)$ is evaluated below for 5 different values
   of x:

x	-3.21	-0.987	0	0.654	3.21
y	-2.38	-0.276	0	0.126	0.413

   Using this same precision 3 arithmetic, construct the inverted difference
   table (3.5.4) and so obtain the partial denominators $q_0, q_1, \ldots, q_4$ of the
   continued fraction (3.5.3) which interpolates these five data pairs, taken
   in order of increasing x.

   (b) The continued fraction interpolating the first four points is
   $q_0 + \overset{3}{\underset{k=1}{K}} (x - x_{k-1}) \downarrow q_k$. Taking the approximate values of q_0, q_1, q_2,
   and q_3 derived as in part (a), the continued fraction is equivalent to the
   rational function:

   $$(x^2 + 34.0412x - 0.411366)/(33.76x + 150.1257) .$$

   Express this in the form $(A + Bx)(x + C)/(x + D)$ where $A, B, C,$ and
   D are constants. What would be the values of these constants if the
   numerical work of part (a) was conducted in infinite precision?

(c) The approximate continued fraction derived in part (a) which interpolates all the data converts into the rational function:

$$(1.01277x^2 - 0.925778x + 0.0112015)/(x^2 + 3.69292x - 4.08792).$$

Express this in the form $A(x + B)(x + C)/[(x + D)(x + E)]$ and hence show that it embodies a spurious closely-spaced zero and pole. Why is this?

(d) Had the ordinate at $x = 3.21$ been quite different (say, equal to unity) what would have been the rational function of order $2/2$ you would have expected (in theory) to interpolate the five data, had all the arithmetic been exact? In fact, with $q_4 = -0.0834$ (corresponding to the last data ordinate being unity), the continued fraction produced as in part (a) can be reduced to the form given in part (c) with $A = 0.9366$, $B = 0.01205$, $C = -3.76893$, $D = 4.54033$, and $E = -3.41891$. Compare this approximating representation of $y(x)$ with the theoretical one by sketching a graph over $x \in (-4, 4)$.

7. (a) Construct and graph over $x \in (-3, 3)$, the rational function

$$y(x) = x(A + Bx^2)/(C + x^2)$$

with A, B, and C constant, which interpolates the data given in problem 4 of section 3.2.3, and compare with previous graphs of this series. (Hint: interpolate y/x versus x^2.)

(b) List the values of $y'(x)$ at the support points, and find the maximum value of $y(x)$ in $(-2.5, 2.5)$ together with its position.

8. (a) The successive convergents of the continued fraction (3.5.5) can be represented by $\overset{n'}{\underset{j=1}{K}}(x - x_{j-2}) \downarrow q_{j-1} = A_n/B_n$, where A_n and B_n satisfy the recurrence relations (2.7.5). Show that A_n and B_n are polynomials whose leading terms (of highest power in x) are given by

$$A_{2m} = (q_1 + q_3 + \ldots + q_{2m+1})x^{m-1} + \ldots; \quad A_{2m+1} = x^m + \ldots;$$

$$B_{2m} = x^m + \ldots; \quad B_{2m+1} = (q_0 + q_2 + \ldots + q_{2m})x^m + \ldots$$

for any integer $m \geqslant 0$, and with $A_0 = 0$. (Hint: prove by induction.)

(b) If $\rho_k = \overset{r}{\underset{j=0}{\sum}} q_{k-2j}$, where $r = \text{entier } (k/2)$, show that ρ_{n-1} is a symmetrical function of the n data (x_j, y_j) for $j = 0, 1, \ldots, (n-1)$. The quantities ρ_k are called **reciprocal differences**: this property of symmetry recommends their use in certain applications in preference to inverted differences.

(c) Determine an explicit expression for $\rho_2 = q_0 + q_2$ in terms of the three data (x_j, y_j) for $j = 0, 1$, and 2, and hence confirm that it is invariant with respect to any permutation of the data (and so is a symmetric function of these data).

9. A backward recurrence relation similar to (2.8.2) for the evaluation of (3.5.3) by nested division is given by

$$r_N = q_N, \quad r_{N-k} = q_{N-k} + (x - x_{N-k})/r_{N-k+1} \colon (k = 1, 2, \ldots, N)$$

where $N = \mu + \nu$. The value of r_0 is the required value of $y(x)$. Write an instruction sequence based on this recurrence which assigns r equal to $y(x)$, given N, x, and the values of x_k, y_k, q_k for $k = 0, 1, \ldots, N$. Assume that an infinite inverted difference has a value $\inf$, and that any $0/0$ indeterminacy (at a node) is dealt with by supposing $y(x)$ to be equal to the data ordinate at that node.

10. (a) Let $\nu(m+1) = \mu(m) = entier\ (m/2)$, $\sigma(m+1) = -\sigma(m)$ with $\sigma(0) = \pm 1$, for any integer m. Denote by $\mathbf{S}$ any subset of $\{0, 1, \ldots, n\}$, by $\mathbf{S}_j$ the set $\mathbf{S}$ augmented by an element $j \in \mathbf{S}$, and by ϕ the empty set. If $|\mathbf{S}| \geqslant 1$, place $\pi(\mathbf{S}; x) = \prod_{k \in \mathbf{S}} (x - x_k)$, and suppose that $P(\mathbf{S}; x), Q(\mathbf{S}; x)$ are polynomials of degree $\mu(|\mathbf{S}|), \nu(|\mathbf{S}|)$ respectively, defined so that $P(\mathbf{S}; x_k) = y_k^\sigma Q(\mathbf{S}; x_k) \ \forall\, k \in \mathbf{S}$, where $\sigma \equiv \sigma(|\mathbf{S}|)$. Further define $\pi(\phi; x) \equiv P(\phi; x) \equiv 1, Q(\phi; x) \equiv 0$. Then prove that:

  (i) $C_j(\mathbf{S}; x) \equiv P(\mathbf{S}_j; x) P(\mathbf{S}; x) - Q(\mathbf{S}_j; x) Q(\mathbf{S}; x) = \hat{P}(\mathbf{S}_j) \hat{P}(\mathbf{S}) \pi(\mathbf{S}; x)$, where $\hat{P}(\mathbf{S})$ is the leading coefficient (of x^μ) in $P(\mathbf{S}; x)$; and

  (ii) $(x - x_i) C_i(\mathbf{S}_j; x) C_i(\mathbf{S}; x) = (x - x_j) C_j(\mathbf{S}_i; x) C_j(\mathbf{S}; x)$.

(b) Define $R(\mathbf{S}; x) \equiv [P(\mathbf{S}; x)/Q(\mathbf{S}; x)]^\sigma$, where $\sigma = \sigma(|\mathbf{S}|)$. Then verify that

  (i) if $\sigma(0) = 1$, then $R(\phi; x) = \infty$, and $R(\mathbf{S}; x)$ is a rational function of order $\mu(|\mathbf{S}|)/\nu(|\mathbf{S}|)$ for $|\mathbf{S}| \geqslant 1$;

  (ii) if $\sigma(0) = -1$, then $R(\phi; x) = 0$, and $R(\mathbf{S}; x)$ is a rational function of order $\nu(|\mathbf{S}|)/\mu(|\mathbf{S}|)$ for $|\mathbf{S}| \geqslant 1$; and

  (iii) if $\sigma(0) = \pm 1$, then in general $R(\mathbf{S}; x_k) = y_k \ \forall\, k \in S$, and placing $\mathbf{S}_{ij} \equiv \mathbf{S}_i \cup \mathbf{S}_j$ where $i = j$, use the result (ii) of part (a) to show that:

$$(x - x_i) [R(\mathbf{S}_{ij}; x) - R(\mathbf{S}_j; x)] [R(\mathbf{S}_i; x) - R(\mathbf{S}; x)]$$
$$= (x - x_j) [R(\mathbf{S}_{ij}; x) - R(\mathbf{S}_i; x)] [R(\mathbf{S}_j; x) - R(\mathbf{S}; x)] .$$

(c) Confirm from the results of part (b) that if A is either zero or infinite, the recurrence

$$R_{j, -1} \equiv A, \quad R_{j,0} \equiv y_j, \quad \text{for all } n \geqslant j \geqslant 0 ,$$

and $R_{j,k} = R_{j,k-1} + (x - x_j)/[(x_j - x_{j-k}) (R_{j,k-1} - R_{j-1,k-1})^{-1}$
$$- (x - x_{j-k}) (R_{j,k-1} - R_{j-1,k-2})^{-1}]$$

for all $n \geqslant j \geqslant k \geqslant 1$, defines a family of rational functions such that $R_{j,k} \equiv R_{j,k}(x)$ in general interpolates (x_i, y_i) for all $i = j - k, j - k + 1$, $\ldots, j$, and has order μ/ν where $\mu + \nu = k$, with $\mu + 1 \geqslant \nu \geqslant \mu$ if $A = 0$, or with $\nu + 1 \geqslant \mu \geqslant \nu$ if $A = \infty$. (The tableau of values of R_{jk} is analogous to that of the Neville algorithm which generates polynomials $a_{jk} = y_{j-k,j-k+1,...,j}$ and evidently leads to $R_{n,n}$ interpolating all the $(n+1)$ data.)

(d) As an example of the failure of the recurrence of part (c) *always* to perform correctly, try interpolating the points $(-1,-1), (0,0), (1,1)$, with $A = 0$, which ought to lead to the rational function $R_{2,2}(x) \equiv x$. Verify that it does so if the first two points are reversed in order.

Section 3.6

1. (a) Three points taken in order along a curve have (x, y) Cartesian coordinates $(-4, 3), (0, 0), (4, 3)$. Find the quadratic polynomials $x(t), y(t)$ such that the curve C given parametrically by $x = x(t)$, $y = y(t)$ interpolates all three data points, assuming that the successive parameter values are $-1, 0, 1$ corresponding to each successive point on C. (This is an example of cardinal parametric interpolation.)

(b) Eliminating t between these equations $x = x(t)$, $y = y(t)$, verify that C is the parabola $y = 3x^2/16$.

(c) A further point with coordinates $(8.8, 14.52)$ is now prescribed following the other three: confirm that this lies on the curve C. The curve C' constructed by using cardinal parametric (cubic) polynomial interpolation through all four points might be expected to be identical with C: is it? If not, is there *any* fourth point which would leave the interpolating curve unchanged?

2. Repeat the analysis of the last problem using the chord length between successive points as parameter increment, as in equation (3.6.2).

3. (a) A common method of constructing a curve of continuous tangent through a sequence of given points $P_0, P_1, \ldots, P_n$ uses a succession of (generally different) circular arcs joining successive pairs of points $P_0 P_1, P_1 P_2, \ldots, P_{n-1} P_n$. If γ_k is the angle made by the direction of the chord $P_{k-1} P_k$ with some given datum line $(0 \leqslant \gamma_k < 2\pi)$, and if θ_k denotes the angle made with the same datum by the tangent at P_k to the piecewise continuous curve, verify that $\theta_k + \theta_{k-1} = 2\gamma_k$, so that if θ_0 is given, then $\theta_1, \theta_2, \ldots, \theta_n$ are determined.

(b) If the curve to be constructed is closed, the points P_0 and P_n being coincident, prove that this curve cannot (in general) belong to class C^1 if $n = 2m$ is an even integer. (Hint: note that $\theta_{k+2} - \theta_k = 2\Delta\gamma_{k+1}$, which is twice the angle between the successive chord lines which join at P_{k+1}).

(c) If the curve is closed, the points P_0 and P_{2m+1} being coincident (so that $n = 2m + 1$ is an odd integer), show that the curve belongs to class C^1 if

$$\theta_0 - \gamma_1 = \sum_{j=1}^{m} \Delta\gamma_{2j} - \pi.$$

4. (a) Using the expression (3.4.2) for a quadratic piece $y = \pi_k(x)$ of a quadratic spline in the interval $[x_{k-1}, x_k]$, express the values of $\pi'_k(x_{k-1})$ and $\pi'_k(x_k)$ in terms of the data and the value of μ_k.

   (b) Show that there is in general no solution corresponding to a periodic spline of continuous slope unless the number of distinct data pairs (n) in each full period (ω) is odd, and then with $n = 2m + 1$, say,

$$y'_0 = \sum_{j=1}^{2m+1} (-1)^{j+1} \hat{y}_{j-1,j} .$$

   [Hint: note that $x_n - x_0 = \omega$, and use the expression (3.4.4) for μ_n.]

   (c) Find the general expression for the tangent (y'_k) of the slope of the periodic quadratic spline at each node $x = x_k$ for $k = 1, 2, \ldots, n$.

5. A periodic cubic spline interpolating the data $(x_k, \sin x_k)$ where $x_k = 2k\pi/n$ for $k = 0, 1, 2, \ldots, n$ is given by cubic pieces $\pi_k(x)$ as in (3.4.6) with

$$M_k = -(6/h^2)[1 - \cos(h)] \sin(kh)/[2 + \cos(h)]$$

where $h = 2\pi/n$. If $n = 6$, find the value of the ordinate of the spline at $x = \pi/2$.

6. (a) Write down the system of n linear equations connecting the values of $y''(x_k) = M_k$ for the periodic cubic spline interpolating the n distinct data pairs (x_k, y_k) for $k = 0, 1, \ldots, (n-1)$ with $x_0 \leqslant x_k < x_0 + \omega$ which describe a variation of fundamental period ω. (Hint: it will be found notationally convenient to define $x_{k+n} = x_k + \omega$ and $y_{k+n} = y_k$ for all integers k, so that for example $x_n = x_0 + \omega$ and $y_n = y_0$.)

   (b) If the nodes are equi-spaced, so that $x_k = k\omega/n$, the $n \times n$ diagonally dominant matrix $\mathbf{C}$ (say) of the linear system for the values of $M_0, M_1, \ldots, M_{n-1}$ has a particular form such that the element in the jth row and kth column of $\mathbf{C}$ is given by

$$c_{jk} = \gamma_{k-j} \quad \text{if } k \geqslant j$$

$$= \gamma_{k-j+n} \quad \text{if } k < j .$$

   Scaling the equations so that the diagonal elements $c_{kk} = \gamma_0 = 4$, identify the values of $\gamma_1, \gamma_2, \ldots, \gamma_{n-1}$. (A matrix of this form is called **circulant**.)

   (c) A circulant matrix is readily inverted, and it can be shown that the equations are solved by

$$(\omega^2/6n^2) M_j = (1 + \sqrt{3}) y_j - \sqrt{3}[1 - (-\alpha)^n]^{-1} \sum_{k=0}^{n-1} (-\alpha)^k (y_{j+k} + y_{j-k})$$

   for $j = 0, 1, \ldots, (n-1)$, where $\alpha = 2 - \sqrt{3}$, and where y_{i+n} is defined equal to y_i for any integer i. Verify that this solution is correct for $n = 3$.

7. (a) Using the identity that if $\sin \psi \neq 0$, then

$$\{\sin[\phi + (2k + 1)\psi] - \sin[\phi + (2k - 1)\psi]\}/\sin \psi = 2 \cos(\phi + 2k\psi),$$

show that

$$\sum_{k=0}^{n} \cos(\phi + 2k\,\psi) = \cos(\phi + n\psi)\sin[(n+1)\psi]/\sin\psi \quad \text{if } \sin\psi \neq 0,$$
$$= (n+1)\cos\phi \qquad\qquad\qquad \text{if } \sin\psi = 0.$$

(b) Show that, if j and k are non-negative integers $\leqslant n/2$, and $\alpha = 2\pi/n$, then

$$\sum_{r=0}^{n-1} \cos(rj\alpha)\cos(rk\alpha) = (n/2)(\delta_{j,k} + \delta_{j+k,0} + \delta_{j+k,n})$$

$$\sum_{r=0}^{n-1} \sin(rj\alpha)\sin(rk\alpha) = (n/2)(\delta_{j,k} - \delta_{j+k,0} - \delta_{j+k,n})$$

$$\sum_{r=0}^{n-1} \sin(rj\alpha)\cos(rk\alpha) = 0 ,$$

where $\delta_{p,q}$ is the Kronecker delta (unity if $p = q$, else zero). (Hint: express the products as sums/differences of sines/cosines.)

(c) Verify that the coefficients of the interpolating trigonometric polynomial (3.6.3) are given by (3.6.4).

8. What is the shape of the closed curve through the three data points $(-1, 1)$, $(-1, -1)$, $(2, 0)$ which is given by cardinal parametric interpolation, using trigonometric polynomials to represent $x(t)$ and $y(t)$?

Section 3.7.1

1. If $y = p_{2n+1}(x)$ is the Hermitian polynomial interpolate which makes osculatory contact with $y = f(x)$ at the $(n+1)$ points $x_0, x_1, \ldots, x_n$ in $[a, b]$, then prove that

$$f(x) = p_{2n+1}(x) + f^{(2n+2)}(\xi_x)[\pi_n(x)]^2/(2n+2)!$$

for some ξ_x in $[a, b]$.

Section 3.7.2

1. (a) Verify that, for any positive integer ν, $T_\nu(x)$ is a polynomial of genuine degree ν, and that the coefficient of its leading term, $\hat{T}_\nu(x)$, is equal to $2^{\nu-1}$.

(b) Prove that $T_\nu(-x) = (-1)^\nu T_\nu(x)$ for any positive integer ν.

2. Show that the shape function $s_{nk}(x)$ associated with the support point $x_k = \cos[(k + \frac{1}{2})\pi/(n+1)] = \cos\theta_k$, say, of nth degree Tchebyshev interpolation is given by $s_{nk}(x) = C_{nk}T_{n+1}(x)/(x - x_k)$; and find the value of C_{nk} in terms of n and θ_k.

3. (a) Given as in problem 7 of section 3.6 that

$$\sum_{k=0}^{n} \cos(\phi + 2k\,\psi) = \cos(\phi + n\psi)\sin[(n+1)\psi]/\sin\psi, \quad \text{if } \sin\psi \neq 0,$$
$$= (n+1)\cos\phi, \qquad\qquad\qquad \text{if } \sin\psi = 0,$$

show that if $x_0, x_1, \ldots, x_n$ are the Tchebyshev abscissae of nth degree interpolation and $S_n(v) = \sum\limits_{k=0}^{n} T_v(x_k)$, then $S_n(0) = (n+1)$, but $S_n(v) = 0$ for all $v = 1, 2, \ldots, (2n+1)$.

(b) If i and j are non-negative integers $\leqslant n$, verify that

$$\sum_{k=0}^{n} T_i(x_k) T_j(x_k) = 0 \qquad \text{if } i \neq j ,$$

$$= (n+1)/2 \quad \text{if } i = j \neq 0 ,$$

$$= (n+1) \qquad \text{if } i = j = 0 .$$

(Hint: use the identity: $2 \cos A \cos B = \cos(A+B) + \cos(A-B)$.)

(c) Show that if $f(x_k) = f_k$ for $k = 0, 1, \ldots, n$ where x_k are the Tchebyshev abscissae of nth degree interpolation, then the interpolating polynomial is expressable as

$$p_n(x) = (c_0/2) + \sum_{j=1}^{n} c_j T_j(x)$$

where

$$c_i = 2 \sum_{k=0}^{n} f_k T_i(x_k)/(n+1) \quad \text{for} \quad i = 0, 1, \ldots, n .$$

(d) Find the quadratic Tchebyshev interpolate over $[-1, 1]$ of $f(x) = \ln(1 + x/3)$ in the form shown in part (c).

4. (a) Either using the result of problem 3(d), or by constructing a divided difference table, find the quadratic Tchebyshev interpolate over $[-1, 1]$ of $f(x) = \ln(1 + x/3)$, and express it in power form (i.e., as $p_2(x) = a_0 + a_1 x + a_2 x^2$).

(b) Use the remainder term of (3.7.1) to establish bounds on the truncation error $f(x) - p_2(x)$.

(c) Find the two values of $x \in [-1, 1]$ for which $[f(x) - p_2(x)]$ has a stationary value, and calculate the actual error at these positions and at $x = \pm 1$.

5. (a) **Clenshaw's technique** for the evaluation of the truncated Tchebyshev series $p_n(x) = (c_0/2) + \sum\limits_{j=1}^{n} c_j T_j(x)$ at any given x consists of the formation of the sequence $a_n, a_{n-1}, \ldots, a_0$ by the recursion: $a_{n+2} = a_{n+1} = 0$, $a_k = c_k + 2x a_{k+1} - a_{k+2}$, $(k = n, n-1, \ldots, 0)$. Relating the c_j's in the expression for $p_n(x)$ to the a_j's, show that $p_n(x) = (a_0 - a_2)/2$.

(b) Show that Clenshaw's technique can also be used to evaluate $T_n(x)$, and that then the generated sequence is such that $(a_{n-k} - a_{n-k+2})/2 = T_k(x)$ for $k = 1, 2, \ldots, n$.

Section 3.8

1. (a) Let $f(x) \in C^2[a, b]$ and assume $f''(x)$ does not change sign in $[a, b]$. If $p_1(x) = a_0 + a_1 x$ is the minimax linear polynomial approximating $f(x)$

over $[a, b]$, show that

$$a_1 = [f(b) - f(a)]/(b - a) ,$$

$$2a_0 = [bf(a) - af(b)]/(b - a) + f(z_1) - a_1 z_1$$

where z_1 is the unique solution of $f'(z_1) = a_1$.
  (b) Interpret this result geometrically.
  (c) What is the value of $E_1^*(f)$?
  (d) Construct the minimax linear polynomial approximating $\ln(1 + x/3)$ over $[-1, 1]$ and find its maximum error in this interval.

2. (a) Verify that for $n \geqslant 1$, $t_{n-1}(x)$ satisfies the necessary and sufficient conditions of the Tchebyshev alternation theorem as a minimax polynomial of degree n approximating x^{n+1} over $[-1, 1]$.
  (b) Since $t_{n-1}(x)$ has genuine degree $(n - 1)$, does it also satisfy these conditions as a minimax polynomial of degree $(n - 1)$?

3. (a) Suppose $p_n(x)$ satisfies the equi-oscillation condition (3.8.2): prove that there can exist no other polynomial $q_n(x)$ of degree n which produces a smaller error magnitude $E_n(f) < |\epsilon|$. (This proves the sufficiency of the conditions of the alternation theorem.)
  (b) Show that if $f \in C^{n+1}$ and $[f(x) - p_n(x)]$ has at least $(n + 1)$ stationary values in $[a, b]$, then $f^{(n+1)}(x)$ would have at least one zero in (a, b); and hence if $f^{(n+1)}(x)$ does not change sign in (a, b) then $z_0 = a$ and $z_{n+1} = b$ in (3.8.2).

4. Illustrate the fact (a) that the Tchebyshev alternation theorem cannot apply to discontinuous functions by trying to find a minimax approximation to $f(x) = \operatorname{sgn}(x)$; and (b) that the number of points of equi-oscillation may exceed $(n + 2)$ by considering the minimax linear polynomial approximating $f(x) = \sin(2N\pi x)$ over $[-1, 1]$, where N is a positive integer.

5. (a) Show that the minimax polynomial approximations of degree $2m$ and $2m + 1$ to an *even* function $f(x)$ of over $[-b, b]$ are identical linear combinations of even powers of x having at least $(2m + 3)$ points of equi-oscillation of error.
  (b) What is the analogous observation that may be made if $f(x)$ is an odd function of x?

6. Let $p_n^*(x)$ be the minimax polynomial approximation of degree n to $f(x)$ over $[-1, 1]$. Verify the following propositions:
  (i) if $f(x) \equiv q_{n+1}(x)$ is a polynomial of degree $\leqslant (n + 1)$, and $\hat{q}_{n+1}$ is the coefficient of x^{n+1}, then $p_n^*(x) \equiv q_{n+1}(x) - (q_{n+1}/2^n)T_{n+1}(x)$;
  (ii) if $f(x) \equiv T_{n+1}(x)$, then $p_n^*(x) \equiv 0$;
  (iii) if $f(x) \equiv T_{n+m}(x)$, where $m > 1$ is an integer, then $p_n^*(x) \equiv 0$; and
  (iv) if $f(x) = 2xT_{n+1}(x)$, then $p_n^*(x) \equiv T_n(x)$.

Section 3.8.1

1. Using the methods listed below, construct a nearly-minimax quadratic polynomial approximating $f(x) = (x^2 + 0.5625)^{-1}$ over $[-1, 1]$, listing the positions and the values of maximum/minimum error, and the values of the error at $x = \pm 1$, and hence determining $E_2(f)$ in each instance:
   (a) by finding the Tchebyshev interpolate;
   (b) by economising the Taylor polynomial $(\alpha^2 - \alpha^4 x^2 + \alpha^6 x^4)$ where $\alpha = 4/3$ (why does this give a poor approximation?);
   (c) by truncating the Tchebyshev series representing this function: namely

$$f(x) = (8/15)[2 + \sum_{n=1}^{\infty} (-1)^n 4^{1-n} T_{2n}(x)] \ ;$$

   (d) by forced oscillation of the error at the points of equi-oscillation of $T_4(x)$ (why T_4 and not T_3?);
   (e) by constructing the minimax linear polynomial approximating $f(z^{\frac{1}{2}})$ over $z \in [0, 1]$, and then putting $x^2 = z$.

Section 3.8.2

1. (a) Suppose $f(x) = \ln(1 + x/3)$ is approximated over $[-1, 1]$ by two minimax linear polynomial segments over $[-1, 0]$ and $[0, 1]$: what would be the maximum error? (Hint: find the linear polynomials by the method of problem 1, section 3.8.)
   (b) Estimate (roughly) the number of equal-length segments into which the interval $[-1, 1]$ should be divided to achieve the same actual error as that of the unsegmented minimax quadratic polynomial, $E_2^*(f) = 0.0034$.

2. (a) Prove that if $f(x) \in C^2[a, b]$, and neither $f(x)$ nor $f''(x)$ changes sign in $[a, b]$, then $[f(x)/f'(x) - x]$ is a monotone function of x over $[a, b]$, and that therefore $p_1(x)/f(x)$ can have no more than one stationary value in $[a, b]$, where $p_1(x)$ is any given linear polynomial.
   (b) Show that the maximum relative error magnitude $\max_{x \in [a,b]} |1 - p_1(x)/f(x)|$ is least and equal to $|\epsilon|$ if

$$p(x) = (1 - \epsilon)[(x - a)f(b) + (b - x)f(a)]/(b - a)$$

   where $(1 + \epsilon) = p_1(z_1)/f(z_1)$ and $x = z_1$ is the (unique) root in $[a, b]$ of

$$[(x - a)f(b) + (b - x)f(a)]f'(x) = [f(b) - f(a)]f(x) \ .$$

   (c) Find the linear polynomial with minimax relative error which approximates $f(x) = x^{1/2}$ over $[1/4, 1]$, and determine its accuracy.

3. If the minimax nth degree polynomial approximation of $\ln(1 + x/3)$ for $x \in [-1, 1]$ is $p_n(x)$, and that of $\log_{10}(t)$ for $t \in [1, 2]$ is $q_n(t)$, show how the latter is related to the former.

Section 3.10

1. (a) It can be shown (from, for instance, the summation formula of problem 3, section 3.7.2) that, if k is any non-negative integer $\leqslant (n+1)$ and $x_j = \cos[(j+1)\pi/(n+2)]$, then

$$\sum_{j=0}^{n+1} (-1)^j [T_k(x_j) - T_k(x_{j-1})] = 0 .$$

Hence confirm that, if $\sigma_{n+1}(x) = (-1)^j$ for $x \in (x_j, x_{j-1})$ and $j = 0, 1, \ldots, (n+1)$ and $q_n(x)$ is any arbitrary polynomial of degree $\leqslant n$, then

$$\int_{-1}^{+1} q_n(x)\sigma_{n+1}(x)\mathrm{d}x = 0 .$$

(Hint: suppose that $q_n(x) = q'_{n+1}(x)$, where $q_{n+1}(x)$ is an arbitrary polynomial of degree $n+1$.)

(b) If $f(x)$ is continuous in $[-1, 1]$ and $p_n(x)$ interpolates $f(x)$ at $x = x_0, x_1, \ldots, x_n$, as defined above, and only at these points, so that $\operatorname{sgn}[f(x) - p_n(x)] = c\sigma_{n+1}(x)$, where c is either plus or minus unity and $\sigma_{n+1}(x_j)$ is defined as zero for $j = 0, 1, \ldots, n$, then verify that

$$\int_{-1}^{+1} |f(x) - p_n(x)|\mathrm{d}x = \int_{-1}^{+1} [f(x) - q_n(x)]\operatorname{sgn}[f(x) - p_n(x)]\mathrm{d}x$$

for any arbitrary polynomial $q_n(x)$ of degree $\leqslant n$. Hence prove that $p_n(x)$ is that polynomial of degree $\leqslant n$ which minimises $\int_{-1}^{+1} |f(x) - p_n(x)|\mathrm{d}x$. (Hint: show that the integral $\leqslant \int_{-1}^{+1} |f(x) - q_n(x)|\mathrm{d}x$.)

2. (a) The polynomials $U_n(x)$ which are developed by defining $U_0(x) = 1$, $U_1(x) = 2x$, and $U_{\nu+1}(x) = 2x U_\nu(x) - U_{\nu-1}(x)$ for $\nu = 1, 2, 3, \ldots$ are called **Tchebyschev polynomials of the second kind**. Evidently they satisfy the same recurrence relation as the $T_n(x)$'s. Confirm that for $|x| \leqslant 1$, their values are given by $U_\nu(\cos\theta) = \sin[(\nu+1)\theta]/\sin\theta$.

(b) Verify that $U_n(x) - U_{n-2}(x) = 2 T_n(x)$. (Hint: place $x = \cos\theta$, and verify for $|x| \leqslant 1$.)

(c) Prove that $p_n(x) = x^{n+1} - U_{n+1}(x)/2^{n+1} = u_{n-1}(x)$, say is that polynomial of degree $\leqslant n$ which approximates $f(x) = x^{n+1}$ for $x \in [-1, 1]$ with minimum unweighted absolute error norm.

(d) Show that the choice $\xi_j = \cos[(j+1)\pi/(n+2)]$ minimises the value of $\int_{-1}^{+1} |(x - \xi_0)(x - \xi_1) \ldots (x - \xi_n)|\mathrm{d}x$, with respect to all possible sequences $\xi_0 \geqslant \xi_1 \geqslant \ldots \geqslant \xi_n$, and that $U_{n+1}(x)/2^{n+1}$ is that polynomial with leading term x^{n+1} whose unweighted absolute norm over $[-1, 1]$ is least. (It will be recalled that $T_{n+1}(x)/2^n$ has the same property *vis-a-vis* the maximum norm.)

PROBLEMS: CHAPTER 4

Section 4.1.1

1. The following functions satisfy the condition $f(0)f(1) < 1$: what value of x

is located by the bisection method applied to the interval $[0, 1]$ with each function $f(x)$? Is the point a root of $f(x) = 0$? If not, why not?
(a) $f(x) \equiv \cos 10x$; (b) $f(x) \equiv (x - 0.3)^{-1}$; (c) $f(x) = \text{sgn}(x - 0.2)$.

2. If the initial interval $[a, b]$ is such that $b - a = 2s$, show that the least number of iterations of the bisection method needed to provide an estimate of the root with error magnitude $\leqslant xtol$ is equal to ceil$[\ln(s/xtol)/\ln 2]$.

3. (a) Using correctly rounded decimal arithmetic of precision 2, and an initial bracketing interval of $[0, 1]$, write down the sequence of root bounds $[a, b]$ given by the instruction sequence for the bisection method quoted in the text applied to the location of the roots of $f(x) \equiv 3x - 0.91 = 0$, assuming that $xtol$ is zero. What is the terminating value of xi?
   (b) Repeat part (a) but use chopping instead of correct rounding.
   (c) Repeat part (a) but use correct rounding and $f(x) \equiv 1.7x - 0.91 = 0$.
   (d) Repeat part (c) but replace the assignment to xi by $xi: = (b + a)/2$. Note that $(b + a)$ is rounded before division by 2. (This should indicate why $xi: = a + (b - a)/2$ does not give numerically the same value, but is sometimes preferable, as here).
   (e) Repeat part (d) using chopping instead of correct rounding.

4. Modify the instruction sequence of the text in such a manner that if $f(xi) = 0$ is zero, then both a and b are set equal to xi.

Section 4.1.2

1. Prove that, if $f(a)f(b) < 0$, then the value of c given by (4.1.1) lies in (a, b).

2. (a) Show that as $a \to \xi$, where $f(\xi) = 0$, then $f(a)/(\xi - a) \to -f'(\xi)$.
   (b) Verify the approximate relation (4.1.2) for small $(\xi - a)$.

Section 4.1.3

1. (a) If $x = x_{n+3}$ is a root estimate, then show that the same root estimate is derived as the intersection with the x-axis of the straight line joining the points $(x_n, \beta f_n), (x_{n+2}, f_{n+2})$ where $\beta = (f_{n+2}/f_n)(x_{n+3} - x_n)/(x_{n+3} - x_{n+2})$.
   (b) The rational function $y = (x - p)/(qx + r)$ interpolating (x_k, f_k) for $k = n, n + 1$, and $n + 2$, intersects the x-axis at $x = p$, where

   $$(f_{n+2}/f_n)(p - x_n)/(p - x_{n+2}) = \hat{f}_{n+1,n+2}/\hat{f}_{n,n+1}$$

   in terms of divided differences $\hat{f}_{j,k} = (f_k - f_j)/(x_k - x_j)$. Given that x_{n+2} is the intersection with the x-axis of the straight line joining the points $(x_n, f_n), (x_{n+1}, f_{n+1})$, show that the estimate $x_{n+3} = p$ is given by taking $\beta = 1 - (f_{n+2}/f_{n+1})$, as in (4.1.5).

2. (a) Choose any two numerical 'estimates' $x_0 > 0$ and $x_1 \in (-1, 0)$ bracketing the zero root of the equation $x(1 + x) = 0$. Find the first iterated root estimate x_2 by the method of *regula falsi*, confirming that this also lies in the interval $(-1, 0)$. Then verify that the modification (4.1.6) applied to the value of f_0 implies that the second iterated root estimate x_3 is (correctly) zero.

  (b) As they stand, assuming that the flag *mark* is initialised as zero and that x_0, x_1 are the initial bracketing root estimates, the modifications proposed (in the text) at labels UBMOD and LBMOD have no effect until x_4 comes to be estimated. Suggest a change in the program sequence forcing a modification to be applied to the root estimate x_3, and show that the estimates at iteration 2 in Table 4.1 would all thereby be improved.

3. (a) In the notation of (4.1.6), show that $\alpha^2 - \gamma_0 = (\gamma_0 - \gamma_1 - 1)^2/4 - \gamma_1$ and hence confirm that $\beta < 1$. (Hint: note that γ_1 is positive.)

  (b) Similarly confirm that $\alpha^2 - \gamma_0 = (\gamma_0 + \gamma_1 - 1)^2/4 - \gamma_1 \gamma_0$, and that β given by (4.1.6) is therefore larger than the value $\beta = 1 - \gamma_1$ given by (4.1.5).

4. (a) Suppose that successive estimates of the root $x = \xi$ of the equation $f(x) = 0$ by *regula falsi* are x_0, x_1, x_2, with x_0 on one side of the root and x_1, x_2 on the other. One of the modifications by 3-point interpolation is then used to obtain a new root estimate, x_3. Provided that all estimates are sufficiently close to the root, it can be shown that

$$\epsilon_3 \simeq -A\epsilon_0 \epsilon_1 \epsilon_2/6 \qquad \text{if (4.1.6) is used,}$$
$$\simeq (3B^2 - 2A)\epsilon_0 \epsilon_1 \epsilon_2/12 \quad \text{if (4.1.5) is used,}$$

where $\epsilon_k = \xi - x_k$ is the error of the kth estimate, and $A = f'''(\xi)/f'(\xi)$, $B = f''(\xi)/f'(\xi)$. What conditions on the derivatives of $f(x)$ at $x = \xi$ must be met if x_3 is to replace x_0 as a new bound to the root, and either (4.1.5) or (4.1.6) is used?

  (b) Are these conditions met for the equation $f(x) \equiv \exp(x^2) - 50 = 0$ used in Table 4.1?

  (c) Verify that $x_3 = \xi$ precisely if $f(x)$ is any rational function of order $1/1$ and the modification of (4.1.5) is used.

5. The equation $f(x) \equiv x^2 - \sin(x) = 0$ has a root in $[0.85, 0.9]$; if successive root estimates by *regula falsi* are 0.9 and 0.85 (in that order) confirm that the next is 0.875950. What value of β would give the true root ($x = 0.876726$) at the next iteration? and what values of β would be applied to form the next root estimate by (a) the Illinois method, (b) the Pegasus method, (c) use of hyperbolic interpolation, and (d) use of parabolic interpolation? In each example, note

whether or not the value of β achieves a replacement of the upper bound (0.9). Which method gives the smallest upper bound?

Section 4.2.1

1. Supposing that the error ϵ_n obeys a relationship of the form given by (4.2.4), confirm that (4.2.2) implies that the order of convergence of the secant rule is equal to $p = (1 + \sqrt{5})/2$,

2. (a) If b is the base of the floating point representation in use, and one writes $\ln|\epsilon_n| = -m_n \ln(b)$, so that m_n can be interpreted as the number of accurate (or significant) digits in the current approximation of ξ by x_n, show that an order of convergence equal to p implies that the number of significant digits ultimately increases by a factor of approximately p at each iteration.
   (b) If the convergence is linear, show that m_n increases in linear progression, as $n \to \infty$, provided $|q| < 1$.

3. (a) Show that if the secant rule is applied to the location of the roots of $f(x) \equiv x^2 - c = 0$, then $x_{n+1} = (x_{n-1}x_n + c)/(x_{n-1} + x_n)$.
   (b) Confirm that the error $\epsilon_n = \mathrm{sqrt}(c) - x_n$ implied by this recurrence is consistent with (4.2.2), and find the value of C_n. Verify also that the limit C_∞ is consistent with (4.2.3).
   (c) Demonstrate by calculation that if the secant rule is used to find the value of $\mathrm{sqrt}(2)$ by this recurrence, then the 'waltzing' pattern of error is present unless both x_0 and x_1 exceed $\mathrm{sqrt}(2)$. (Take x_0 and x_1 both positive.)

Section 4.2.2

1. (a) If Newton's method is applied to the solution of $f(x) \equiv x^{-1} - c = 0$, one obtains an iterative method of finding the reciprocal of c without using division: find the recurrence relation for the root estimates.
   (b) Verify that the error $\epsilon_n = c^{-1} - x_n$ of the root estimates obeys the recurrence $\epsilon_{n+1} = c\epsilon_n^2$.
   (c) Show that $\epsilon_n = (c\epsilon_0)^m/c$ where $m = 2^n$ is consistent with the recurrence of part (b), and hence prove that the recurrence of part (a) converges if and only if $0 < cx_0 < 2$.

2. (a) Find a recurrence for the generation of the cube root of c, as given by the root of $f(x) \equiv x^3 - c = 0$.
   (b) By taking $f(x) \equiv x^{3-m} - cx^{-m}$, find the value of m which will give third-order convergence by Newton's method, and write down the appropriate recurrence relation.

3. (a) Suppose that $x_0(c)$ is a (close) approximation to $\mathrm{sqrt}(c)$ for $c \in I$, and that $x_1 \equiv x_1(c)$ is given in terms of $x_0 \equiv x_0(c)$ by Heron's rule. If

$|\text{sqrt}(c) - x_1(c)| \leqq \delta\, \text{sqrt}(c)$ for all $c \in I$, where δ is a required relative error, show that the relative error of $x_0(c)$ must not exceed $(2\delta)^{1/2}$ in magnitude. (Hint: take $C_0 \cong C_\infty$ in (4.2.6).)

(b) The rational function of order $1/1$ which approximates $\text{sqrt}(c)$ for $1/2 \leqslant c \leqslant 1$ with minimax relative error has a maximum relative error magnitude of $10^{-3.49}$. If this function is chosen as $x_0(c)$, what will be the maximum relative error of $x_1(c)$ over the same range of c? Express the answer as 2^{-p}, and so establish the maximum binary precision of floating-point number for which this approximation is suitable.

4. Given that $\text{erfc}(x) \doteq 2\pi^{-1/2}\int_x^\infty \exp(-t^2)dt$ may be evaluated with sufficient accuracy as $\text{erfc}(x) = 0.794\exp(-x^2)/(x + 0.789)$, find the root of the equation $\text{erfc}(x) = 0.25$ correct to 3 significant decimal digits.

5. The Newton–Raphson recurrence may also be applied to the generation of a sequence of complex root estimates. Find the recurrence relation for the root estimates $z_0, z_1, z_2, \ldots,$ of the equation $f(z) \equiv z^2 + 1 = 0$, and separate the relation into real and imaginary parts by taking $z_n = x_n + iy_n$. (Assume $y_0 \neq 0$.)

Section 4.2.3

1. (a) The equation $f(x) \equiv \cos(x) - 0.1x = 0$ has a finite number of real roots. Establish interval brackets enclosing each such root.
   (b) If it was preferred to use an unbracketed method of locating these roots, what is the steep-side of $y = f(x)$ in the neighbourhood of each root?
   (c) Find the largest (positive) root of $f(x) = 0$ by the Newton–Raphson method.

2. (a) The equation $f(x) \equiv \tan(x) - x = 0$ has an infinity of real roots which, for large $|x|$, are each individually close to the values $a_m = (m + \frac{1}{2})\pi$, for integer m. Show that, if m is large, the root close to a_m is given approximately (for large m) by $\xi_m = a_m - 1/a_m$. (Hint: place $\xi_m = a_m - \delta$ and solve for δ.)
   (b) Establish bracketing values for the roots $\xi_m (m \geqslant 0)$, and identify the 'steep-side' of each root.
   (c) Find the value of the root close to $x = 99$ using the Newton–Raphson method.

3. (a) Show that any non-zero starting value for the Newton–Raphson iterations applied to the equation $f(x) \equiv x^{1/3} = 0$ fails to converge to the root at $x = 0$.
   (b) Similarly, consider the convergence of the iterations as applied to the equation $f(x) \equiv \text{sgn}(x)|x|^{1/2} = 0$.
   (c) For what range of choice of x_0 do the Newton–Raphson iterations converge to the (zero) root of $f(x) \equiv x(1 + x^2)^{-1/2} = 0$? (Note that $x = 0$ is a point of inflection and there is no 'steep-side' of this root.)

Section 4.2.4

1. (a) Suppose that $a < \varphi(x) < b$ for all $x \in [a, b]$; then show that if $\varphi(x)$ is continuous in $[a, b]$ there is at least one root of $x = \varphi(x)$ in $[a, b]$.
   (b) If additionally $\varphi(x)$ obeys the **Lipschitz condition** $|\varphi(s) - \varphi(t)| \leqslant M |s - t|$ for all $s, t \in [a, b]$ and where $0 \leqslant M < 1$, then prove that there is only one root of $x = \varphi(x)$ in $[a, b]$.
   (c) If $x = \xi$ is this (unique) root, and $x_{n+1} = \varphi(x_n)$, then confirm that $\lim_{n \to \infty} x_n = \xi$ for any choice of $x_0 \in [a, b]$.
   (d) If $\varphi(x) \in C^1[a, b]$ and $\max_{a \leqslant x \leqslant b} |\varphi'(x)| = m$, then show that $\varphi(x)$ obeys the Lipschitz condition over $[a, b]$ with $M = m$.

2. (a) Suppose $x = \xi$ is a solution of $x = \varphi(x)$; then if $\varphi'(x) \in C^1[a, b]$ and $\max_{a \leqslant x \leqslant b} \varphi'(x)| = m$ for some interval $|x - \xi| \leqslant \delta$, where $0 \leqslant m < 1$, prove that $a < \varphi(x) < b$ for all $x [a, b]$, where $a = \xi - \delta$ and $b = \xi + \delta$. Hence show that ξ is the only root in $[a, b]$, and if $x_{n+1} = \varphi(x_n)$, then $\lim_{n \to \infty} x_n = \xi$ for any choice of $x_0 \in [a, b]$.
   (b) Verify that if $\epsilon_n = \xi - x_n$ then $\epsilon_{n+1}/\epsilon_n \to \varphi'(\xi)$ as $n \to \infty$.

3. (a) If $\varphi(x) \in C^p[a, b]$ and $x = \xi$ is a root of $x = \varphi(x)$ where $\xi \in [a, b]$, and further if $\varphi^{(j)}(\xi) = 0$ for $j = 1, 2, \ldots, (p - 1)$, but $\varphi^{(p)}(\xi) \neq 0$, then show that if the iteration $x_{n+1} = \varphi(x_n)$ converges, its order of convergence is p.
   (b) Use the result of part (a) to verify that the Newton-Raphson method has local quadratic convergence in the neighbourhood of a simple root.
   (c) If $u(x) = f(x)/f'(x)$ and $f(x) \in C^3$, show that if any of the following iterations converge, then their order of convergence is at least 3:

   (i) $x_{n+1} = x_n - \frac{1}{2} u(x_n)[3 - u'(x_n)]$ (**Tchebyshev's method**);

   (ii) $x_{n+1} = x_n - 2u(x_n)/[1 + u'(x_n)]$ (**Halley's method**);

   (iii) $x_{n+1} = x_n - \frac{1}{2} u(x_n)[1 + 1/u'(x)]$.

   (Their global convergence properties are, however, different from those of the Newton-Raphson method.)

4. **Steffenson's method** of locating roots of the equation $f(x) = 0$ consists of the iterative scheme:

$$x_{2i+1} = x_{2i} + f_{2i}, \quad x_{2i+2} = x_{2i} - (f_{2i})^2/(f_{2i+1} - f_{2i}),$$
$$i = 0, 1, 2, \ldots.$$

Show that this is equivalent to alternate steps of fixed-point iteration and the secant rule. (Its order of convergence averaged over successive pairs of iterations is $\sqrt{2}$.)

5. (a) If b and $c > 0$ are constants, find sufficient conditions on x_0 for the iteration $x_{n+1} = (bx_n + c/x_n)/(b+1)$ to converge to the positive square root of c.

   (b) Taking $x_0 = b = c = 2$, determine the values of x_k for $k = 1, 2, \ldots, 10$.

   (c) What is the asymptotic value of the error factor $\epsilon_{n+1}/\epsilon_n$ for this iterative scheme? Compare this with the actual values of the error factor obtained in part (b).

6. (a) If ξ is the limit of the sequence $\{x_k\}$, and any three successive values of $\epsilon_k = \xi - x_k$ for $k = n-1$, n and $n+1$ are assumed to be in geometric progression with common ratio r_{n+1}, then two equations connect x_{n-1}, x_n and x_{n+1} with r_{n+1} and with an estimate $\xi = \xi_{n+1}$ (say) of the limit. Show that

$$r_{n+1} = (x_{n+1} - x_n)/(x_n - x_{n-1})$$

   and

$$\xi_{n+1} = (x_{n+1} x_{n-1} - x_n^2)/(x_{n+1} + x_{n-1} - 2x_n)$$

$$= x_{n+1} - (\Delta x_n)/\delta^2 x_n \ .$$

   If the sequence $\{x_k\}$ converges linearly, then the transformed sequence $\{\xi_k\}$ will converge more rapidly: this is an example of convergence acceleration called **Aitken's δ^2-process**.

   (b) Verify that $\xi - \xi_{n+1} = (\epsilon_n^2 - \epsilon_{n+1}\epsilon_{n-1})/\delta^2 x_n$.

   (c) If the sequence $\{x_k\}$ is convergent and is generated by the iteration $x_{n+1} = \varphi(x_n)$, then show that $[\epsilon_{n+1}/\epsilon_n - \varphi(\xi)] \to -\frac{1}{2}\epsilon_n\varphi''(\xi)$, as $n \to \infty$. Hence demonstrate that the sequence $\{\xi_k\}$ converges linearly to ξ with asymptotic error ratio r^2, where $r = \varphi'(\xi)$. In other words, it converges twice as fast as $\{x_k\}$.

   (d) Given the sequence $\{x_k : k = 0, 1, \ldots, 10\}$ of problem 5(b), determine $\{\xi_k : k = 2, \ldots, 10\}$.

7. (a) Re-apply the Aitken δ^2-process to the sequence of values $\{1.375, 1.40655738, 1.41308921, 1.41407334, 1.41419731\}$ obtained in problem 6(d).

   (b) Another convergence accelerator of any sequence $\{x_k\}$ having linear convergence, is the ϵ-**algorithm**. In this, for $j = 0, 1, 2, \ldots$ in turn, one generates a triangular array $x_{j,k}$ from the recurrence:

$$x_{j,-1} = 0, \quad x_{j,0} = x_j, \quad x_{j,k+1} = x_{j-1,k-1} + 1/(x_{j-1,k} - x_{j,k})$$

   for $k = 0, 1, \ldots, (j-1)$. (Note that this is another example of a rhombus rule.) Only the results in the even-numbered columns are meaningful. It may be shown that if $x_j = a + \sum_{j=1}^{m} b_i \beta_i^j$ where $\beta_i \neq 1$, then $x_{j,2m} = a$ for all $j \geq m$. Furthermore, if $1 > |\beta_1| \geq |\beta_2| \geq \ldots \geq |\beta_m|$, so that the sequence $\{x_j\}$ converges linearly to the limit a, and if $k < m$, then

$\{x_{j,2k}\}$ tends to a at the rate of $|\beta_{k+1}|^j$. Check the first part of this theorem by taking $x_j = 5 - 4(-1)^j + 3(2)^j$, and showing that $x_{4,4} = 5$.

(c) Confirm that $\{x_{j,2}\}$ is the sequence $\{\xi_j\}$ generated from $\{x_j\}$ by Aitken's δ^2-process.

(d) Apply the ϵ-algorithm to the sequence $\{x_k: k = 0, 1, \ldots, 6\}$ of problem 5(b) to generate $x_{j,4}$ and compare the errors $\sqrt{2} - x_{j,4}$ with those of the sequence generated in part (a). About how rapidly are the two sequences converging?

Section 4.2.5

1. (a) Use the reversed Newton form of the quadratic interpolating polynomial to establish (4.2.10).

   (b) Use the error formula (3.2.1) to show that if Muller's method converges, then $\epsilon_{n+1} = C_n \epsilon_n \epsilon_{n-1} \epsilon_{n-2}$ where $C_\infty = -f'''(\xi)/6f'(\xi)$.

   (c) Confirm that this recurrence expression for the error implies that the order of convergence is the positive root $p = 1.84$ of $p^3 - p^2 - p - 1 = 0$.

   (d) Treating both f_n and $(x_n - x_{n-1})$ as small quantities show that (4.2.10) becomes

   $$x_{n+1} \cong x_n - f_n/(\hat{f}_{n-1,n} - f_{n-1}\hat{f}_{n-2,n-1,n}/\hat{f}_{n-1,n})$$

   which is almost identical with (4.2.11).

2. (a) Verify the root estimate $x = x_{n+1}$ of (4.2.11), using the result quoted in problem 1(b) of section 4.1.3.

   (b) Show that, if the points (x_k, f_k) for $k = n-2, n-1, n$ tend to coalesce (as in osculating interpolation) then (4.2.11) is equivalent to Halley's recurrence

   $$x_{n+1} = x_n - 2u(x_n)/[1 + u'(x_n)] \quad \text{where} \quad u(x) = f(x)/f'(x) \ .$$

   (See problem 3(c) of section 4.2.4.)

3. (a) By direct (rather than inverse) interpolation one can construct the osculating cubic polynomial $p_3(x)$ which interpolates $f(x)$ and $f'(x)$ at $x = x_{n-1}, x_n$; the value of $p_3''(x_n)$ can then be used as an approximation to the value of $f''(x_n)$ in the expression $u'(x) = 1 - f(x)f''(x)/[f'(x)]^2$ required in the single-point iteration formulae having cubic convergence (given in problem 3(c) of section 4.2.4), such as that of Tchebyshev's method. Show that

   $$p_3''(x_n) = 2(2f_n' + f_{n-1}' - 3\hat{f}_{n-1,n})/(x_n - x_{n-1}).$$

   (b) Another simpler approximation is given by

   $$f''(x_n)/f'(x_n) = [1 - u'(x_n)]/u(x_n) \cong 2(1 - \hat{u}_{n-1,n})/(u_{n-1} + u_n)$$

where $u_k = u(x_k)$. Obtain a two-point iteration formula relating x_{n+1} to x_n, u_n, x_{n-1}, and u_{n-1} by applying this approximation in Tchebyshev's method. Starting with $x_0 = 3$ and $x_1 = x_0 - u_0$, use this iteration formula to locate the root of the equation $f(x) \equiv \exp(x^2) - 50 = 0$. (This iteration, if it converges, does so with order $1 + \sqrt{2}$, whereas the approximation of part (a) would give an order of convergence of $1 + \sqrt{3}$.)

4. Consider the iterative scheme, starting with given initial x_0, represented for $i = 0, 1, 2, \ldots$ by

$$x_{2i+1} = x_{2i} - f_{2i}/f'_{2i}, \quad x_{2i+2} = x_{2i+1} - (x_{2i+1} - x_{2i})f_{2i+1}/(2f_{2i+1} - f_{2i}) .$$

This is Newton-Raphson iteration with the derivative evaluation omitted at each second step. Since its local rate of convergence can be shown to be quadratic just as for Newton-Raphson's method, it is more efficient than that method. Apply the scheme to the solution of $\exp(x^2) - 50 = 0$, starting at $x_0 = 3$.

Section 4.3

1. (a) The 'multiplicity-independent' form of Newton-Raphson's iteration is that applied to the location of the simple roots of $u(x) = 0$, where $u(x) = f(x)/f'(x)$. Express the resulting recurrence formula for x_{n+1} in terms of x_n, f_n, f'_n, and f''_n.

   (b) Likewise the 'multiplicity independent' form of the secant rule is that applied to the location of the roots of $u(x) = 0$. Write down the relation for x_{n+1} for this form in terms of values of x, f, and f'.

   (c) Supposing that $f(x)$, $f'(x)$, and $f''(x)$ all require roughly equivalent computational work, which of the above two 'multiplicity independent' forms would be more efficient?

2. (a) Show that in the neighbourhood of a root $x = \xi$ of multiplicity m, the iteration $x_{n+1} = x_n - mf(x_n)/f'(x_n)$ converges quadratically.

   (b) Apply the formulae (4.3.2) and (4.3.3) to the location of the (double) root of $f(x) = \sin(x) - 1 = 0$ starting with $x_0 = m_0 = 1$.

   (c) Apply the method to the location of a root close to zero of $x^4 - 8x^3 + 18x^2 - 8x + 1 = 0$.

3. Using a calculator or computer, explore the behaviour of $u(x) = f(x)/f'(x)$ where $f(x) = x^4 - 8x^3 + 18x^2 - 8x + 1$, close to the double root of $f(x)$ at $x = 2 + \sqrt{3}$. (The method of problem 2 will most likely fail to achieve a very accurate location of this root. Why is this?)

4. (a) Apply the secant rule to the location of the zero of $(x - 1)^2 = 0$, generating the sequence of root estimates up to x_{10}, starting with $x_0 = 0$, $x_1 = 0.1$.

   (b) The asymptotic error factor $r = \epsilon_{n+1}/\epsilon_n$ for the secant rule near a double root should be the positive root of $r(1 + r) = 1$. Do the numerical results bear this out?

(c) Generate the sequence $\{\xi_k: k = 2, \ldots, 10\}$ by applying Aitken's δ^2-process (problem 6, section 4.2.4) to the sequence $\{x_k\}$ of root estimates. Estimate its rate of convergence.

Section 4.4

1. (a) If $z^n = \pm(1 + b_1 z + b_2 z^2 + \ldots + b_{n-1} z^{n-1})$ and $B = \max_{1 \leqslant j \leqslant n-1} |b_j|$ is non-zero, show that any root of this equation satisfies $|z| < 1 + B$. (Hint: show that $|z| \geqslant 1 + B$ implies a contradiction.)

   (b) If the restriction of non-zero B is relaxed in part (a), show that the proposition remains true if any root is asserted to satisfy $|z| \leqslant 1 + B$.

   (c) Verify that any zeros of $p_n(x)$ as given by (4.4.1) satisfy the condition $|x| \leqslant c$, where c is given by (4.4.2). (Hint: convert the equation to a form consistent with that of part (a).)

2. Show that any zeros of $p_n(x)$ as given by (4.4.1) satisfy the condition $\mu^2/c \leqslant |x| \leqslant c$, where c is given by (4.4.2) and $\mu = |a_0/a_n|^{1/n}$.

3. (a) If $\sum_{j=0}^{n-1} |a_j| \leqq |a_n|$, prove from first principles that any zero of $p_n(x)$ as given by (4.4.1) has modulus $|x| \leqslant 1$.

   (b) If $\alpha = \sum_{j=0}^{n-1} |a_j|/|a_n| > 1$, then prove that any zero of $p_n(x)$ satisfies $|x| < \alpha$.

   These two results provide another simple method of bounding the zeros of a polynomial.

Section 4.4.1

1. (a) Verify that the sequence $\{(-1)^k \cos[(m - k)x]: k = 0, 1, \ldots, m\}$ satisfies the conditions of a Sturm sequence in $[0, \pi]$.

   (b) Confirm that Sturm's theorem applied to $\cos(mx)$ correctly evaluates the number of its real roots in $[0, \pi]$.

2. Show that $\{x^2 - 2a_1 x + a_0, x - a_1, a_1^2 - a_0\}$ where $a_0 \neq a_1^2$ represents a Sturm sequence for any real x.

3. Given that $\{f_k(x): k = 0, 1, \ldots, m\}$ is a Sturm sequence in $[a, b]$ and that $\{\varphi_k(x): k = 0, 1, \ldots, m\}$ is a sequence of functions which are all positive (or all negative) and non-zero in $[a, b]$, then verify that $\{\varphi_k(x)f_k(x): k = 0, 1, \ldots, m\}$ is also a Sturm sequence in $[a, b]$.

4. A sequence $\{g_k(x): k = 0, 1, \ldots, m\}$ meets all conditions for a Sturm sequence in $[a, b]$, save only that $g_0'(t)g_1(t) < 0$ for all zeros $x = t$ of $g_0(t)$. Confirm that $\{(-1)^k g_k(x): k = 0, 1, \ldots, m\}$ is a Sturm sequence in $[a, b]$.

5. (a) A sequence of polynomials $p_0(x), p_1(x), p_2(x), \ldots$ has the following properties:

   (i) $p_n(x)$ is a polynomial of genuine degree n;

(ii) there exist numbers A, B, and C dependent (possibly) on n but independent of x, such that

$$p_{n+1}(x) = (A+Bx)p_n(x) - Cp_{n-1}(x) ;$$

and

(iii) there exist numbers P, Q, and R dependent (possibly) on n but independent of x, and a quadratic polynomial $q(x)$ which is independent of n, such that

$$q(x)p_n'(x) = (P+Qx)p_n(x) - Rp_{n-1}(x) .$$

Verify that the Tchebyshev polynomials $T_0(x)$, $T_1(x)$, $T_2(x)$, ... form such a sequence, by identifying the values of A, B, C, P, Q, and R, given that $q(x) \equiv 1 - x^2$. (All sequences of what are called the classical orthogonal polynomials have these properties.)

(b) What conditions (if any) have to be met by A, B, C, P, Q, and R, and by $q(x)$, in order that $\{p_{m-k}(x): k = 0, 1, \ldots, m\}$ forms a Sturm sequence for all real x? Confirm that $p_n(x) \equiv T_n(x)$ meets these conditions.

(c) Show that the Tchebyshev polynomials of the second kind $U_0(x)$, $U_1(x)$, $U_2(x)$, ... where $U_n(\cos\theta) = \sin[(n + 1)\theta]/\sin\theta$, also form a sequence with the properties listed in part (a), and confirm that $\{U_{m-k}(x): k = 0, 1, \ldots, m\}$ is a Sturm sequence for all real x.

(d) Show that $\{T_m(x), T_{m-k}'(x): k = 0, 1, \ldots, (m-1)\}$ is a Sturm sequence for all real x. (Hint: $T_n'(x) = nU_{n-1}(x)$.)

Section 4.4.2

1. Show that the recurrence (4.4.5) terminates with a polynomial $f_m(x)$ of degree ≥ 1 if and only if this polynomial is a common factor of the polynomials $f_0(x) = p_n(x)$ and $f_1(x) = p_n'(x)$.

2. Find Sturm sequences for the following polynomials, and roughly locate any real roots: (a) $x^3 + x^2 - 4x + 1$; (b) $x^4 - 2x^2 + 8x - 22$.

3. Find the Sturm sequence for

$$p_4(x) = x^4 + 1.33x^3 + 0.667x^2 + 1.15x + 0.346$$

using correctly rounded floating-point arithmetic with 3 significant decimal digits.

4. (a) In any suitable programming language, write the body of a recursive procedure setbnds(a, b, va, vb, j), whose parameters are all called by value, which assigns suitable values for the bounds bracketing each real zero of a polynomial in the interval [a, b] to the non-local arrays *upper* and *lower* beginning at their jth elements. It should be assumed that the number of sign changes in the Sturm sequence for this polynomial at any

value of x is given by the value of the function $v(x)$, and that va and vb are the integer values of $v(a)$ and $v(b)$.

(b) What would be a suitable call on this procedure which assigns to the arrays values bracketing all real zeros, if it were known that they have magnitude less than the value assigned to $xlim$? What values would be assigned if $xlim = 6$, and $v(x)$ equals 3 for $x < -2$, equals 2 for $x < 2.06$ and is zero for $x > 2.69$.

5. Show that the recurrence (4.4.5) provides the partial denominators $q_j(x)$ of a continued fraction representation

$$f_1(x)/f_0(x) = \overset{m}{\underset{j=1}{K'}} (-1) \downarrow q_j(x)$$

where $q_m(x)$ is defined as $f_{m-1}(x)/f_m(x)$ and $f_{m+1}(x) = 0$. (In general, with $m = n$ and all the $q_J(x)$ linear in x, this is a J-fraction representation of the rational function f_1/f_0.)

Section 4.4.3

1. (a) If δ_0, δ_1, and δ_2 are all quantities small compared with unity, verify that the roots of $p_2(x) \equiv (1 + \delta_2)\beta^2 x^2 + 2(1 + \delta_1)\alpha\beta x + (1 + \delta_0)\alpha^2 = 0$ are approximately $-(\alpha/\beta)[1 \pm (2\delta_1 - \delta_0 - \delta_2)^{1/2}]$.

   (b) If forced cancellation is used in forming the numerical Sturm sequence for $p_2(x)$, show that the roots would be discriminated as double if $|2\delta_1 - \delta_0 - \delta_2| < N\epsilon$, approximately.

2. Find the number of sign changes of the Sturm sequence for $p_4(x) = x^4 - x^3 - 3x^2 + x + 2$, given in the text, at $x = -\infty, 0, 1.5, \infty$.

3. (a) If $\{f_k(x): k = 0, 1, \ldots, m\}$ is the sequence generated by the Euclidean algorithm with $f_0(x) = p_n(x)$ and $f_1(x) = p_n'(x)$, then show that $\{\phi_k(x) = f_k(x)/f_m(x): k = 0, 1, \ldots, m\}$ is a Sturm sequence.

   (b) Verify that the Sturm theorem applied to $\{f_k(x): k = 0, 1, \ldots, m\}$ counts the zeros of $\phi_0(x)$.

Section 4.4.4

1. (a) Let $p_2(x) = x^2 - (\alpha_1 + \alpha_2)x + \alpha_1\alpha_2$ where $\alpha_1\alpha_2 \neq 0$. If a factor $(x - \xi)$ is extracted by synthetic division, verify that the deflated polynomial is $(x + \xi - \alpha_1 - \alpha_2)$.

   (b) If $\xi = \alpha_1(1 - \beta\epsilon)$, show that the relative error in determining the other root of $p_2(x)$ is equal to $-(\alpha_1/\alpha_2)\beta\epsilon$ (and is therefore likely to be smaller if $|\alpha_1| < |\alpha_2|$).

2. (a) Let $p_3(x) = x^3 - (\alpha_1 + \alpha_2 + \alpha_3)x^2 + (\alpha_1\alpha_2 + \alpha_2\alpha_3 + \alpha_3\alpha_1)x - \alpha_1\alpha_2\alpha_3$, where $\alpha_1\alpha_2\alpha_3 \neq 0$. If the factor $(x - \xi)$ with $\xi = \alpha_1(1 - \beta\epsilon)$ is extracted

by synthetic division, verify that the deflated polynomial is approximately
$$p_2(x) = x^2 - (\alpha_2 + \alpha_3 + \alpha_1 \beta \epsilon) x + \alpha_2 \alpha_3 + (\alpha_2 + \alpha_3 - \alpha_1) \alpha_1 \beta \epsilon.$$

(b) Show that if $\alpha_2 \neq \alpha_3$, the zeros of this deflated polynomial are given (approximately) by

$$\alpha_2 + (\alpha_3 - \alpha_1) \alpha_1 \beta \epsilon / (\alpha_3 - \alpha_2), \quad \alpha_3 + (\alpha_2 - \alpha_1) \alpha_1 \beta \epsilon / (\alpha_2 - \alpha_3).$$

(c) If a cubic has 3 real zeros ξ_1, ξ_2, ξ_3 with $|\xi_1| \ll |\xi_2| \ll |\xi_3|$, and $\alpha_1, \alpha_2, \alpha_3$ represents some permutation of these zeros, then show that the relative errors induced in α_2 and α_3 due to relative error $\beta \epsilon$ in α_1 are least if $\alpha_1 \equiv \xi_1$ and greatest if $\alpha_1 \equiv \xi_3$.

(d) If $\alpha_2 = \alpha_3$, what are the zeros of the deflated polynomial $p_2(x)$ of part (a)?

3. The equation $10x^3 - 990x^2 + 550x - 1 = 0$ has roots $\{98.4413, 0.556874, 0.00182417\}$. Using correctly rounded floating-point arithmetic of precision 3, find these roots in (a) increasing and (b) decreasing order of magnitude, using deflation after each root is evaluated.

4. (a) If the roots of the equation $p_n(x) = 0$ are $\xi_1, \xi_2, \ldots, \xi_n$, then those of the reciprocal equation $q_n(x) \equiv z^n p_n(z^{-1}) = 0$ are $\xi_1^{-1}, \xi_2^{-1}, \ldots, \xi_n^{-1}$, so that the root of largest magnitude of the former equation is the reciprocal of the root of smallest magnitude of the latter. Find the roots of $z^3 - 550z^2 + 990z - 10 = 0$ in increasing order of magnitude, using deflation after each root is found, and employing correctly rounded arithmetic of decimal precision 3. Hence infer the roots of the cubic equation in problem 3.

(b) The process of synthetic division represented by the recurrence $a_0' = a_0, a_k' = a_k + a_{k-1}'/\xi$: $k = 1, 2, \ldots, n$ is called **backward deflation** of the polynomial $p_n(x) \equiv a_n x^n + a_{n-1} x^{n-1} + \ldots + a_0$ by a factor $(1 - x/\xi)$, and $p_n(x) - p_n(\xi)(x/\xi)^n = (1 - x/\xi)(a_{n-1}' x^{n-1} + a_{n-2}' x^{n-2} + \ldots + a_0')$, with $a_n' = p_n(\xi)\xi^{-n}$. Use this process in finding the roots of the cubic equation of problem 3 in decreasing order of magnitude, employing correctly rounded arithmetic of decimal precision 3.

(c) Assuming that all roots are precisely determined, show algebraically that the deflated polynomial of the reciprocal equation is identical with that obtained by backward deflation of the (original) equation. (Numerically, backward deflation is the preferable approach, and is stable, if the roots are evaluated in decreasing order of magnitude.)

5. The equation $x^3 - 20x^2 + 100x - 1 = 0$ has roots $\{10.3114, 9.67856, 0.0100201\}$. Using correctly rounded floating-point arithmetic of precision 3 find the roots (a) in increasing order of magnitude, using deflation after each root is evaluated, and (b) in decreasing order of magnitude using backward deflation (as described in problem 4(b) above).

6. (a) If the process of zero-suppression (as opposed to deflation) is applied in finding a zero of $f(x)$ after the roots $x = \xi_1, \xi_2, \ldots, \xi_m$ have been obtained, show that Newton's iterative method takes the form

$$x_{k+1} = x_k - f(x_k)/[f'(x_k) - f(x_k) \sum_{j=1}^{m} (x_k - \xi_j)^{-1}] .$$

(b) Apply this iteration to find the small zero $x = \xi_2$ of $x^3 - 20x^2 + 100x - 1$ after the approximate zero $\xi_1 \cong 10$ has been found. (Note that deflation would produce the quadratic polynomial $x^2 - 10x$ and $\xi_2 = 0$.)

Section 4.5.2

1. (a) Write an instruction sequence (in any programming language) which locates the minimum of a unimodular function $f(x)$ in an initial interval $[a, b]$ by the method of golden section, iterating whilst the span of the contracted interval is larger than some (positive) value *eps*. (Ignore the possibility that the function values at the two interval points of any interval may be equal, and assume that $b > a$.)

(b) Will the iterations terminate if *eps* is zero? If not, include a condition which will none the less force termination. Is such a condition unnecessary if *eps* is positive?

(c) Use the algorithm to find the first 3 contracted intervals in locating the minimum of $\max(2 - x^2, x - 4)$ in the initial interval $[1, 6]$.

(d) Use the algorithm to find the first 3 contracted intervals in locating the minimum of $\mathrm{abs}(x)$ in $[-1, 1]$. (Observe thereby what happens if the function values at the two internal points of an interval are equal.)

2. (a) Given three data (x_k, y_k) for $k = 0, 1, 2$, show that a quadratic polynomial interpolating these data has a stationary value at

$$x = \tfrac{1}{2}(x_0 + x_2 - \hat{y}_{0,2}/\hat{y}_{0,1,2}) .$$

(b) If $x_0 < x_1 < x_2$, and the minimal condition is satisfied by $y = f(x)$ at the point $x = x_1$, show that the position of this stationary value lies in the interval $[(x_0 + x_1)/2, (x_1 + x_2)/2]$.

PROBLEMS: CHAPTER 5

Section 5.1

1. (a) If $y = p_2(x)$ is the quadratic arc interpolating the points (x_0, y_0), $(x_1, y_1), (x_2, y_2)$, where $x_0 < x_1 < x_2$, as given by equation (3.2.2), verify that the integral mean of $p_2(x)$ over $[x_0, x_2]$ is

$$(x_2 - x_0)^{-1} \int_{x_0}^{x_2} p_2(x)dx = [(2h_1 - h_2)h_2 y_0 + (h_1 + h_2) y_1$$
$$+ (2h_2 - h_1)h_1 y_2]/(6h_1 h_2)$$

where $h_k = x_k - x_{k-1}$.

(b) If $h_1 = h_2$, confirm that the integral mean is given by Simpson's rule, namely $(y_0 + 4y_1 + y_2)/6$. What correction term has to be added to this value if $h_1 \neq h_2$?

(c) Putting $y(x) \equiv p_2(x)$, confirm that $\int_0^1 y(x)dx = [3y(1/3) + y(1)]/4$, and so prove that the 'three-eighths' rule:

$$\int_0^1 y(x)dx = [y(0) + 3y(1/3) + 3y(2/3) + y(1)]/8$$

is true for any quadratic integrand.

(d) Putting $y(x) \equiv p_2(x)$, find a formula for the value of $\int_0^1 y(x)dx$ depending on $y(0)$, $y'(0)$, and $y(1)$; hence deduce a symmetric formula that depends on the value of $y'(1)$ as well, and is true for any quadratic integrand.

Section 5.1.1

1. (a) Show by repeated integration by parts that

$$\int_a^b (b-x)(x-a)f''(x)dx = (b-a)[f(a) + f(b)] - 2\int_a^b f(x)dx$$

and thereby establish the expression (5.1.3) for the error of the trapezoidal rule.

(b) Verify the expression (5.1.6) for the error of the mid-ordinate rule (Hint: use integration by parts.)

2. (a) The quadrature formula (5.1.11) is said to be *symmetrical* if $\lambda_k^{(n)} = \lambda_{n-k}^{(n)}$ and $x_k + x_{n-k} = a + b$. Show that this property ensures that any integrand which is an odd function of $(x - c)$ is correctly integrated, where $x = c = (b + a)/2$ is the mid-point of the range of integration.

(b) Show that the order of error of a symmetric formula must be an even integer.

3. The Cotes formulae should converge with increase in n when applied to an integrand like $\sin(x)$, since its derivatives are all bounded in magnitude. Apply the Cotes (closed-type) formulae for $n = 1, 2, \ldots, 6$ to the evaluation of $(2/\pi)\int_0^{\pi/2} \sin(x)dx$ and list the actual error, and the implied value of ξ which would have to be used in (5.1.12) to give this error.

4. Repeat the calculations of problem 3, using the interlaced (open-type) point distribution, having weights as listed in Table 5.2, for $n = 0, 1, \ldots, 5$.

5. (a) Find the values of the weights in the following Cotes open-type formula:

$$\int_0^1 f(x)dx \cong \lambda_1 f(1/4) + \lambda_2 f(1/2) + \lambda_3 f(3/4)$$

by arranging that the formula is precise if $f(x) \equiv x^k$ for $k = 0, 1, 2$. (This approach is called the **method of undetermined coefficients**.) Check that the formula remains precise if $f(x) \equiv x^3$.

(b) This formula has an error depending on $f^{iv}(\xi)$ for some $\xi \in (0,1)$. Find the expression for the error by placing $f(x) \equiv x^4$.

6. (a) A sequence of functions $\{g_k(t): k = 0, 1, 2, ...\}$ is developed by the recurrence

$$g_{k+1}(t) = \int_{-1}^{t} [g_k(s) - \bar{g}_k] ds, \quad \text{where} \quad \bar{g}_k = (1/2) \int_{-1}^{1} g_k(t) dt$$

for $k = 0, 1, 2, ...$ Show that if $g_0(t)$ is an even function of t, integrable over $[-1, 1]$, then $g_1(t)$ is an odd function, and $g_2(t)$ an even function, of t and that $g_1(-1) = g_1(1) = g_2(-1) = g_2(1) = 0$. Hence prove by induction, that if k is an odd/even integer, then $g_k(t)$ is an odd/even function of t which (at least for $k > 0$) vanishes at $t = \pm 1$.

(b) If $g_0(t) = (1 - t^2)/2$ then $\{g_k(t)\}$ becomes a sequence of polynomials of ascending degree (related to the **Bernoulli polynomials**); calculate expressions for $g_1(t)$ and $g_2(t)$ and values for $\bar{g}_0, \bar{g}_1, \bar{g}_2$.

(c) Show by successive integration by parts that if $F(t) \in C^\infty[-1, 1]$, then with $g_k(t)$ defined as in part (a),

$$\int_{-1}^{+1} g_0(t) F''(t) dt$$

$$= \sum_{k=0}^{m-1} [F^{(2k+1)}(1) - F^{(2k+1)}(-1)] \bar{g}_{2k} + \int_{-1}^{+1} g_{2m}(t) F^{(2m+2)}(t) dt$$

for any positive integer m.

7. (a) If $F(t) = \frac{1}{2}\cos(xt)/\sin\alpha$, where α is a constant and $0 < \alpha < \pi$, verify that $F^{(2k+1)}(1) = -F^{(2k+1)}(-1) = \frac{1}{2}(-1)^{k+1}\alpha^{2k+1}$.

(b) Putting $g_0(t) = \frac{1}{2}(1 - t^2)$, verify that $\int_{-1}^{+1} g_0(t) F''(t) dt = \cot(x) - (1/\alpha)$, and so from the result of problem 6(c) above, there exists a series expansion:

$$\alpha \cot(\alpha) = 1 + \sum_{k=0}^{\infty} (-1)^{k+1} \bar{g}_{2k} \alpha^{2k+2} .$$

(c) Given that

$$\alpha \cot(\alpha) = 1 + 2\alpha^2 \sum_{n=1}^{\infty} (n^2 \pi^2 - \alpha^2)^{-1}$$

show that

$$\bar{g}_{2k} = (-1)^{k+1}(2/\pi^{2k+2}) \sum_{n=1}^{\infty} n^{-(2k+2)}$$

and hence express $\bar{g}_{2k}$ in terms of the Bernoulli numbers given by (2.5.5).

8. (a) Use (5.1.3) and the result of problem 6(c) above, with $g_0(t) = \frac{1}{2}(1 - t^2)$, to verify that $-E_T = \sum_{j=1}^{m-1} [f^{(2j-1)}(b) - f^{(2j-1)}(a)](b-a)^{2j}(\bar{g}_{2j-2}/2^{2j}) + R_m$, which is an expression for the error of the trapezoidal rule.

(b) If it is true that, for $k = 0$, $g_{2k}(t)$ does not change sign in $(-1, 1)$ and $g_{2k+1}(t)$ changes sign in $(-1, 1)$ only at $t = 0$, then prove that the same is true for $k = 1, 2, ...$

(c) Using the result (b) show that the remainder term R_m in (a) is expressable as

$$R_m = f^{(2m)}(\xi)(b-a)^{2m+1}(\bar{g}_{2m-2}/2^{2m})$$

for some $\xi \in (a, b)$.

Section 5.1.2

1. (a) Using the expression for the error E_T as given by the result of problem 8(a) of section 5.1.1, and summing over each segment $[x_{j-1}, x_j]$ for $j = 1, 2, \ldots, N$, show that $-E_{N,1} = [f'(b) - f'(a)](b-a)\bar{g}_0/(2N)^2 + [f'''(b) - f'''(a)](b-a)^3\bar{g}_2/(2N)^4 + \ldots$.

   (b) Identifying the values of $\bar{g}_{2k}$ with the Bernoulli numbers B_{2k+2} as in problem 7(c) of section 5.1.1, establish the Euler-Maclaurin summation formula, (5.1.23).

2. The expression $s_m(\xi) = m^{-1}\sum_{j=0}^{m-1} f[(j+\xi)/m]$ for $0 \leq \xi \leq 1$ is called a **Riemann sum**, and functions which are Riemann integrable over $[0,1]$ are such that $\lim_{m\to\infty} \{s_m(\xi)\} = \int_0^1 f(x)dx = I$. Prove that any $(n+1)$-point composite quadrature rule of N equal segments applied to the estimation of this integral converges to the value I as $N \to \infty$ provided merely that $f(x)$ is Riemann integrable. (This result of course tells us nothing about the rate of convergence.)

Section 5.1.3

1. Using the Romberg method, estimate the value of $(2/\pi)\int_0^{\pi/2} \sin(x)dx$ after completing 4 or 5 rows of the tableau, and compare the actual accuracy of this estimate with its *inferred* accuracy – that is, the error inferred from the tabulated values alone without knowledge of the precise value.

2. Another way of determining the convergence of the Romberg method is to require that two consecutive elements on a diagonal differ by less than some prescribed tolerance, i.e. $|I_{j,k+1} - I_{j-1,k}| < \epsilon$. Show that this implies that $|I_{j,k} - I_{j-1,k}| < \epsilon$, and that it is therefore a (slightly) more severe test.

3. (a) Verify that Richardson extrapolation applied to Simpson's rule leads to Milne's formula.

   (b) The nine-point Cotes formula (corresponding to $n = 8$) has 10th-order error, but negative weights. Does the nine-point rule in the Romberg scheme have negative weights? and what is its order of error?

4. Apply Richardson extrapolation to the mid-ordinate rule to derive 4th-order single-element formulae of the following forms, and find the weighting coefficients:

$$\int_0^1 f(x)dx \cong \lambda_1 f(1/4) + \lambda_2 f(1/2) + \lambda_3 f(3/4)$$

$$\cong \mu_1 f(1/6) + \mu_2 f(1/2) + \mu_3 f(5/6) .$$

5. (a) In estimating the error $E_{N,1}$ incurred by the composite trapezoidal rule in evaluating $(2b)^{-1}\int_{-b}^{b}(1+x^2)^{-1}dx$ for large $b \gg 1$, the derivatives of $f(x) \equiv (1+x^2)^{-1}$ at $x = \pm b$ are approximately those of x^{-2} at $x = \pm b$. Show that, with this approximation,

$$E_{N,1} \cong (1/b^2)[B_1(2/N)^2 - B_2(2/N)^4 + B_3(2/N)^6 - + \ldots]$$

and that this series is semi-convergent. (Hint: use equation (2.5.6).)

(b) Show that in evaluating $(2/\pi)\int_0^{\pi/2}\sin x\, dx$, the error $E_{N,1}$ incurred by the composite trapezoidal rule is a convergent series in $(1/N^2)$ for $N \geq 1$.

6. (a) Accepting the development of $E_{N,1}$ as a series in $1/N^2$ given by (5.1.21), show that the lowest order term in the error $E_{N,k}$ of the estimates in the kth column of the Romberg tableau is $a_k/(2^{k-1}N^2)^k$. (Hint: truncate the series to k terms and apply (3.7.1).)

(b) Show that the values of $(2^{k-1}N^2)^k$ increase monotonically with k, if the value of N is that appropriate to estimates in the jth row of the Romberg tableau. (Hence if a_k decreases rapidly with k, so that this lowest order term in the series for $E_{N,k}$ is numerically dominant — which as is clear from problem 5 is not always true — the error decreases as k is increased.)

7. In an instruction sequence for the Romberg method, the elements $I_{j,k}$ are mapped onto a vector r in such a way that $r_{j-k+1} = I_{j,k}$, so that rows are over-written on r in reverse order. Assuming that the $(j-1)$th row has been calculated and the results assigned to $r_1, r_2, \ldots, r_{j-1}$, and that $I_{j,1}$ has been assigned to the variable I, write an instruction sequence that assigns the jth row to r for $l = j, \ldots, 2, 1$, jumping however to a label *CONVERGED* if $|I_{j,k} - I_{j-1,k}| \leq eps$, with the value of I currently assigned equal to $I_{j,k}$ (or $I_{j,k+1}$).

8. Using the analogue of the Neville algorithm for the evaluation of an interpolating continued fraction (see problem 10 of section 3.5) show that with this form of interpolation, the recurrence for $I_{j,k}$ becomes the rhombus rule:

$$I_{j,k+1} = I_{j,k} + [(4^k - 1)(I_{j,k} - I_{j-1,k})^{-1} - 4^k(I_{j,k} - I_{j-1,k-1})^{-1}]^{-1},$$

with $I_{j,0}$ defined as 0 or ∞ (depending on the sequence of orders of rational function interpolation required). Why, in general, may this make only a small difference to the tableau of values (as compared with those using polynomial interpolation)?

Section 5.1.5

1. Verify the expression (5.1.25) for the integral of a cubic piece, and the values of the end-derivatives given by (5.1.28).

2. (a) Show that Simpson's rule is exact when applied to the integration over $[x_{k-1}, x_{k+1}]$ of two consecutive pieces of a quadratic spline with equispaced nodes $x_k = x_{k-1} + h$. Hence the composite Simpson's rule can

be regarded as derived by assuming that the integrand values are inter-polated by *any* quadratic spline (having an even number of pieces) with equi-spaced nodes at the points of integrand evaluation.

(b) In order to extend this result to a quadratic spline $y = y(x)$ with an *odd* number of pieces ($n = 2m + 1$, say), it is necessary to introduce some assumption about the end-condition to be satisfied by the spline. Assuming that this is chosen to minimise the variance of $y''(x)$ over $[x_0, x_{2m+1}]$, then it is found that

$$h(y'_{2m+1} - y'_0) = [(2m + 1)/m(m + 1)] \sum_{k=0}^{2m} (-1)^k (k - m) \Delta y_k .$$

What is the form of quadrature formula which precisely integrates such a quadratic spline over $[x_0, x_{2m+1}]$? Confirm that it corresponds with the three-eighths rule if $m = 1$ (i.e. $n = 3$), and evaluate the weights for $m = 2$ (i.e. $n = 5$), checking that it correctly integrates $y(x) \equiv x^j$ for $j = 0, 1, 2$, and 3 (and so is of fourth order).

3. (a) What quadrature rules result from the truncated Gregory formula (5.1.29) appropriate to Bessel interpolation if the number of points is either 3 or 4?

(b) Truncating the Gregory formula (5.1.30) at the fourth difference terms, confirm that the resulting quadrature formula is identical with the Cotes rules (Table 5.1) for $n = 4$ and 5.

4. (a) The following table lists the acceleration $a(t)$ recorded at time t on a vertically ascending sounding rocket:

t (secs.)	0	1	2	3	4	5	6	7	8
$a(t)$ (m/s²)	150	166	181	200	223	257	314	422	662.

Estimate the speed achieved after 8 seconds by (i) the composite trapezoidal rule, (ii) the composite Simpson's rule, and (iii) the truncated Gregory formula (5.1.29) appropriate to Bessel interpolation.

(b) Tabulate the values of $a'(t)$ used in Bessel piecewise cubic interpolation and hence, using (5.1.25), estimate the speed $v(t)$ achieved at $t = 1, 2, \ldots, 8$ seconds after launch.

(c) To find the height, it is necessary to form the repeated integral of $a(t)$, that is, $\int_0^t v(t) dt$. Show that the height reached at $t = t_n$ can be expressed as $\int_0^{t_n} (t_n - t) a(t) dt$. Evaluate this integral by the same three methods which are used in part (a), so as to estimate the height reached after 8 seconds.

(d) Since values of $v(t)$ and $v'(t) = a(t)$ are tabulated in part (b), another way to estimate height is to apply Hermitian piecewise cubic interpolation to these data and to integrate $v(t)$ using (5.1.25). Tabulate the height at 1-second intervals by this method.

5. (a) Express the end-conditions for a cubic spline $y(x)$ corresponding to (i) quadratic end-pieces, and (ii) the 'not-a-knot' condition as equations connecting values of y_k' with the data, and specialise these for equi-spaced nodes. (Hint: see problem 3 of section 3.4.)

   (b) Solve the system of equations to find the cubic spline fitting the acceleration data of problem 3, using one or the other of these end-conditions, and estimate the speed of the rocket at $t = 8$ secs.

   (c) Compare the values of $a'(t)$ at $t = 0$ and 8 secs obtained from this solution with the 'rough' values for the end-derivatives of cubic splines given in the text.

Section 5.2.1

1. Derive from first principles (a) the 3-point (6th-order) Gauss-Legendre formula, (b) the 4-point (6th-order) Lobatto formula, and (c) the 4-point (6th-order) Tchebyshev formula.

2. Assuming that the error is equal to $cf^{(m)}(\xi)/m!$ for some $|\xi| < 1$ where c is a constant, find the error in each of the three 6th-order formulae of problem 1. (Hint: integrate $f(x) \equiv x^6$.)

3. Use the 3-point (6th-order) Gauss-Legendre formula to estimate the value of $(2/\pi) \int_0^{\pi/2} \sin\theta \, d\theta$, and find the value of ξ in the expression for $E_G^{(2)}$ corresponding to the error of this estimate.

4. (a) If $n \gg 1$, verify from (5.2.5) that $E_G^{(n)} \cong (\pi/2) f^{(2n+2)}(\xi)/[2^{2n+2}(2n+2)!]$.

   (b) The error of the quadrature rule in the mth column of the Romberg scheme applied to $(1/2) \int_{-1}^{+1} f(x) dx$ is equal to $(-1)^m B_m f^{(2m)}(\xi)/[2^{m(m-3)}(2m)!]$. Show that this is smaller than the error of the Gaussian formula of the same order if $m \gg 1$ (assuming the value of ξ is the same). Of course the rule employs many more function evaluations (equal to $2^{m-1} + 1$).

5. Using the recurrence formulae (5.2.6) for Legendre polynomials, find the optimal points of integrand evaluation for the 4-point Gauss-Legendre and the 5-point Lobatto formulae.

6. The following table lists the magnitudes of the errors $|E_G^{(n)}|$ of Gaussian quadrature, and those of the estimate I_{nn} in the Romberg method, applied to the integral $I = \pi^{-1} \int_0^\pi \exp(x) \cos(x) dx$. Supposing that the calculation proceeds by increasing n by unit increments, graph the error against the running total of number of integrand evaluations in each method, and determine the more efficient technique —

n	2	3	4	5	6	7
$|E_G^{(n)}|$	8.5×10^{-2}	1.8×10^{-2}	5.0×10^{-5}	5.7×10^{-6}	4.7×10^{-9}	3.6×10^{-10}
$|I - I_{nn}|$	1.5×10^{-1}	1.9×10^{-2}	2.4×10^{-5}	2.8×10^{-7}	2.2×10^{-11}	

Section 5.2.2

1. Find the points of integrand evaluation used in the 7-point Patterson formula.

2. (a) A sequence of polynomials $\{p_n(x): \ n = 0, 1, 2, ...\}$ of successively increasing degree is said to be **orthogonal** on an interval $[a, b]$ with weight $w(x)$, where $w(x)$ is positive over (a, b), if

$$\int_a^b w(x)\, p_n(x)\, p_m(x)\, \mathrm{d}x = 0 \quad \text{for} \quad n \neq m \ .$$

The Legendre polynomials $P_n(x)$ of genuine degree n are orthogonal on $[-1, 1]$ with unit weight. Show that the $(n + 1)$ points of evaluation of the Gaussian quadrature formula are the $(n + 1)$ zeros of $P_{n+1}(x)$. (Hint: use (5.2.8) with $i = 0$.)

   (b) If $p_n(x)$ is a polynomial of degree n from a sequence orthogonal on $[a, b]$ with weight $w(x)$, and $q_m(x)$ is a polynomial of degree $m < n$, show that $\int_a^b w(x)\, p_n(x)\, q_m(x)\mathrm{d}x = 0$.

   (c) The n zeros of a polynomial $p_n(x)$ of degree n from a sequence orthogonal on $[a, b]$ are all real and distinct, and all lie in (a, b). Prove this by supposing that $p_n(x)$ has only $m < n$ changes of sign in (a, b) and that $q_m(x)$ is a polynomial of degree m which likewise changes sign at the same points in (a, b); then showing that this leads to a contradiction.

3. (a) Confirm that if $x_0, x_1, \ldots, x_n$ are the $(n + 1)$ zeros of $P_{n+1}(x)$, then the $(n + 1)$ conditions of (5.2.8) are satisfied if no points are fixed (i.e. $i = 0$).

   (b) If $x_0 = -x_n = -1$ are regarded as fixed, then verify that if $x_1, x_2, \ldots, x_{n-1}$ are selected as the $(n - 1)$ zeros of $P_n'(x)$, the conditions of (5.2.8) are satisfied with $r = 0, 1, \ldots, (n - 2)$.

4. (a) Given that the points of integrand evaluation in (5.2.3) are the $(n + 1)$ zeros of $P_{n+1}(x)$, show that the Christoffel numbers are

$$\lambda_k = [2P_{n+1}'(x_k)]^{-1} \int_{-1}^{+1} (x - x_k)^{-1} P_{n+1}(x)\, \mathrm{d}x \ .$$

   (b) If $p_n(x)$ is any polynomial of degree n which has a zero at $x = x_\nu$, say, prove that $[p_n(x) - p_n'(x_\nu)(x - x_\nu)]/(x - x_\nu)^2$ is a polynomial of degree $(n - 2)$.

   (c) Show further that $\lambda_k = \frac{1}{2}[P_{n+1}'(x_k)]^{-2} \int_{-1}^{+1} (x - x_k)^{-2} [P_{n+1}(x)]^2 \mathrm{d}x$, and therefore that the Christoffel numbers are all positive. (Hint: use part (a) and the orthogonality property of the Legendre polynomial.)

5. The **Weierstrasse approximation theorem** states that: if $f(x)$ is continuous on a finite interval $[a, b]$ then, given any $\epsilon > 0$, there exists an integer $N \equiv N(\epsilon)$ and a polynomial $p_N(x)$ of degree N such that $|f(x) - p_N(x)| < \epsilon$ for all x in $[a, b]$. Use this to show that the fact that the Christoffel numbers are all positive (see problem 3) ensures that the error $E_G^{(m)}$ of Gaussian quadrature converges

to zero with increase of n if $f(x) \in C^0[-1, 1]$. (No general theorems are known guaranteeing the existence of the nodes x_j for Patterson's method as $n \to \infty$, nor can it be proved that the weights are all positive, although they appear to be so from calculation. Thus this result cannot be extended, other than tentatively, to Patterson's method.)

6. Use the result of problem 1 of section 3.7.1 to show that if $f(x) \in C^{2n+2}[-1, 1]$ then the error of the Gauss-Legendre formula is given by

$$E_G^{(n)} = \tfrac{1}{2} f^{(2n+2)}(\xi) \int_{-1}^{+1} [P_{n+1}(x)]^2 d x / [(2n + 2)! \hat{P}_{n+1}^2]$$

for some $\xi \in [-1, 1]$, where $\hat{P}_{n+1}$ is the coefficient of x^{n+1} in the Legendre polynomial $P_{n+1}(x)$.

7. (a) Define $q_0(x) \equiv 1$ and $q_{i+1}(x) = (x - x_i) q_i(x)$ for $i = 0, 1, \ldots, n$ so that $q_{n+1}(x) = \prod_{i=0}^{n} (x - x_i)$. Suppose further that $q_i(x) = \sum_{m=0}^{} q_{i,i-m} P_m(x)$ where $P_m(x)$ is the Legendre polynomial of degree m. Show from (5.2.6) that the array of coefficients $q_{i,j}$ can be generated by placing $q_{i,-1} = 0$, $q_{i,0} = 2^i (i!)^2 / (2i)!$ for $i = 0, 1, \ldots, (n+1)$ and using the recurrence $q_{i+1,j} = [(i - j + 2) q_{i,j-2}/(2i - 2j + 5)] - x_i q_{i,j-1} + [(i - j + 1) q_{i,j}/(2i - 2j + 1)]$ for $j = 1, 2, \ldots, (i + 1)$, and $i = 0, 1, \ldots, n$. (Values of $q_{i,i+1}$ can be arbitrary, but finite, as they only occur multiplied by zero.)

   (b) Write an instruction sequence which assigns the value of $q_{n+1,j}$ to a vector q_j for $j = 0, 1, \ldots, (n + 1)$, given the vector $x_0, x_1, \ldots, x_n$ and the value of n.

   (c) If the values of $x_0, x_1, \ldots, x_n$ are symmetrically placed about $x = 0$ so that $x_k = -x_{n-k}$, then certain coefficients $q_{n+1,j}$ of the expansion for $q_{n+1}(x)$ should vanish. Which are these coefficients?

   (d) If $q_n^{(k)}(x) = q_{n+1}(x)/(x - x_k)$ for any $0 \leqslant k \leqslant n$, so that $q_n^{(k)}$ is a continued product omitting the factor $(x - x_k)$, and $q_n^{(k)}(x) = \sum_{m=0}^{n} q_{n-m}^{(k)} P_m(x)$, verify that $q_0^{(k)} = q_{n,0}$ and defining $q_{-1}^{(k)} = 0$, $q_j^{(k)} = (2n - 2j + 1)[q_{n+1,j} + x_k q_{j-1}^{(k)} - (n - j + 2) q_{j-2}^{(k)}/(2n - 2j + 5)]/(n - j + 1)$ for $j = 1, 2, \ldots, n$.

8. (a) The weighting coefficients of any quadrature formula based on polynomial interpolation of the integrand are given by (5.1.1) in terms of integrals of the shape functions s_{nk} which are expressed by (3.2.5). If the range of integration $[a, b]$ is supposed to be mapped on to $[-1, 1]$ as in (5.2.1), and one places $s_{nk}(x) = \sum_{m=0}^{n} r_{n-m}^{(k)} P_m(x)$, say, where $P_m(x)$ is the Legendre polynomial of degree m, then show that the weights λ_k in the expression for the integral mean are given by $\lambda_k = r_n^{(k)}$.

   (b) Using the results of problem 6(d), write an instruction sequence which assigns the values of $r_n^{(k)}$ to a vector l_k for $k = 0, 1, \ldots, n$, given the

value of n, the vector $x_0, x_1, \ldots, x_n$ of the points of integrand evaluation, and the vector of coefficients $q_0, q_1, \ldots, q_n$ of the expansion
$$\prod_{n=0}^{n} (x - x_i) = \sum_{m=0}^{n+1} q_{n+1-m} P_m(x).$$ (This is the vector derived in problem 6(b).)

(c) If $x_0, x_1, \ldots, x_n$ are the $(n+1)$ zeros of $P_{n+1}(x)$ used in Gauss-Legendre quadrature and $n = 4$, what is the weight associated with the integrand value at $x = 0$?

Section 5.2.3

1. If $I(n)$ denotes the estimate of the integral by the transformation method using n segments, show that for $n \geq 1$,

$$I(4n) = \tfrac{1}{2} \{I(2n) + (2n)^{-1} \sum_{j=1}^{n} \phi'(t_{2j-1})[f(x_{2j-1}) + f(x_{4n-2j+1})]\}$$

where $t_k = (\tfrac{1}{2} k/n) - 1$ and $x_k = \phi(t_k)$.

2. Using correctly rounded floating-point arithmetic to base b and precision p, what is the largest value of $|t|$ in $[-1, 1]$ for which the value of $\phi(t)$ as given by (5.2.10) is representable as unequal to unity? Evaluate this limit if $c = 2, b = 2$, and $p = 24$. What is the value of $\phi'(t)$ at that limit?

3. (a) If $x = \phi(t) = \tanh[mt/(1 - t^2)^{1/2}]$ and $f(x) \sim A(1 - x)^{\nu-1}$ as $x \to 1$, where $\nu > 0$, then transform $\int_{-1}^{+1} f(x)dx$ into $\int_{-1}^{+1} g(t)dt$, and show that $g(t)$ and all its derivatives vanish at $x = 1$.

(b) Confirm that this transformation leads to the quadrature formula

$$(1/2)\int_{-1}^{+1} f(x)\,dx = (m/n) \sum_{j=1}^{n-1} (1 - t_j^2)^{-3/2}(1 - x_j^2)f(x_j) + \epsilon_n$$

where $t_j = (2j - n)/n$ and $x_j = \phi(t_j)$.

(c) Evaluate the weights and points of integrand evaluation (x_j) of the formula corresponding to $n = 16, m = 4$; and verify that the sum of the weights is unity correct to the 6th decimal place. (Note the curious distribution of points of integrand evaluation.)

4. Verify that $\int_0^{\infty} x^{\alpha-1}\exp(-px)dx = p^{-\alpha}\int_0^1 [\ln(1/t)]^{\alpha-1}dt$ where p and α are positive constants, and apply a transformation method to evaluate this integral as a function of p for $\alpha = 0.25$.

Section 5.3

1. If the error in the integral-mean over each segment of the range is less in magnitude than some prescribed value, show that the resulting aggregated integral mean over the entire range is less than the same prescribed value.

2. Apply the adaptive trapezoidal rule to the evaluation of $\int_0^{+1}\exp[(-16t^2/(1 - t^2)]dt = 0.2122 \ldots$ with required accuracy $\epsilon = 0.005$.

3. (a) Show that the second central difference $\delta^2 f(x) = f(x + h) + f(x - h) - 2f(x)$ is approximately equal to $f''(x)h^2$. What would be the approximate value of the 4th central difference $\delta^4 f(x)$?

  (b) What limit should apply to the fourth difference $\delta^4 f(x)$ if the error of the integral mean estimated by the composite Simpson rule applied to the 2 segments $[x - 2h, x]$, $[x, x + 2h]$ is to be less than $\epsilon/(b - a)$?

  (c) In the adaptive Simpson's rule, one has access to 5 successive integrand values $f_0, f_1, \ldots, f_4$ at the end-points of 4 equal segments. If the error of the integral mean is *not* acceptable, it is proposed either to bisect the current interval and recursively estimate the integral in each bisected sub-interval, or to apply the Romberg method to the current interval. The latter is to be preferred if $|I_S(2) - I_S(1)|$ is less than some stipulated positive fraction $(k < 2/3)$ of $|I_T(4) - I_T(2)|$, where $I_S(n)$ and $I_T(n)$ are the composite Simpson and trapezoidal rule estimates applied to n segments of the current interval. Express this criterion as an inequality between the integrand values.

4. Let f_0, f_1, f_2 be the integrand values at $x = x_1 - \sqrt{3/5}h, x_1, x_1 + \sqrt{3/5}h$, regarded each as the mid-points of three segments of the interval $[x_1 - h, x_1 + h]$. Show that if Richardson extrapolation is applied to the values of the integral mean f_1, and that derived from the composite mid-ordinate rule for these three segments, the extrapolated estimate is the 3-point Gauss-Legendre formula.

5. (a) In Robinson's method, the integration interval $[x_1 - h, x_1 + h]$, say, is partitioned into 3 contiguous sub-intervals each centred about one of the 3 points $x_1 - \sqrt{3/5}h, x_1, x_1 + \sqrt{3/5}h$ at which the integrand is evaluated in the Gauss-Legendre 3-point quadrature formula (applied to the un-segmented interval). This 3-point formula is then applied to each of the 3 sub-intervals: if the nine points of integrand evaluation thereby involved are expressed as $x_1 + \alpha_k h$ for $k = 0, 1, \ldots, 8$ with $\alpha_0 < \alpha_1 < \ldots < \alpha_8$, what are the values of α_k?

  (b) Suppose that $I^{(6)}$ is the estimate by the 3-point Gauss-Legendre formula of the integral mean of some stipulated integrand $f(x)$ over the unsegmented interval, and let I_1, I_2, I_3 be the integral means of $f(x)$ over the component sub-intervals (arranged in order of increasing x, say) estimated by the same formula. Since $I^{(6)}$ is obtained from a sixth-order formula it follows that, on regarding $f^{vi}(x)$ to be constant over the unsegmented interval, a more accurate estimate of the integral mean over $[x_1 - h, x_1 + h]$ can be constructed by Richardson extrapolation, which is expressible as $I^{(8)} = \mu_0 I^{(6)} + \mu_1 I_1 + \mu_2 I_2 + \mu_3 I_3$, say. Find the values of $\mu_0, \mu_1, \mu_2, \mu_3$ and confirm (by, for instance, integrating successive integer powers of x) that $I^{(8)}$ is an eighth-order estimate.

  (c) A tenth-order estimate of the overall integral mean can be derived by integrating the polynomial interpolating the integrand $f(x)$ at the nine points $x_1 + \alpha_k h$, for $k = 0, 1, \ldots, 8$. This estimate can be expressed as $I^{(10)} = \sum\limits_{k=0}^{8} \lambda_k f(x_1 + \alpha_k h)$ where it is found that $\lambda_0 = \lambda_8 = 0.06169388063$, $\lambda_1 = \lambda_7 = 0.10838422910$, $\lambda_2 = \lambda_6 = 0.03984636033$, $\lambda_3 = \lambda_5 = 0.17520903531$, $\lambda_4 = 0.22973298923$. If the error of $I^{(8)}$ is judged as $I^{(10)} - I^{(8)}$ and expressed as $\sum\limits_{k=0}^{8} \epsilon_k f(x_1 + \alpha_k h)$, what are the values of ϵ_k?

(d) Find the errors of $I^{(6)}$, $I^{(8)}$ and $I^{(10)}$ as applied to the estimation of the integral mean of $(1 + x^2)^{-1}$ over $[-1, 1]$. What are the errors in $I^{(6)}$ and $I^{(8)}$ which would be judged by accepting $I^{(8)}$ and $I^{(10)}$, respectively, as the 'precise' integral mean?

Section 5.4

1. (a) If the matrix C of the bilinear form (5.4.3) is skew-symmetric, then show that its value can be expressed as $\sum\limits_{j=0}^{n} \sum\limits_{k=j+1}^{n} c_{jk}(y_j x_k - x_j y_k)$, in terms of the coefficients c_{jk} above the diagonal (that is, for $k > j$).

   (b) Show that if C is skew-symmetric, then the bilinear form $y^T C x$ is invariant to any rotation of the (x, y) axes.

   (c) Verify that $J = I - \frac{1}{2}(x_n y_n - x_0 y_0) = \frac{1}{2}\int_\gamma [y(s) x'(s) - y'(s) x(s)] ds$ is also invariant under any rotation of the axes; and that if $r = r(\theta)$ is the polar equation of the curve γ, then $J = -(1/2)\int_\gamma r^2(\theta) (d\theta/ds) ds$.

2. (a) Let P_i denote the point with Cartesian coordinates (x_i, y_i). Then show that $\frac{1}{2}(y_j x_k - x_j y_k)$ is the expression for the area of the triangle $OP_j P_k$ where O is the origin of the coordinate system, being positive if O, P_j, P_k are ordered in the clockwise sense round the triangle. (This result gives a meaning to each component term $c_{jk}(y_j x_k - x_j y_k)$ of the bilinear form as some fraction of a triangular area.)

   (b) Show that the value of $K = J - \frac{1}{2}(y_0 x_n - x_0 y_n)$ represents the area between the curve γ and the straight line joining its end-points.

   (c) If K is expressed as a bilinear form $y^T C' x$. say, and c'_{jk} are the coefficients of the matrix C', then verify that all the coefficients are the same as those of C except that $c'_{0n} = c_{0n} - (1/2), c'_{n0} = c_{n0} + (1/2)$, so that its value (i.e., the area K) is also invariant to any rotation of the axes. This particular area K is also unaffected by the position of the origin of the axis system: what condition must therefore be satisfied by the coefficients c'_{jk}?

3. (a) Let P_0, P_1, and P_2 denote any three consecutive coordinates of γ: then, using the result of problem 2(b) above, interpret (5.4.6) as implying that the area between the arc $P_0 P_1 P_2$ of γ and the straight line $P_0 P_2$ is $(4/3)$ times the triangular area $P_0 P_1 P_2$.

   (b) Since this area is invariant under any change of axes, let the origin O be the mid-point of $P_0 P_2$ and let the y-axis be the line OP_1. Show from

Simpson's rule that the area between the quadratic $y = p_2(x)$ interpolated through P_0, P_1, P_2 and the line $P_0 P_2$ is $(4/3)$ times the triangular area $P_0 P_1 P_2$. (The curve $y = p_2(x)$ is of course a parabola.)

(c) Is Brun's rule the same as that estimate obtained by applying Lagrangian interpolation to the integrand (see problem 1 of section 5.1)? If not, why not?

4. (a) Confirm that the expression for $C^{(3)}$ can be obtained by Richardson extrapolation of the composite trapezoidal rule (5.4.4), assuming the latter to have second-order error.

(b) Interpreting Selmer's rule, given by the matrix $C_1^{(4)}$ of (5.4.7), as implying that the area between the arc $P_0 P_1 P_2 P_3$ through any 4 consecutive points on γ and the straight line $P_0 P_3$ is $(9/8)$ times the area of the quadrilateral $P_0 P_1 P_2 P_3$, show that this also follows from the trapezoidal rule by the application of Richardson extrapolation.

5. Show that if $x_{k+1} - x_k = h$, a constant, then the value of I given by using the matrix of (5.4.8) is identical with the truncated Gregory formula appropriate to Bessel interpolation.

6. (a) If the points (x_k, y_k) for $k = 0, 1, \ldots, n$ are chosen as n distinct equi-spaced points on the circumference of a circle γ of radius r, ordered clockwise, with (x_0, y_0) coincident with (x_n, y_n), show that the estimate of the area within the circle γ is given by (5.4.9) as $r^2 n \sum_{k=1}^{m} b_k^{(m)} \sin(2\pi k/n) = A_{mn}$ say.

(b) What is the estimate A_{mn}, if $n = 8$ and m is taken as 1, 2, 3, and 4?

(c) Find the approximate errors in $A_{1,n}$ and $A_{2,n}$ if n is large compared with (2π), showing that these are of second and fourth order in $(2\pi/n)$.

7. (a) The latitude (θ) and longitude (ψ) of 16 coastal locations on the continent of S. America are converted to the coordinates $R\sin\theta$, $R\psi$ of an equal-area projection, with the earth radius R taken as 6360 km. The results are listed below, starting at the southern extremity and proceeding clockwise:

$R\psi$	7104	8214	7974	7837	8470	9015	8547	8323
$R\sin\theta$	5199	4324	3464	2567	1524	493	-440	-1214

$R\psi$	7433	6111	4927	3900	4264	4806	6218	7219 (7104)
$R\sin\theta$	-1152	-661	283	654	1413	2475	3643	4354 (5199)

Estimate the area of the continent using the approximation (5.4.9) with $m = 1$, 2, and 3.

(b) The shape is better modelled with a discontinuity in slope at the southern extremity, so that a formula such as the composite Brun's rule, or the

Bessel interpolate (5.4.8), would be preferable. Re-estimate the area using one or other of these, with the southern extremity as the common end point.

8. Write an instruction sequence which calculates the area within a polygon whose vertices are successive positions of a pen on a graphics pad, accessed interactively by *read (x, y)* where *(x, y)* are Cartesian coordinates, the polygon being closed at any time by the line joining the currently recorded vertex with the first recorded vertex. The expression *read (x, y)* has a Boolean value which is *true* if the actual parameters have been assigned equal to the current pen coordinates, and a value *false* if the pen has been switched-off (in which event the actual parameter values are undefined). It may be assumed that the values of the coordinates (x, y) are non-negative integers less than $b^{p/2}$, where b is the base and p the precision of the floating-point arithmetic in use; why is this assumption important?

Section 5.5

1. (a) If the mid-ordinate rule is applied to the estimation of $\int_0^h \ln(1/x)\,dx$, show that the error is equal to $h \ln(e/2)$, where $e = \exp(1)$.

    (b) If the composite mid-ordinate rule is applied to the estimation of $\int_0^{Nh} \ln(1/x)\,dx$ using N equal segments, show that the error is approximately $h \ln(\sqrt{2})$ if N is large. (Hint: use Stirling's formula (2.5.7).)

    (c) In estimating the value of $\int_e^1 \ln(1/x)\,dx$ by the composite mid-ordinate rule using a large number of equal segments of span $h = 1/N$, show that as $N \to \infty$, the error will be proportional to $(1/N^2)$ if ϵ is fixed independently of N, but it will be proportional to $(1/N)$ if ϵ is taken as equal to $(1/N)$. (The same result would be true if ϵ bore any fixed proportion to $1/N$.)

2. Suggest substitutions which eliminate the singularity (if any) in the integrand of the following integrals:

    (a) $\int_0^1 x^{-1/2} \cos(x)\,dx$;   (b) $\int_0^1 x^{-1} \sin(x)\,dx$;

    (c) $\int_0^1 (1-x^2)^{1/2} \sin(\pi x)\,dx$;   (d) $\int_0^1 [x(1-x^2)]^{-1/2}\,dx$.

3. Use integration by parts to eliminate the logarithmic singularity in:

    (a) $\int_0^1 \exp(-2x) \ln(x)\,dx$;   (b) $\int_0^\pi x \ln[\sin(x)]\,dx$.

4. Suggest a method by which $\int_0^\infty \exp(-x) \ln(x)\,dx$ might be evaluated numerically.

5. Use the property of orthogonality of Tchebyshev polynomials $T_k(x)$, together with the expression for the Tchebyshev interpolate over $-1 \leqslant x \leqslant 1$ of a function $f(x)$ as a truncated Tchebyshev series (problem 3, section 3.7.2), to show that

$$\int_{-1}^{+1}(1 - x^2)^{-1/2} f(x)\,dx \cong [\pi/(n+1)] \sum_{k=0}^{n} f(x_k) ,$$

where $x_k = \cos[(2k + 1)\pi/(2n + 2)]$ for $k = 0, 1, \ldots, n$ are the Tchebyshev abscissae of degree n.

6. (a) It is suggested that the Gauss-Laguerre quadrature formula:

$$\int_0^\infty \exp(-x)f(x)\,dx = \sum_{k=1}^{n} \lambda_k^{(n)} f(x_k)$$

with weights $\lambda_k^{(n)}$ and abscissae x_k known for some particular n, could be applied to the evaluation of $\mathrm{erfc}(x) = 2\pi^{-1/2}\int_x^\infty \exp(-t^2)\,dt$. Find the resulting expression for $\mathrm{erfc}(x)$. Would this be an appropriate method to use for the evaluation of $\mathrm{erfc}(x)$?

(b) Suggest a better method for the evaluation of $\mathrm{erfc}(x)$, using numerical quadrature.

7. (a) Show by integration by parts that if m is a non-zero integer

$$\int_0^1 \sin(m\pi x)f(x)\,dx = (m\pi)^{-1}[(-1)^{m-1}f(1) + f(0)]$$
$$- (m\pi)^{-2}\int_0^1 \sin(m\pi x)f''(x)\,dx$$

and hence establish the series expansion

$$\int_0^1 \sin(m\pi x)f(x)\,dx =$$
$$\sum_{k=0}^{N-1}(-1)^k[(-1)^{m-1}f^{(2k)}(1) + f^{(2k)}(0)]/(m\pi)^{2k+1} + R_N .$$

(b) Supposing that $f(x) = x/(4 - x^2)$, confirm that the series expansion is semi-convergent, and hence estimate the value of $\int_0^1 [x/(4 - x^2)]\sin(10\pi x)\,dx$ correct to 6 significant decimal digits.

8. (a) Suppose that $f(x)$ and $g(x)$ are analytic over $[a, b]$, and that $g(x)$ has one simple zero at some interval point $x = \xi$ of this interval. Verify that

$$g'(\xi)\int_a^b [f(x)/g(x)]\,dx = f(\xi)\ln|g(b)/g(a)| + \int_a^b [h(x)/g(x)]\,dx$$

where $h(x) = g'(\xi)f(x) - f(\xi)g'(x)$, and find the value of $[h(x)/g(x)]$ at $x = \xi$. (This is a generalisation of the result for $g(x) \equiv (x - \xi)$ quoted in the text.)

(b) Suggest how the improper integral $\int_0^{\pi/2} (x - \cos x)^{-1}\,dx$ might appropriately be rearranged so that its value could be estimated numerically.

9. (a) If $f(x) \to 0$ as $x \to \infty$, and m is integer, confirm the formal expansion:

$$(-1)^m\int_m^\infty f(x)\sin(\pi x)\,dx = \sum_{k=0}^{\infty}(-1)^k f^{(2k)}(m)/\pi^{2k+1}.$$

(b) Evaluate $-\int_1^\infty [\sin(\pi x)/\ln(x)]\,dx$ correct to 2 decimal places.

Hints and Solutions

SECTION 1.2

1. (a) 0.10×10, 1.0×10^0, 10×10^{-1}
   (b) 0.50×10^{-1}, 5.0×10^{-2}, 50×10^{-3}
   (c) 0.10×10^{-5}, 1.0×10^{-6}, 10×10^{-7}.

2. (a) $(1111)_2 = 15 \geqslant |m| \geqslant (1000)_2 = 8$.
   (b) From (1.2.3) $(10)_2 = 2 > |m| \geqslant 1 \therefore |m| = (1.d_1 d_2 ...)_2$ where the d's are binary digits.

3. (a) With $x = 1/16$, sum $= [1 - (1/16)]^{-1} = 16/15$.
   But $(0.\dot{1})_{16} = 1/16 + (1/16)^2 + ... = x + x^2 + ... = 16/15 - 1 = 1/15$.
   (b) $(1/5) = 3/15 = 3 \times (0.\dot{1})_{16} = (0.\dot{3})_{16}$
   $(1/3) = 5/15 = 5 \times (0.\dot{1})_{16} = (0.\dot{5})_{16}$
   $(4/5) = 12/15 = 12 \times (0.\dot{1})_{16} = (0.\dot{C})$, since hex digit C $= 12$.
   (c) $(1/5) = (0.\dot{3})_{16} \simeq (0.333)_{16} = (333)_{16} \times 16^{-3} = 819/4096$.

5. (a) Digit 1 followed by p zeros: precisely representable as $(b^{p-1} \times b^1)$.
   (b) Digit 1 followed by $(p-1)$ zeros, followed by digit k: *not* representable.
   (c) Digit $(b/2)$ followed by $(p-2)$ zeros, followed by digit k: precisely representable.
   (d) Digit 1 followed by $(p-2)$ zeros, followed by digits 1 and 0: precisely representable, as $(b^{p-1} + 1) \times b$.

SECTION 1.3

1. (a) If $x = n + f$ where n is integer and $0 \leqslant f < 1$, then by definition, entier $(x) = n$, and since $-n - 1 < -x \leqslant -n$, therefore also ceil $(-x) = -n$.
   (b) If $f = |y| -$ entier $(|y|)$, then $0 \leqslant f < 1$, and ceil $(|y| - 0.5)$ equals entier $(|y|)$ if $0 \leqslant f \leqslant 0.5$, but equals entier $(|y|) + 1$ if $0.5 < f < 1$. Difference arises only if $f = 0.5$.

2. (a) From (1.3.3), for entier $(|y| + 0.5) = b^p$; i.e. $b^p \leqslant |y| + 0.5 < b^p + 1$, and so from (1.3.1) for $b^p - 0.5 \leqslant |y| < b^p$.

   (b) From (1.3.4), for ceil $(|y|) = b^p$; i.e. $b^p - 1 < |y| \leqslant b^p$, and so from (1.3.1) for $b^p - 1 < |y| < b^p$.

3. No. (The least value of $|m|$ is also b^{p-1}).

4. (a) 1.5×10^{-11},   (b) 7×10^{-15},   (c) 1.2×10^{-7},   (d) 10^{-9},   (e) 10^{-6},
   (f) 1.5×10^{-8}.

5. (a) Exact mantissa unaltered: only exponent (i) changed. (b) $c = 0$.

6. Effect on output of method of rounding used:

method:	chopping	correct rounding	boosting
output:	*entier* $(frac)$	*entier* $(frac + 0.5)$	*ceil* $(frac)$.

SECTION 1.4

2. $m: =$ entier $(y + 0.5)$.

4. With $b = 2$, $p = 10$, and $n = 100$, the sample mean $\bar{a} = 0.385$ (say), so that from Table 1.1, $|\bar{a} - \alpha| = 0.038$ whereas $\sigma_\alpha n^{-1/2} = 0.0216$, so that $|\bar{a} - \alpha| n^{1/2}/\sigma_\alpha = 1.75$. Tables of the normal distribution show that values of $\bar{a}$ as far from α as this occur with probability 0.08.

5. (a) 24 binary digits gives $\epsilon/(b + 1) = 4 \times 10^{-8}$; (b) 8 octal digits (each of 3 bits) gives $\epsilon/(b + 1) = 5 \times 10^{-8}$; and (c) 6 (4-bit) hexadecimal digits, $\epsilon/(b + 1) = 6 \times 10^{-8}$.

6. (a) 1.6×10^{-12}, (b) 2.3×10^{-15}, (c) 4×10^{-8}, (d) 10^{-10}, (e) 6×10^{-8},
   (f) 5×10^{-9}.

SECTION 1.5

1. (i) x; (a) $|x| \geqslant 1$; (b) $|x| \leqslant 1$.
   (ii) $1/\ln|x|$; (a) $|x| \simeq 1$; (b) $|x| \geqslant 1$ and $|x| \leqslant 1$.
   (iii) $x \cot(x)$; (a) $|x| \simeq n\pi$, $(n \geqslant 1$ integer) and $|x| \geqslant 1$; (b) $|x| \simeq (n - \frac{1}{2})\pi$.
   (iv) $-x \tan(x)$; (a) $|x| \simeq (n - \frac{1}{2})\pi$ and $|x| \geqslant 1$, (b) $|x| \simeq n\pi$ and $|x| \leqslant 1$.
   (v) $2x/\sin(2x)$; (a) $|x| \simeq n\pi/2$ and $|x| \geqslant 1$; (b) never well-conditioned.
   (vi) $-x(x^2 + 1)^{-1/2}$; (a) never ill-conditioned; (b) $|x| \simeq 0$.
   (vii) $-x(x^2 - 1)^{-1/2}$; (a) $|x| \simeq 1$; (b) never well-conditioned.

2. With decimal precision 3, and since $1 \leqslant x < 10$, maximum input error magnitude is 0.005, and input relative error is $|\beta\epsilon| < 0.005/x$.
   (a) cond$\{\sin(x)\} = x \cot x = -22.7 = \phi$; $\therefore |\phi\beta\epsilon| < 0.038$.
   (b) cond$\{(x^2 - 10)^{1/2}\} = x^2/(x^2 - 10) = 16.9 = \phi \therefore |\phi\beta\epsilon| < 0.026$.
   (c) cond$\{x^x\} = $ cond$\{\exp(x\ln x)\} = x(1 + \ln x) = 19.3 \therefore |\phi\beta\epsilon| < 0.014$.

3. (a) Use the result $d(\ln|t|)/dt = 1/t$ to show that:
$$d(\ln|y|)/d(\ln|x|) = (dy/y)/(dx/x) = (x/y)\,(dy/dx) = xf'(x)/f(x).$$
   (b) Here $\mathrm{cond}\{f^{-1}(y)\} = \mathrm{cond}\{x\} = d(\ln|x|)/d(\ln|y|)$
$$= 1/[d(\ln|y|)/d(\ln|x|)]$$
$$= 1/\mathrm{cond}\{f(x)\} = 1/\phi(x).$$

4. (a) $\phi(x) = \mathrm{cond}\{x^n\} = n; \therefore \mathrm{cond}\{y^{1/n}\} = 1/n.$
   (b) $\phi(x) = \mathrm{cond}\{\exp(x)\} = x; \therefore \mathrm{cond}\{\ln(|y|)\} = 1/x = 1/\ln|y|.$
   (c) $\phi(x) = \mathrm{cond}\{\tan(x)\} = x\sec^2(x)/\tan(x) = x[1 + \tan^2(x)]/\tan(x).$
   $\therefore \mathrm{cond}\{\arctan(y)\} = \tan x/[x(1 + \tan^2 x)] = y/[(1 + y^2)\arctan(y)].$

5. (a) Let $y = f(x) = f_1(x)f_2(x)\cdots f_k(x)$. Then $\ln|y| = \sum\limits_{j=1}^{k}\ln|f_j(x)|.$
   Thus: $\mathrm{cond}\{f(x)\} = d(\ln|y|)/d(\ln|x|) = \sum\limits_{j=1}^{k} d[\ln|f_j(x)|]/d(\ln|x|) = \sum\limits_{j=1}^{k}\mathrm{cond}\{f_j(x)\}.$
   (b) Here $f_j(x) = x + j$ and $\mathrm{cond}\{f_j(x)\} = x/(x + j)$, for $j = 1, 2, \ldots, 20.$
   Thus $\mathrm{cond}\{f(x)\} = x\sum\limits_{k=1}^{20}(x + j)^{-1} = 20.15\sum\limits_{k=1}^{20}(20.15 - j)^{-1} = 201.5.$

6. (a) $dx_i/dx_k = (dx_i/dx_j)(dx_j/dx_k) = f_{ij}'(x_j)f_{jk}'(x_k)$. Hence, $\phi_{ik} = \mathrm{cond}\{f_{ik}(x_k)\} = x_k f_{ik}'(x_k)/f_{ik}(x_k) = x_k f_{ij}'(x_j)f_{jk}'(x_k)/x_i = [x_k f_{jk}'(x_k)/f_{jk}(x_k)][x_j f_{ij}'(x_j)/f_{ij}(x_j)] = \phi_{ij}\phi_{jk}$, since $x_j = f_{jk}(x_k).$
   (b) Suppose $\phi_{j1} = \prod\limits_{i=1}^{j-1}\phi_{(i+1)i}$. Then by part (a), with $i = j + 1$, and $k = 1,$
   $\phi_{(j+1)1} = \phi_{(j+1)j}\phi_{j1} = \phi_{(j+1)j}\prod\limits_{i=1}^{j-1}\phi_{(i+1)i} = \prod\limits_{i=1}^{j}\phi_{(i+1)i}$. Thus the proposition
   is also true of the relation for $\phi_{(j+1)1}$. But it is true for $j = 2$. Hence by
   induction it is true for $j = 3, 4, \ldots.$

SECTION 1.6

2. (a) (i) $2x^2/[(1 + 2x)(1 + x)]$;   (ii) $2x^2/[(1 + x^2)^{1/2} + (1 - x^2)^{1/2}]$;
        (iii) $2\sin^2(x/2)$; (iv) $2\sin(x/2)\cos(1 + x/2).$
   (b) (i) 0.002, 0.000294, 0.000292;   (ii) 0, 0.000152, 0.000151;
        (iii) 0, 0.0000756, 0.0000756; (iv) 0.006, 0.00654, 0.00658.

3. Expand $\sin x$ as power-series: $x - \sin x = x - [x - (1/6)x^3 + (1/120)x^5 - \ldots] = (x^3/6)[1 - (1/20)x^2 + \ldots]$. At $x = 0.0123$, the first term $(x^3/6)$ provides adequate accuracy and evaluates as $3.10 \times 10^{-7}.$

4. (a) $\ln(1 + z) = z - \frac{1}{2}z^2 + \frac{1}{3}z^3 - + \ldots$ and $(1 + z)^{-1} = 1 - z + z^2 - z^3 + \ldots$. Therefore $f(x) = -(1/3)z^3 + (1/2)z^4 - (3/5)z^5 + \ldots$
$$= -(1/3)(x - 1)^3 + (1/2)(x - 1)^4 - + \ldots$$
   i.e., $f(x) = \sum\limits_{k=3}^{\infty}(-1)^k(1 - 2/k)(x - 1)^k = \sum\limits_{k=0}^{\infty}f^{(k)}(1)(x - 1)^k/k!$ Thus
   $f(1) = f'(1) = f''(1) = 0$ and $f^{(k)}(1) = (-1)^k(k - 2)(k - 1)!$ if $k > 2.$

(b) $\text{cond}\{f(x)\} = -(x-1)^2/[xf(x)] \to 3/(x-1) \to \infty$ as $x \to 1$ by part (a). First 2 terms of series give: $f(1.010) = -0.328 \times 10^{-6}$.

(c) $\text{cond}\{f(x)\} \simeq 300$ by part (b) and input error bounds are ± 0.0005. Thus inherent error bounds are $\pm 15\%$.

5. $\text{cond}\{\arccos(x)\} = -x(1-x^2)^{-1/2}/\arccos(x)$ $\therefore$ ill-conditioned for small $(1-x^2)$, particularly so for x close to 1, when $\arccos(x) \to 0$. These are also ranges in which formulation involves cancellation error in evaluation of $1-x^2$ and $\pi/2 - \arctan(..)$. Better would be: $\arccos(x) = 2\arctan\{\text{sqrt}\,[(1-x)/(1+x)]\}$ for all x, except $x = -1$, when both this and given formula cause overflow.

6. The evaluations $x_2 = x_1 \uparrow N$, $x_0 = x_2 \uparrow (1/N)$ involved in calculating z_1 have condition numbers $\phi_1 = N$, $\phi_2 = 1/N$, interchanged in the calculation of z_2. Thus error involved by first exponentiation is attenuated (by factor $1/N$) when z_1 is calculated, but magnified (by N) in the calculation of z_2.

7. (a) $f(x \pm \frac{1}{2}h) = f(x) \pm \frac{1}{2}hf'(x) + (h^2/8)f''(x) \pm (h^3/48)f'''(x) + \ldots$ and so $\delta f(x)h = f'(x) + (h^2/24)f'''(x) + \ldots$. Thus $C = -f'''(x)/24$.

(b) $\delta\tilde{f}(x) = \tilde{f}(x + \frac{1}{2}h) - \tilde{f}(x - \frac{1}{2}h) = f(x + \frac{1}{2}h)(1 - \alpha_1\epsilon) - f(x - \frac{1}{2}h)(1 - \alpha_2\epsilon) = \delta f(x) + [\alpha_2 f(x - \frac{1}{2}h) - \alpha_1 f(x + \frac{1}{2}h)]\epsilon \cong \delta f(x) + (\alpha_2 - \alpha_1)f(x)\epsilon$. Hence α can be identified with $(\alpha_1 - \alpha_2)$. But correct rounding implies $|\alpha_1|, |\alpha_2| \leq \frac{1}{2}$, so that $|\alpha| \leq 1$; that is, $A = 1$.

(c) In the quoted instance, putting $A = 1$, $|f'(x) - \delta\tilde{f}(x)/h| = |C|h^2 + |f(x)|\epsilon h^{-1}$. Differentiating with respect to h and equating the derivative to zero to find a stationary value, it follows that this expression is minimum for $2|C|h = |f(x)|\epsilon h^{-2}$ i.e., for $h = |12\epsilon f(x)/f'''(x)|^{1/3}$. In quoted computation, $\alpha = 2^{-23}$ and h is about 0.01.

SECTION 1.7

3. $n_1 * n_2$ has precision $2p$ or $2p - 1$ i.e., max $= 2p$.

4. Using a calculator with $b = 10$, $p = 13$, $i = 7$, some results (indicating chopping) are as follows:

k_1	67	20	18	17	10	11	1		2	20	17	20	67
k_2	3	10	11	6	10	9	1		-1	-5	-6	-10	-3
$k_1 k_2$	201	200	198	102	100	99	1		-2	-100	-102	-200	-201
n_0	200	200	100	100	100	0	0		-100	-100	-200	-200	-300

6. Superimposed on the growth of relative error with n (if chopping is employed) is a slower oscillation as the mantissa of the sum goes from its lower to higher limit. This is more difficult to observe if correct rounding is used in summation, because the relative error may fluctuate in sign (within bounds roughly proportional to $\sqrt{n}$).

7. (a) (i) Let $n_3 = n_1 \circledast n_1$, $n_4 = n_2 \circledast n_2$, $n_0 = n_3 \simeq n_4$. From (1.7.2) if $\beta_1 = \beta_2 = 0$, then $\beta_3 = \alpha_1$, $\beta_4 = \alpha_2$ (say), and so $\beta_0 = (x_3\beta_3 - x_4\beta_4)x_0^{-1} + \alpha_3$. Therefore generated error (ignoring α_3) is $(x_1^2 \alpha_1 - x_2^2 \alpha_2)/(x_1^2 - x_2^2)$ and if $\alpha_1 \neq \alpha_2$ this tends to ∞ as $x_1 \to x_2$.

    (ii) Let $n_3 = n_1 \tilde{+} n_2$, $n_4 = n_1 \simeq n_2$, $n_0 = n_3 \circledast n_4$. With $\beta_1 = \beta_2 = 0$, then $\beta_3 = \alpha_1$, $\beta_4 = \alpha_2$ as before, but $\beta_0 = \beta_3 + \beta_4 + \alpha_3$, so that generated error is simply $(\alpha_1 + \alpha_2)$, which is bounded (whether or not $x_1 \to x_2$).

  (b) (i) mean = $\bar{\alpha}$, variance = $(x_1^4 + x_2^4)\sigma_\alpha^2/(x_1^2 - x_2^2)^2$

    (ii) mean = $2\bar{\alpha}$, variance = $2\sigma_\alpha^2$.

  (c) With correct rounding $\bar{\alpha} = 0$, and so error of (ii) can be expected to be greater than of (i) if its variance is greater than that of (i). From part (b), if $(x_1/x_2)^2 = R$, this implies $2 > (R^2 + 1)/(R-1)^2$. But $(R^2 + 1)/(R-1)^2 = 2$ if $R = 2 \pm \sqrt{3}$.

8. There are $(i-1)$ multiplications to form 2^i. Each non-zero binary digit of k (expressed as a binary integer of i digits) implies a multiplication of an accumulator to form $x \uparrow k$. There is at least 1, at most i, of these, so total multiplications are between i and $2i - 1$.

9. Let $n_3 = \tilde{\ln}(n_1)$, $n_4 = n_2 \circledast n_3$, $n_0 = \tilde{\exp}(n_4)$. From (1.5.4) since cond$\{\ln(x)\} = 1/\ln(x)$, $\therefore \beta_3 = \beta_1[\ln(x_1)]^{-1} + \alpha_1$. From (1.7.2), $\beta_4 = \beta_2 + \beta_3 + \alpha_2$, and since cond$\{\exp(x)\} = x$, $\beta_0 = x_4\beta_4 + \alpha_3$. Substituting expressions for β_3, β_4 and $x_4 = x_2 x_3 = x_2 \ln(x_1)$, therefore $\beta_0 = [(\beta_2 + \alpha_1 + \alpha_2) \ln x_1 + \beta_1]x_2 + \alpha_3$. The component of generated error is $(\alpha_1 + \alpha_2)x_2 \ln(x_1) = (\alpha_1 + \alpha_2)\ln(x_0)$.

10. Correct value if difference has same (or larger) floating-point exponent as $f(x)$: i.e., for $f(x) \geqslant b + 1$ or $f(x) \leqslant 0$.

11. (c) Let stated relation for $\bar{a}_n$ be true for $n = m - 1$; then $m\bar{a}_m = \sum_{k=1}^{m} a_k = a_m + (m-1)\bar{a}_{m-1} = \delta_m + m\bar{a}_{m-1}$, i.e., $\bar{a}_m = (\delta_m/m) + \bar{a}_{m-1} = (\delta_m/m) + \sum_{k=1}^{m-1} (\delta_k/k) = \sum_{k=1}^{m} (\delta_k/k)$ so that result is true for $n = m$. But it is true for $n = 1$, since $\bar{a}_1 = a_1$. Again, let stated relation for σ_n^2 be true for $n = m - 1$; then $(m-1)\sigma_m^2 = a_m^2 + \sum_{k=1}^{m-1} a_k^2 - m(\bar{a}_m)^2 = (\bar{a}_{m-1} + \delta_m)^2 - m[\bar{a}_{m-1} + (\delta_m/m)]^2 + \sum_{k=1}^{m-1} a_k^2 = [(m-1)\delta_m^2/m] + \sum_{k=1}^{m-1} a_k^2 - (m-1)(\bar{a}_{m-1})^2 = [(m-1)\delta_m^2/m] + (m-2)\sigma_{m-1}^2$, i.e., $\sigma_m^2 = (\delta_m^2/m) - [\sigma_{m-1}^2/(m-1)] + \sigma_{m-1}^2 = \sum_{k=2}^{m} [\delta_k^2/k) - \sigma_{k-1}^2/(k-1)]$ so that result is true for $n = m$. But it is true for $n = 2$, since $\sigma_2^2 = (a_2 - a_1)^2/2 = \delta_2^2/2$.

SECTION 1.8

1. (a) Fails if x_1^N overflows. Better: $x_1 * (1 + (x_2/x_1) \uparrow N) \uparrow (1/N)$.

(b) Fails if $\exp(-x)$ overflows. Better as $\exp(\ln(\mathrm{abs}(x)) * N - x)$.

(c) Fails (zero-divide) if $x = -1$. Better for $0 > x \geqslant -1$ as $\pi/2 - \arctan(\mathrm{sqrt}((1 + x)/(1 - x)))$.

(d) Fails (zero-divide) if $x = 0$. Test for $x = 0$, and in this event, evaluate as 0.

2. Let Op 1 be first operand, Op 2 second operand. Let $x_1, x_2 \neq 0, VR$.

Operation: Op 1 Op 2 =	add or subtract			multiplication			division		
	x_2	0	VR	x_2	0	VR	x_2	0	VR
$x_1 \neq 0$	$x_1 \pm x_2$	x_1	VR	$x_1 * x_2$	0	VR	x_1/x_2	VR	0
0	x_2	0	VR	0	0	u/d	0	u/d	0
VR	VR	VR	VR	VR	u/d	VR	VR	VR	u/d

u/d = undefined operation $(0/0, 0 * VR \text{ or } VR * 0, VR/VR)$.

3. Let M and N be max. & min. exponent values respectively. Then form (1.2.4) the largest magnitude representable is $(b^P - 1)b^M$ whilst least is b^{P-1+N}. Their product is $(1 - b^{-P})b^{2P-1+M+N}$, and their quotient is $(1 - b^{-P})b^{M+1-N}$. Taking $(1 - b^{-P}) \simeq 1 \therefore M + N = 1 - 2p$, and $M - N = 2R - 1$ whence $M = R - p$, $N = 1 - R - p$.

4. (a) Using formulae $(1 - \sin x)\tan^2 x = \sin^2 x/(1 + \sin x)$ and $\arctan x = \pi/2 - \arctan(1/x)$, arrange computation as: $\arctan[2s^2/(1 + s)]$ if $s \geqslant -0.5$, and $(\pi/2) - \arctan[(1 + s)/(2s^2)]$ if $s < -0.5$, where $s = \sin x$. This avoids cancellation error and overflow (of $\tan x$).

(b) Value to be found is $\ln|2 \sinh(x/2)/x|$, which is an even function of x. If base is b, precision p, then let $N > \frac{1}{2}p \ln b$, and arrange computation as $\ln[\sinh(y)/y]$ if $y \leqslant N$, $y - \ln[\mathrm{abs}(x)]$ if $y > N$, where $y = \mathrm{abs}(x)/2$. This avoids cancellation error (at $x = 0$) and underflow or overflow of $\exp(..)$.

5. (a) Determine the C's by noting that $gd(0) = 0$. Thus: $C_1 = 0$, $C_2 = 0$, $C_3 = -2 \arctan(1) = -\pi/2$, $C_4 = -\arccos(1) = 0$.

(b) (i) $\arctan(\sinh x)$ would overflow for moderately large $|x|$ in calculating $\sinh(x)$, even though $gd(x)$ is bounded.

(ii) $\arcsin(\tanh x)$ has no overflow problems, because $\tanh|x| \to 1$ as $|x| \to \infty$.

(iii) $2 \arctan(e^x) - \pi/2$ has the same overflow problem as (i), and suffers from loss of precision near $x = 0$.

(iv) $\arccos(\mathrm{sech}\, x)$ would overflow for moderately large $|x|$ if $\mathrm{sech}\, x$ is computed as $1/\cosh x$, and it loses precision near $x = 0$ (because $\arccos(..)$ is badly conditioned for argument values close to 1). Moreover the evaluation of $\arccos(..)$ would usually provide the positive principle part, whereas $gd(x)$ is negative for $x < 0$, necessitating a multiplication by $\mathrm{sgn}(x)$. Therefore choose (ii).

SECTION 2.2

1. (a) $(-m)_k = (-m)(-m+1)(-m+2)\ldots(-m+k-1) = (-1)^k(m-k+1)$
   $\ldots (m-1)m = (-1)^k m!/(m-k)!$. Note that $(-m)_k = 0$ for all integer $k > m$: therefore:
   $$(1-z)^m = m! \sum_{k=0}^{\infty} (-z)^k/[(m-k)!k!] .$$

   (b) $(m)_k = m(m+1) \ldots (m+k-1) = (m+k-1)!/(m-1)!$ Thus
   $(1-z)^{-m} = \sum_{k=0}^{\infty} (m+k-1)! z^k/[(m-1)!k!]$. If $m=1$, $(1)_k = k!$ so
   that $(1-z)^{-1} = \sum_{k=0}^{\infty} z^k$ (the geometric series).

2. (a) Even function of z ∴ odd powers of z in two series cancel and
   $$(1-z)^{-c} + (1+z)^{-c} = 2\sum_{k=0}^{\infty} (c)_{2k} z^{2k}/(2k)!$$

   (b) $(1+z)(1-z)^{-c} = \sum_{k=0}^{\infty} (c)_k z^k/k! + \sum_{k=0}^{\infty} (c)_k z^{k+1}/k! = 1 + \sum_{k=0}^{\infty}$
   $[(c)_k + k(c)_{k-1}] z^k/k! = 1 + \sum_{k=0}^{\infty} (c+2k-1)(c)_{k-1} z^k/k!$

3. (a) $(1-z)^{-c} = 2^c \sum_{k=0}^{\infty} (c)_k (2z-1)^k/k! = \sum_{k=0}^{\infty} (c)_k 2^{k+c}(z-\frac{1}{2})^k/k!$

   (b) $(1-z)^{-c} = [1-\alpha-(z-\alpha)]^{-c} = (1-\alpha)^{-c}[1-(z-\alpha)/(1-\alpha)]^{-c} =$
   $\sum_{k=0}^{\infty} (c)_k (1-\alpha)^{-k-c}(z-\alpha)^k/k!$

4. (a) $\theta = \pi/3$: $(1-z+z^2)^{-1} = 1 + z - z^3 - z^4 + z^6 + z^7 - \ldots$
   $\theta = \pi/2$: $(1+z^2)^{-1} = 1 - z^2 + z^4 - \ldots$
   $\theta = 2\pi/3$: $(1+z+z^2)^{-1} = 1 - z + z^3 - z^4 + z^6 - z^7 + \ldots$

   (b) $\lim_{\theta \to 0} \{[\sin(k+1)\theta]/\sin\theta\} = (k+1)$; ∴ $(1-2z+z^2)^{-1} = \sum_{k=0}^{\infty} (k+1)z^k =$
   $(1-z)^{-2}$. Confirm by placing $m=2$ in answer 1(b) above.

SECTION 2.3

1. (a) Assume $\Gamma(z+k) = (z)_k \Gamma(z)$: then $\Gamma(z+k+1) = (z+k)\Gamma(z+k) =$
   $(z+k)(z)_k \Gamma(z) = (z)_{(k+1)} \Gamma(z)$.
   (b) $(z)_0 = \Gamma(z)/\Gamma(z) = 1$ and $(z)_{(-m)} = \Gamma(z-m)/\Gamma(z) = 1/[\Gamma(z)/\Gamma(z-m)]$
   $= 1/(z-m)_m$.
   (c) $(z)_{(-m)} = 1/[(z-m)(z-m+1)\ldots(z-1)]$.

2. (a) $(z+k)_{(m+1)} - (z+k-1)_{(m+1)} = \Gamma(z+k+m+1)/\Gamma(z+k) - \Gamma(z+k+m)/$
   $\Gamma(z+k-1) = [(z+k+m)-(z+k-1)]\Gamma(z+k+m)/\Gamma(z+k) =$
   $(m+1)(z+k)_m$.

(b) From question 1(c), sum is $\sum_{k=0}^{\infty}(x+n+1+k)_{(-n-1)} = \sum_{k=1}^{\infty}(x+n+k)_{(-n-1)}$.

Thus from part (a), with $z = x + n$, $m = -n - 1$, sum is $(-n)^{-1}$ $[(x + n + \infty)_{(-n)} - (x + n)_{(-n)}] = (x + n)_{(-n)}/n = [(x)_n n]^{-1}$.

(c) Series $= 2[(1/2)+(1/6)+(1/12)+(1/20)+...] = 2\sum_{k=0}^{\infty}[(k+1)(k+2)]^{-1}$
$= 2$ from part (b) with $x = n = 1$.

3. This series cannot be deranged, as it is not absolutely convergent: the two component series $1 + (1/2) + (1/3) + ...$ diverge, and $(\infty - \infty) \neq 0$.

4. Let S_L be the series on the left, and S_R that on the right.
   (a) $S_R = (1 + 2^{-2} + 3^{-2} + ...) = S_L + (2^{-2} + 4^{-2} + 6^{-2} + ...)$
   $= S_L + (1 + 2^{-2} + 3^{-2} + ...)/4 = S_L + S_R/4 \therefore S_L = (3/4)S_R$.
   (b) $S_R = (1 - 3^{-1} + 5^{-1} - 7^{-1} + 9^{-1} - + ...) = S_L - (3^{-1} - 9^{-1} + 15^{-1} - ...)$
   $= S_L - (1 - 3^{-1} + 5^{-1} + ...)/3 = S_L - S_R/3 \therefore S_L = (4/3)S_R$.
   This derangement of a non-absolutely convergent series is permissible since both component series are convergent.

Section 2.3.1

1. (a) Suppose integer i is chosen so that $b^p > |\tilde{S}_n|b^i \geqslant b^{p-1}$, i.e. $\tilde{S}_n b^i$ is the integer mantissa of the machine representation of S_n. If $\tilde{S}_{n+1} = \tilde{S}_n$, and correct rounding is in use, then $|\tilde{t}_{n+1}b^i| \leqslant 1/2$, so that $|\tilde{t}_{n+1}/\tilde{S}_n| \leqslant 1/(2|\tilde{S}_n|b^i) \leqslant 1/(2b^{p-1}) = \epsilon/2$.
   (b) If chopping is in use and $\tilde{S}_{n+1} = \tilde{S}_n$, then $\text{sgn}(\tilde{t}_{n+1}) \neq -\text{sgn}(\tilde{S}_n)$, and $|\tilde{t}_{n+1}b^i| < 1$. Whence $0 \leqslant (\tilde{t}_{n+1}/\tilde{S}_n) < \epsilon$.

2. Suppose (as in question 1), $b^p > |\tilde{S}_n|b^i \geqslant b^{p-1}$, where $|\tilde{S}_n|b^i$ is integer. Then $|\tilde{t}_{n+1}| \leqslant (b^{-p}/2)|\tilde{S}_n| \Rightarrow |\tilde{t}_{n+1}b^i| \leqslant (b^{-p}/2)|\tilde{S}_n b^i| < (1/2)$. This implies that $|\tilde{S}_{n+1}|b^i = \text{entier}(|\tilde{S}_n|b^i \pm \tilde{t}_{n+1}b^i + 0.5) = |\tilde{S}_n|b^i$, i.e. $|\tilde{S}_{n+1}| = |\tilde{S}_n|$.

3. Since $S_\infty - S_n = \mu t_{n+1}$ where $0 < \mu < 1$, then $|S_\infty - S_e| < |t_{n+1}/2|$. Relative error magnitude is $|S_\infty - S_e|/|S_\infty| < |t_{n+1}/2S_\infty| \simeq |t_{n+1}/2S_e| \leqslant |e/2|$.

4. (a) From question section 2.2(1), with $(-z)$ replacing z, $t_k = (c)_k(-z)^k/k!$
   i.e. $t_k/t_{k-1} = (c)_k(-z)^k(k-1)!/[(c)_{k-1}(-z)^{k-1}k!] = -(c+k-1)z/k$.
   (b) For example: assigning initially $k = 0$, and $sum = term = 1$, the summing loop could be:

```
LOOP: k := k + 1; term := (1 - k - c) * term * z/k;
         if abs(term) > abs(sum * e) then
            begin oldsum := sum; sum := sum + term;
                  if oldsum ≠ sum then goto LOOP
            end
         else sum := sum + term/2;
```

5.   1   (True sum $= 27/64 = .421875$)

$-.1667$	.4167								
.9676	.4005	.4086							
$-.0616$	.4530	.4267	.4176						
.8389	.3886	.4208	.4238	.4207					
.0684	.4537	.4212	.4210	.4224	.4215				
.7177	.3930	.4234	.4223	.4216	.4220	.4218			
.1766	.4472	.4201	.4217	.4220	.4218	.4219	.4218		
.6237	.4002	.4237	.4219	.4218	.4219	.4219	.4219	.4219	

6. (a) $S'_j = (S_j + S_{j-1})/2 = S_{j-1} + t_j/2$. Therefore $t'_{j+1} = S'_{j+1} - S'_j = S_j - S_{j-1} + (t_{j+1} - t_j)/2 = (t_{j+1} + t_j)/2$. Also $S'_1 = S_0 + (t_1/2) = t_0 + (t_1/2) = S'_0 + (t_1 + t_0)/2 \therefore S'_0 = t_0/2$.

   (b) Since $|t'_{j+1}| < |t_j| \therefore \operatorname{sgn}(t'_{j+1}) = \operatorname{sgn}(t_j)$ and the averaged series is alternating in sign.

Section 2.3.3

1. (a) $\tau_n/t_{n+1} = 1 + r_{n+2} + r_{n+2} r_{n+3} + r_{n+2} r_{n+3} r_{n+4} + \ldots$
$$< 1 + r_{n+2} + r_{n+2}^2 + r_{n+2}^3 + \ldots = 1/(1 - r_{n+2}).$$

   (b) Here $r_k = (z/2k)^2$ and provided $n + 2 > z/2$ so that $r_{n+2} < 1$, $r_n/t_{n+1} < 4(n+2)^2/[4(n+2)^2 - z^2] = 4(n+2)^2/[2n+4-z)(2n+4+z)]$.

2. (a) $\tau_n = \sum_{k=0}^{\infty} (n+1+k)^{-m-1} > \sum_{k=0}^{\infty} [(n+k+1)(n+k+2) \ldots (n+k+m+1)]^{-1}$
$= [m(n+1)_m]^{-1}$.

   (b) $\tau_n = \sum_{k=0}^{\infty} (n+1+k)^{-3} < \sum_{k=0}^{\infty} (n+1+k)^{-1} [(n+1+k)^2 - 1]^{-1}$
$= \sum_{k=0}^{\infty} [(n+k)(n+k+1)(n+k+2)]^{-1} = [2n(n+1)]^{-1}$.

   (c) From part (a) $\tau_n > [2(n+1)(n+2)]^{-1}$, so that if $S_e = S_n + [2(n+1)(n+2)]^{-1}$, then truncation error $E_n = \tau_n - [2(n+1)(n+2)]^{-1}$, and so from (a) and (b): $0 < E_n < [n(n+1)(n+2)]^{-1}$.

3. $[m(n+1-m/2)_m]^{-1} = \sum_{k=0}^{\infty} [(n+k+1-m/2)(n+k+2-m/2) \cdots$
$(n+k+1+m/2)]^{-1}$. If $m = 2p$ is even, $[2p(n+1-p)_{2p}]^{-1} = \sum_{k=0}^{\infty} \{[(n+k+1)^2 - p^2]$
$[(n+k+1)^2 - (p-1)^2] \cdots [(n+k+1)^2 - 1](n+k+1)\}^{-1} > \sum_{k=0}^{\infty} (n+k+1)^{-2p-1} =$
$\sum_{k=n+1}^{\infty} k^{-2p-1} = \tau_n$. If $m = 2p - 1$ is odd, $[(2p-1)(n-p+3/2)_{(2p-1)}]^{-1} =$
$\sum_{k=0}^{\infty} \{[(n+k-1)^2 - (p-1/2)^2] \cdots [(n+k+1)^2 - (1/2)^2]\}^{-1} > \sum_{k=0}^{\infty} (n+k+1)^{-2p} =$
τ_n. Whence true for all m.

4. (a) From the binomial theorem $(1 + a^{-1})^k = 1 + (k/a) + k(k-1)/(2a^2) + \ldots$
$> 1 + k/a$.

(b) Noting that $\tau_n = t_{n+1} \sum_{k=0}^{\infty} [(n + 3/2)/(n + k + 3/2)] x^k$ and using the result of part (a) with $a = n + 3/2$, $(1 - \mu x)^{-1} = \sum_{k=0}^{\infty} (\mu x)^k < (\tau_n/t_{n+1}) < \sum_{k=0}^{\infty} x^k = (1 - x)^{-1}$ where $\mu = (n + 3/2)/(n + 5/2)$.

(c) If $S_e = S_n + (1 - \mu x)^{-1}$ then $0 < (S_\infty - S_e) < [(1-x)^{-1} - (1 - \mu x)^{-1}] t_{n+1}$
$= x[1 + (n + 3/2)(1 - x)]^{-1} (1 - x)^{-1} t_{n+1}$.

Section 2.3.4

1. (a) Here $t(u) = u^{-m-1}$ so that $t''(u) = (m + 1)(m + 2) u^{-m-3} > 0$ for all $u > 0$. Thus (2.3.16) applies and since $\int_v^\infty u^{-m-1} du = v^{-m}/m$, therefore $(n + 1)^{-m-1}(n + 1 + m/2) < m\tau_n < (n + 1/2)^{-m}$. Quoted result follows after multiplication throughout by $(n + 1)^m$.

   (b) Comparison series gives $(n + 1)/(n + 2) < 2(n + 1)^2 \tau_n$ whereas (with $m = 2$) part (a) gives $1 + (n + 1)^{-1} < 2(n + 1)^2 \tau_n$. But $(n + 1)/(n + 2) < 1 < 1 + (n + 1)^{-1}$, so lower limit is smaller than that of (2.3.16). Again, the comparison series gives $2\tau_n < (n^2 + n)^{-1}$, whereas from part (a), $2\tau_n < (n + 1/2)^{-2} = (n^2 + n + 1/4)^{-1}$. Thus the upper limit is larger than that of (2.3.16).

2. (a) $t(u) = u^{-1} - \ln(u+1) + \ln u$; and so $\int_v^\infty t(u) du = \{(u+1)\ln[u/(u+1)]\}_v^\infty = (v + 1)\ln[(v + 1)/v] - 1$ since $(u + 1)\ln(1 + u^{-1}) \to (u + 1)/u \to 1$ as $u \to \infty$.

   (b) Expand $\ln(1 + k^{-1})$ as series $k^{-1} - (k^{-2}/2) + (k^{-3}/3) - + \ldots$ so that $t_n = (n^{-2}/2) - (n^{-3}/3) + (n^{-4}/4) - + \ldots$. Similarly $t(u) = (u^{-2}/2) - (u^{-3}/3) + \ldots$ and $\int_v^\infty t(u) du = (v^{-1}/2) - (v^{-2}/6) + (v^{-3}/12) - + \ldots = -\sum_{j=1}^{\infty} (-v)^{-j}/[j(j+1)]$. Whence $t_{100} = 0.5000 \times 10^{-4} - 0.0033 \times 10^{-4} = 0.4967 \times 10^{-4}$ and $\int_{100.5}^\infty t(u) du = 0.4975 \times 10^{-2} - 0.0017 \times 10^{-2} = 0.4958 \times 10^{-2}$.

3. Here $t(u) = (u^3 + 1)^{-1/2}$, $t'(u) = -(3u^2/2)(u^3 + 1)^{-3/2}$ and $t''(u) = (3/4)(5u^4 - 4u)(u^3 + 1)^{-5/2}$ so that $t''(u) > 0$ if $u \geqslant 1$. Therefore use estimate (2.3.17) with truncation error $|E_n| < |t'(n + 1/2)/8| < 10^{-3}$. This implies that $n = 8$ is large enough, and $S_e = S_8 + \int_{8.5}^\infty t(u) du$. But $\int_{8.5}^\infty t(u) du = \int_{8.5}^\infty (u^3 + 1)^{-1/2} du = \int_{8.5}^\infty u^{-3/2}(1 + u^{-3})^{-1/2} du = \int_{8.5}^\infty u^{-3/2}[1 - (u^{-3}/2) + \ldots] du = [-2u^{-1/2} + (u^{-7/2}/7) + \ldots]_{8.5}^\infty$ integrating term-by-term. and this integral equals therefore 0.686 (second and subsequent terms of series being negligible). Summing, $S_8 = 2.608$ and so $S_e = 3.294$.

4. (a) $t(u) = (u + 1/2)^{-1} \exp(-pu)$, and differentiating twice, $t''(u) = [p^2(u + 1/2)^2 + p(2u + 1) + 2](u + 1/2)^{-3} \exp(-pu) > 0$ for all $u > 0$ and $p > 0$ (i.e. $|x| < 1$). Thus (2.3.16) applies.

   (b) $\int_{n+1/2}^\infty t(u) du = \int_{n+1/2}^\infty (u + 1/2)^{-1} \exp(-pu) du = \exp(p/2) \int_{(n+1)p}^\infty s^{-1} e^{-s} ds$ where $s = (u + 1/2)p$. But $\exp(p/2) = |x|^{-1/2}$: whence quoted result.

5. (a) Averaged series is $(1/2)+(1/3)-(1/15)+(1/35)-+\ldots=1/2+\sum_{k=0}^{\infty}(-1)^k$
$[(2k+1)(2k+3)]^{-1}$. From (2.3.24), $|E_n|<[(2n+3)(2n+5)]^{-1}$. Thus
$n=49$.

(b) Liebnitz sum is $(2/3)+(2/35)+\ldots = 2\sum_{k=0}^{\infty}[(4k+1)(4k+3)]^{-1}$.
The term magnitude function and its derivative are: $t(u)=$
$2[(4u+1)(4u+3)]^{-1} = (4u+1)^{-1}-(4u+3)^{-1}$. $t'(u)=$
$4[-(4u+1)^{-2}+4(4u+3)^{-2}] = 32(2u+1)(4u+1)^{-2}(4u+3)^{-2} =$
$32(2u+1)[4(2u+1)^2-1]^{-2}$. From (2.3.18), therefore $|E_n|<8(n+1)$
$[16(n+1)^2-1]^{-2}$. This implies $|E_n|<10^{-4}$ if $n=6$.

SECTION 2.4

1. Abbreviated results:

			S_k	$\sum_{j=25-k}^{24} t_j$	S_k	Spill	Kahan's device	
k	$k*k$	$1/(k*k)$	$(p=2)$	$(p=2)$	$(p=4)$	register	spill	sum
1	1	1	1	.0017	1	0	1	1
2	4	.25	1.3	.0036	1.250	−.050	.25	1.3
3	9	.11	1.4	.0057	1.360	−.040	.06	1.4
4	16	.063	1.5*	.0080	1.423	−.077	.023	1.4
5	25	.040		.011	1.463	−.037	.063	1.5
6	36	.028		.014	1.491	−.009	−.009	1.5
7	49	.020		.017	1.511	.011	.011	1.5
11	120	.0083		.033	1.557	.057	.057	1.6
12	140	.0071		.039	1.564	.064	−.036	1.6
13	170	.0059		.046	1.570	.070	−.030	1.6
17	290	.0034		.092	1.586	.086	−.014	1.6
18	320	.0031		.11	1.589	.089	−.011	1.6
19	360	.0028		.14	1.592	.092	−.0082	1.6
23	530	.0019		.60	1.601	.10*	.0006	1.6
24	580	.0017		1.6	1.603		.0023	1.6

*Summand is saturated from here onwards.

(a) *sum* saturated and equal to 1.5;
(b) *sum* = 1.6;
(c) accumulator rounds to single length 1.6;
(d) either (*sum* + *spill*) saturated and equal to 1.6, or (by Kahan's device)
sum = 1.6.
(N.B. correct *sum* is 1.60572 ...).

2. Abbreviated results:

k	Chopped values $k*k$	$1/(k*k)$	S_k $(p=2)$	$\sum\limits_{j=25-k}^{24} t_k$ $(p=2)$	S_k $(p=4)$	Spill register	Kahan's device: spill	sum
1	1	1	1	.0017	1	0	1	1
2	4	.25	1.2	.0036	1.250	.05	.25	1.2
3	9	.11	1.3*	.0056	1.360	.06	.16	1.3
4	16	.062		.0078	1.422	.12	.12	1.4
5	25	.040		.010	1.462	.16	.060	1.4
6	36	.027		.012	1.489	.18	.087	1.4
7	49	.020		.015	1.509	.20	.10	1.5
11	120	.0083		.031	1.554	.23*	.045	1.5
12	140	.0071		.037	1.561		.052	1.5
13	160	.0062		.044	1.567		.058	1.5
17	280	.0035		.089	1.583		.074	1.5
18	320	.0031		.10	1.586		.077	1.5
19	360	.0027		.12	1.588		.079	1.5
23	520	.0019		.58	1.595		.086	1.5
24	570	.0017		1.5	1.596		.087	1.5

*Summand is saturatee from here onwards.

   (a) *sum* saturated and equal to 1.3;
   (b) *sum* $= 1.5$;
   (c) accumulator converts to single length 1.5;
   (d) either (*sum* + *spill*) saturated and chops to 1.5, or (by Kahan's device) *sum* $= 1.5$.

3. Ration of successive terms $t_k/t_{k-1} = -x^2/(4k^2) = -(x/2k)^2$. Thus if $x \simeq 20$, terms t_{10} and t_9 are largest and approximately cancel. The magnitude of $t_9 = (x^2/4)^k/(k!)^2 \simeq 7.6 \times 10^6$. This has to be calculated correctly to 4th decimal place, implying a need for a precision of (at least) 11 decimal digits.

4. (a) In (1.7.5) the β's are ignored, and the vast majority of the partial sums are approximately the final sum, so that error is nearly the sum of the α's.
   (b) $\bar{\alpha} \simeq 0$, since as many of the α's would be negative as positive.
   (c) The error in the jth term is $\beta_j \epsilon$ and contributes a relative error $\beta_j t_j \epsilon/S_N$ to the sum S_N. The expected relative error would therefore equal
$$(\bar{\beta}\,\epsilon/S_N)\sum_{j=1}^{N} jt_j.$$

(d) Here the generated error would be $(\overline{\beta}\,\epsilon/S_N)\sum_{j=1}^{N}(n-j)t_j = \overline{\beta}\,\epsilon\sum_{j=1}^{N}S_j/S_N$ and as in part (a) this is roughly $N\overline{\beta}\,\epsilon$, whereas the generated error in summation is given by placing $s_j = S_N - S_{N-j}$ and $\sum_{j=1}^{N}(S_N - S_{N-j}) = \sum_{j=1}^{N}jt_j$, so that emphasis in source of error is reversed, as compared with (c).

SECTION 2.5

1. Let $I_n = \int_x^\infty t^{-n-1}e^{-t}dt$ so that $E_1(x) = I_0$. Then integrating by parts $I_n = [-e^{-t}/t^{n+1}]_x^\infty - (n+1)\int_x^\infty t^{-n-2}e^{-t}dt = (e^{-x}/x^{n+1}) - (n+1)I_{(n+1)}$. Thus $E_1(x) = I_0 = (e^{-x}/x) - I_1 = (e^{-x}/x) - (e^{-x}/x^2) + 2I_2 = (e^{-x}/x) - (e^{-x}/x^2) + (2e^{-x}/x^3) - 6I_3$, and so on. Whence stated result with $(e^{-x}/x)E_n = (-1)^{n+1}(n+1)!\,I_{n+1}$, i.e., $(e^{-x}/x)\,|E_n| = (n+1)!\int_x^\infty t^{-n-2}e^{-t}dt < (n+1)!\,x^{-n-2}\int_x^\infty e^{-t}dt = (n+1)!\,x^{-n-2}e^{-x}$. Infinite series is divergent, since terms (ultimately) diverge in magnitude.

2. (a) Term-ratio is $t_{k+1}/t_k = -(k+1)/x$. Thus minimum term is t_{n+1} where n is largest value for which $(n+1) \leqslant x$, i.e., $n = \text{entier}(x-1) = m-1$. Thus $|t_m| < m!\,x^{-m} \leqslant m!/m^m$.

   (b) Use (2.5.7), noting that $E_0 < 0$.

3. Assuming that $x > 2$, then the series will sum to a value between 1 and 1/2 (as follows by placing $n = 0$ in question 1). Hence for truncation error to be compatible with rounding error, $|E_n| < 2^{-25}$. From question 2, we find this bound implied by $x > 19.74$ and $m = 19$ (i.e., 19 terms of series required).

4. $\ln\{[1+(a/N)]^{bN+c}\} = (bN+c)\ln[1+(a/N)] = (bN+c)[(a/N) - (a^2/2N^2) + \ldots] = ab[1 + (c/bN)][1 - (a/2N) + \ldots] \to ab$, as $N \to \infty$. Thus required limit is $\exp(ab)$. Therefore from (2.5.7), as $N \to \infty$:

$$N!/(N+a)^{N+c} \simeq (2\pi)^{1/2}N^{N+1/2}\exp(-N)/[N^{N+c}(1+a/N)^{N+c}]$$
$$\simeq (2\pi)^{1/2}N^{1/2-c}\exp(-N-a) \ .$$

5. (a) Series is: $-(1/12z) - (1/360z^3) - (1/1260z^5) - (1/1680z^7) - (1/1180z^9) - \ldots$. Thus $t_2 = t_1$ if $z^2 = 1/30$: i.e. $t_2 = t_1 = -0.4564$ at $z = 0.1826$, and $t_3 = t_2$ if $z^2 = 2/7$: i.e. $t_3 = t_2 = -0.0182$ at $z = 0.5345$. It follows that t_2 is the smallest term (i.e. $t_2 < t_1$ and $t_2 < t_3 < t_4 \ldots$), and upper bound to $|E_n|$ is rendered least by choice of $n=1$, if $0.1826 < z < 0.5345$, the bound decreasing from 0.4654 to 0.0182 over this range. Similarly:

Optimal n =	←1→		←2→	←3→	←4→	←5→
From/to z =	0.1826	0.5345	0.8660	1.189	1.509	1.828
Error bound =	0.456	1.82×10^{-2}	1.63×10^{-3}	1.77×10^{-4}	2.07×10^{-5}	2.51×10^{-6}

   (b) Using (2.5.6), $t_j \simeq \pi^{-1}(2j-2)!(2\pi z)^{1-2j}$, so that $t_{j+1}/t_j = j(j-1/2)/(\pi z)^2$. Hence $t_{j+1} = t_j$ if $j^2 - (j/2) - \pi^2 z^2 = 0$, i.e. $j = (1/4) + [\pi^2 z^2 + (1/16)]^{1/2}$.

But, using binomial theorem, $[\pi^2 z^2 + 1/16]^{1/2} = \pi z[1 + (1/32\pi^2 z^2) + ...]$. Whence stated result since $\pi z \to \infty$ if $j \to \infty$. At this value of z, least upper bound is given by choosing $n = (j-1)$ or j and then $|E_n| < |t_{j+1}| \simeq \pi^{-1}(2j)!/(2\pi z)^{2j+1} = \pi^{-1}(2j)!/(2j-1/2)^{2j+1}$. Putting $N = 2j$, $a = -1/2$ and $c = 1$ in result of question 4, therefore $|E_n| < (e/\pi j)^{1/2}\exp(-2j)$. For $j = 6$, $z = 5.75/\pi = 1.830$ and $|E_n| < \exp(-11.5)/(6\pi)^{1/2} = 2.33 \times 10^{-6}$.

6. (a) Differentiating repeatedly, $t^{(k)}(u) = (-1)^k(k+1)!\, u^{-k-2}$, whence quoted result for b_{nj}. Series only semi-convergent, since for $j \gg 1$ the term ratio is $|b_{n,j+1}/b_{n,j}| = (B_{j+1}/B_j)(n+1)^{-2} \simeq (2j+1)(2j+2)/[2\pi(n+1)]^2$ by (2.5.6), and so becomes greater than 1 for all $j \geqslant \pi(n+1)$.

(b) $|b_{2,j}|$ decreases with increasing j up to $j = 9$. Thus least upper bound on $E_{2,m}$ is provided by taking $m = 8$, and $|E_{2,8}| < |b_{2,9}| \simeq 5 \times 10^{-8}$.

(c) $|E_{5,5}| < |b_{5,6}| = B_6/6^{13} \simeq 2 \times 10^{-11}$. Further $S_5 + t_6/2 = 1.4775$ precisely; $\int_6^\infty u^{-2}\, du = 1/6$; and $\sum_{j=1}^{6} b_{nj} = 7.6740018 \times 10^{-4}$; therefore $S_\infty = 1.64493406685 + E_{5,5}$. (Correct value is $\pi^2/6$, correct to 12 significant digits as quoted.)

7. Since $\tau_n = S_\infty - S_n$, and (with $m = 0$) it is given that $|E'_{n,0}| < |b'_{n,1}|$, modified formula shows that the sign of $[\tau_n - \int_{n+1/2}^\infty t(u)du]$ is the same as the sign of $-b'_{n,1}$, which in turn is that of $t'(n+1)$ and therefore negative: that is, $\tau_n < \int_{n+1/2}^\infty t(u)du$. Similarly, the summation formula of problem 6 shows that the sign of $[\tau_n - (t_{n+1}/2) - \int_{n+1}^\infty t(u)du]$ is that of $(b_{n,1} + E_{n,0})$, which is the sign of $b_{n,1}$ (since $|E_{n,0}| < |b_{n,1}|$) and so that of $-t'(n+1)$; whence $\tau_n > (t_{n+1}/2) + \int_{n+1}^\infty t(u)du$.

8. $\zeta(1.01) = 100.5779433 \pm |E'_{3,3}|$. But $|E'_{3,3}| < |b'_{3,4}| \simeq 2 \times 10^{-7}$. If by direct summation, truncation error were not to be greater, then τ_n must be $< 2 \times 10^{-7}$. Estimating τ_n as approximately $\int_{n+1/2}^\infty t(u)\, du = [-100/u^{0.01}]_{n+1/2}^\infty \simeq 100/n^{0.01}$, this implies $n^{0.01} > 5 \times 10^8$, i.e., $n > 10^{870}$ (approx). That's a lot of terms!

9. Here $t(u) = u^{-1}$, so that $t^{(k)}(u) = (-1)^k k!\, u^{-k-1}$. Thus $b_{nj} = (-1)^{j+1}(B_j/2j)(n+1)^{-2j}$, and $|E_{5,5}| < 4095\,|b_{5,6}| \simeq 4 \times 10^{-8}$. Further $S_5 + (t_6/2) = 0.7$ precisely, and $\sum_{j=1}^{6} (4^j - 1)b_{nj} = 6.85281 \times 10^{-3}$. Since $\text{sgn}(t_6) = -1$, therefore $S_\infty = 0.69314719 + E_{5,5}$. (Correct value is $\ln 2 = 0.69314718$).

SECTION 2.6

1. (a) Using recursion:

```
gamma: = if z ⩾ N then strlg(z)
            else if z ⩾ 0.5 then gamma (z+1)/z
            else pi/sin(pi*z)/gamma (1–z);
```

Using iteration:

```
real d, x;
if z ⩾ 0.5 then begin d: = 1; x: = z end
          else begin d: = pi/sin(pi*z); x: = 1−z end failing at a pole;
while x < N do begin d: = d * x; x: = x+1 end;
gamma: = if z ⩾ 0.5 then strlg(x) d else d/strlg(x);
```

(b) $\Gamma(-1.5) = \pi/s$. Since $\Gamma(1/2)^2 = \pi = (4s/3)^2$, therefore $s = 3\pi^{\frac{1}{2}}/4$.

2. (a) Both sides of consistency equation are equal to: $f(z) + f(1-z) + f(1/z) + f[1/(1-z)] + f[(z-1)/z] + f[z/(z-1)]$.

   (b) $f(z) = A(z) - B(1-z) + f[(1/(1-z)]$,   (C)
   $= B(z) - A(1/z) + f[(z-1)/z]$.   (D)
   $= E(z) - f[z/(z-1)]$,   (E)
   where $E(z) = A(z) - B(1-z) + A[z/(z-1)] = B(z) - A(1/z) + B[(z-1)/z]$, in conformity with consistency equation of part (a).

   (c) Range: $z < -1$ $-1 \leqslant z < 0$ $1/2 < z \leqslant 1$ $1 \leqslant z \leqslant 2$ $2 \leqslant z$
       Relation: (C)   (E)   (A)   (D)   (B)

3. (a) $L_2(0) = 0$, by definition; $L_2(-1) = -c$, by (B); $L_2(1/2) = c - (\ln 2)^2/2$, by (A); $L_2(1) = 2c$, by (A) or (B); and $L_2(2) = 3c$. From integral, $dL_2(z)/dz = -(\ln|1-z|)/z = 0$ at $z = 2$. Also $L_2(z) = 0$ for some $z > 2$, since $L_2(2) > 0$ and $L_2(z) = -(\ln z)^2/2 < 0$ for large $z \to \infty$.

   (b) The series in w is alternating and converges much more rapidly (only terms up to $k = 3$ required even at $x = 0.5$), but formation of $w = \ln(1-x)$, which may not otherwise be needed, is computationally time-consuming, and loses significance for $x \leqslant 0.25$, particularly for very small x. Latter objection would be met by forming w as equal instead to -2 argtanh $[x/(2-x)]$. The series in x, on the other hand, converges sufficiently rapidly to allow summing to saturation: only 16 terms are required at $x = 1/2$; and its coefficients are readily computed.

SECTION 2.7

1. Suppose (A_j/B_j) for $j = 1, 2, \ldots k$ are first k convergents. Then the $(k+1)$th differs from the kth only by q_k being replaced by $q_k + (p_{k+1}/q_{k+1})$, and q_k only affects the value of the kth convergent: $(A_k/B_k) = (q_k A_{k-1} + p_k A_{k-2})/(q_k B_{k-1} + p_k B_{k-2})$. Hence the $(k+1)$th convergent is a quotient with numerator $[q_k + (p_{k+1}/q_{k+1})]A_{k-1} + p_k A_{k-2} = A_k + (p_{k+1}/q_{k+1})A_{k-1}$, and a denominator formed by replacing A by B; hence it is $(q_{k+1} A_k + p_{k+1} A_{k-1})/(q_{k+1} B_k + p_{k+1} B_{k-1}) = (A_{k+1}/B_{k+1})$. Proof follows by induction since $A_1/B_1 = p_1/q_1$ is 1st convergent.

2. (a) Use (2.7.1) with $p_1 = 1, p_{j+1} = r_j z, c_0 = 1/z, c_j = z$ for $j = 1, 2, \ldots$

(c) Use results of parts (a) and (b). $s_{2j+1} = (2j)!/(2^j j!)^2$. $s_{2j+2} = 2^{2j+1}(j!)^2/(2j+1)!$, for $j = 0, 1, \ldots$

3. (a) $[1 + (r_2 + r_3)z]/[1 + (r_1 + r_2 + r_3)z + r_1 r_3 z^2]$.
   (b) $r_1 = -a, r_2 = (a^2 + 1)/a, r_3 = -1/a$.
   (c) First derivative zero at $z = 0$.

4. (a) Follows from definition
   (b) If K denotes converged value, then

   $$\overset{\infty}{\underset{j=1}{K}}\ p \downarrow q = \overset{\infty}{\underset{j=2}{K}}\ p \downarrow q = K \text{ and from part (a), } K = p/(q + K). \text{ Solve for } K.$$

   (c) Placing $c_j = -1$ for $j = 0, 1, 2 \ldots$ in (2.7.1), we see that K changes sign

   with q: i.e. $\overset{\infty}{\underset{j=1}{K}}\ p \downarrow q = -\overset{\infty}{\underset{j=1}{K}}\ p \downarrow (-q)$. Hence $\overset{\infty}{\underset{j=1}{K}}\ p \downarrow q = (q/2)\,[\pm(1 + 4p/q^2)^{\frac{1}{2}} - 1]$. But if p and q are positive, the converged value must be positive; hence positive square-root must be chosen.

5. (a) In (2.7.6), $r_{j-1} = 1$ for all j. Hence A_{j+1} and B_j are solutions of same difference equation for any j, and $B_0 = B_1 = 1$ whereas $A_1 = A_2 = 1$. Hence $B_k = A_{k+1}$. For $z = 2$: $A_k/B_k = A_k/A_{k+1} = (1/2)\,[1-(-1)^k 2^{-k}]/[1 + (-1)^k 2^{-k-1}] = (1/2)\,[1-(-1)^k 2/2^{k+1}]\,[1-(-1)^k/2^{k+1} + \ldots]$ using binomial theorem. Whence stated result if $2^k \gg 1$.
   (b) $A_k/B_k = A_k/A_{k+1} = 2k/(k + 1)$. Note less rapid convergence at $z = -1/4$ than at $z = 2$.

6. (a) Eliminate A_{k+1} and A_{k-1} from the three equations $A_j = A_{j-1} + b_j A_{j-2}$ formed by placing $j = k, k + 1$ and $k + 2$.
   (b) Place $\bar{A}_j = A_{2j}, \bar{B}_j = B_{2j}$, and identify successive convergents of the continued fraction on the r.h.s. with $\bar{A}_j/\bar{B}_j$; note that $\bar{A}_0 = 0, \bar{A}_1 = A_2 = 1$ and $\bar{B}_0 = 1, \bar{B}_1 = B_2 = 1 + b_2$.
   (c) Put $b_j = r_{j-1}z^{-1}$ for $j = 2, 3, \ldots$ in result of (b) and then apply similarity transform with $c_0 = z^{-1}$, and $c_j = z$ for all $j = 1, 2, \ldots$

7. Note that the reciprocal of the augmented continued fraction is equal to

$$B'_n/A'_n = \overset{n}{\underset{j=0}{K'}}\ p_j \downarrow q_j = \overset{n+1}{\underset{j=1}{K'}}\ p_{j-1} \downarrow q_{j-1} = A''_{n+1}/B''_{n+1}, \text{ say, where } A''_j, B''_j \text{ obey}$$

(2.7.5) except that p_{j-1}, q_{j-1} replace p_j, q_j respectively (and p_0 is taken as unity). Then identify B'_{j-1} with A''_j and A'_{j-1} with B''_j.

8. (a) Place $A'_j = A_{2j+1}, B'_j = B_{2j+1}, a_j = b_j$ in result of problem 6(a), and identify successive convergents of augmented continued fraction on the r.h.s. with A'_j/B'_j. Note that $A'_{-1} = A_{-1} = 1, A'_0 = A_1 = a_1, B'_{-1} = B_{-1} = 0, B'_0 = B_1 = 1$.
   (b) Match the partial numerators and denominators of the continued fraction on the r.h.s. of the identity in part (a) with $p_1, q_1, p_2, q_2, \ldots$ in problem 6(b).

9. Assume that $\overset{k}{\underset{j=0}{K'}} r_j z \downarrow 1 = A_{k+1}/B_{k+1}$ where A_j, B_j are given by (2.7.6) and propose that A_j is of degree entier $[(j-1)/2]$ and B_j of degree entier $(j/2)$ for all positive $j \leqslant k$. Confirm from (2.7.6) that proposition is true for $j = k + 1$, and is true by definition for $k = 2$.

SECTION 2.8

1. Divide the equation $B_k = q_k B_{k-1} + p_k B_{k-2}$ throughout by B_{k-1} and place $\beta_k = B_{k-1}/B_k$ to show that $\beta_k^{-1} = q_k + p_k \beta_{k-1}$. Next eliminate q_k between the equations for A_k and B_k to show that $(A_k - p_k A_{k-2})/A_{k-1} = (B_k - p_k B_{k-2})/B_{k-1}$. Multiply both sides of this equation by A_{k-1}/B_k and place $t_k = (A_k/B_k) - (A_{k-1}/B_{k-1})$.

2. (a) Use (2.8.1) with $j = k + 1$ and substitute for t_k, $t_{k-1}, \ldots t_1$.

(b) As $z \to 0$, $\beta_j \to 1$ for $j = 1, 2, \ldots$ Thus $t_{n+1} = O(z^n)$ as $z \to 0$ by part (a). But t_{n+1} is difference between nth and $(n + 1)$th convergents, and evidently $d^k t_{n+1}/dz^k = O(z^{n-k})$ for $k = 1, 2, \ldots (n - 1)$.

3. 9th convergent is $(e^{-x}/x) \overset{9}{\underset{j=1}{K'}} entier (j/2) x^{-1} \downarrow 1 = 0.91303628(e^{-x}/x) = 6.095635 \times 10^{-6}$. Truncation error is between 0 and $t_{10} = -9.63 \times 10^{-8}(e^{-x}/x)$. Relative error is therefore less in magnitude than 1.1×10^{-7}, much smaller than that of series.

4. With $p_1 = 1$, $p_j = entier\,(j/2) x^{-1}$ for $j = 2, 3, \ldots$ and $q_j = 1\;\forall\,j$, use similarity transform with $c_0 = c_1 = 1$, and $c_{2j} = (j - 1)!/j! = 1/j$, $c_{2j+1} = j!/(j!\,x^{-1}) = x$, $(j = 1, 2, \ldots)$ to transform S-fraction to $\overset{\infty}{\underset{j=1}{K}} 1 \downarrow c_j$. But clearly $\overset{\infty}{\underset{j=1}{\Sigma}} c_j$ diverges.

5. $\overset{n}{\underset{n\;\;j=0}{K'}}(j/2) \downarrow 1.$ Repeated averages

n					
4	.73684					
5	.76923	.75304				
6	.75145	.76034	.75669			
7	.76167	.75656	.75845	.75757		
8	.75556	.75861	.75759	.75802	.75779	
9	.75933	.75744	.75803	.75781	.75791	.75785
10	.75694	.75813	.75779	.75791	.75786	.75788

It appears that correct value to 4 decimal places is 0.7579. (Without averaging this same accuracy is achieved with $n > 20$, precise value is 0.7578722. .).

6. (a) Solve equation $\beta = 1/(1 + p\beta)$ and select positive root (since $\beta_j > 0$ if $p_j, q_j > 0$). Then $t_{j+1}/t_j \to -p\beta^2 = -2p/[1 + 2p + \sqrt{(1 + 4p)}]$.

(b) Converting to an S-fraction by similarity transform, given continued fraction is equivalent to $\overset{\infty}{\underset{j=0}{K'}}$ $[j^2 x^2/(4j^2 - 1)] \downarrow 1$, so that partial numerators $\to (x/2)^2$ as $j \to \infty$. Thus from part (a), the asymptotic term-ratio in its series representation has magnitude $x^2/[x^2 + 2 + 2\sqrt{(1 + x^2)}] < x^2/4$ clearly less than that of power series $(= x^2)$.

(c) 2nd convergent is $(1 + x^2/3)^{-1}$, and 2nd partial sum is $(1 - x^2/3)$ which is less (since $1 - x^4/9 < 1$). Both however are too small, since next term of alternating series is positive. Thus $S_2 < K_2 < S_\infty = K_\infty$, and so $|K_\infty - K_2| < |S_\infty - S_2|$.

7. (b) Put $r_j = t_j/t_{j-1} \simeq - [(j - 1)/j]^\alpha$, and so $b_j = 1 + r_j \simeq 1 - [(j - 1)/j]^\alpha = 1 - (1 - j^{-1})^\alpha \simeq \alpha/j$, as $j \to \infty$. Therefore $a_j = - r_j/b_j b_{j-1} \simeq - r_j j(j - 1)/\alpha^2 \simeq j^2/\alpha^2$ as $j \to \infty$.

8. (a) If $C_{m,n} = S_{m,n}/S_{m-1,n}$ for some $m = k \leqslant n$. Then $C_{k-1,n} = - r_{k-1}/(1 + r_{k-1} + C_{k,n}) = - r_{k-1}/(1 + r_{k-1} - S_{k,n}/S_{k-1,n}) = - S_{k-1,n}/[S_{k-1,n} + (S_{k-1,n} - S_{k,n})/r_{k-1}] = - S_{k-1,n}/(S_{k-1,n} + t_{k-2})$ and so relation is true for $m = k - 1$. But it is true for $m = n$, and so true for all $m \leqslant n$.

(b) $t_1 \overset{n}{\underset{j=1}{K'}} (-r_j) \downarrow (1 + r_j) = t_1/(1 + C_{2,n}) = t_1/(1 - S_{2,n}/S_{1,n}) = S_{1,n}$.

(c) Applying similarity transform $\overset{n}{\underset{j=1}{K}}$ $p_j \downarrow q_j = (1/\beta_0) \overset{n}{\underset{j=1}{K}} \beta_j \beta_{j-1} p_j \downarrow \beta_j q_j$. Place $\beta_1 q_1 = 1$, $t_1 = p_1 \beta_1 = p_1/q_1$, $r_j = - \beta_{j-1} \beta_j p_j \, \forall j \geqslant 2$, where $\beta_j q_j + \beta_j \beta_{j-1} p_j = 1$, i.e., $\beta_j = (q_j + \beta_{j-1} p_j)^{-1}$.

10. (a)

k	A_k	B_k	A_k/B_k	β_k	t_k	$\overset{k}{\underset{j=1}{\Sigma}} t_j$	C_{4-k}
0	0	1	0	-	-	-	0
1	1	1	1	1	1	1	0.0123
2	1	−0.0100	−100	−100	−101	−100	−1.00
3	1.01	0.0023	439	−4.35	540	440	∞

Backward recursion fails "zero-divide". Correct answer 440.

(b)

0	0	1	0	-	-	-	0
1	1	1	1	1	1	1	−1.01
2	1	1.01	0.990	0.990	−0.0122	0.988	−1.23
3	−0.01	0	∞	∞			−4.35

Forward recursion and summation fails "zero-divide". Correct answer −4.35.

11. (a) Augmented form evaluates as Z_{-1}/Z_0.

   (b) Here $p_1 = q_1 = q_2 = \ldots = q_N = 1$, and $p_{k+1} = r_k z$ for all $0 < k < N$.
Thus, for example:

$$a_0 := b_0 := 1;$$
for $j := 1$ **step** 1 **until** entier $(N/2)$ **do** $a_j := b_j := 0;$
for $k := N-1$ **step** -1 **until** 1 **do**
    for $j :=$ entier $((N+1-k)/2)$ **step** -1 **until** 1 **do**
    begin $a_j := b_j; b_j := b_j + r_k{}^* a_{j-1}$
    end

SECTION 2.9

1. S-fraction is $\dfrac{x^{-1}}{1+}\ \dfrac{x^{-1}}{1+}\ \dfrac{x^{-1}}{1+}\ \dfrac{2x^{-1}}{1} = (1 + 3x^{-1})/(1 + 4x^{-1} + 2x^{-2}) =$
$(1 + 3x^{-1})[1 - 4x^{-2}(1 + x^{-1}/2) + 16x^{-2}(1 + x^{-1}/2)^2 - 64x^{-3}(1 + x^{-1}/2)^3 +$
$\ldots] = (1 + 3x^{-1})(1 - 4x^{-1} + 14x^{-2} - 48x^{-3} + \ldots) = (1 - x^{-1} + 2x^{-2} -$
$6x^{-3} + \ldots).$

2. (a) Using the Q-D algorithm, table of r_{jk} is:

      1
      0     $1/3 = r_1$
      0     $7/5$ $16/15 = r_2$
      0   $155/49$ $432/245$ $81/35 = r_3$
      0   $889/155$ $19536/7595$ $2035/441$ $256/63 = r_4$

   (b) $r_j = j^4/(4j^2 - 1)$ for $j = 1, 2, \ldots$ Converges for $z > 0$.

   (c) $\overset{3}{\underset{j=0}{K'}}\ r_j \downarrow 1 = 115/144 = .7986$, $\overset{4}{\underset{j=0}{K'}}\ r_j \downarrow 1 = 3019/3600 = .8386$ whereas
$\pi^2/12 = 0.8225.$

3. (a) Note from problem (9) of § 2.7 that

$$C_{m+1,m+2\ell}(z) = a_{m+1}\ \overset{2\ell-1}{\underset{k=0}{K'}}\ r_{m+1+k,k}z \downarrow 1 \text{ is a rational function of order}$$

$(\ell - 1)/\ell$, and that both the stated rational functions have the given
expansion about $z = 0$, by hypothesis.

   (b) Identify $zC_{m+1,m+2\ell}(z)$ with $\overset{2n}{\underset{j=1}{K}}\ b_j \downarrow 1$ by placing $n = \ell, b_1 = a_{m+1}z$, and
$b_j = r_{m+j,j-1}z$; and $C_{m,m+2\ell}(z)$ with $\overset{2n+1}{\underset{j=1}{K}}\ \alpha_j \downarrow 1$ by placing $\alpha_1 = a_m$ and
$\alpha_j = r_{m+j-1,j-1}z$ for $j = 2, 3, \ldots (2\ell + 1)$. Defining $r_{j,0} = 0\ \forall\ j > 0$, the
recurrence relations follow.

   (c) If $n = 2\ell$, then order is $(m + \ell)/\ell$, and if $n = 2\ell + 1$, it is $(m + \ell)/(\ell + 1)$.

4. (a) $r_{jj} = r_j$, $r_{j,0} = 0$.

$r_{jk} = r_{j,k+1} + r_{j-1,k} - r_{j-1,k-1}$ $(k$ odd$)$ $\Big\}$ $k = j - 1, j - 2, \ldots 1$
$\quad\ = r_{j,k+1} r_{j-1,k}/r_{j-1,k-1}$ $\quad (k$ even$)$

for $j = 1, 2, 3, \ldots$ The terms $r_{j,1}$ are then equal to $- a_j/a_{j-1}$, and $a_0 = 1$.

(b) The terms $r_{j,1} = (2j - 1)/(j + 1)$ and the series is that for $[(1 + 2z)^{\frac{1}{2}} - 1]/z$.
i.e., $1 - (1/2)z + (1/2)z^2 - (5/8)z^3 + (7/8)z^4 - (21/16)z^5 + \ldots$

(c) $1/\phi(z) = 1 + \overset{\infty}{\underset{j=1}{K}} \rho_j z \downarrow 1 = 1 + \rho_j z \overset{\infty}{\underset{j=0}{K'}} \rho_{j+1} z \downarrow 1 = 1 + \overset{\infty}{\underset{j=0}{\Sigma}} \rho_i a_j z^{j+1}$.

5. (a) Starting from row j and column 2ℓ the array has elements:

$$\begin{array}{ccccc} \epsilon & c_1 \epsilon & & & \\ c_2/\epsilon & c_2/\epsilon & & & \\ & -c_2/\epsilon & -c_2/\epsilon & * & * & \cdots \\ & c_3 \epsilon & & * & * & \cdots \end{array}$$

where $c_1 = a_{j-1,2\ell-1}/a_{j-1,2\ell}$, $c_2 = a_{j,2\ell-1} a_{j+1,2\ell}$, and asterisks denote values which will lose significance in computation due to cancellation error; e.g., $a_{j+2,2\ell+4} = c_2/\epsilon - a_{j+1,2\ell+3} - c_2/\epsilon$.

(b) QD-algorithm applied to the two functions gives:

1				1		
0 4				0 4		
0 5 1				0 5	1	
0 5 0 0				0 5	0	0
0 5 0 * *				0 49/10	$-1/10$	∞	∞
0 5 0 * * *				0 24/5	$-1/10$	49/10	$-\infty$ *

Asterisks denote indeterminate values. The *column* of zeros distinguishes the terminated S-fraction (on the left)

6. $e_0^{n+1} = 0$, $q_1^n = - a_{n+1}/a_n$, $e_m^n = e_{m-1}^{n+1} + q_m^{n+1} - q_m^n$, $q_{m+1}^n = q_m^{n+1} e_m^{n+1}/e_m^n$ for all $n \geqslant 0$ and $m \geqslant 1$.

7. (a) $F(a, c; c; z) = F(c, a; c, z) = \overset{\infty}{\underset{m=0}{\Sigma}} (a)_m z^m/m! = (1 - z)^{-a}$ as in problem 1, § 2.2.

(b) Place $a = c = 1/2$, $b = - 1/2$ and $z = x^2$ in Gauss's result, noting that $F(1/2, - 1/2; 1/2; x^2) = (1 - x^2)^{1/2}$, to obtain $\arcsin(x)/[x(1 - x^2)^{1/2}] =$

$$\cfrac{1}{1 -} \cfrac{2x^2}{3 -} \cfrac{2x^2}{5 -} \cfrac{12x^2}{7 -} \cfrac{12x^2}{9 -} \cfrac{30x^2}{11 -} \ldots$$ (Half fractions have been cleared

by a similarity transform.)

(c) Place $a = p$, $b = - q$, $c = p$ and $z = x$ in Gauss's result, noting that $F(p, - q; p; x) = (1 - x)^q$, to obtain: $B_x(p, q)/[x^p(1 - x)^q] =$

$$\cfrac{1}{p -} \cfrac{p(p + q)}{1 + p -} \cfrac{(1 - q)}{2 + p -} \cfrac{(1 + p)(1 + p + q)}{3 + p -} \cfrac{2(2 - q)}{4 + p -} \ldots$$

8. (a) Place $b = 0$ in Gauss's continued fraction: then since $F(a, 0; c; z) \equiv 1$,

$F(a, 1; c + 1; z) = c \underset{j=0}{\overset{\infty}{K'}} g_j z \downarrow (j + c)$ where $g_{2j+1} = -(j + a)(j + c)$ and

$g_{2j} = -j(j + c - a)$. (Note that the factor c cancels with c in first partial denominator and in $g_1 = -ac$.)

(b) (i) $a = c = 1, z = -x, x^{-1} \ln(1 + x) = \underset{j=1}{\overset{\infty}{K'}} [entier (j/2)]^2 x \downarrow j$.

(ii) $a = c = 1/2, z = x^2, x^{-1} \operatorname{argtanh}(x) = \underset{j=0}{\overset{\infty}{K'}} (-j^2 x^2) \downarrow (2j + 1)$.

(iii) $a = c = 1/2, z = -x^2, x^{-1} \arctan(x) = \underset{j=0}{\overset{\infty}{K'}} j^2 x^2 \downarrow (2j + 1)$.

(iv) $a = 1 - \alpha, c = 1, z = -x \quad (\alpha x)^{-1} \quad [(1 + x)^\alpha - 1] =$

$$= \frac{1}{1 +} \frac{(1 - \alpha)x}{2 +} \frac{(1 + \alpha)x}{3 +} \frac{2(2 - \alpha)x}{4 +} \cdots$$

9. (a) Put $b = \alpha, c = \beta$ and $z = x/a$ in Gauss's result and let $a \to \infty$. This gives first result. With $a = \alpha, c = \beta$ and $z = x/b$, then the second result follows by considering the limit, $b \to \infty$.

(b) $[\exp(x) - 1]/x = \underset{k=0}{\overset{\infty}{\Sigma}} x^k/(k + 1)! = \Phi(1, 2; x)$, and since $\Phi(0, \beta; x) \equiv 1$,

$[\exp(x) - 1]/x = \Phi(1, 2; x)/\Phi(0, 1; x) = \underset{j=1}{\overset{\infty}{K'}} (-1)^{j+1} entier (j/2) x \downarrow j$.

(c) From (2.6.2), $\operatorname{erf}(x) = x \Phi(1/2, 3/2; -x^2)$ so that since $\Phi(\alpha, \alpha; z) = \exp(z)$, $x^{-1} \exp(x^2) \operatorname{erf}(x) = \Phi(1/2, 3/2; -x^2)/\Phi(1/2, 1/2; -x^2)$. Again from (2.6.3), $\operatorname{erf}(x) \equiv x \exp(-x^2) \Phi(1, 3/2; x^2)$ so that $x^{-1} \exp(x^2) \operatorname{erf}(x) = \Phi(1, 3/2; x^2)/\Phi(0, 1/2; x^2)$. From part (a), both continued fraction

representations are identical: $x^{-1} \exp(x^2) \operatorname{erf}(x) = \underset{j=0}{\overset{\infty}{K'}} (-1)^j 2j x^2 \downarrow (2j + 1)$.

10. (a) Use similarity transform to obtain $F(a, b + 1; c + 1; z)/F(a, b; c; z) = \underset{j=0}{\overset{\infty}{K'}} G_j z/[(c + j - 1)(c + j)] \downarrow 1$, the G_j's being given as in problem 3.

(b) Since $F(a, 0; c; z) \equiv 1$ for any a, c or z, $\underset{m=0}{\overset{\infty}{\Sigma}} (\alpha)_m (-x)^{-m} = \underset{c \to \infty}{\lim}$ $[F(\alpha, 1; c + 1; -c/x)/F(\alpha, 0; c, -c/x)]$. Thus in part (a), with $a = \alpha$, $b = 0, z = -c/x$ and $c \to \infty$, $G_{2j+1} z/[(c + 2j)(c + 2j + 1)] = (j + \alpha)$ $(j + c)cx^{-1}/[(c + 2j)(c + 2j + 1)] = (j + \alpha)x^{-1}$. $G_{2j} z/[(c + 2j - 1)$ $(c + 2j)] = j(j + c - \alpha)cx^{-1}/[(c + 2j - 1)(c + 2j)] = jx^{-1}$. Hence $\underset{m=0}{\overset{\infty}{\Sigma}}$

$(\alpha)_m (-x)^m = \underset{j=0}{\overset{\infty}{K'}} r_j x^{-1} \downarrow 1$, where $r_{2j} = j$, and $r_{2j+1} = j + \alpha$.

11. (a) $J_\nu(z)/J_{\nu-1}(z) = (z/2\nu) \lim\limits_{\alpha\to\infty} \Phi(\alpha, \nu+1; -z^2/4\alpha)/\Phi(\alpha, \nu; -z^2/4\alpha) =$

$(z/2\nu) \lim\limits_{\alpha\to\infty} \left\{\nu \overset{\infty}{\underset{j=0}{K'}} \alpha(-z^2/4\alpha) \downarrow (j+\nu)\right\} = (z/2) \overset{\infty}{\underset{j=0}{K'}} (-z^2/4) \downarrow (j+\nu) =$

$z \overset{\infty}{\underset{j=0}{K'}} (-z^2) \downarrow 2(j+\nu).$

(b) With $\nu = 1/2$, and $z = iz$, $z^{-1}\tanh(z) = \overset{\infty}{\underset{j=0}{K'}} z^2 \downarrow (2j+1)$. Taking recipro-

cal and subtracting 1, $z \coth(z) - 1 = \overset{\infty}{\underset{j=1}{K}} z^2 \downarrow (2j+1)$.

SECTION 2.10

1. (b) The denominator of $C_1 = x/(1-C_2)$ is either small, or zero-divide occurs, because C_2 should be very close to 1.

(c) Here C_3 becomes very close to 3 so that $C_2 = x^2/(3-C_3)$ has either a large magnitude, or (possibly) a zero-divide error occurs. If C_2 is large, then C_1 is small. In general, if $x = n\pi/2$ then calculation should reveal that $C_2 \simeq 1$ if n is odd or $C_3 \simeq 3$ if n is even (unless $n = 0$, when all C's

are zero). These results are accounted for by the fact that $C_2 = \overset{\infty}{\underset{j=2}{K}} (-x^2) \downarrow$

$(2j-1)$ is the continued fraction for $1 - x \cot x$ and $C_3 = \overset{\infty}{\underset{j=3}{K}} (-x^2) \downarrow$

$(2j-1)$ is that for $3 - x^2(1 - x \cot x)^{-1}$.

3. (a) Let $A_k = c_1 X_k + c_2 Y_k$ where c_1, c_2 are constants. Then relations $c_1 = -Y_0, c_2 = X_0$ follow since $A_0 = 0$ and $A_1 = p_1$. Similarly for B_k, since $B_0 = 1, B_1 = q_1$, we find that $B_k = [(Y_1 - q_1 Y_0) X_k + (X_1 - q_1 X_0) Y_k]/p_1$. But by recurrence relation $Y_1 - q_1 Y_0 = p_1 Y_{-1}$, and similarly $X_1 - q_1 X_0 = p_1 X_{-1}$.

(b) Value of continued fraction is, from part (a), $\lim\limits_{n\to\infty} (A_n/B_n) =$

$\lim\limits_{n\to\infty} [(X_0 Y_n - Y_0 X_n)/(Y_{-1}X_n - X_{-1}Y_n)] = -X_0/X_1$ if $X_n/Y_n \to 0$.

(c) Let K be value of continued fraction. Then as in part (b), if $\varsigma = \lim\limits_{n\to\infty} (X_n/Y_n)$, it follows that $(X_0 - \varsigma Y_0)/(Y_{-1}\varsigma - X_{-1}) = K$, or

$\varsigma = (X_0 + KX_{-1})/(Y_0 + KY_{-1})$. Placing $X'_k = X_k - \varsigma Y_k$ and $\varsigma' = \lim\limits_{n\to\infty}$

(X'_n/Y_n), we find that $\varsigma' = (X'_0 + KX'_{-1})/(Y_0 + KY_{-1})$ as before, and $X'_0 + KX'_{-1} = (X_0 + KX_{-1}) - \varsigma(Y_0 + KY_{-1}) = 0$, so that $\varsigma' = 0$.

4. (a) From (2.10.5), and given continued fraction: $-Z_n/Z_{n-1} = \overset{\infty}{\underset{j=n+1}{K}} (-x^2) \downarrow$

$2(j+\nu) = \overset{\infty}{\underset{j=0}{K}} (-x^2) \downarrow 2(j+\nu+n+1) = x J_{\nu+n+1}(x)/J_{\nu+n}(x)$. But $Z_{-1} =$

$J_\nu(x) \therefore Z_0 = x J_{\nu+1}(x), Z_1 = x^2 J_{\nu+2}(x), \dots Z_n = x^{n+1} J_{\nu+n+1}(x)$.

(b) See next part.

(c) Use above continued fraction to calculate $(-Z_9/Z_8)$ and assume (say) $Z_9 = 1$ giving Z_8. Backward recursion then generates values of Z_k/Z_9, tabulated below for $k = 7, 6, \ldots 0, -1$. Scale all values so that Z_{-1} has defined value: results are tabulated in second column. These are to be compared with results from use of forward recursion (using given Z_{-1}, Z_0 and calculating $Z_1, Z_2, \ldots Z_9$ in turn) shown in last column, displaying numerical instability.

k	Z_k/Z_9	Z_k	Z_k by foward recursion
9	1	$2.1166184 \times 10^{-12}$	-0.234898
8	32.377271	$6.8530319 \times 10^{-11}$	-0.013078
7	943.16384	1.9963180×10^{-9}	-8.19639×10^{-4}
6	24411.505	5.1669841×10^{-8}	-5.87110×10^{-5}
5	552513.17	1.1694596×10^{-6}	-3.75293×10^{-6}
4	10708833	2.2666513×10^{-5}	2.21704×10^{-5}
3	172709360	3.6555981×10^{-4}	3.65497×10^{-4}
2	2222526500	4.7042406×10^{-3}	4.70423×10^{-3}
1	21338159000	4.5164741×10^{-2}	4.5164740×10^{-2}
0	134765460000	0.28524706	0.28524706
-1	402354950000	0.85163191	0.85163191

5. (a) Substitute for Z_n and for Z_{n-1}, which equals $Ac^{n-1} + b/(c-1)$, in the recurrence relation, and verify that it is satisfied for any A. (If $c = 1$, then $Z_n = A + bn$ is a general solution.)

(b) $Z_1 = 12Z_0 - 6 = 12*0.545 - 6 = 0.540$. $Z_2 = 12Z_1 - 6 = 12*0.540 - 6 = 0.480$. Similarly $Z_3 = -0.24$, $Z_4 = -8.88$. The inherent error in Z_0 causes Z_1 to be unequal to Z_0, and this excites the unwanted term $12^n A$ of the general solution, which increases so rapidly that it dominates subsequent Z's.

(c) Placing $n = 0$ in general solution: $Z_n = 12^n A + 6/11$, we see that $Z_n = 12^n Z_0 - 6(12^n - 1)/11$. Condition number of Z_n with respect to Z_0 is: $\phi_n = (Z_0/Z_n)(d Z_n/d Z_0) = 12^n Z_0/Z_n = [1 - 6(1 - 12^{-n})/11Z_0]^{-1}$. If $Z_0 = 6/11$ precisely, then $\phi_4 = 12^4 \simeq 2 \times 10^4$; for Z_0 close, but unequal, to $6/11$ we see that $\phi_4 \simeq [1 - 6/(11Z_0)]^{-1}$ remains large. Thus Z_4 (and in general Z_n) is very poorly conditioned in its dependence on Z_0.

6. (a) $I_n + I_{n-1} = \int_0^{1/3} (x^n + x^{n-1})(x+1)^{-1} dx = \int_0^{1/3} x^{n-1} dx = [x^n/n]_0^{1/3}$ and $I_0 = \int_0^{1/3} (x+1)^{-1} dx = [\ln(1+x)]_0^{1/3} = \ln(4/3)$. With decimal precision 3, for example, this gives $I_0 = 0.285$, and using the recursion

$I_n = (1/3^n n) - I_{n-1}$, the sequence $\{I_n : n = 1, 2, \ldots 5\}$ becomes $\{0.048,$ $7.60 \times 10^{-3}, 4.70 \times 10^{-3}, -1.61 \times 10^{-3}, 2.43 \times 10^{-3}\}$ Clearly these are in error as all I_n's should be positive, and should decrease monotonically with n.

(b) $I_n = \int_0^{1/3} x^n (1 - x + x^2 - + \ldots) \, dx = [(x^{n+1}/(n + 1) - x^{n+2}/(n + 2) +$ $\ldots)]_0^{1/3} = (1/3^{n+1})[(n + 1)^{-1} - 3^{-1}(n + 2)^{-1} + \ldots] = (1/3^{n+1}) \sum_{k=0}^{\infty}$ $[(-3)^k (n + k + 1)]^{-1}$. With $n = 5$, this sums to 1.78×10^{-6}. This is the correct answer. That of (a) is affected by numerical instability, as the required particular solution of the recurrence relation decreases rapidly with increasing n, whereas the general solution of $I_n + I_{n-1} = 0$ is $I_n = A(-1)^n$ and does not decrease in magnitude. In effect, part (b) uses reverse recurrence assuming $I_{N+1} = 0$ for some large N and working back to $n = 10$.

(c) From part (b), $(-1)^n I_n = \sum_{k=n+1}^{\infty} (-1)^{k+1}/(3^k k) = \tau_n = S_\infty - S_n = S_\infty -$ $\sum_{k=1}^{n} (-1)^{k+1}/(3^k k)$ where $S_\infty = I_0 = \ln(4/3)$. Part (b) evaluates $I_5 = -\tau_5$ by summation of the tail, part (a) on the other hand calculates it as $(S_\infty - S_n)$ which involves cancellation error, because $|\tau_5| << S_\infty$.

SECTION 3.2

1. (a) From (3.2.2), collecting terms multiplying $(x - x_1)^2$, $(x - x_1)$ and independent of x together, and writing $D = (x_0 - x_1) (x_1 - x_2)$ $(x_2 - x_0)$, $a_2 = [y_0(x_2 - x_1) + y_1(x_0 - x_2) + y_2(x_1 - x_0]/D$, $a_1 = [-y_0(x_2 - x_1)^2 + y_1(x_2 - x_0) (x_2 - 2x_1 + x_0) + y_2(x_1 - x_0)^2]/D$, $a_0 = y_1$.

   (b) Noting that $D = h_1 h_2(h_1 + h_2)$, $p_2'(x_1) = a_1 = [(y_2 - y_1)h_1^2 +$ $(y_1 - y_0)h_2^2]/D$, $p_2''(x_1) = 2a_2 = 2[(y_2 - y_1)h_1 - (y_1 - y_0)h_2]/D$.

   (c) $p_2'(x_1) = (y_2 - y_0)/(2h)$; $p_2''(x_1) = (y_2 + y_0 - 2y_1)/h^2$. Further $p_2'(x) = a_1 + 2a_2(x - x_1) = 0$ for $|x - x_1| \leqslant h$ if $|2a_2 h| \geqslant |a_1|$: that is if $|y_2 + y_0 - 2y_1| = |\Delta^2 y_0| \geqslant \frac{1}{2}|y_2 - y_0|$.

2. (a) Using (3.2.5): $s_{30} = (x - x_1) (x - x_2) (x - x_3)/[(x_0 - x_1) (x_0 - x_2) (x_0 - x_3)]$, $s_{31} = (x - x_0) (x - x_2) (x - x_3)/[(x_1 - x_0) (x_1 - x_2) (x_1 - x_3)]$, $s_{32} = (x - x_0) (x - x_1) (x - x_3)/[(x_2 - x_0) (x_2 - x_1) (x_2 - x_3)]$, $s_{33} = (x - x_0) (x - x_1) (x - x_2)/[(x_3 - x_0) (x_3 - x_1) (x_3 - x_2)]$.

   (b) With $x_k = kh$, and writing $x = \zeta h$, $s_{30} = -(\zeta - 1) (\zeta - 2) (\zeta - 3)/6 = -(\zeta^3 - 6\zeta^2 + 11\zeta - 6)/6$, $s_{31} = \zeta(\zeta - 2) (\zeta - 3)/2 = (\zeta^3 - 5\zeta^2 + 6\zeta)/2$. Stationary values of s_{30} at $\zeta = 2 \pm \sqrt{1/3}$, and of s_{31} at $\zeta = (5 \pm \sqrt{7})/3$.

3. (a) From (3.2.5), if $\Phi_{nk}(x) = \Phi(x)/(x - x_k)$, then $s_{nk} = \Phi_{nk}(x)/\Phi_{nk}(x_k)$.
But $\Phi_n'(x) = \Phi_{nk}(x) + (x - x_k)\Phi_{nk}'(x_k)$ so that $\Phi_{nk}(x_k) = \Phi_n'(x_k)$.

(b) $ds_{nk}/dx = [1/\Phi_n'(x_k)] \, d[\Phi_n(x)/(x - x_k)] = [(x - x_k)\Phi_n'(x) - \Phi_n(x)] /$
$[(x - x_k)^2 \Phi_n'(x_k)]$. But at $x = x_i \neq x_k$, $\Phi_n(x_i) = 0$, whence stated result.
At $x = x_k$, the derivative is indeterminate, but can be evaluated by
l'Hôpital's rule to give $\lim_{x \to x_k} \{ [x - x_k)\Phi_n''(x)]/[2(x - x_k)\Phi_n'(x_k)]\} =$
$\Phi_n''(x_k)/[2\Phi_n'(x_k)]$.

(c) $\Phi_n'(x_j) = \Phi_j(x_j) = \prod_{k=0}^{j-1} (x_j - x_k) \prod_{k=j+1}^{n} (x_j - x_k) = \prod_{k=0}^{j-1} (j - k) \prod_{k=j+1}^{n}$
$(j - k) h^n = [j(j - 1) \ldots 1] [(-1)(-2)\ldots(-n + j)] h^n = (-1)^{n-j} j!$
$(n - j)! h^n$.

(d) From (c) in (b), $(ds_{nk}/dx) |_{x=x_0=0} = \Phi_n'(x_0)/[-kh\Phi_n'(x_k)] = (-1)^{k+1}$
$(n/k)n!/[k!(n - k)!]$. With $n = 12$, $k = 6$, value is $-(2.12!)/(6!)^2 =$
-1848. (Whence very rapid rise in value of $|s_{12,6}|$ close to $x = x_0$).

4. Each identity results from using (3.2.3) and interpolating the values of the
function on the right-hand-side at $x = x_k \, \forall k$.

5. (a) $v_{jk} = x_{j-1}^{k-1}$ (NB: $v_{j1} = 1 \; \forall j$).

(b) Since $x_1 = 0 \therefore a_0 = y_1$. Other two equations are $a_1 x_0 + a_2 x_0^2 = (y_0 - y_1)$
and $a_1 x_2 + a_2 x_2^2 = (y_2 - y_1)$, so that with $D = x_0 x_2(x_2 - x_0)$, $a_1 =$
$[(y_1 - y_0)x_2^2 + (y_2 - y_1)x_0^2]/D$ and $a_2 = [(y_1 - y_2)x_0 + (y_0 - y_1)x_2]/D$.

(c) Note that det (V_n) vanishes if $x_n = x_j \; \forall \; j = 0, 1, \ldots (n - 1)$ since last
row of determinant is identical with $(j + 1)$th row, so $(x_n - x_j)$ is factor
of det (V_n).

Section 3.2.1

1. The cubic is being interpolated through $(0, 0)$, $(1, 1)$, $(2, 0)$, $(3, 0)$ and its
value is required at $x = 0.7847$.

			Aitken's scheme			Neville's scheme		
j	x_j	a_{j0}	a_{j1}	a_{j2}	a_{j3}	a_{j1}	a_{j2}	a_{j3}
0	0	0						
1	1	1	0.7847			0.7847		
2	2	0	0	0.9537		1.2153	0.9537	
3	3	0	0	0.8692	1.0563	0	1.3460	1.0563

2. For example:

(a) **for** j: = 0 **step** 1 **until** n **do** d_j: = y_j;
if $n > 0$ **then for** k: = 0 **step** 1 **until** $n-1$ **do**
begin s: = $x - x_k$;
        for j: = $k+1$ **step** 1 **until** n **do** d_j: = $d_k + (d_j - d_k)*s/(x_j - x_k)$
end:

(b) **for** $j:=0$ **step** 1 **until** n **do**
    begin $c_j:=y_j$; $s=x-x_j$;
        if $j>0$ **then for** $\ell:=j-1$ **step** -1 **until** 0 **do**
$$c_\ell:=c_{\ell+1}+(c_{\ell+1}-c_\ell)*s/(x_j-x_\ell)$$
    end;

In this instruction sequence, the values of $(j-k)$ are assigned to ℓ, and the outer loop is left with $c_\ell=a_{n,n-\ell}$ so that $p_n(x)=c_0$.

(c) **for** $m:=n$ **step** -1 **until** 0 **do**
    begin $d_m:=y_m$; $s:=x-x_m$;
        if $m<n$ **then for** $j:=m+1$ **step** 1 **until** n **do**
$$d_j:=d_{j-1}+(d_{j-1}-d_j)*s/(x_m-x_j)$$
    end;

In this algorithm values of $(n-j+k)$ are assigned to m, and $a_{j,k}$ is assigned to d_j.

3. (a) From (3.2.8) in (3.2.6), $y_{0,1,\ldots k,j}=[(x-x_k)y_{0,1,\ldots k-1,j}-(x-x_j)\\ y_{0,1,\ldots k}]/(x_j-x_k)$, so comparing with (3.2.7), S is the set $\{(x_i,y_i):\\ i=0,1,\ldots k-1\}$.

(b) S is the set $\{(x_i,y_i): i=(j-k),(j-k+1),\ldots(j-1)\}$.

4. (a) Substitute for a_{jk} in terms of a'_{jk} in (3.2.6), and note that $\sum_{i=0}^{k} a'_{ii} - \sum_{i=0}^{k-1} a'_{ii}=a'_{kk}$.

Section 3.2.2

1. (a) For example:

    for $j:=n$ **step** -1 **until** 0 **do**
        begin $a_j:=d_j$;
            if $j<n$ **then for** $k:=j$ **step** 1 **until** $n-1$ **do** $a_k:=a_k-a_{k+1}*x_j$
    end

(b)

j	3	2		1		0			Final
k			2		1	2	0	1	2	Values
a_0							1	1		1
a_1				2	7				7	7
a_2		3	-5			-9			-9	-9
a_3	4									4

(c) $n(n+1)$, half of these multiplications, the other half subtractions.
(d) Replace x_j by $(x_j-\xi)$.

(e) The power form in x gives a bad answer due to generated (cancellation) error.

3. $\hat{y}_{j-k-1,j-k,\ldots j} = (\hat{y}_{j-k,j-k+1,\ldots j} - \hat{y}_{j-k-1,j-k,\ldots j-1})/(x_j - x_{j-k-1})$.

4. (a) See problem (3) of § 3.2.

(b) From (3.2.3) replacing n by m, then $\hat{y}_{0,1,\ldots m} = \sum\limits_{k=0}^{m} y_k \hat{s}_{mk}(x_0, x_1, \ldots$

$x_m; x) = \sum\limits_{k=0}^{m} [y_k/\Phi'(x_k)]$. If the ordering of the $(m + 1)$ data pairs is

permuted, this sum is evidently unchanged.

(c) $\hat{s}_{m-1,k}(x_1, x_2, \ldots x_m; x) = \prod\limits_{\substack{i=1 \\ i \neq k}}^{m} (x_k - x_i)^{-1}; \hat{y}_{1,2,\ldots m} = \sum\limits_{k=1}^{m} y_k \hat{s}_{m-1,k}$

$(x_1, x_2, \ldots x_m; x)$.

(d) $(x_k - x_0)\hat{s}_{m,k}(x_0, x_1, \ldots x_m; x) = \hat{s}_{m-1,k}(x_1, x_2 \ldots x_m; x); \hat{y}_{1,2,\ldots m} -$

$\hat{y}_{0,1,\ldots m-1} = \sum\limits_{k=1}^{m} y_k \hat{s}_{m-1,k}(x_1, \ldots x_m; x) - \sum\limits_{k=0}^{m-1} y_k \hat{s}_{m-1,k}$

$(x_0, \ldots x_{m-1}; x) = \sum\limits_{k=0}^{m} y_k [(x_k - x_0) \hat{s}_{m,k}(x_0, \ldots x_m; x) - (x_k - x_m)$

$\hat{s}_{m,k}(x_0, \ldots x_m; x)] = (x_m - x_0) \sum\limits_{k=0}^{m} y_k \hat{s}_{m,k}(x_0, x_1, \ldots x_m; x) =$

$(x_m - x_0)\hat{y}_{0,1,\ldots m}$. (Note that this is a particular example of the identity (3.2.14), and of the recursion formula given in problem 3, with $k = j - 1$ and $j = m$.)

5. $1 - 3(x + 1) + 4x(x + 1) - 4x(x + 1)(x - 1) = -8$ at $x = 2$.

6. $(((y_3 - d_0)/(x_3 - x_0) - d_1)/(x_3 - x_1) - d_2)/(x_3 - x_2)$. Verity from the Newton form (3.2.11) with $n = 3$, $x = x_3$, and $p_3(x) = y_3$.

7. (a) Differentiate the Newton form $p_2(x) = y_i + \hat{y}_{ij}(x - x_i) + \hat{y}_{012}(x - x_i)$
$(x - x_j)$ giving $p_2'(x) = \hat{y}_{ij} + \hat{y}_{012}(2x - x_i - x_j)$.

(b) From part (a), $p_2'(x_1) = \hat{y}_{01} + \hat{y}_{012}(x_1 - x_0) = \hat{y}_{12} + \hat{y}_{012}(x_1 - x_2)$. Result follows from elimination of $\hat{y}_{012}$.

8. (b) In terms of the original ordering, put $\hat{p}(S; x) = \hat{y}_{0,1,\ldots(m-1)} = d_{m-1}$ and $j = m + 1$, $k = m$. Then $d_m = \hat{p}(S_k; x)$, $d_{m+1} = \hat{p}(S_j \cup S_k; x)$ and so $d_m' = \hat{p}(S_j; x) = d_m + (x_{m+1} - x_m) d_{m+1}$.

(c) Roll the mth point down to the nth position and then omit it. Thus the new coefficients are given (in terms of the original ordering) by $d_k' = d_k + (x_{k+1} - x_m) d_{k+1}$, for $k = m, (m + 1) \ldots (n - 1)$ and $d_n' = 0$.

9. New coefficients are given (in terms of the original ordering) by $c'_k = c_k + (x_{k-1} - x_m) c_{k-1}$ for $k = m, (m-1), \ldots 1$ and $c'_0 = 0$.

Section 3.2.3

1. (a) .46947 15628

.03052 84372

.50000 00000 $-.0_3 60\ 91730$

.02991 92642 $-.0_4 3\ 64519$

.52991 92642 $-.0_3 64\ 56249$ $.0_6 7863$

.0.2927 36393 $-.0_4 3\ 56656$ $.0_7 440$

.55919 29035 $-.0_3 68\ 12905$ $.0_6 8303$

.02859 23488 $-.0_4 3\ 48353$

.58778 52523 $-.0_3 71\ 61258$

.02787 62230

.61566 14753

(b) y_4
y_2 $h\hat{y}_{24}$ $2h^2\hat{y}_{024}$
y_0 $h\hat{y}_{02}$ $2h^2\hat{y}_{012}$ $6h^3\hat{y}_{0124}$
y_1 $h\hat{y}_{01}$ $2h^2\hat{y}_{012}$ $6h^3\hat{y}_{0123}$ $24h^4\hat{y}_{01234}$
       $h\hat{y}_{01}$ $2h^2\hat{y}_{013}$ $6h^3\hat{y}_{0123}$ $24h^4\hat{y}_{01235}$ $120h^5\hat{y}_{012345}$
y_3 $h\hat{y}_{13}$ $2h^2\hat{y}_{135}$ $6h^3\hat{y}_{0135}$
y_5 $h\hat{y}_{35}$

But Newton polynomial is $p_5(x) = y_0 + \hat{y}_{01}(x - x_0) + \hat{y}_{012}(x - x_0)(x - x_1) + \ldots$ the $(m+1)$th coefficient being $a_m = \hat{y}_{0,1,\ldots m}$. Identifying these from (a) and scaling (with $h = 2$), $d_0 = .5299192642$. $d_1 = .01463681965$, $d_2 = -.0_4 8070311$, $d_3 = -.0_6 74303$, $d_4 = .0_8 20477$, $d_5 = .0_{10} 1146$.

(c) Terms of Newton polynomial and interpolated values for $m = 1, 2, \ldots 5$:
.0018069154, .0000186957, .0000003655, .0000000039, .0000000001
.5317261796, .5317448753, .5317452408 .5317452447, .5317452448
(Last value is in fact correct value of $\sin(x)$ at $x = 32.12345$).

2. (a) From (3.2.20), $\Delta^n y_j = n! h^n \hat{y}(x)$, a constant for all $j = 0, 1, 2, \ldots$

(b) Third differences are constant (and equal 6). therefore a cubic.

(c) Result follows by applying binomial theorem to $y_\nu = (1 + \Delta)^\nu y_0$. Verification follows from (3.1.20), noting that with $x = x_0 + \nu h$, $h^k(\nu - k + 1)_k = h^k(\nu - k + 1)(\nu - k + 2) \ldots (\nu - 1)\nu = (x - x_{k-1})(x - x_{k-2}) \ldots (x - x_0)$.

3. (a) The error propagates in a triangular pattern and grows quickly; the sign alternates and one recognises the binomial coefficients in the different columns: see example at left below.

```
0                                                    −ε
    0                                          2ε
0        0                               ε            −4ε
    0        ε                               −2ε              8ε
0        ε        −4ε                    −ε            4ε           −16ε
    ε        −3ε        10ε                   2ε              −8ε
ε        −2ε        6ε           −20ε    ε            −4ε           16ε
    −ε        +3ε        −10ε                  −2ε              8ε
0        ε        −4ε           15ε    −ε            4ε           −16ε
    0        −ε        5ε                   2ε              −8ε
0        0        ε           −6ε    ε            −4ε           16ε
    . . .                            . . .
```

Example on right shows worst combination of error with error of kth difference equal in magnitude to $2^{k+1}\epsilon$.

(b) Error of kth difference is (say) $\Delta^k \epsilon_j = \epsilon_j - k\epsilon_{j+1} + k(k-1)\epsilon_{j+2}/2 - \ldots + (-1)^k \epsilon_{j+k}$ where the terms involve the binomial coefficients. If the errors are random with mean μ, then the mean of $\Delta^k \epsilon_j = \mu[1 - k + k(k-1)/2 - \ldots + (-1)^k] = [1 + (-1)]^k = 0$ and the variance is $\sigma^2[1 + k^2 + k^2(k-1)^2/4 + \ldots + 1]$. For $k = 1, 2, \ldots$ this gives the sequence of values $2\sigma^2$, $6\sigma^2$, $20\sigma^2$, $70\sigma^2$, .−. whose general term is $\sigma^2(2k)!/(k!)^2$. From (2.5.7), this is approximately $(2^k \sigma)^2/(\pi k)^{1/2}$ if k is large.

4. (b) Coefficients of Newton form: $d_0 = -3, d_1 = -0.33, d_2 = 0.94, d_3 = 0.13/6, d_4 = -4.15/24$ and $d_5 = 8.3/120$. Power form of polynomial is $p_5(x) = x(3.7664 - 0.8429x^2 + 0.0692x^4)$. (Values at $x = 1, 2, 3$ are 2.99, 3.00, and 5.36. It is an odd function of x.) Alternatively, taking account of the asymmetry of the data, regard y/x as quadratic in x^2 interpolating the 3 data: $(0.25, 3.5 + 6)$, $(2.25$ $2.22)$, and $(6.25, 1.2)$ giving $y = x[3.46 - 0.67(x^2 - 0.25) + 0.0692(x^2 - 0.25)(x^2 - 2.25)]$.

(c)

x	−2.5	−1.5	−0.5	0.5	1.5	2.5
$p_5'(x)$	1.48	−0.17	3.16	3.16	−0.17	1.48
$p_5''(x)$	−8.98	2.92	2.36	−2.36	−2.92	8 98

Max. at $x = 1.44$ where $p_5(x) = 3.34$.

Section 3.2.4

2. (a) $y(-1) = 1, y(0) = 3, y'(0) = 5, y''(0) = 14, y(1) = 29, y'(1) = 78$.

(b) Six. The three groups of data at $x = -1, 0$ and 1 may be permuted, retaining the ordering of each group.

(c) Divided difference table:

(-1)	1				
(0)	3	2			
(0)	3	5	3		
(0)	3	5	7	4	
(1)	29	26	21	14	5
(1)	29	78	52	31	17	6

$p_5(x) = 1 + 2(x+1) + 3(x+1)x + 4(x+1)x^2 + 5(x+1)x^3 + 6(x+1)x^3$
$(x-1) = 29 + 78(x-1) + 52(x-1)^2 + 31(x-1)^2 x + 17(x-1)^2 x^2 + 6(x-1)^2 x^3$.

3. Use values of $E_1'(x) = -x^{-1} \exp(-x)$ at $x = 0.03$ and 0.04 to obtain the divided difference table:

$(.03)$	2.9591			
$(.03)$	2.9591	-32.35		
$(.04)$	2.6813	-27.78	457	
$(.04)$	2.6813	-24.02	376	-8100

and so $p_3(x) = 2.9591 - 32.35(x - 0.03) + 457(x - 0.03)^2 - 8100(x - 0.03)^2$
$(x - 0.04)$ which gives $p_3(0.0357) = 2.7907$.

4. (a) For instance:

```
for j : = 0 step 1 until n do
    begin cⱼ : = yⱼ;
        if j > 0 then for ℓ: = j−1 step −1 until 0 do
            if xⱼ = xℓ then begin t : =cℓ ; cℓ : = cℓ₊₁/(j−ℓ);
                                cℓ₊₁: = t
                            end
                        else cℓ: = (cℓ₊₁−cℓ)/(xⱼ−xℓ)
    end
```

(b) For example, in the above algorithm, after "**if** $x_j = x_\varrho$ **then begin**" insert the instruction "**if** $x_\varrho \neq x_{\varrho+1}$ **then goto** *FAIL*;"

5. $u_{nk}(x_j) = v_{nk}'(x_j) = \delta_{jk}$, $u_{nk}'(x_j) = v_{nk}(x_j) = 0$ for all $j, k = 0, 1, \ldots n$. Each polynomial obeys $(2n + 2)$ conditions and has therefore degree $\leq (2n + 1)$. each must moreover have double zeros at the $2n$ values of $x_j \neq x_k$, so that if $p_n(x) = \prod_{\substack{j=1 \\ j \neq k}}^{n} (x - x_j)^2$, then they must each have the form $(ax + b)p_n(x)$ for some a and b.

Thus $u_{nk}(x) = [p_n(x_k) - (x - x_k)p_n'(x_k)]p_n(x)/[p_n(x_k)]^2$ and $v_{nk}(x) = (x - x_k)p_n(x)/p_n(x_k)$.

SECTION 3.3

1. (a) For example:

$$k: = \text{entier } ((\xi - x_0)/h) + 1;$$
$$\text{if } k \leqslant 0 \text{ then } k: = 1 \text{ else if } k > n \text{ then } k: = n$$

(b) Search by bisection of the interval of possible k values. For example:

$$k: = 1; \, kmax: = n;$$
while $k < kmax$ do
$$\text{begin } j: = \text{entier } ((k + kmax)/2);$$
$$\text{if } \xi < x_j \text{ then } kmax: = j \text{ else } k: = j + 1$$
end

(Here k is the minimum possible value of k currently determined).

2. (a) From (3.3.1), $A = (\hat{y}_{k-1,k} - y'_k)/h_k$, $B = (y'_{k-1} - \hat{y}_{k-1,k})/h_k$, where $\hat{y}_{k-1,k} = (y_k - y_{k-1})/h_k$ is a divided difference.

(b) Substitute for $y'_k = y'(x_k)$ and $y'_{k-1} = y'(x_{k-1})$ from (3.3.3) in result of part (a), and note that $h_k + h_{k+1} = x_{k+1} - x_{k-1}$.

(c) Here $\hat{y}_{k-1,k\,k+1} = \hat{y}_{k-2,k-1,k} = \hat{p}_2(x)$ is a constant independent of k, so that from part (b), $\pi_k(x)$ is a quadratic such that $\pi_k''(x) = p_2''(x) = 2\hat{p}_2(x)$ and as $\pi_k(x)$ has the same value as $p_2(x)$ at $x = x_{k-1}, x_k$, therefore $\pi_k(x) = p_2(x)$.

3. (a) $y(x_j) = \delta_{jk}$ and from (3.3.3) $y'(x_j) = [(\delta_{jk} - \delta_{j-1,k})(h_{j+1}/h_j) + (\delta_{j+1,k} - \delta_{j,k})(h_j/h_{j+1})]/(h_j + h_{j+1})$. That is, $y'(x_j) = 0$ for all j except $j = k - 1$, $k, k + 1$, for which $y'(x_{k-1}) = h_{k-1}/[h_k(h_{k-1} + h_k)]$, $(2 \leqslant k \leqslant n)$. $y'(x_k) = (h_{k+1} - h_k)/(h_k h_{k+1})$, $(1 \leqslant k \leqslant n - 1)$. $y'(x_{k+1}) = -h_{k+2}/[h_{k+1}(h_{k+1} + h_{k+2})]$, $(0 \leqslant k \leqslant n - 2)$.

(b) $s_k(x) = (1 - |\xi|)(1 + |\xi| - 3\xi^2/2)$ $0 \leqslant |\xi| \leqslant 1$
$$= (1 - |\xi|)(2 - |\xi|)^2/2 \quad\quad 1 \leqslant |\xi| \leqslant 2$$
$$= 0 \quad\quad\quad\quad\quad\quad\quad\quad\quad 2 \leqslant |\xi|$$

(c) Equal minima are reached at $|\xi| = 4/3$ where $s_k(x) = -2/27$.

4. (a) The interpolate is an odd function of x: for $x > 0$; adopting the form of expression given in problem 2, we find

k	x_{k-1}	x_k	y_{k-1}	y_k	A	B	$\pi_k(x)$
3	−0.5	0.5	−1.78	1.78	1.005	−1.005	$x(4.06 - 2.01x^2)$
4	0.5	1.5	1.78	3.33	0.94	1.005	$0.227 + 3.674x - 1.168x^2$
							$+ 0.065x^3$
5	1.5	2.5	3.33	3.0			$0.3 + 3.43x - 0.94x^2$

(Values at $x = 1, 2$, and 3 are 2.80, 3.40 and 2.13).

(b)

x	-2.5	-1.5	-0.5	0.5	1.5	2.5
$y'(x)$	-1.27	0.61	2.55	2.55	0.61	-1.27

Max. at $x = 1.82$ where $y(x) = 3.43$.

SECTION 3.4

1. (a) Differentiating (3.4.2), $\pi_k'(x) = \hat{y}_{k-1,k} + \mu_k(2x - x_{k-1} - x_k)/h_k$, so that $\pi_k'(x_{k-1}) = \hat{y}_{k-1,k} - \mu_k$ and $\pi_k'(x_k) = \hat{y}_{k-1,k} + \mu_k$.

   (b) As in part (a), $\mu_k = \pi_k'(x_k) - \hat{y}_{k-1,k} = y_k' - \hat{y}_{k-1,k}$, and so on substituting in the recurrence relation for $(\mu_k + \mu_{k+1})$ in (3.4.3), the required relation follows if k is replaced by $(j - 1)$. Applying the trapezoidal rule to the integral, it would be evaluated as $(h_j/2)(y_{j-1}' + y_j')$.

   (c) Substitute for μ_k and μ_{k+1} from (3.4.4) in (3.4.3), and verify that (3.4.4) is also the appropriate expression for μ_1.

   (d) From (3.4.4), $\mu_0 = \mu_1 = \ldots = \mu_{m-1} = 0$, $\mu_m = 1/h$, $\mu_{m+1} = -3/h$, and $\mu_{m+2} = -\mu_{m+3} = \ldots (-1)^{n-m} \mu_n = 4/h$. Substituting in (3.4.2):

$$s_{nm}(x) = 0 \qquad\qquad \text{for } x_0 \leqslant x \leqslant x_{m-1}$$

$$= (x - x_{m-1})^2/h^2 \qquad\qquad \text{for } x_{m-1} \leqslant x \leqslant x_m$$

$$= (x_{m+1} - x)(3x - 3x_m + h)/h^2 \qquad \text{for } x_m \leqslant x \leqslant x_{m+1}$$

$$= 4(-1)^{k+m+1}(x - x_{k-1})(x_k - x) \qquad \text{for } x_{k-1} \leqslant x \leqslant x_k,$$

where $k = m + 2, m + 3, \ldots n$.

2. (a) Differentiate $\sum\limits_{k=1}^{n} \mu^2{}_k/h_k$ w.r.t. y_0' and equate to zero to find stationary value: differentiate again to show that it is a minimum.

   (b) Substitute for μ_k from (3.4.4) in the equation $\sum\limits_{k=1}^{n} (-1)^k \mu_k = 0$ to obtain an expression for y_0' in terms of divided differences. Then replace $\hat{y}_{k-1,k}$ by $(y_k - y_{k-1})/h$ to interpret y_0' in terms of data ordinates. For instance, with $n = 3$: $-\mu_1 = -\hat{y}_{0,1} + y_0'$, $\mu_2 = \hat{y}_{1,2} - 2\hat{y}_{0,1} + y_0'$, $-\mu_3 = -\hat{y}_{2,3} - 2\hat{y}_{0,1} + 2\hat{y}_{1,2} + y_0'$, whence $-\mu_1 + \mu_2 - \mu_3 = -5\hat{y}_{0,1} + 3\hat{y}_{1,2} - \hat{y}_{2,3} + 3y_0' = 0$; that is, $y_0' = (y_3 - 4y_2 + 8y_1 - 5y_0)/(3h)$. Similarly for $n = 2$, $y_0' = (-y_2 + 4y_1 - 3y_0)/(2h)$, and for $n = 4$, $y_0' = (-y_4 + 4y_3 - 8y_2 + 12y_1 - 7y_0)/(4h)$.

3. (a) Differentiating (3.4.6): $6h_k\pi_k'(x) = 6h_k\hat{y}_{k-1,k} - (x - x_{k-1})(x_k - x)(M_k - M_{k-1}) + (2x - x_{k-1} - x_k)[(h_k + x_k - x)M_{k-1} + (h_k + x - x_{k-1})M_k]$; $6h_k\pi_k''(x) = 2(2x - x_{k-1} - x_k)(M_k - M_{k-1}) + 2(h_k + x_k - x)M_{k-1} + 2(h_k + x - x_{k-1})M_k = 2(3x - 2x_{k-1} - x_k + h_k)M_k + 2(h_k + 2x_k + x_{k-1} - 3x)M_{k-1} = 6(x - x_{k-1})M_k +$

$6(x_k - x)M_{k-1}$. Thus $\pi_k''(x)$ is linear and interpolates (x_{k-1}, M_{k-1}), (x_k, M_k).

(b) From above expression for $\pi_k'(x)$, we see that $\pi_k'(x_{k-1}) = \hat{y}_{k-1,k} - (2M_{k-1} + M_k)h_k/6$, $\pi_k'(x_k) = \hat{y}_{k-1,k} + (M_{k-1} + 2M_k)h_k/6$. The system (3.4.7) is then the rearranged form of the equations $\pi_1'(x_0) = y_0'$, $\pi_k'(x_k) = \pi_{k+1}'(x_k)$ for $k = 1, 2, \ldots (n-1)$, $\pi_n'(x_n) = y_n'$.

(c) From part (b), $2M_{k-1} + M_k = (6/h_k)(\hat{y}_{k-1,k} - y_{k-1}')$, $M_{k-1} + 2M_k = (6/h_k)(y_k' - \hat{y}_{k-1,k})$, and quoted result follows by solving these equations for M_{k-1} and M_k. Placing $k = j + 1$ in equation for M_{k-1}, and $k = j$ in equation for M_k and equating the two expressions for M_j, we find $h_{j+1}y_{j-1}' + 2(h_j + h_{j+1})y_j' + h_jy_{j+1}' = 3(h_{j+1}\hat{y}_{j-1,j} + h_j\hat{y}_{j,j+1})$ for $j = 1, 2, \ldots (n-1)$. Result quoted follows after dividing through by $(h_j + h_{j+1})$.

(d) Here $\lambda_j = 1/2$, and $\hat{y}_{j-1,j} = (y_j - y_{j-1})/h$, so that multiplying through by $2h/3$, the system becomes $(h/3)(y_{j-1}' + 4y_j' + y_{j+1}') = y_{j+1} - y_{j-1}$.

4. (a) Integrating by parts: $\sum\limits_{k=1}^{n} \int_{x_{k-1}}^{x_k} y''(x)\eta''(x)dx = \sum\limits_{k=1}^{n} \left\{ [y''(x)\eta'(x)]_{x_{k-1}}^{x_k} - \int_{x_{k-1}}^{x_k} y'''(x)\eta'(x)dx \right\} = [y''(x)\eta'(x)]_{x_0}^{x_n} - \sum\limits_{k=1}^{n} \int_{x_{k-1}}^{x_k} y'''(x)\eta'(x)dx$ since $y''(x)\eta'(x)$ is continuous over $[x_0, x_n]$.

(c) Because $y = y(x) + \eta(x)$ interpolates the same data, then $\eta(x_0) = \eta(x_1) = \ldots \eta(x_n) = 0$, and since it has the same tangent, $\eta'(x_0) = \eta'(x_n) = 0$. Whence the conditions of parts (a) and (b) are satisfied.

(d) $\sum\limits_{k=1}^{n} \int_{x_{k-1}}^{x_k} [y''(x) + \eta''(x)]^2 dx = \int_{x_0}^{x_n} [y''(x)]^2 dx + \sum\limits_{k=1}^{n} \int_{x_{k-1}}^{x_k} [\eta''(x)]^2 dx$ using the result of part (c).

(e) Here $\eta'(x_0)$ and $\eta'(x_n)$ are not necessarily zero, but the result of part (a) nonetheless still applies, since $y''(x)$ is zero at $x = x_0, x_n$. (The rest is unchanged.)

5. (a) $B_k(x) = A(x - x_{k-2})^3$ for $x \in [x_{k-2}, x_{k-1}]$, where A is some constant, since $B_k(x) = B_k'(x) = B_k''(x) = 0$ at $x = x_{k-2}$. Further, $B_k(x) = A(x - x_{k-2})^3 + C(x - x_{k-1})^3$ for $x \in [x_{k-1}, x_k]$ where C is a constant, since again $B_k(x)$, $B_k'(x)$ and $B_k''(x)$ are continuous at $x = x_{k-1}$. But $B_k(x_k) = 1$, by definition, and $B_k'(x_k) = 0$ by symmetry. Thus if h is the distance between nodes: $B_k(x_k) = 8Ah^3 + Ch^3 = 1$ and $B_k'(x_k) = 12Ah^2 + 3Ch^2 = 0$, giving $C = -4A$ and $A = 1/(4h^3)$. Therefore: placing $\xi = (x - x_k)/h$, we find $B_k(x) = 1 - 3\xi^2(2 - |\xi|)$ for $|\xi| \leq 1$, $= (2 - |\xi|)^3$ for $1 \leq |\xi| \leq 2$, and it is zero for $|\xi| \geq 2$, i.e. outside (x_{k-2}, x_{k+2}).

(b)

$j =$	$k - 2$	$k - 1$	k	$k + 1$	$k + 2$
$B_k(x_j) =$	0	1/4	1	1/4	0
$hB_k'(x_j) =$	0	3/4	0	-3/4	0
$h^2 B_k''(x_j) =$	0	3/2	-3	3/2	0

(Values of $B_k(x)$ at all other nodes are zero.)

(c) Referring to the values given in the part (b) and comparing values of $y(x)$
at $x = x_j$, we see that for $j = 0, 1, \ldots n$, $(1/4)\, c_{j-1} + c_j + (1/4)\, c_{j+1} = y_j$,
which is a set of $(n + 1)$ linear equations connecting the $(n + 3)$ coefficients. The various pairs of end-conditions are given by:

(i) $-c_{-1} + c_1 = 4y_0' h/3, -c_{n-1} + c_{n+1} = 4y_n' h/3$;

(ii) $\Delta^2 c_{-1} = c_{-1} - 2c_0 + c_1 = 0, \nabla^2 c_{n+1} = c_{n-1} - 2c_n + c_{n+1} = 0$, (so
that $c_0 = 2y_0/3, c_n = 2y_n/3$); and since $h^3 B_k'''(x)$ has the value $3/2$,
$-9/2, 9/2, -3/2$ in successive intervals $(x_{k-2}, x_{k-1}), \ldots (x_{k+1},$
$x_{k+2})$, therefore:

(iii) $\Delta^3 c_{-1} = -c_{-1} + 3c_0 - 3c_1 + c_2 = 0, \nabla^3 c_{n+1} = -c_{n-2} + 3c_{n-1} -$
$3c_n + c_{n+1} = 0$;

(iv) $\Delta^4 c_{-1} = c_{-1} - 4c_0 + 6c_1 - 4c_2 + c_3 = 0, \nabla^4 c_{n+1} = c_{n-3} - 4c_{n-2}$
$+ 6c_{n-1} - 4c_n + c_{n+1} = 0$.

Using the equations connecting the coefficients with the data, (iv) can
also be shown to imply, more simply, that $c_1 = (8y_1 - y_0 - y_2)/9$,
$c_{n-1} = (8y_{n-1} - y_{n-2} - y_n)/9$.

(d) Uniqueness implies that there is *one* function belonging to $C^2 [x_0, x_n]$
which reduces to cubic polynomials on each subinterval (x_{k-1}, x_k) of
$[x_0, x_n]$ and which satisfies the interpolation conditions. Since each
B-spline defined on this set of nodes $\in C^2 [x_0, x_n]$ and reduces to cubics
on each subinterval, therefore so does the linear combination of B-splines:
hence if such a combination satisfies the interpolation conditions, it will
be identical with the cubic spline. The diagonal dominance of the coefficient matrix for the coefficients $c_{-1}, c_0, \ldots c_{n+1}$ ensures the existence
(and uniqueness) of such a combination.

6. (a) The "not-a-knot" conditions gives $M_1 = - M_4 = \delta^2 y_1 = 1.88$; because
$y(x)$ is an odd function of x, therefore $M_2 = -M_3$ and (3.4.9) shows that
$M_1 + 3M_2 = 6\delta^2 y_2 = 12.06$ whence $M_2 = 3.393$. Then (3.4.8) gives
$M_0 = - M_5 = 0.367$, and so for $x > 0$:
$y(x) = \pi_3(x) = x(3.843 - 1.131x^2)$ for $|x| \leqslant 0.5$
$\quad = \pi_4(x) = \pi_5(x) = -0.173 + 4.879x - 2.074x^2 + 0.252x^3$ for $x \geqslant 0.5$
(Values at $x = 1, 2$ and 3 are $2.88, 3.305$, and 2.60.)

(b)

x	-2.5	-1.5	-0.5	0.5	1.5	2.5
$y'(x)$	-0.77	0.36	2.99	2.99	0.36	-0.77
$y''(x)$	0.37	1.88	3.39	-3.39	-1.88	-0.37

Max at $x = 1.71$ where $y(x) = 3.37$.

(c) From problem 5, part (c) (iv), $c_4 = -c_1 = 2.429$; other values then found
as $c_6 = -c_{-1} = 1.410, c_5 = - c_0 = 2.040, c_3 = - c_2 = 1.564$.

7. (a) Using (3.4.9) and (3.4.10), $M_0 = M_1 = - M_4 = - M_5 = 1.556, M_2 =$
$- M_3 = 3.501$. Thus for $x > 0$:

$$y(x) = \pi_3(x) = x(3.851 - 1.167x^2) \qquad \text{for } |x| \leqslant 0.5$$
$$= \pi_4(x) = -0.186 + 4.969x - 2.236x^2 + 0.324x^3 \quad \text{for } 0.5 \leqslant x \leqslant 1.5$$
$$= \pi_5(x) = 0.908 + 2.782x - 0.778x^2 \qquad \text{for } x \geqslant 1.5$$

(Values at $x = 1, 2$, and 3 are 2.87, 3.36, and 2.25.)

(b)

x	-2.5	-1.5	-0.5	0.5	1.5	2.5
$y'(x)$	-1.11	0.45	2.98	2.98	0.45	-1.11
$y''(x)$	1.56	1.56	3.50	-3.50	-1.56	-1.56

Max at $x = 1.79$ where $y(x) = 3.39$.

8. Here $n = 2$ and $h_1 = h_2 = 1$. For the natural cubic spline, $M_0 = M_2 = 0$, and from (3.4.9), since $\lambda_1 = 1/2$, therefore $2M_1 = 6\hat{y}_{0,1,2} = 3\delta^2 y_1$; that is, $M_1 = 3(1 - 0.425) = 1.725$. Thus from (3.4.6), $\pi_2(x) \equiv x + 0.425(1 - x) - 0.2875x$ $(1 - x)(2 - x) = 0.425 + 0.2875(3 - x)x^2$ and by symmetry $y(-x) = y(x) = \pi_2(x)$ for $0 \leqslant x \leqslant 1$. Hence:

$|x|$	0	0.147	0.261	0.363	0.458	0.546	0.630	0.709	0.858	1
$y(x)$	0.425	0.443	0.479	0.525	0.578	0.635	0.695	0.756	0.878	1
$r(x)$	0.580	0.737	0.972	1.291	1.717	2.273	3.036	4.120	9.163	∞

and $y'(1) = \pi_2'(1) = 0.8625$, so that inclination of tangent to y-axis is $49.2°$.

SECTION 3.5

1. If $b_0 = b_1 = \ldots = b_\nu = 0$ then from (3.5.2) $\sum\limits_{k=0}^{\mu} x_j^k a_k = A_\mu(x_j) = 0$ for $j = 0, 1, \ldots (\mu + \nu)$. Thus $A_\mu(x)$ has $\mu + \nu + 1 > \mu$ zeros and would be identically zero; hence $a_0 = a_1 = \ldots = a_\mu = 0$. But the solution is not trivial, and so this is a contradiction. The assumption is therefore incorrect.

2. (a) Note that $Q_k(x) = q_k + (x - x_k)/Q_{k+1}(x)$.
   (b) $q_{j0} = Q_0(x_j) = y(x_j) = y_j$, from (3.5.3). Further $q_{kk} = Q_k(x_k) = q_k$ from part (a), which also shows that $q_{j,k+1} = Q_{k+1}(x_j) = (x_j - x_k)/[Q_k(x_j - q_k] = (x_j - x_k)/(q_{jk} - q_{kk})$.
   (c) The orders of $Q_0(x), Q_1(x), \ldots Q_{\mu+\nu}(x)$ are respectively $\mu/\nu, \nu/(\mu-1)$, $(\mu - 1)/(\nu - 1), (\nu - 1)/(\mu - 2), \ldots 0/0$.

3. (a) The triangular arrays for the inverted differences are:

   (i) 1
       3 1/2
       6 2/5 −10
       10 1/3 −12 −1/2

   $$y(x) = 1 + \cfrac{x}{1/2+} \cfrac{x-1}{-10+} \cfrac{x-2}{(-1/2)}$$
   $$= (x + 1)(x + 2)/2;$$

   (ii) 1
        1/3 −3/2
        1/6 −12/5 −10/9
        1/10 −10/3 −12/11 99/2

   $$y(x) = \cfrac{1}{1+} \cfrac{x}{-3/2+} \cfrac{x-1}{-10/9+} \cfrac{x-2}{(99/2)}$$
   $$= 12(4x + 3)/(x^2 - 9x + 36).$$

   (b) Adding another line to the triangular arrays above: (i) 15, 2/7, −14, −1/2, ∞; (ii) 1/15, −30/7, −14/13, 117/2, 1/9. The continued fraction (i) is

unchanged (since $q_4 = \infty$), but the other is $\dfrac{1}{1+} \dfrac{x}{-3/2+} \cdots \dfrac{x-2}{99/2+} \dfrac{x-3}{(1/9)} =$
$(x + 1)(x + 2)/2$.

4. It is necessary to interchange the datum $(1, 2)$ with, say, $(3, 3.5)$ to avoid q_1 being infinite; the inverted difference table derived from (3.5.3) then becomes

x_k	y_k			
0	$2 = q_0$			
3	3.5	$2 = q_1$		
1	2	∞	$0 = q_2$	
4	4.4	5/3	-3	$-1 = q_3$

$$y(x) = 2 + \dfrac{x}{2+} \dfrac{x-3}{0+} \dfrac{x-1}{-1} = (x^2 + x + 2)/(x + 1).$$

5. (b) Defining $(\infty \pm \infty) = \infty$, then the continued fraction is only undefined at $x = x_1$ owing to occurrence of indeterminacy $0/0$.

6. (a) The inverted difference table derived from (3.5.3) is:

x_k	y_k				
-3.21	$-2.38 = q_0$				
-0.987	-0.276	$1.06 = q_1$			
0	0	1.35	$3.40 = q_2$		
0.654	0.126	1.54	3.42	$32.7 = q_3$	
3.21	0.413	2.30	3.39	-321	$-0.00723 = q_4$

(b) $A = 1.0087$, $B = 0.02962$, $C = -0.0121$, $D = 4.447$. Should be $A = 1$, $B = 0$, $C = 0$, and $D = 4.56$.

(c) $1.013(x - 0.01226)(x - 0.9018)/[(x + 4.585)(x - 0.8917)]$ Zero at $x \simeq 0.9$ and pole at $x \simeq 0.89$ are spurious: this approximates to an "arbitrary" degenerate interpolate, because the representation of the data by a rational function of order 2/2 should be non-unique (and therefore arbitrarily degenerate) had all the numerical working been of infinite precision, since the data are exactly interpolated by a rational function $x/(4.56 + x)$ of order 1/1.

(d) Since all but one of the data (namely that at $x = 3.21$) are interpolated by a rational function of order 1/1, the interpolating rational function of order 2/2 should be $x(x - 3.21)/[(x + 4.56)(x - 3.21)]$. Values are compared below with the quoted result using arithmetic of decimal precision 3:

x	-4	-3	-1	1	2	2.5	3	3.5	4
$x/(4.56 + x)$	-7.143	-1.923	-0.281	0.180	0.305	0.354	0.397	0.434	0.467
precision = 3	-7.239	-1.915	-0.282	0.196	0.359	0.461	0.687	-1.357	0.175

Note how the theoretical degeneracy at $x = 3.21$ is "spread" by the numerical imprecision.

7. (a) The table of inverted differences, derived from (3.5.3) is

$$x_k^2 \quad y_k/x_k$$

0.25   3.56 $= q_0$

2.25   2.22 $-1.493 = q_1$

6.25   1.20 -2.542 $-3.81 = q_2$

i.e. $y(x)/x = 3.56(x^2 - 0.25)/[1.493 + (x^2 - 2.25)/3.81]$ and so $y(x) = x(13.2 - 0.25x^2)/(3.44 + x^2)$. (Values at $x = 1, 2$ and 3 are 2.91, 3.28 and 2.64).

(b)

x	-2.5	-1.5	-0.5	0.5	1.5	2.5
$y'(x)$	-0.67	0.27	3.04	3.04	0.27	-0.67

Maximum at $x = 1.66$, where $y(x) = 3.35$.

8. (a) Since $A_0 = 0, B_0 = 1, A_1 = 1, B_1 = q_0$ the quoted results are true for $m = 0$. If true for $m = 0, 1, \ldots k$ (say) then from (2.7.5) $A_{2k+2} = q_{2k+1} A_{2k+1} + (x - x_{2k}) A_{2k} = q_{2k+1}[x^k + \ldots] + (x - x_{2k})[(q_1 + q_3 + \ldots + q_{2k-1})x^{k-1} + \ldots] = (q_1 + q_3 + \ldots + q_{2k+1})x^k + \ldots$ and $B_{2k+2} = q_{2k+1} B_{2k+1} + (x - x_{2k}) B_{2k} = q_{2k+1}[(\ldots)x^k \ldots] + (x - x_{2k})(x^k + \ldots) = x^{k+1} + \ldots$, and a similar consideration of A_{2k+3}, B_{2k+3} establishes the proof by induction for all m.

(b) A_n/B_n is the (generally) unique rational function of order r/s where $r = \text{entier } [(n - 1)/2]$ and $s = \text{entier } (n/2)$ which interpolates the n data (x_k, y_k) for $k = 0, 1, \ldots (n - 1)$, and so is unaffected by the ordering of these data. Hence in particular the ratio of the leading coefficients in A_n and B_n is unaffected by that ordering: this ratio is equal to ρ_{n-1} which therefore must depend symmetrically upon these data. (If $\rho_{n-1}(\mathbf{S})$ depends on the set of n data $\mathbf{S}$, then in the notation of (3.1.7) there exists a recursion: $\rho_{n+1}(\mathbf{S}_j \cup \mathbf{S}_k) - \rho_{n-1}(\mathbf{S}) = (x_j - x_k)/[\rho_n(\mathbf{S}_j) - \rho_n(\mathbf{S}_k)]$ with $\rho_{-1} = 0$, and $\rho_0(\mathbf{S}_i) = y_i$, by which the reciprocal differences can be constructed.)

(c) $q_0 + q_2 = y_0^{-1} + (x_2 - x_1)/[(x_2 - x_0)/(y_2^{-1} - y_0^{-1}) - (x_1 - x_0)/ (y_1^{-1} - y_0^{-1})] = y_0^{-1} + (x_2 - x_1)(y_2^{-1} - y_0^{-1})(y_1^{-1} - y_0^{-1})/Q_{012} = P_{012}/Q_{012}$, say, where $Q_{012} = x_0(y_2^{-1} - y_1^{-1}) + x_1(y_0^{-1} - y_2^{-1}) + x_2(y_1^{-1} - y_0^{-1})$, $P_{012} = y_0^{-1}Q_{012} + (x_2 - x_1)(y_2^{-1} - y_0^{-1})(y_1^{-1} - y_0^{-1}) = x_0 y_0^{-1}(y_2^{-1} - y_1^{-1}) + x_1 y_1^{-1}(y_0^{-1} - y_2^{-1}) + x_2 y_2^{-1}(y_1^{-1} - y_0^{-1})$. The values of P_{012} and Q_{012} are unchanged for any even permutation of the data, and both are reversed in sign for any odd permutation: thus $\rho_2 = P_{012}/Q_{012}$ is invariant to any permutation of the data.

9. For example:

$r := q_N$;

  if $N > 0$ **then for** $j := N-1$ **step** -1 **until** 0 **do**

    if $r \neq 0$ **then if** $r \neq \inf$ **then** $r := q_j + (x - x_j)/r$

        else $r := q_j$

$$\text{else if } x \neq x_j \text{ then } r: = \inf$$
$$\text{else begin } r: = y_j; \text{ goto FINISH}$$
$$\text{end};$$

FINISH: **comment** r assigned equal to $y(x)$;

10. (a) (i) Trivial if $|\mathbf{S}| = 0$. If $|\mathbf{S}| = m > 0$, then $C_j(\mathbf{S}; x)$ is a polynomial of degree $\mu(m + 1) + \mu(m) = m$, with leading coefficient (of x^m) equal to $\hat{P}(\mathbf{S}_j)\,\hat{P}(\mathbf{S})$. By definition of P and Q, its m zeros are the values of $x = x_k \; \forall \, k \in \mathbf{S}$, because $\sigma(m) + \sigma(m + 1) = 0$.

   (ii) Trivial if $i = j$. If $i \neq j$ and $\mathbf{S}_{ij} \equiv \mathbf{S}_j \cup \mathbf{S}_k$, then (i) shows that both sides of equation equal $\hat{P}(\mathbf{S}_{ij})\,\hat{P}(\mathbf{S}_i)\,\hat{P}(\mathbf{S}_j)\,\hat{P}(\mathbf{S})\,\pi(\mathbf{S}_{ij}; x)\,\pi(\mathbf{S}; x)$.

(b) (iii) If $\sigma\,(|\mathbf{S}|) = 1$, then stated recurrence is obtained by dividing both sides of the result of (a) (ii) by $P(\mathbf{S}_{ij}; x)\,Q(\mathbf{S}_i; x)\,Q(\mathbf{S}_j; x)\,P(\mathbf{S}; x)$. If $\sigma(|\mathbf{S}|) = -1$, divide the same result by $Q(\mathbf{S}_{ij}; x)\,P(\mathbf{S}_i; x)\,P(\mathbf{S}_j; x)\,Q(\mathbf{S}; x)$.

(c) Place $\mathbf{S} \equiv \{ j - k + 1, j - k + 2, \ldots j - 1 \}$ and $i = j - k$ in the result of part (b) (iii), and write $R_{j,k} \equiv R(\mathbf{S}_{ij}; x)$, so that $R_{j,k-1} \equiv R(\mathbf{S}_i; x)$, $R_{j-1,k-1} \equiv R(\mathbf{S}_j; x)$ and $R_{j-2,k-2} \equiv R(\mathbf{S}; x)$.

SECTION 3.6

1. (a) $x(t) = 4t, \; y(t) = 3t^2$.

   (c) New point corresponds to $t = 2$. Interpolating through (x, y) values in turn $x(t) = 4t + A(t^2 - 1)t, \; y(t) = 3t^2 + B(t^2 - 1)t$, with $A = 2/15$, $B = 0.42$, so that $y(t) - 3[x(t)/4]^2 = t(t^2 - 1)(2 - t)(t^2 + 2t + 63)/300$, and clearly C' coincides with C only at $t = -1, 0, 1, 2$. However, C and C' would coincide if fourth point (corresponding to $t = 2$) were $(8, 12)$.

2. Parameter values of points, according to (3.6.2), starting at $(-4, 3)$ are $0, 5$, 10, and 22.48; a linear transformation of parameter does not affect the interpolation, so take parameter as $t' = (t - 5)/5$, with values $-1, 0, 1$, and 3.496. Then (a) and (b) follow exactly as before. In (c), $A = -0.1321$, $B = -0.5645$, and again C' and C do not coincide; nor can any distinct fourth point be added which leaves the interpolating curve unchanged.

3. (b) To belong to class C^1, then $\displaystyle\sum_{j=1}^{m} (\theta_{2j} - \theta_{2j-2}) = \theta_{2m} - \theta_0 = 2\pi$; but in general $\displaystyle\sum_{j=1}^{m} \Delta\gamma_{2j-1}$ is not equal to π.

   (c) If curve belongs to class C_1, then $2 \displaystyle\sum_{j=1}^{m} \Delta\gamma_{2j} = \sum_{j=1}^{m} (\theta_{2j+1} - \theta_{2j-1}) =$
   $$\theta_{2m+1} - \theta_1 = 2\pi + \theta_0 - \theta_1 = 2\pi + \theta_0 - (2\gamma_1 - \theta_0) = 2(\pi + \theta_0 - \gamma_1).$$

4. (a) $\pi'_k(x_k) = \hat{y}_{k-1,k} + \mu_k, \; \pi'_k(x_{k-1}) = \hat{y}_{k-1,k} - \mu_k$.

(b) From (3.4.4) and part (a) $y'_n = \hat{y}_{n-1,n} + \mu_n = 2 \sum_{j=1}^{n} (-1)^{j+n} \hat{y}_{j-1,j} +$

$(-1)^n y'_0$ and since $x_n = x_0 + \omega$, therefore $y'_n = y'_0$ and there is no solution if n is even.

(c) $y'_k = \sum_{j=1}^{k} (-1)^{j+k} \hat{y}_{j-1,j} + \sum_{j=k+1}^{2m+1} (-1)^{j+k+1} \hat{y}_{j-1,j}.$

5. Relevant piece of spline is $\pi_2(x)$, and since $M_1 = M_2$, $h = \pi/3$ and $\pi/2 = x_1 + h/2$, therefore $\pi_2(\pi/2) = (y_1 + y_2)/2 - (h^2/8)M_1 = \{1 + (3/4)[1 - \cos(h)] / [2 + \cos(h)]\} \sin(h) = 1.15 \sin(\pi/3) = 0.9959.$

6. (a) Since period is $\omega = (x_n - x_0)$ therefore $y(x_0) = y_0 = y(x_n) = y_n, y'(x_0) = y'_0 = y'(x_n) = y'_n$, and $y''(x_0) = M_0 = y''(x_n) = M_n$. Eliminating $y'_0 = y'_n$ between the first and last equations of (3.4.7): $(2M_0 + M_1)h_1 + (M_{n-1} + 2M_n)h_n = 6(\hat{y}_{0,1} - \hat{y}_{n-1,n})$ i.e. $2(h_0 + h_1)M_0 + h_1M_1 + h_0M_{n-1} = 6(\hat{y}_{0,1} - \hat{y}_{-1,0})$ where $h_0 = x_0 - x_{-1} = x_0 + \omega - x_{n-1}$. Placing $\lambda_k = h_k/(h_k + h_{k+1})$, the system therefore becomes $2M_0 + (1 - \lambda_0)M_1 + \lambda_0 M_{n-1} = 6\hat{y}_{-1,0,1}$, $\lambda_k M_{k-1} + 2M_k + (1 - \lambda_k)M_{k+1} = 6\hat{y}_{k-1,k,k+1}$ for $k = 1, 2, \ldots (n - 2)$, $(1 - \lambda_{n-1})M_0 + \lambda_{n-1}M_{n-2} + 2M_{n-1} = 6\hat{y}_{n-2,n-1,n}.$

(b) For equispaced nodes, $\lambda_k = 1/2$ for all k, and $6\hat{y}_{k-1,k,k+1} = 3\delta^2 y_k/h^2$. Thus $4M_0 + M_1 + M_{n-1} = 6\delta^2 y_0/h^2$, $M_{k-1} + 4M_k + M_{k+1} = 6\delta^2 y_k/h^2$, for $k = 1, 2, \ldots (n - 2)$, $M_0 + M_{n-2} + 4M_{n-1} = 6\delta^2 y_{n-1}/h^2$. Here $\gamma_0 = 4$, and all other γ's are zero except $\gamma_{n-1} = \gamma_1 = 1.$

(c) Putting $(\omega^2/6n^2)M_j = (h^2/6)M_j = x_j$ and $n = 3$ in above equations $4x_0 + x_1 + x_2 = \delta^2 y_0 = y_{-1} - 2y_0 + y_1 = -2y_0 + y_1 + y_2; x_0 + 4x_1 + x_2 = \delta^2 y_1 = y_0 - 2y_1 + y_2; x_0 + x_1 + 4x_2 = \delta^2 y_2 = y_0 + y_1 - 2y_2$. Evaluating the given expression for $x_0 = (\omega^2/6n^2)M_0$, $x_0 = 2.732y_0 - 1.699[2y_0 - 0.268(y_1 + y_{-1}) + 0.072(y_2 + y_{-2})] = -0.667y_0 + 1.699 \times (0.268 - 0.072)(y_1 + y_2) = -0.667y_0 + 0.333(y_1 + y_2)$. More precisely, $x_0 = (-2y_0 + y_1 + y_2)/3$ and similarly $x_1 = (y_0 - 2y_1 + y_2)/3$ and $x_3 = (y_0 + y_1 - 2y_2)/3$, which can be verified as compatible with the above system.

7. (a) Stated result is a form of the identity: $\sin[(A + B)/2] - \sin[(A - B)/2] = 2 \cos(A/2)\sin(B/2)$ (*). It follows that, if $\sin \psi \neq 0$, $\sum_{k=0}^{n} \cos(\phi + 2k\psi) = \sum_{k=0}^{n} \{\sin[\phi + (2k + 1)\psi] - \sin[\phi + (2k - 1)\psi]\}/(2\sin\psi) = \{\sin[\phi + (2n + 1)\psi] - \sin(\phi - \psi)\}/(2\sin\psi)$, and result follows from (*). If $\sin\psi = 0$, then ψ is a multiple of π, and $\cos(\phi + 2k\psi) = \cos\phi.$

(b) Use result of part (a) noting that $2\cos(rj\alpha)\cos(rk\alpha) = \cos[r(j - k)\alpha] + \cos[r(j + k)\alpha]$; $2\sin(rj\alpha)\sin(rk\alpha) = \cos[r(j - k)\alpha] - \cos[r(j + k)\alpha]$; $2\sin(rj\alpha)\cos(rk\alpha) = \sin[r(j + k)\alpha] + \sin[r(j - k)\alpha] = \cos[(\pi/2) - r(j + k)\alpha] + \cos[(\pi/2) - r(j - k)\alpha].$

(c) If $n = 2m$, then from (3.6.3) for $0 \leqslant j \leqslant m$ since $y_r = y(r\alpha)$, $\displaystyle\sum_{r=0}^{n-1} y_r \cos(jr\alpha)$

$$= \sum_{r=0}^{n-1} \left\{ (A_0/2) + \sum_{k=1}^{m-1} [A_k \cos(kr\alpha) + B_k \sin(kr\alpha)] + (A_m/2)\cos(mr\alpha) \right\}$$

$\cos(jr\alpha)$. Interchanging the order of summation with respect to k and r,

and using part (b), $\displaystyle\sum_{r=0}^{n-1} y_r \cos(jr\alpha) = (n/2)\left\{ A_0 \delta_{j,0} + \sum_{k=0}^{m} A_k \delta_{j,k} + (A_m/2) \right.$

$(\delta_{j,m} + \delta_{j+m,2m})\big\}$, and the right-hand-side equals $(nA_j/2)$ for $0 \leqslant j \leqslant m$

as in (3.6.4). Similarly $\displaystyle\sum_{r=0}^{n-1} y_r \sin(jr\alpha) = (n/2) B_k \delta_{j-k,0} = (n/2) B_j$ for

$0 \leqslant j \leqslant m$, and similar reasoning confirms (3.6.4) for $n = 2m + 1$.

8. Cardinal parametric interpretation implies equally spaced values of t_k, or $t_k = 2\pi k/3$ for $k = 0, 1, 2$. Interpolating $x(t)$, $y(t)$ using (3.6.3) and (3.6.4) we find $x(t) = - (\cos t + \sqrt{3} \sin t)$, $y(t) = \cos t - (1/\sqrt{3}) \sin t$. Expressing $\cos t$ and $\sin t$ in terms of x and y, and eliminating t by using $\cos^2 t + \sin^2 t = 1$, it follows that $x^2 + 3y^2 = 4$ which is the equation of an ellipse.

Section 3.7.2

1. (a) True for $\nu = 1$, since $T_1(x) = x$. If true for $\nu = n$, say, then from (3.7.3) $\hat{T}_{n+1}(x) = 2\hat{T}_n(x) = 2^n$, and so true for $\nu = n + 1$. Hence true for all $\nu = 1, 2, \ldots$

(b) Use (3.7.3) and inductive proof, or from (3.7.1), with $x = \cos\theta$, $T_\nu(-x) = T(-\cos\theta) = T_\nu[\cos(\pi-\theta)] = \cos[\nu(\pi-\theta)] = \cos(\nu\pi)\cos\nu\theta = (-1)^\nu \cos\nu\theta = (-1)^\nu T_\nu(x)$.

2. Both the shape function, and $T_{n+1}(x)/(x - x_k)$, are polynomials of degree n having the value δ_{jk} at $x = x_j$ for $j = 0, 1, \ldots n$. Therefore they are proportional, and since $s_{nk}(x_k) = 1$, therefore $c_{nk} + \lim_{x \to x_k} [(x - x_k)/T_{n+1}(x)] = \lim_{\theta \to \theta_k} \{(\cos\theta - \cos\theta_k)/\cos[(n + 1)\theta]\} = \sin(\theta_k)/\{(n + 1)\sin[(n + 1)\theta_k]\}$. But by (3.7.4) $\sin[(n + 1)\theta_k] = \sin[(k + 1/2)\pi] = (-1)^k$, and so $c_{nk} = (-1)^k \sin(\theta_k)/(n + 1)$.

3. (a) Placing $\theta_k = \arccos(x_k) = (k + 1/2)\pi/(n + 1)$ then from (3.7.2), $S(\nu) = \displaystyle\sum_{k=0}^{n} \cos(\nu\theta_k) = \sum_{k=0}^{n} \cos(\phi + 2k\psi)$ with $\phi = \psi = \nu\pi/(2n + 2)$. Thus from quoted result, if $\nu \neq 0$ and since $\nu < (2n + 2)$, $S(\nu) = \cos(\nu\pi/2) \sin(\nu\pi/2)/\sin[\nu\pi/(2n + 2)] = \sin(\nu\pi)/\{2\sin[\nu\pi/(2n + 2)]\} = 0$, whereas $S(0) = (n + 1)$.

(b) $\displaystyle\sum_{k=0}^{n} T_i(x_k)T_j(x_k) = \sum_{k=0}^{n} \cos(i\theta_k)\cos(j\theta_k) = \{\cos[(i + j)\theta_k] + \cos[|i - j| \theta_k]\}/2 = \{S(i + j) + S(|i - j|)\}/2$. and stated results follow.

(c) Since $p_n(x_k) = f_k$ for $k = 0, 1, \ldots, n$, therefore $\displaystyle\sum_{k=0}^{n} f_k T_i(x_k) = \sum_{k=0}^{n}$

$$[c_0/2 + \sum_{j=1}^{n} c_j T_j(x_k)] T_i(x_k) = (c_0/2) \sum_{k=0}^{n} T_i(x_k)T_0(x_k) + \sum_{j=1}^{n} c_j \sum_{k=0}^{n}$$

$$T_i(x_k)T_j(x_k) = [(n + 1)/2](c_0\delta_{i,0} + \sum_{j=1}^{n} c_j\delta_{ij}) = (n + 1)c_i/2, \text{ from part}$$

(b).

(d) For quadratic interpolate, $\cos\theta_k = \cos[(2k + 1)\pi/6]$, i.e. $x_0 = \sqrt{3}/2$, $x_1 = 0$, and $x_2 = -\sqrt{3}/2$: whence from part (c), $c_0/2 = c_2 = (1/3)$ $\ln(11/12)$ and $c_1 = (1/\sqrt{3})\ln[(6 + \sqrt{3})/(6 - \sqrt{3})]$, i.e. $p_2(x) = 0.343085 T_1(x) - 0.029004[1 + T_2(x)]$.

4. (a) $p_2(x) = (0.343 - 0.058x)x$.

   (b) Since $f'''(x) = 2/(3 + x)^3$, therefore remainder term is $E = \pi_2(x)/[3(3 + \xi_x)^3]$ with $|\xi_x| < 1$. But $|\pi_2(x)| \leqslant 1/4$ over $[-1, 1]$, and placing $\xi_x = -1$ we see that $|E| < 1/96$.

   (c) Stationary values at $x = -0.524, 0.481$ where error is respectively 0.00374 and -0.00289. Errors at $x = \pm 1$ are -0.00437 (at $x = -1$) and 0.00261 (at $x = 1$).

5. (a) $\displaystyle p_n(x) = (a_0 - 2xa_1 + a_2)/2 + \sum_{j=1}^{n} (a_j - 2xa_{j+1} + a_{j+2})T_j(x) = (a_0 - $

   $2xa_1 + a_2)/2 + a_1 T_1(x) - a_2 T_0(x) + \displaystyle\sum_{j=2}^{n} a_j[T_j(x) - 2xT_{j-1}(x) + $

   $T_{j-2}(x)]$. But the sum is zero since the square-bracketed quantity is zero by (3.7.3). Further, since $T_0(x) = 1$ and $T_1(x) = x$, therefore $p_n(x) = (a_0 - a_2)/2$.

   (b) Here $c_n = 1, c_{n-1} = c_{n-2} \ldots = c_0 = 0$, and so $a_n = 1, a_{n-1} = 2x, a_{n-2} = 4x^2 - 1$. Thus placing $b_k = (a_{n-k} - a_{n-k+2})/2$, we see that $b_1 = x = T_1(x)$ and $b_2 = 2x^2 - 1 = T_2(x)$. But from Clenshaw's recursion formula, $b_k = 2xb_{k-1} - b_{k-2}$, and so from (3.7.3), $b_k = T_k(x)$ for all integers $k > 0$.

SECTION 3.8

1. (a) Since $f''(x)$ does not change sign in $[a, b]$ therefore the points of equi-oscillation are at $x = z_0, z_1, z_2$ where $z_0 = a, z_2 = b$ and $z_1 \in (a, b)$. Relations for a_0 and a_1 follow by solving the system of equations (3.2.7), namely $a_0 + z_k a_1 + (-1)^k \epsilon = f(z_k)$ $(k = 0, 1, 2)$, whereas $f'(z_1) = a_1$ follows since error must be stationary at the interior node $x = z_1$.

   (b) Let L be the line joining $(a, f(a)), (b, f(b))$, and T the tangent to the curve $y = f(x)$ parallel to L. Then $y = p_1(x)$ is locus of points equi-distant from L and T (i.e., it is a line parallel to L and half-way between L and T).

(c) $2\epsilon = 2a_0 + 2z_1a_1 - 2f(z_1) = [bf(a) - af(b)]/(b - a) + a_1z_1 - f(z_1)$
     $= \pm 2E_1^*(f)$.

(d) $a_1 = (\ln 2)/2$ and $f'(z_1) = (3 + z_1)^{-1} = a_1$ so that $z = 2/(\ln 2) - 3$; this gives
     $2a_0 = [\ln(8/9)]/2 + \ln[2/(3 \ln 2)] - 1 + (3 \ln 2)/2 = \ln[16/(9 \ln 2)] - 1$.
     Whence $p_1(x) = 0.347x - 0.029$ and $E_1^*(f) = 0.030$. (NB $f''(x) = -(3 + x)^{-2} < 0 \; \forall \, x$).

2. (a) The error is $T_{n+1}(x)/2^n$ which has $(n + 2)$ points of equi-oscillation
        (with $\epsilon = 2^{-n}$).

    (b) Yes. More points of equi-oscillation than those presented by the theorem
        are not precluded.

3. (a) Prove by *reductio ad absurdum.* If such a polynomial existed then
        $r_n(x) = q_n(x) - p_n(x)$ is also a (non-vanishing) polynomial or degree n,
        and has the sign of $(-1)^k$ at the $(n + 2)$ points $x = z_k$, because $r_n(x_k) =$
        $[f(x_k) - p_n(x_k)] - [f(x_k) - q_n(x_k)] = (-1)^k \epsilon - \lambda\epsilon = (-1)^k \epsilon (1 \pm \lambda)$,
        where $|\lambda| < 1$, since the error $|f(x_k) - q_n(x_k)| < |\epsilon|$. Thus $r_n(x)$ has
        $(n + 2)$ changes in sign and therefore $(n + 1)$ zeros, which is inconsistent
        with it being of degree n.

    (b) First part of proposition follows by repeated applications of Rolle's
        theorem, noting that the $(n + 1)$th derivative of $p_n(x)$ is zero. Second
        part follows by implying that $t(x) - p_n(x)$ has (at most) n stationary
        values.

4. (a) The error at $x = \pm \delta$ as $\delta \to 0$ is inevitably $\pm\frac{1}{2}$ (or more) whatever the
        degree of the polynomial whereas elsewhere it can certainly be less in
        magnitude than a half.

    (b) Minimax polynomial is $p_1(x) = 0$ with $E_1^*(f) = 1$ and its error has $4N$
        points of equi-oscillation.

5. (a) Place $x^2 = z$, and let $q_m(z)$ be the minimax polynomial of degree m
        approximating $f(z^{1/2})$ over $z \in [0, b^{1/2}]$. Then $f(z^{1/2}) - q_m(z)$ has at
        least $m + 2$ points of equi-oscillation (including possibly $z = 0$). Hence
        $p_{2m}(x) = q_m(x^2)$ is a polynomial of degree $\leqslant 2m$ with at least $(2m + 3)$
        points of equi-oscillation about $f(x)$: by Tchebyshev's alternation
        theorem it is therefore the (unique) minimax polynomial approximation
        of degree $2m + 1$ (as well as that of degree $2m$).

    (b) The minimax polynomials of degree $(2m - 1)$ and $2m$ are identical linear
        combinations of odd powers of x having at least $(2m + 2)$ points of equi-
        oscillation.

6. (i) Substitute in (3.7.1), noting that $f^{(n+1)}(\xi) = (n + 1)! \, \hat{p}_{n+1}$, where $\hat{p}_{n+1}$ is
        the coefficient of x^{n+1} in $p_{n+1}(k)$.
    (ii) Follows from (i), with $\hat{p}_{n+1} = 2^n$.
    (iii) By (ii), $p_{n+m-1}^*(x) \equiv 0$ so that we can place $p_\nu^*(x) \equiv 0$ for any $\nu < n + m$.

(iv) Here $f(x) = T_{n+2}(x) + T_n(x)$, so that by (i), $p^*_{n+1}(x) \equiv T_n(x)$ is of degree n. Hence we can place $p^*_n(x) \equiv p^*_{n+1}(x)$.

Section 3.8.1

1. Each polynomial has the form $p_2(x) = a_0 + a_2 x^2$; each has 3 stationary values of error at $x = z_1, z_2, z_3$, say, where $z_3 = -z_1$ and $z_2 = 0$; in the table below, the equal error at $x = \pm 1$ is e_0, that at $x = z_1 = -z_3$ is e_1, and the error at $x = 0$ is e_2.

	a_0	a_2	$-z_1$	e_0	e_1	e_2	$E_n(f)$
(a)	1.7778	−1.3545	0.5447	0.217	−0.212	0	0.217
(b)	1.0754	2.4582	−	−2.893	−	0.702	2.893
(c)	1.6000	−1.0667	0.6370	0.107	−0.134	0.178	0.178
(d)	1.6439	−1.1378	0.6124	0.134	−0.151	0.134	0.151
(e)	1.6356	−1.1378	0.6124	0.142	−0.142	0.142	0.142

Note: (b) fails because the Taylor expansion about $x = 0$ converges only for $|x| < 3/4$; and in (d) the forced oscillation at the four points of equi-oscillation of $T_3(x)$ would leave the values of a_0, a_2, and ϵ indeterminate, and in fact $T_4(x)$ is the first neglected term of the Tchebyshev series for $f(x)$, not T_3, because $f(x)$ is an even function of x.

Section 3.8.2

1. (a) Since $f'(x) = (3 + x)^{-1}$, stationary value of polynomial $(a_0 + a_1 x)$ in $[-1, 0]$ is at root $x = z_1$ of $(3 + x)^{-1} = a_1 = [f(0) - f(-1)] \doteq \ln(3/2)$, giving $z_1 = [\ln(3/2)]^{-1} - 3 = -0.5337$ and $2E^*_1(f) = |f(0) + a_1 z_1 - f(z_1)| = |[z_1/(3 + z_1)] - \ln(1 + z_1/3)| = 0.0205$ from problem 1(c), § 3.8. Similarly over $[0, 1]$, stationary value is at root $x = z_1$ of $(3 + x)^{-1} = a_1 = [f(1) - f(0)] = \ln(4/3)$, so that $z_1 = 0.4760$ and here $2E^*_1(f) = |f(0) + a_1 z_1 - f(z_1)| = 0.0103$. Thus magnitude of maximum error is 0.01025 (reached in $[-1, 0]$).

   (b) Since error varies (roughly) as $(b - a)^{n+1}$ and here $n = 1$, doubling the number of segments (i.e., to 4) should suffice. (In fact the error of the minimax linear polynomial over the segment $[-1, -1/2]$, where the error is largest, would then be 0.0031. Three *unequal* segments, with the points of partition at $x = -0.48$ and 0.175, would also suffice.)

2. (a) Derivative of $\phi(x) = [f(x)/f'(x) - x]$ is $-f(x) f''(x)/[f'(x)]^2$ which does not change sign in $[a, b]$: hence $\phi(x)$ increases or decreases monotonically with x in $[a, b]$. Derivative of $p_1(x)/f(x) = (a_0 + a_1 x)/f(x)$, say, is $[a_1 f(x) - (a_0 + a_1 x)f'(x)]/[f(x)]^2$ and this must be zero at stationary value, where $f(x)/f'(x) - x = \phi(x) = (a_0/a_1)$. But since $\phi(x)$ is monotone over $[a, b]$ this cannot have more than one root in $[a, b]$.

(b) By part (a), there is only one stationary value of error at $x = z_1$ (say); thus other two points of equi-oscillation must be at $x = a, b$. Whence

$$p_1(a) + \epsilon f(a) = f(a), \text{ i.e., } p_1(a) = (1 - \epsilon)f(a), \tag{A}$$
$$p_1(b) + \epsilon f(b) = f(b), \text{ i.e., } p_1(b) = (1 - \epsilon)f(b), \tag{B}$$
$$\text{and} \quad p_1(z_1) - \epsilon f(z_1) = f(z_1), \text{ i.e., } (1 + \epsilon) = p_1(z_1)/f(z_1), \tag{C}$$

From (A) and (B) one obtains given expression for $p_1(x)$ by linear interpolation. Equation for z_1 follows from $\phi(x) = a_0/a_1$ by identifying a_0 and a_1 in expression for $p_1(x)$.

(c) Since $f'(x) = x^{-1/2}/2 = f(x)/(2x), x = z_1$ is (unique) root of $[(x - 1/4) + (1 - x)/2] = [1 - 1/2](2x)$, i.e., $(2x + 1)/4 = x$ giving $z_1 = 1/2$. Thus $(1 + \epsilon)/(1 - \epsilon) = [(z_1 - a)f(b) + (b - z_1)f(a)]/[(b - a)f(z_1)] = 2\sqrt{2}/3$ so that $\epsilon = -(3 - 2\sqrt{2})^2$, and substituting in the expression for $p_1(x)$ given in part (b): $p_1(x) = -(1 - \epsilon)(2x + 1)/3$.

3. $q_n(t) = [\ln(3/2) + p_n(2t - 3)]/\ln 10$.

SECTION 3.10

1. (a) The Tchebyschev polynomials $T_0(x), T_1(x), \ldots, T_{n+1}(x)$, form a basis for the space S_{n+1} of polynomials of degree $\leqslant (n + 1)$ and $q_{n+1}(x) \in S_{n+1}$; thus from given identity $0 = \sum\limits_{j=0}^{n+1} (-1)^j [q_{n+1}(x_j) - q_{n+1}(x_{j-1})] = \sum\limits_{j=0}^{n+1} \int_{x_{j-1}}^{x_j} q'_{n+1}(x)\,\sigma_{n+1}(x)\mathrm{d}x = -\int_{-1}^{1} q'_{n+1}(x)\,\sigma_{n+1}(x)\mathrm{d}x$.

(b) Since $|z| = z\,\mathrm{sgn}(z)$ for any real z, therefore from part (a) $\int_{-1}^{+1} |f(x) - p_n(x)|\,\mathrm{d}x = c \int_{-1}^{+1} [f(x) - p_n(x)]\,\sigma_{n+1}(x)\mathrm{d}x = c \int_{-1}^{+1} f(x)\,\sigma_{n+1}(x)\mathrm{d}x = c \int_{-1}^{+1} [f(x) - q_n(x)]\,\sigma_{n+1}(x)\mathrm{d}x = \int_{-1}^{+1} [f(x) - q_n(x)]\,\mathrm{sgn}[f(x) - p_n(x)]\,\mathrm{d}x = \int_{-1}^{+1} |f(x) - q_n(x)|\,\mathrm{sgn}[f(x) - q_n(x)]\,\mathrm{sgn}[f(x) - p_n(x)]\,\mathrm{d}x \leqslant \int_{-1}^{+1} |f(x) - q_n(x)|\mathrm{d}x$.

2. (a) By inspection, $U_0(\cos\theta) = 1$, and $U_1(\cos\theta) = \sin(2\theta)/\sin\theta = 2\cos\theta$ conforming with definition; further for $v \geqslant 1$, $U_{v+1}(\cos\theta) + U_{v-1}(\cos\theta) = \{\sin[(v + 2)\theta] + \sin(v\theta)\}/\sin\theta = 2\cos\theta\,\sin[(v + 1)\theta]/\sin\theta = 2\cos\theta\,U_v(\cos\theta)$ in conformity with recurrence relation.

(b) $U_n(\cos\theta) - U_{n-2}(\cos\theta) = \{\sin[(n + 1)\theta] - \sin[(n - 1)\theta]\}/\sin\theta = 2\sin\theta\,\cos(n\theta)/\sin\theta = 2\cos(n\theta) = 2T_n(\cos\theta)$.

(c) From part (b), $U_{n-1}(x) = x^{n+1} - [T_{n+1}(x)/2^n] - U_{n-1}(x)/2^{n+1} = t_{n-1}(x) - U_{n-1}(x)/2^{n+1}$ is a polynomial of degree $\leqslant (n - 1)$. Further $f^{(n+1)}(x) = (n + 1)!$ has no zeros in $[-1, 1]$, so it is sufficient that $p_n(x)$ interpolates $f(x)$ at $x = x_j = \cos\theta_j$ for $j = 0, 1, \ldots, n$, where $\theta_j = (j + 1)\pi/(n + 2)$. But since $U_{n+1}(x_j) = U_{n+1}(\cos\theta_j) = \sin[(n + 1)\theta_j]/$

$\sin \theta_j = \sin[(j + 1)\pi]/\sin \theta_j = 0$ for $j = 0, 1, \ldots, n$, therefore $u_{n-1}(x_j) = x_j^{n+1}$, and so $p_n(x) = u_{n-1}(x)$ is the required polynomial with minimum L_1-norm.

(d) Let $\pi_n(x) = (x - \xi_0)(x - \xi_1) \ldots (x - \xi_n) = x^{n+1} - p_n(x)$, say: then minimise $\int_{-1}^{+1} |x^{n+1} - p_n(x)| \, dx$, giving $p_n(x) = x^{n+1} - U_{n+1}(x)/2^{n+1}$ as in part (c). Hence $\pi_n(x) = U_{n+1}(x)/2^{n+1}$, having zeros at $x = \xi_j = \cos[(j + 1)\pi/(n + 2)]$ for $j = 0, 1, \ldots, n$.

Section 4.1.1

1. (a) $x = 2.5\pi$ (a root).
   (b) $x = 0.3$, not a root because $f(x)$ is discontinuous at $x = 0.3$.
   (c) $x = 0.2$, a root, despite discontinuity.

2. Initially, maximum error magnitude of mid-point root estimate is s; after n iterations it is $s/2^n$. Thus magnitude is $\leqslant xtol$ if $2^n \geqslant (s/xtol)$, whence result.

3. (a) b 1.0   0.50   0.50   0.38   0.32   0.32   0.31   0.31
       a 0   0   0.25   0.25   0.25   0.29   0.29   0.30 $(xi = 0.31 = b)$
   (b) b 1.0   0.50   0.50   0.37   0.31   0.31   0.31   0.31
       a 0   0   0.25   0.25   0.25   0.28   0.29   0.30 $(xi = 0.30 = a)$
   (c) b 1.0   1.0   0.75   0.63   0.57   0.54   0.54   0.54
       a 0   0.50   0.50   0.50   0.50   0.50   0.52   0.53 $(xi = 0.54 = b)$
   (d) b 1.0   1.0   0.75   0.65   0.60   0.55
       a 0   0.50   0.50   0.50   0.50   0.50 $(xi = 0.55 = b)$
   Note 'adjacent' values of xi are now 0.05 apart, not 0.01 as in (c).
   (e) b 1.0   1.0   0.75   0.60   0.55
       a 0   0.50   0.50   0.50   0.50 $(xi = 0.50 = a)$

4. In place of the single conditional statement **if** $msgn = asqn$ **then** $\ldots$, use (say):

$$\text{if} \quad msgn \neq asqn \text{ then } b := xi;$$
$$\text{if} - msgn \neq asqn \text{ then } a := xi;$$

Section 4.1.2

1. From (4.1.1), $c - a = (b - a)f(a)/[f(a) - f(b)] < b - a$, so that $c < b$. Similarly show that $b - c < b - a$ so that $c > a$.

2. (b) From (4.1.1), $c - a = -(b - a)f(a)/[f(b) - f(a)] \simeq -(b - \xi)f(a)/f(b)$, if $(\xi - a)$ is small. Result follows after dividing through by $(\xi - a)$ and using part (a).

Section 4.1.3

1. (a) From (4.1.1) $x_{n+3} = (x_n f_{n+2} - x_{n+2}\beta f_n)/(f_{n+2} - \beta f_n)$.
   (b) From (4.1.1), $x_{n+2} - x_{n+1} = -f_{n+1}/\hat{f}_{n,n+1}$, and so $\hat{f}_{n+1,n+2}/\hat{f}_{n,n+1} = (f_{n+2} - f_{n+1})/[(x_{n+2} - x_{n+1})\hat{f}_{n,n+1}] = (f_{n+1} - f_{n+2})/f_{n+1}$.

2. (a) For example $x_0 = 1$, $x_1 = -1/2$, gives $x_2 = -1/3$ and so $\beta = 1/3$. Then $\beta f_0 = 2/3$, and since $f_2 = -2/9$, therefore $x_0 f_2 = x_2 \beta f_0$.

   (b) For instance, immediately following the assignment to fx, insert

   if *mark* $= 0$ **then** *mark* $:=$ sgn $(fx * fb)$;

   Entries in Table 4.1, for iteration 2, then become

   1.023264   1.034720   1.034704   2.624270   1.157001

3. (a) Put $\sqrt{(\alpha^2 - \gamma_0)} = D$, and $1 - \beta = A - D$, say. Then $A = (1 - \gamma_1 - \gamma_0)/2$ and D are both essentially positive, and $A^2 - D^2 = \gamma_1 > 0$. Therefore $A > D$ and so $\beta < 1$.

   (b) Put $\beta - (1 - \gamma_1) = C + D$, say, where $C = (\gamma_0 + \gamma_1 - 1)/2$. Then $D^2 - C^2 = -\gamma_1 \gamma_0 > 0$, and so $D > |C|$. Therefore $C + D > 0$, and so $\beta > 1 - \gamma_1$.

4. (a) Since it is given that $\epsilon_1 \epsilon_2 > 0$ and it is required that $\epsilon_3 \epsilon_0 > 0$, therefore A must be negative if (4.1.6) is used, or $A < 3B^2/2$ if (4.1.5) is used.

   (b) Differentiation yields $A = 2(3 + 2\xi^2)$, $B = (1 + 2\xi^2)/\xi$. Evidently A is essentially positive, so the condition is not met by parabolic interpolation. However, $3B^2 - 2A = (3 + 4\xi^4)/\xi^2$ is essentially positive, so it is met by hyperbolic interpolation.

   (c) Take $f(x) \equiv a + b(x + c)^{-1}$ (a, b, and c constants). Then $A = 6/(\xi + c)^2$, $B = 2/(\xi + c)$.

5. From problem 1(a), since $x_n = 0.9$, $x_{n+2} = 0.875950$, then to give $x_{n+3} = 0.876726$, the value of β should equal 0.971178 (and only larger values of β would fail to achieve an overestimate of the true root). (a) $\beta = 0.5$; (b) $\beta = 0.970864$; (c) $\beta = 0.969990$; and (d) $\beta = 0.970959$. All give upper bounds, and (d) gives the smallest.

Section 4.2.1

1. For $n \to \infty$, from (4.2.2) and (4.2.3) in (4.2.4), $|\epsilon_{n-1}|/|\epsilon_n|^{p-1} \to |q/C_\infty|$ and so $|\epsilon_{n+1}|^{p-1}/|\epsilon_n| \to |C_\infty/q|$. Using (4.2.4) again, therefore $|\epsilon_n|^{p(p-1)-1} \to |C_\infty/q|$. But since $\epsilon_n \to 0$ as $n \to \infty$, it follows that $p^2 - p - 1 = 0$; whence quoted result, noting that $p > 0$.

2. (a) From (4.2.4), as $n \to \infty$, $\ln(|\epsilon_{n+1}|/|\epsilon_n|^p) = (pm_n - m_{n+1}) \ln(b) \to \ln(q)$ so that $m_{n+1} \cong pm_n - \ln(q)/\ln(b)$. But for large n, pm_n becomes much larger than $\ln(q)/\ln(b)$, so that $m_{n+1} \cong pm_n$.

3. (b) $C_n = (x_n + x_{n-1})^{-1} \to 1/[2 \text{ sqrt}(c)]$ as $n \to \infty$.

Section 4.2.2

1. (a) $x_{n+1} = (2 - cx_n)x_n$.

   (c) Substitute $\epsilon_{n+1} = (c\epsilon_0)^{2m}/c$ and $\epsilon_n = (c\epsilon_0)^m/c$ in the expression $\epsilon_{n+1} = c\epsilon_n^2$; the recurrence (a) converges if and only if $|c\epsilon_0| < 1$.

2. (a) $x_{n+1} = (2x_n^3 + c)/(3x_n^2)$.

   (b) Take $m = 1$: then $f''(x) = 2 - 2cx^{-3} = 2f(x)/x^2$ and $x_{n+1} = (x_n^3 + 2c)x_n/(2x_n^3 + c)$.

3. (a) Taking $C_0 = C_\infty$ in (4.2.6), it follows that $\delta \,\mathrm{sqrt}(c) \geqslant |\epsilon_1| = \epsilon_0^2/[2\,\mathrm{sqrt}(c)]$ so that $|\epsilon_0| \leqslant (2\delta)^{1/2}\,\mathrm{sqrt}(c)$.

   (b) $(2\delta)^{1/2} = 10^{-3.49} = 2^{-11.59}$, so that $\delta = 2^{-24.18}$. The approximation is therefore adequate for a binary mantissa of 24 bits.

4. With $f(x) \equiv \mathrm{erfc}(x) - 0.25 = 0$, and $f'(x) = -2\pi^{-1/2}\exp(-x^2)$, Newton-Raphson's iteration formula gives $x_{n+1} = x_n + [\pi^{1/2}\exp(x_n^2)/2][\mathrm{erfc}(x_n) - 0.25] = x_n + 0.704(x_n + 0.789) - 0.222\exp(x_n^2)$. The root is 0.823.

5. As in (4.2.7) with $c = -1$, $z_{n+1} = (z_n^2 - 1)/(2z_n) = (z_n\bar{z}_nz_n - \bar{z}_n)/(2z_n\bar{z}_n) = [(x_n^2 + y_n^2)(x_n + iy_n) - (x_n - iy_n)]/[2(x_n^2 + y_n^2)]$ so that separating real and imaginary parts, $x_{n+1} = (x_n/2)[1 - (x_n^2 + y_n^2)^{-1}]$, $y_{n+1} = (y_n/2)[1 + (x_n^2 + y_n^2)^{-1}]$.

Section 4.2.3

1. (a) Since $|\cos(x)| \leqslant 1$, all real roots lie in the interval $|x| \leqslant 10$. Superimposing sketch curves of $y = 0.1x$ and $y = \cos(x)$ will show that negative roots lie in each negative half-lobe, and positive roots in each positive half-lobe, of $y = \cos(x)$. Whence checking that $f(-10) > 0$ and $f(-3\pi) < 0$ one establishes brackets for 7 real roots: $[-10, -3\pi]$, $[-3\pi, -5\pi/2]$, $[-3\pi/2, -\pi]$, $[-\pi, -\pi/2]$, $[0, \pi/2]$, $[3\pi/2, 2\pi]$, $[2\pi, 5\pi/2]$.

   (b) Since $f'(x) = -\sin(x) - 0.1$, the steep-sides of these 7 roots $\xi_0, \xi_1, \ldots \xi_6$, where $|f'(x)|$ is larger than $|f'(\xi_k)|$, are $[-10, \xi_0]$, $[\xi_1, -5\pi/2]$, $[-3\pi/2, \xi_2]$, $[\xi_3, -\pi/2]$, $[\xi_4, \pi/2]$, $[3\pi/2, \xi_5]$, $[\xi_6, 5\pi/2]$.

   (c) Iteration formula is $x_{n+1} = x_n + [\cos(x_n) - 0.1x_n]/[\sin(x_n) + 0.1]$ and a starting value of $x_0 = 5\pi/2$ (or somewhat less) will converge to ξ_6; one finds $\xi_6 = 7.06889 = 2.25010\pi$ in 3 iterations.

2. (a) With $\xi_m = a_m - \delta$, $\tan(\xi_m) = \cot(\delta) = \xi_m = a_m - \delta$, and if $a_m \gg 1$, then $\cot(\delta) \cong 1/\delta$ is large and $\delta \cong 1/a_m$.

   (b) Root $\xi_0 = 0$; others are $\xi_m \in [a_m - \pi/4, a_m)$ or $\delta \in (0, \pi/4)$; steep-sides are $[\xi_m, a_m]$.

   (c) Since $99 = 31.51\pi$, root ξ_{31} is closest. Use iteration formula $x_{n+1} = x_n - [\tan(x_n) - x_n]/[\sec^2(x_n) - 1] = x_n\,\mathrm{cosec}^2(x_n) - \cot(x_n)$ and choose x_0 between $(a_m - 1/a_m) = 98.95006$ and $a_m = 98.96017$ to find $\xi_{31} = 98.9500628244$.

3. (a) $x_{n+1} = x_n - 3x_n = -2x_n \therefore x_n = (-2)^n x_0$ diverges with n (if $x_0 \neq 0$).

   (b) $x_{n+1} = x_n - 2x_n = -x_n \therefore x_n = (-1)^n x_0$ oscillates indefinitely.

   (c) $x_{n+1} = x_n - x_n(1 + x_n^2) = -x_n^3$ converges if and only if $|x_0| < 1$, oscillates if $|x_0| = 1$, and diverges if $|x_0| > 1$.

Section 4.2.4

1. (a) $\varphi(a) - a > 0$ and $\varphi(b) - b < 0$ so that since $\varphi(x)$ is continuous at least one root is bracketed.

(b) Let ξ_1, ξ_2 be two distinct roots in $[a, b]$: then $|\xi_1 - \xi_2| = |\varphi(\xi_1) - \varphi(\xi_2)| \leqslant M|\xi_1 - \xi_2| < |\xi_1 - \xi_2|$ which is a contradiction.

(c) $|x_{n+1} - \xi| = |\varphi(x_n) - \varphi(\xi)| \leqslant M|x_n - \xi|$, i.e. $|x_n - \xi| \leqslant M^n|x_0 - \xi| \to 0$, as $n \to \infty$.

(d) By mean value theorem, $\varphi(s) - \varphi(t) = f'(r)(s - t)$ for r between s and t: whence if $s, t \in [a, b]$, then $r \in [a, b]$ and $|\varphi(s) - \varphi(t)| \leqslant m|s - t|$.

2. (a) By mean value thoerem, as in 1(d), $|\varphi(x) - \xi| = |\varphi(x) - \varphi(\xi)| \leqslant m|x - \xi| \leqslant m\delta < \delta$. Evidently conditions of problem 1(a) and 1(d) are met, so remaining results follow as in problems 1(b) and (c).

3. (a) By Taylor's theorem, $x_{n+1} = \varphi(x_n) = \varphi(\xi) + (x_n - \xi)\varphi'(\xi) + \ldots + (x_n - \xi)^p\varphi^{(p)}(\xi_n)/p!$ for ξ_n between ξ and x_n. Thus if $\xi - x_n = \epsilon_n$, then $\epsilon_{n+1} = -(-\epsilon_n)^p\varphi^{(p)}(\xi_n)/p!$ and result follows from (4.2.4).

(b) Here $\varphi(x) = x - u(x)$, where $u(x) = f(x)/f'(x)$ and $u'(x) = 1 - f(x)f''(x)/[f'(x)]^2$. Thus $u(\xi) = 0$ and $u'(\xi) = 1$. But $\varphi'(\xi) = 1 - u'(\xi) = 0$ and $\varphi''(\xi) = -u''(\xi) = f''(\xi)/f'(\xi) \neq 0$ at a simple root.

(c) As in (b), $u(\xi) = 0$, $u'(\xi) = 1$, but $u''(\xi) \neq 0$ at a simple root. Thus at $x = \xi$:

(i) $\varphi'(x) = 1 - (3 - u')u'/2 + uu''/2 = 0$; $\varphi''(x) = 1 + 3(u' - 1)u''/2 + uu'''/2 = 0$;

(ii) $\varphi'(x) = (1 - u')(1 + u')^{-1} + 2uu''(1 + u')^{-2} = 0$; $\varphi''(x) = 2[(u' - 1)u'' + uu'''](1 + u')^{-2} - 4u(u'')^2(1 + u')^{-3} = 0$;

(iii) $\varphi'(x) = [1 - u' + uu''(u')^{-2}]/2 = 0$; $\varphi''(x) = [u''(u')^{-1} - u'' + uu'''(u')^{-2} - u(u'')^2(u')^{-3}]/2 = 0$.

4. x_{2i+1} is derived by fixed point iteration with $x_{2i+1} = \varphi(x_{2i}) \equiv x_{2i} + f(x_{2i})$. The next iteration is the secant rule, (4.2.1): $x_{2i+2} = x_{2i+1} - (x_{2i+1} - x_{2i})f_{2i+1}/(f_{2i+1} - f_{2i}) = x_{2i} + f_{2i} - f_{2i}f_{2i+1}/(f_{2i+1} - f_{2i}) = x_{2i} - f_{2i}^2/(f_{2i+1} - f_{2i})$.

5. (a) With $\varphi(x) = (bx + c/x)/(b + 1)$, $\varphi'(\sqrt{c}) = (b - 1)/(b + 1)$ and $|\varphi'(\sqrt{c})| < 1$ if $b > 0$. With $b > 0$, $|\varphi'(x)| < 1$ for all $x^2 > c/(1 + 2b)$ so that $x_0^2 > c/(1 + 2b)$ is sufficient. (In fact, $x_0 > 0$ is necessary and sufficient.)

(b) With $\varphi(x) = (2/3)(x + 1/x)$ and $x_0 = 2$ one finds successive x_k: 1) 1.666667, 2) 1.511111, 3) 1.448584, 4) 1.425942, ... 8) 1.414360, 9) 1.414262, 10) 1.414230 correct to 4 decimal places.

(c) $\varphi'(\xi) = \varphi'(\sqrt{c}) = (b - 1)/(b + 1) = 1/3$. In part (b), $(\sqrt{2} - x_{10})/(\sqrt{2} - x_9) = 0.333345$.

6. (a) Two equations are $(\xi_{n+1} - x_{n+1})/(\xi_{n+1} - x_n) = (\xi_{n+1} - x_n)/(\xi_{n+1} - x_{n-1}) = r_{n+1}$.

(c) By Taylor's theorem, $\epsilon_{n+1} = -[\varphi(x_n) - \varphi(\xi)] = -[(x_n - \xi)\varphi'(\xi) + (x_n - \xi)^2 \varphi''(\bar{\xi})/2]$ where $\bar{\xi}$ is between x_n and ξ. Whence stated result, and in (b) $\xi - \xi_{n+1} = \epsilon_n \epsilon_{n-1}[(\epsilon_n/\epsilon_{n-1}) - (\epsilon_{n+1}/\epsilon_n)]/\delta^2 x_n \cong \epsilon_n \epsilon_{n-1}$ $(\epsilon_n - \epsilon_{n-1})\varphi''(\xi)/(2\delta^2 x_n) = \epsilon_n \epsilon_{n-1}(x_{n-1} - x_n)\varphi''(\xi)/(2\delta^2 x_n) = \frac{1}{2} \epsilon_n \epsilon_{n-1} \varphi''(\xi)/(1 - r_{n+1}) \cong \frac{1}{2} \epsilon_n \epsilon_{n-1} \varphi''(\xi)/(1 - r)$ as $n \to \infty$. Thus $(\xi - \xi_{n+1})/(\xi - \xi_n) = \epsilon_n/\epsilon_{n-2} \to r^2$ as $n \to \infty$.

(d) Values of ξ_k for $k = 2, 3, \ldots 10$ are 2) 1.375, 3) 1.40655738, 4) 1.41308921, $\ldots$ 8) 1.41421336, 9) 1.4141354, 10) 1.41421356 correct to 8 decimal places.

7. (a) $\{1.41479406, 1.41424792, 1.41421518\}$

(b)
j	$x_{j,0} = x_j$	$x_{j,1}$	$x_{j,2}$	$x_{j,3}$	$x_{j,4}$
0	4				
1	15	$-1/11$			
2	13	$1/2$	$173/13$		
3	33	$-1/20$	$163/11$	$-35/216$	
4	49	$-1/16$	113	$-13/216$	5

(c) $x_{j,1} = 1/\nabla x_j$ and $x_{j,2} = x_{j-1} + 1/(1/\nabla x_j - 1/\nabla x_{j-1}) = x_{j-1} - \nabla x_{j-1}$ $\nabla x_{j-1}/\delta^2 x_{j-1} = \xi_j$.

(d)
j	$x_{j,1}$	$x_{j,2} = \xi_j$	$x_{j,3}$	$x_{j,4}$
2	6.42857143	1.375		
3	15.9930314	1.40655738	-25.2597403	
4	44.16627392	1.41308921	-137.103442	1.41549843
5	128.4205893	1.41407334	-971.956738	1.41428702
6	381.0768767	1.41419731	-7938.13163	1.41421689

Correct limit is $\sqrt{2} = 1.41421356$ and $x_{j,4}$ has errors $\{-1.28 \times 10^{-3}, -7.34 \times 10^{-5}, -3.33 \times 10^{-6}\}$ decreasing by a factor $\cong 1/20$. Sequence in part (a) has errors $\{-5.80 \times 10^{-4}, -3.44 \times 10^{-5}, -1.62 \times 10^{-6}\}$ decreasing at about the same rate.

Section 4.2.5

1. (a) $p_2(x) = f_n + \hat{f}_{n-1,n}(x - x_n) + \hat{f}_{n-2,n-1,n}(x - x_n)(x - x_{n-1})$ is interpolating polynomial, and $p_2(x_{n+1}) = 0$ implies that $\hat{f}_{n-2,n-1,n}(x_{n+1} - x_n)^2 + \omega(x_{n+1} - x_n) + f_n = 0$. Choose root giving the smaller magnitude of $(x_{n+1} - x_n)$, if real.

(b) From (3.2.1), $p_2(\xi) = -\epsilon_n \epsilon_{n-1} \epsilon_{n-2} f'''(\bar{\xi})/6$ where $\bar{\xi}$ is in the interval containing x_{n-2}, x_{n-1}, x_n and ξ. If the method converges, then $\bar{\xi} \to \xi$, and $p_2(\xi) \to p_2(x_{n+1}) + (\xi - x_{n+1})p_2'(x_{n+1}) = \omega\epsilon_{n+1} \to f'(\xi)\epsilon_{n+1}$, so that $C_\infty = -f'''(\xi)/6f'(\xi)$.

(d) $x_{n+1} \cong x_n - 2f_n \, \mathrm{sgn}(\omega)/[\, |\omega| + |\omega|(1 - 2f_n\hat{f}_{n-2,n-1,n}/\omega^2)] = x_n - f_n/(\omega - f_n\hat{f}_{n-2,n-1,n}/\omega) \cong x_n - f_n/(\omega - f_n\hat{f}_{n-2,n-1,n}/\hat{f}_{n-1,n}) = x_n - f_n/\{\hat{f}_{n-1,n} + [(x_n - x_{n-1})\hat{f}_{n-1,n} - f_n]\hat{f}_{n-2,n-1,n}/\hat{f}_{n-1,n}\}$ whence quoted result.

2. (a) Placing $p = x_{n+3}$ and replacing n by $(n-2)$ throughout, $(x_{n+1} - x_n)/(x_{n+1} - x_{n+2}) = f_n f_{n-2,n-1}/(f_{n-1,n}f_{n-2})$ whence $x_{n+1} = x_n - (x_n - x_{n-2})f_n/(f_n - f_{n-2}\hat{f}_{n-1,n}/\hat{f}_{n-2,n-1})$.

   (b) If the points coalesce, then $\hat{f}_{n-2,n-1} \to \hat{f}_{n-1,n} \to f'(x_n)$ and $\hat{f}_{n-2,n-1,n} \to f''(x_n)/2$.

3. (a) From problem 2, § 3.1.3, $p''(x_n) = 2(B - 2A)$ and result follows.

   (b) Iteration formula is $x_{n+1} = x_n - u_n[1 + u_n(1 - \hat{u}_{n-1,n})/(u_{n-1} + u_n)]$. Successive root estimates are 1) 2.834362, 2) 2.567782, 3) 2.290965, 4) 2.060028, 5) 1.979235 and estimate is correct to 6 decimal places on 6th iteration. (Compare Table 4.2 and note better global convergence of this method.)

4. Successive root estimates are 1) 2.834362, 2) 2.581593, 3) 2.400263, 4) 2.155880, 5) 2.035076, 6) 1.980511, 7) 1.977899 and estimate is correct to 9 decimal places at the 8th iteration. (Compare Table 4.2.)

SECTION 4.3

1. (a) Since $u'(x) = 1 - f(x)f''(x)/[f'(x)]^2$ and $x_{n+1} = x_n - u(x_n)/u'(x_n)$ therefore $x_{n+1} = x_n - f_n f'_n/[(f'_n)^2 - f_n f''_n]$.

   (b) From (4.2.1), $x_{n+1} = x_n - (x_n - x_{n-1})u_n/(u_n - u_{n-1})$, where $u_k = u(x_k)$. Thus $x_{n+1} = x_n - (x_n - x_{n-1})f_n f'_{n-1}/(f_n f'_{n-1} - f_{n-1}f'_n)$.

   (c) Method (a) has quadratic convergence at each step and requires f, f' and f'' at each step: and method (b) converges with order $(1 + \sqrt{5})/2$ but needs only f and f' at each step. Three steps of (b) converge with order $(1 + \sqrt{5})^3/8 = 4.24$, whereas two steps of (a) require equivalent work but converge with order $2^2 = 4$. Thus (b) is slightly more efficient.

2. (a) $x_{n+1} = x_n - mf(x)/f'(x) = x_n - g(x)/g'(x)$ where $g(x) = [f(x)]^{1/m}$, and $g(x)$ has a simple zero at $x = \xi$.

   (b) Successive estimates are 1.293408, 1.572589 and 1.570796 with m_1, m_2, and m_3 evaluated as 2.

   (c) Starting with $x_0 = 0$, $m_0 = 1$, successive estimates are 0.125, 0.2625, 0.267941 and 0.267949 with $m_1, m_2, \ldots = 2$. (The equation has in fact a double root at $x = 1/(2 + \sqrt{3})$).

4.

n	2	3	4	5	6	7	8	9	10
x_n	0.52632	0.68966	0.81250	0.88312	0.92800	0.95545	0.97248	0.98299	0.98949
ξ_n	−0.03065	0.79111	1.18516	0.97860	1.00628	0.99864	1.00033	0.99992	1.00002

Theoretical asymptotic error factor $= 2/(\sqrt{5} + 1) = 0.618$; $\epsilon_{10}/\epsilon_9 = 0.618$. Value of $(\xi - \xi_{10})/(\xi - \xi_9) = -0.24$.

Section 4.4

1. (a) $|z|^n \leqslant |1 + b_1 z + b_2 z^2 + \ldots + b_{n-1} z^{n-1}| \leqslant 1 + |b_1||z| + |b_2||z|^2 + \ldots + |b_{n-1}||z|^{n-1} \leqslant 1 + B(|z| + |z|^2 + \ldots + |z|^{n-1}) = 1 + B(|z|^n - |z|)/(|z| - 1)$. Thus $|z| \geqslant 1 + B$ implies $|z|^n \leqslant 1 + |z|^n - |z|$, i.e. $|z| \leqslant 1$, which is absurd.

   (b) If $B = 0$, then $|z| = 1$.

   (c) Place $x = \mu z$ and $a_j = a_0 \, b_j/\mu^j$ for $j = 1, 2, \ldots (n-1)$, where $\mu = (a_0/a_n)^{1/n}$. Then $p_n(x)/a_0 = 0$ converts to the form of part (a).

2. Place $x = \mu/z$, and apply (4.4.2) to the bounding of the roots of $z^n \, p_n(\mu/z) = 0$.

3. (a) $|a_n||x|^n = |a_{n-1}x^{n-1} + \ldots + a_1 x + a_0| \leqslant \sum\limits_{j=0}^{n-1} |a_j||x|^j$. Thus if $|x| > 1$, then $|x|^n \leqslant \sum\limits_{j=0}^{n-1} |a_j|/|a_n| \leqslant 1$, which is a contradiction.

   (b) If $|x| \geqslant \alpha > a$, then $|x|^n < \alpha^{n-1} (\sum\limits_{j=0}^{n-1} |a_j|/|a_n|) = \alpha^n$, again a contradiction.

Section 4.4.1

1. (a) $f_m(x) = (-1)^m \neq 0 \ \forall x$; so condition (i) is satisfied. $f_{k-1}(x)f_{k+1}(x) = \cos[(m-k)x + x]\cos[(m-k)x - x] = \cos^2[(m-k)x]\cos^2 x - \sin^2[(m-k)x]\sin^2 x = -\sin^2 x$, if $f_k^2(x) = \cos^2[(m-k)x] = 0$. Since $\cos[(m-k)x] \neq 0$ if $\sin x = 0$, therefore condition (ii) is satisfied. Finally $f_0'(t)f_1(t) = m \sin(mt)\cos[(m-1)t] = m \sin^2(mt)\sin t$, if $\cos(mt) = 0$, so that condition (iii) is satisfied provided $t \in [0, \pi]$.

   (b) The sign sequence at $x = 0$ is $+ - + - \ldots$, but at $x = \pi$, since $\cos(n\pi) = (-1)^n$, each function of the sequence has the sign of $(-1)^m$. Thus $v(0) = m$ and $v(\pi) = 0$, and $\cos(mx)$ is shown by Sturm's theorem to have m zeros in $[0, \pi]$. In fact, zeros are at $x = (2n-1)(\pi/2m)$ for $n = 1, 2, \ldots, m$.

2. Here $m = 2$ and condition (i) is clearly satisfied. Condition (ii) requires that $f_0(a_1)f_2(a_1) = -(a_1^2 - a_0)^2$ is negative which is clearly true if $a_0 \neq a_1^2$, whilst condition (i) requires that $f_0'(t)f_1(t) = 2(t - a_1)^2$ is positive for any zero of $f_0(t)$. This is satisfied since $f_0(a_1) = a_0 - a_1^2 \neq 0$.

5. (a) From (3.7.3), $A = 0$, $B = 2$, $C = 1$. Further from (3.7.2), $T_n'(x) = n \sin(n\theta)/\sin\theta$, where $x = \cos\theta$: hence $\varphi(x)T_n'(x) = (\sin^2\theta)T_n'(x) = n \sin\theta \sin(n\theta) = -n \cos\theta \cos(n\theta) + n \cos[(n-1)\theta] = -nxT_n(x) + nT_{n-1}(x)$, so that $P = 0, Q = R = -n$.

   (b) Condition (i) for a Sturm sequence is met, since $f_m(x) \equiv p_0(x)$ is a non-zero constant. Condition (ii) is met, iff $C > 0$. Condition (iii) is met iff $Rq(t) < 0$ for every zero $x = t$ of $p_m(x)$. From part (a), if $p_n(x) \equiv T_n(x)$, then $C > 0, R < 0$, and $q(t) > 0$, since all m zeros of $T_m(x)$ lie in $(-1, 1)$.

(c) Show as in part (a) that $A = 0$, $B = 2$, $C = 1$, and with $q(x) = 1 - x^2$, $P = 0$, $Q = -n$ and $R = -(n + 1)$. Then as in part (b), conditions for a Sturm sequence are met, because the m zeros of $U_m(x)$ are at the points $x = \cos\theta_k$ in $(-1, 1)$ where $\theta_k = k\pi/(m + 1)$, for $k = 1, 2, \ldots, m$.

(d) The sequence $T'_m(x)$, $T'_{m-1}(x)$, $\ldots$, $T'_1(x)$ is a Sturm sequence (of m-terms by part (c), and clearly $f'_0(x)f_1(x) \equiv [T'_m(x)]^2 > 0$ at the (simple) zeros of $T_m(x)$. It remains only to show that $f_0(x)f_2(x) \equiv T_m(x)T'_{m-1}(x) < 0$ at the zeros of $T'_m(x)$: these are the values $x = \cos(k\pi/m)$ for $k = 1, 2, \ldots, (m - 1)$. But $T_m(x)T'_{m-1}(x) = \cos(m\theta) \sin[(m - 1)\theta]/\sin\theta = \cos(m\theta) [\sin(m\theta) \cos\theta - \cos(m\theta)]$, and equals (-1) at $\theta = k\pi/m$ for $k = 1, 2, \ldots, (m - 1)$.

Section 4.4.2

1. If $f_m(x)$ terminates the sequence, then $f_{m+1}(x) = f_m(x)q_m(x) - f_{m-1}(x) \equiv 0$, so that $f_{m-1}(x) \equiv f_m(x)q_m(x)$ and $f_m(x)$ is a factor of $f_{m-1}(x)$. It is also therefore a factor of $f_{m-2}(x) = f_{m-1}(x)q_{m-1}(x) - f_m(x)$ and similarly of $f_k(x)$ for $k = m - 3, m - 4, \ldots 0$. Conversely if it is a factor of $f_0(x), f_1(x)$, it must be also a factor of $f_2(x) = f_1(x)q_1(x) - f_0(x)$ and of every successive function of the sequence which is not identically zero.

2. (a) $\{p_3(x), 3x^2 + 2x - 4, 13(2x - 1)/9, 9/4\}$: 3 real roots in intervals $(-3, -2)$, $(0, 1)$ and $(1, 2)$.

   (b) $\{p_4(x), 4(x^3 - x + 2), x^2 - 6x + 22, 52(-x + 10), - 62\}$: 2 real roots in intervals $(-3, -2)$ and $(1, 2)$. Note that sequence could be terminated with $f_2(x) = (x - 3)^2 + 13$ since this is essentially positive.

3. $\{p_4(x), 4x^3 + 3.99x^2 + 1.33x + 1.15, -0.752x - 0.251, -1\}$. Note that the coefficient of x^2 in $f_2(x)$ is insignificant.

4. (a) For instance:

   if $va > vb + 1$ **then**
       begin real c; **integer** vc;
           $c := (a + b)* 0.5; vc := v(c)$;
           $setbnds(a, c, va, vc, j); setbnds(c, b, vc, vb, j + va - vc)$
       end
   else if $va \neq vb$ **then**
       begin *upper* $_j := b;$ *lower* $_j := a$
       end;

   (b) Suitable call would be $setbnds$ $(-xlim, xlim, v(-xlim), v(xlim), 1)$. Bounds assigned are $(-6, 0), (1.5, 2.25), (2.25, 3)$.

5. Suppose $f_{i+1}/f_i = \overset{m}{\underset{j=i+1}{K'}} (-1) \downarrow q_j$ for some $i = k \leqslant (m - 1)$. Then from (4.4.5) this relation is true for $i = k + 1$. But it is true for $i = (m - 1)$, by definition.

Section 4.4.3

1. (b) Algebraically $f_2(x) \equiv \alpha^2 [(1 + \delta_1)^2/(1 + \delta_2) - (1 + \delta_0)] \cong \alpha^2(1 + 2\delta_1 - \delta_2) - \alpha^2(1 + \delta_0)$, and numerically forced cancellation would be applied when the result has a magnitude of less than $N\epsilon$ relative to one or other of the operands i.e., approximately when $|2\delta_1 - \delta_0 - \delta_2| < N\epsilon$.

2. Sign sequences are respectively $+ - + -, + + - +, - - + +, + + + +$.

3. (a) $\phi_m(x) \equiv 1$, so that condition (i) is satisfied, and $\phi_{m-1}(x)\phi_{m+1}(x)$ has the same sign as $f_{m-1}(x)f_{m+1}(x)$ so that condition (ii) is also satisfied. Finally if $\phi_0(t) = 0$, so that $f_0(t)$ is also zero, $\phi_0'(t)\phi_1(t) = [p_n'(t)/f_m(t)]^2$ which is positive, since if $p_n'(t)$ is zero then $f_m(x)$ will have a zero at $x = t$ of the same multiplicity.

   (b) Note that the sequence of signs of $\phi_k(x)$ is the same as that of $\{\phi_k(x) f_m(x)\}$ at any value of x other than those which are zeros of $f_m(x)$.

Section 4.4.4

2. (a) Deflated polynomial has coefficients $b_1 = \xi - \alpha_1 - \alpha_2 - \alpha_3, b_0 = \alpha_1\alpha_2 + \alpha_2\alpha_3 + \alpha_3\alpha_1 + b_1\xi$. Term involving ϵ^2 in b_0 can be ignored.

   (b) The zeros of $p_2(x)$ are approximately $x = \alpha_k - p_2(\alpha_k)/p_2'(\alpha_k)$ for $k = 2$, 3, and terms dependent on ϵ in expression for $p_2'(\alpha_k)$ can be ignored.

   (c) Using results of part (b) with different choices of $\alpha_1, \alpha_2, \alpha_3$

$\{\xi_1, \xi_2, \xi_3\}$	$\{\alpha_1, \alpha_2, \alpha_3\}$	$\{\alpha_3, \alpha_1, \alpha_2\}$	$\{\alpha_2, \alpha_3, \alpha_1\}$
relative error in α_2	$-(\xi_1/\xi_2)\beta\epsilon$	$-(\xi_2/\xi_3)^2\beta\epsilon$	$(\xi_3^2/\xi_1\xi_2)\beta\epsilon$
relative error in α_3	$(\xi_1\xi_2/\xi_3^2)\beta\epsilon$	$-(\xi_2/\xi_1)\beta\epsilon$	$-(\xi_3/\xi_2)^2\beta\epsilon$.

   Both relative errors are $\ll |\beta\epsilon|$ in magnitude only if $\alpha_1 \equiv \xi_1$.

   (d) Approximately $\alpha \pm [\alpha_1(\alpha - \alpha_1)\beta\epsilon]^{1/2}$ where $\alpha_2 = \alpha_3 = \alpha$.

3. (a) Root $0.00182, p_2(x) = 10x^2 - 990x + 548$; root $0.557, p_1(x) = 10x - 984$; root 98.4.

   (b) Root $98.4, p_2(x) = 10x^2 - 6x - 40$; root $2.32, p_1(x) = 10x + 17.2$; root -1.72. (Smaller roots completely inaccurate due to loss of significance in coefficients of $p_2(x)$.)

4. (a) Root $0.0102, q_2(x) = x^2 - 550x + 984$; root $1.79, q_1(x) = x - 548$; root 548. Thus the roots of cubic of problem 3 are inferred as $\{98.0, 0.559, 0.00182\}$.

   (b) Root $98.4, p_2(x) = -984x^2 + 550x - 1$; root $0.557, p_1(x) = 548x - 1$; root $= 0.00182$. (Note that these roots are more accurate than those obtained in part (a), because the rounding error involved in finding the reciprocal is avoided.)

   (c) If $q_n(z) = c_n z^n + c_{n-1} z^{n-1} + \ldots + c_0$, then $c_k = a_{n-k}$, and deflation applied to extract the factor $z - \xi^{-1}$ leads to $q_{n-1}(z) = c_{n-1}' z^{n-1} +$

$c'_{n-2}z^{n-2} + \ldots + c'_0$ where $c'_{n-1} = c_n$, $c'_{n-k-1} = c_{n-k} + \xi^{-1}c'_{n-k}$:
$k = 1, 2, \ldots (n-1)$. With $c'_{n-k-1} = a'_k$, this is equivalent to $a'_0 = a_0$,
$a'_k = a_k + a'_{k-1}/\xi$.

5. (a) Root 0.00100, $p_2(x) = x^2 - 20x + 100$; root 10.0, $p_1(x) = x - 10$;
   root 10.0.
   (b) Root 10.3, $p_2(x) = 10.3x^2 + 99.9x - 1$; root 9.69, $p_1(x) = 99.8x - 1$;
   root 0.00100. (It is preferable to separate the closer-spaced roots first
   rather than last, even although these are large in magnitude.)

6. (a) Apply Newton's method to $F(x) \equiv f(x)/\prod_{j=1}^{m} (x - \xi_j) = 0$. Then differen-

   tiating, $F'(x) = [f'(x) - f(x) \sum_{j=1}^{m} (x - \xi_j)^{-1}] / \prod_{j=1}^{m} (x - \xi_j)$. Iterations then

   take the form $x_{k+1} = x_k - F(x_k)/F'(x_k)$.
   (b) Starting with $x_0 = 0$, successive iterations yield 0.01001001, 0.01002007
   (the correct small root to this accuracy). The imprecision of ξ_1 does not
   impair the accuracy of ξ_2.

Section 4.5.2

1. (a) For instance:
   ```
   p: = a + (b - a)/(sqrt(1.25) + 0.5); fp : = f(p);
   while b - a > eps do
      begin q : = b - p + a; fq : = f(q);
            if fq < fp then
            begin if p < q then a : = p else b : = p;
                  p : = q; fp : = fq
            end
            else if p < q then b : = q else a : = q
   end
   ```
   This continually updates the interval bounds; p and q are the points of
   golden section.
   (b) If eps were zero, the above algorithm would loop indefinitely with a and
   b equal to adjacent floating-point numbers, and with p and q being equal
   to a and b in that arrangement for which $f(q) \geqslant f(p)$. This could (for
   example) be avoided by iterating while $b - a > \max(eps, abs(p - q))$
   with q initialised equal to p (say). This condition may still be required if
   $eps > 0$, were eps to be smaller than the difference between two adjacent
   floating-point numbers within the initial interval.
   (c) First 3 contracted intervals are $[1, 4.09]$, $[1, 2.91]$, $[1.73, 2.91]$.
   (d) Assuming $q = -p$ in the initial iteration, the contracted intervals are
   $[-0.236, 1]$, $[-0.236, 0.528]$, $[-0.236, 0.236]$.

2. (a) The interpolating polynomial is $y = p_2(x) \equiv y_0 + (x - x_0)\hat{y}_{0,2} + (x - x_0)$
$(x - x_2)\hat{y}_{0,1,2}$ and $p_2'(x) \equiv \hat{y}_{0,2} + (2x - x_0 - x_2)\hat{y}_{0,1,2}$ vanishes for value of x stated.

(b) As in part (a), $x - \frac{1}{2}(x_1 + x_2) = \hat{y}_{1,2}/(2\hat{y}_{0,1,2}) = \frac{1}{2}(x_2 - x_0)[\hat{y}_{1,2}/(\hat{y}_{1,2} - \hat{y}_{0,1})]$. But by minimal condition, $\hat{y}_{1,2}$ and $(-\hat{y}_{0,1})$ are both non-negative, but not both zero. Thus square-bracketed term involving divided differences has value in $[0, 1]$.

SECTION 5.1

1. (b) Correction term is $(\hat{y}_{0,1} + \hat{y}_{1,2})(h_2 - h_1)/6$.

(c) Result follows from (a) with $h_1 = 1/3$, $h_2 = 2/3$. Similarly $\int_0^1 y(x)dx = [y(0) + 3y(2/3)]/4$ is also true for quadratic $y(x)$. Adding both results and dividing by 2, three-eighths rule follows.

(d) From result of part (b), letting $h_1 \to 0$, and $y_1 \to y_0$ but $\hat{y}_{0,1} \to y'(0)$, $\int_0^1 y(x)dx = [5y(0) + y(1)]/6 + [y'(0) + y(1) - y(0)]/6 = [2y(0) + y(1)]/3 + y'(0)/6$. Similarly $\int_0^1 y(x)dx = [y(0) + 2y(1)]/3 - y'(1)/6$. Adding and taking mean: $\int_0^1 y(x)dx = [y(0) + y(1)]/2 - [y'(1) - y'(0)]/12$.

Section 5.1.1

1. (a) Integrating once by parts, $\int_a^b (b - x)(x - a)f''(x)dx = \int_a^b (2x - a - b)f'(x)dx$. Result follows upon integrating once more by parts.

(b) Integrating twice by parts, E_m is equal to $\frac{1}{2}\int_a^b [\frac{1}{2}(b - a) - |x - c|]^2 f''(x)dx = \int_a^b [\frac{1}{2}(b - a)\text{sgn}(x - c) - x + c]f'(x)dx = \frac{1}{2}(b - a)[f(b) - 2f(c) + f(a)] - \int_a^b (x - c)f'(x)dx = -(b - a)f(c) + \int_a^b f(x)dx$.

2. (a) The integral over $[a, b]$ of such an odd function of $(x - c)$ is zero: also if $f(x)$ is such an odd function, then $f_k = -f_{n-k}$ so that the weighted sum is also zero if the formula is symmetric.

(b) Substituting $t = x - c$, and putting $(b - a)/2 = h$, the integral of any odd power of t over $[-h, h]$ is correctly evaluated (as zero) by the formula.

3. Correct integral mean is $(2/\pi)$.

n	1	2	3	4	5	6
E_n	0.1366	-1.451×10^{-3}	-6.40×10^{-4}	5.37×10^{-6}	3.02×10^{-6}	-1.645×10^{-8}
ξ	0.727	0.757	0.749	0.764	0.762	0.769

4.

n	0	1	2	3	4	5
E_n	-0.0705	-1.666×10^{-2}	5.64×10^{-4}	2.90×10^{-4}	-3.84×10^{-6}	-2.30×10^{-6}
ξ	0.755	0.705	0.756	0.751	0.766	0.764

5. (a) $k = 0$: $\int_0^1 x^0 dx = 1 = \lambda_1 + \lambda_2 + \lambda_3$; $k = 1$: $\int_0^1 x\, dx = 1/2 = (\lambda_1/4) +$ $(\lambda_2/2) + (\lambda_3/4)$; $k = 2$: $\int_0^1 x^2 dx = 1/3 = (\lambda_1/16) + (\lambda_2/4) + (9\lambda_3/16)$ giving $\lambda_1 = \lambda_3 = 2/3$, $\lambda_2 = -1/3$.

(b) $\int_0^1 f(x)dx = \lambda_1 f(1/4) + \lambda_2 f(1/2) + \lambda_3 f(3/4) + c f^{iv}(\xi)$, say. Thus with $f(x) = x^4$, and so $f^{iv}(\xi) = 4!$, $1/5 = 2/(4^4 3) - 1/(2^4 .3) + (2.3^4)/(4^4 .3) +$ $4!c$ giving $c = 7/23040$.

6. (a) Since $g_0(t)$ is even, $\bar{g}_0 = \int_0^1 g_0(t)dt = \int_{-1}^0 g_0(t)dt$, and it follows that $g_1(t) = \int_0^t [g_0(s) - \bar{g}_0]\,ds$. This is the indefinite integral about zero of an even function, and so is an odd function of t. Consequently $g_1(1) = -g_1(-1)$ and $g_1(-1)$ is clearly zero; further $\bar{g}_1 = 0$ since $g_1(t)$ is an odd function of t. Thus $g_2(t) = \int_0^t g_1(t)dt + \int_{-1}^0 g_1(t)dt$ can be seen to be an even function, since an indefinite integral about zero of an odd function is an even function. Because $g_2(-1)$ is zero, therefore so also will be $g_2(1)$.

(b) $\bar{g}_0 = \int_0^1 (1 - t^2)dt/2 = 1/3$; $g_1(t) = \int_0^t (1 - 3s^2)ds/6 = (1 - t^2)t/6$; $\bar{g}_1 = 0$; $g_2(t) = \int_{-1}^t (s - s^3)ds/6 = -(1 - t^2)^2/24$; $\bar{g}_2 = -\int_0^1 (1 - t^2)^2 dt/$ $24 = -1/45$.

(c) $\int_{-1}^{+1} g_0(t)F''(t)dt = \int_{-1}^{+1} \bar{g}_0 F''(t)dt + \int_{-1}^{+1} g_1'(t)F''(t)dt = [F'(1) - F'(-1)]\bar{g}_0 - \int_{-1}^{+1} g_1(t)F'''(t)dt = [F'(1) - F'(-1)]\bar{g}_0 + \int_{-1}^{+1} g_2(t)F^{iv}(t)dt$, after second integration by parts, since $g_1(1) = g_1(-1) = 0$, $g_2'(t) = g_1(t)$ and $g_2(1) = g_2(-1) = 0$. Series expansion follows by repeated application of same process.

7. (c) For $\alpha^2 < n^2\pi^2$, $(n^2\pi^2 - \alpha^2)^{-1}$ may be expanded as a series in α^2 and

$$\alpha \cot(\alpha) = 1 + 2 \sum_{n=1}^\infty \sum_{k=0}^\infty (\alpha/n\pi)^{2k+2} = 1 + \sum_{k=0}^\infty 2(\alpha/\pi)^{2k+2} \sum_{n=1}^\infty n^{-2k-2}.$$

From (2.5.5), $\bar{g}_{2k} = (-4)^{k+1} B_{k+1}/(2k + 2)!$

8. (a) Substitute $x = c + (b - a)t/2$, where $c = (b - a)/2$, and $f(x) = F(t)$, so that $f^{(k)}(x) = [2/(b - a)]^k F^{(k)}(t)$. Then from (5.1.3) $-E_T = [(b - a)/2]$

$$\int_{-1}^{+1} g_0(t)\, F''(t)dt = [(b - a)/2] \left\{ \sum_{j=1}^{m-1} [F^{(2j-1)}(1) - F^{(2j-1)}(-1)] \right.$$
$$\left. \bar{g}_{2j-2} + \int_{-1}^{+1} g_{2m-2}(t)F^{(2m)}(t)dt \right\} = \sum_{j=1}^{m-1} [f^{(2j-1)}(b) - f^{(2j-1)}(a)]$$
$$(b - a)^{2j}(\bar{g}_{2j-2}/4^j) + R_m.$$

(b) Assume proposition is true for $k = m$. If $g_{2m+2}(t)$ had a zero in $(-1, 1)$ then since it is zero at $t = \pm 1$, it would have two stationary values in $(-1, 1)$: but as $g_{2m+2}'(t) = g_{2m+1}(t)$ has only one zero in $(-1, 1)$, this is impossible. Again, if $g_{2m+3}(t)$ had a zero in $(0, 1)$ then since it is zero at $t = 0$ and $t = 1$, it would have to have a point of inflection in $(0, 1)$: but $g_{2m+3}''(t) = g_{2m+1}(t)$ has no zero in $(0, 1)$, so this is impossible. Therefore proposition is also true for $k = m + 1$. But proposition is true for $k = 0$: so by induction it is true for $k = 1, 2, \ldots$

(c) The proposition (b) applies to the sequence developed from $g_0(t) = (1 - t^2)/2$, as can be confirmed from problem 6(b). Applying the integral mean value theorem to the expression for R_m given in (a), since $g_{2m-2}(t)$ does not change sign in $(-1, 1)$, therefore for some $\tau \in (-1, 1)$: $R_m = \frac{1}{2}(b - a) \int_{-1}^{+1} g_{2m-2}(t) \, F^{(2m)}(t) dt = (b - a) \bar{g}_{2m-2} \, F^{(2m)}(t) = (b - a)^{2m-1} (\bar{g}_{2m-2}/2^{2m}) f^{(2m)} (g)$ for some $\xi \in (a, b)$.

Section 5.1.2

1. (a) Replacing the values of a and b in the result of § 5.1.1, problem 8(a) by x_{j-1} and x_j and summing over $j = 1, 2, \ldots N, - (b - a)E_{N,1} = - \sum_{j=1}^{N} E_T =$

$$\sum_{j=1}^{N} \sum_{k=1}^{m-1} \left\{ f^{(2k-1)}(x_j) - f^{(2k-1)}(x_{j-1})] (x_j - x_{j-1})^{(2k-1)} \bar{g}_{2k-2}/2^{2k} \right\} +$$

$\sum_{j=1}^{N} R_m$. But $(x_j - x_{j-1}) = (b - a)/N$, and $\sum_{j=1}^{N} [f^{(2k-1)}(x_j) - f^{(2k-1)}(x_{j-1})] = [f^{(2k-1)} (x_N) - f^{(2k-1)} (x_0)]$. Whence stated result, since $x_N = b, x_0 = a$.

(b) It remains only to establish the form of the remainder. Replacing a and b in the result of § 5.1.1, problem 8(c) by x_{j-1} and x_j, and ξ by $\xi_j \in (x_{j-1}, x_j)$, then $\sum_{j=1}^{N} R_m = (x_j - x_{j-1})^{2m+1} (\bar{g}_{2m-2}/2^{2m}) \sum_{j=1}^{N} f^{(2m)}(\xi_j)$.

But if $f^{(2m)} (x)$ is bounded above and below in $[a, b]$, then $\sum_{j=1}^{N} f^{(2m)}(\xi_j) = Nf^{(2m)} (\xi)$ for some $\xi \in (a, b)$, and so using result of problem 7(c),

$$§ 5.1.1: - (b - a)^{-1} \sum_{j=1}^{N} R_m = (-1)^m (b - a)^{2m} B_m f^{(2m)} (\xi)/[(2m)! N^{2m}].$$

2. Show that the composite rule is a weighted average of $(n + 1)$ Riemann sums, $s_n(x_k)$ for $k = 0, 1, \ldots n$.

Section 5.1.3

1.

j	$I_M(2^{j-2})$	$I_T(2^{j-1})$	$I_{j,2} = I_s(2^{j-2})$	$I_{j,3}$	$I_{j,4}$	$I_{j,s}$
1.		.5				
2.	.7071067812	.6035533906	.6380711875			
3.	.6532814824	.6284174365	.6367054518	.6366144028		
4.	.6407288619	.6345731492	.6366250535	.6366196936	.6366197776	
5.	.6376435773	.6361083633	.6366201013	.6366197712	.6366197724	.6366197724

After completing 4 rows, best estimate is either $I_{4,4}$ or $I_{4,3} = 0.636620$ with inferred accuracy better than $|I_{4,3} - I_{3,3}| \cong 5 \times 10^{-6}$ (actual error 2×10^{-7}). After 5 rows, best estimate is either $I_{5,5}$ or $I_{5,4} = 0.636619772$ with inferred accuracy better than $|I_{5,4} - I_{4,4}| = 5 \times 10^{-9}$ (actual error 4×10^{-10}).

2. $|I_{j,k+1} - I_{j-1,k}| = [4^k/(4^k - 1)]|I_{j,k} - I_{j-1,k}| > |I_{j,k} - I_{j-1,k}|$.

3. (a) $[16I_s(2) - I_s(1)]/15 = [(16/12)(f_0 + 4f_1 + 2f_2 + 4f_3 + f_4) - (1/6)$
$(f_0 + 4f_2 + f_4)]/15 = (7f_0 + 32f_1 + 12f_2 + 32f_3 + 7f_4)/90 = I_{3,3}$.

   (b) $(64I_{4,3} - I_{3,3})/63 = [32(7f_0 + 32f_1 + 12f_2 + 32f_3 + 14f_4 + 32f_5 + 12f_6$
   $+ 32f_7 + 7f_8) - (7f_0 + 32f_2 + 12f_4 + 32f_6 + 7f_8)]/5670 = I_{4,4}$. No
   negative weights. Error of 8th order.

4. $[4I_M(2) - I_M(1)]/3 = (2/3)[f(1/4) + f(3/4)] - f(1/2)/3$; i.e. $\lambda_1 = \lambda_3 = 2/3$,
$\lambda_2 = -1/3$, as in problem 5(a), § 5.1.1. $[9I_M(3) - I_M(1)]/8 = (3/8)[f(1/6) +$
$f(1/2) + f(5/6)] - f(1/2)/8$; i.e. $\mu_1 = \mu_2 = 3/8$, $\mu_2 = 1/4$, as in Table 5.2 for
$n = 3$.

5. (a) With stated approximation, $f^{(k)}(x) \cong (k + 1)!/(-x)^{k+2}$ and so $f^{(2k-1)}(b)$
   $- f^{(2k-1)}(-b) = -2(2k)!/b^{2k+1}$. Result follows on substituting in
   (5.1.23). Series semi-convergent, since term-ratio is $-(B_j/B_{j-1})(2/N)^2 \cong$
   $-2j(2j - 1)/(\pi N)^2$, from (2.5.6), which is greater than unity in magnitude for $j > \pi N/2$.

   (b) If $f(x) = \sin(x)$, $f^{(2k-1)}(x) = (-1)^{k+1} \cos(x)$, and here $a = 0$, $b = \pi/2$ so
   that $f^{(2k-1)}(b) - f^{(2k-1)}(a) = (-1)^k$. Thus from (5.1.23), $E_{N,1} = (2/\pi)$
   $[B_1(\pi/2N)^2/2! + B_2(\pi/2N)^4/4! + B_3(\pi/2N)^6/6! + \ldots]$. Term ratio is
   $(B_j/B_{j-1})\pi^2/[8N^2j(2j - 1)] \cong (1/4N)^2$, and series converges for $N \geqslant I$.

6. (a) Putting $x = 1/h^2$ the kth column contains estimates of the value of
   $f(x) \equiv I_T(x^{-1/2})$ at $x = 0$ based on interpolation of $I_T(x^{-1/2})$ by a polynomial in x of degree $(k - 1)$, for values of $x = x_\ell = 1/4^{j-k+\ell}$ and $\ell =$
   $0, 1, \ldots (k - 1)$. Assuming $f(x) = I + \sum_{j=1}^{k} (-1)^j a_j x^j$, then from (3.7.1),
   the error in $I_{j,k} = p_{k-1}(0)$ is $f(0) - p_{k-1}(0) = (-1)^k(x_0 x_1 \ldots x_{k-1})$
   $f^{(k)}(\xi)/k! = a_k/(2^{2j-k-1})^k$. But this is the error $E_{N,k}$ with $N = 2^{j-k}$, the
   number of segments of the $(2k)$th-order rule.

   (b) $(2^{k-1}N^2)^k = 2^{(2k-j-1)k}$ as above; but $(2k - j - 1)k$ increases monotonically with k for $k \leqslant j$, having greatest value $j(j - 1)$ for $k = j - 1$ and
   $k = j$.

7. For instance (with I assigned as $I_{j,k+1}$):

   $m := 1; r_j := I$;   **comment** initially $I = I_{j,1}$;
   for $l := j-1$ **step** -1 **until** 1 **do**
       begin $diffce := I - r_l$; **comment** $(I_{j,k} - I_{j-1,k})$ with $k=j-l$;
           $m := 4*m$;   **comment** $4 \uparrow k$;
           $I := I + diffce/(m-1)$; **comment** value of $I_{j,k+1}$;
         $TEST$: **if** abs$(diffce) \leqslant$ eps **then goto** $CONVERGED$;
            $r_l := I$;   **comment** overwrites $I_{j-1,k}$ by $I_{j,k+1}$;
       end

If $I_{j,k}$ were to be assigned to I at label $CONVERGED$, the conditional statement
labelled $TEST$ would be placed immediately following the assignment of $diffce$.

8. Replace $R_{j,k}$ by $I_{j+1,k+1}$ and x_j by 4^{-j}, and put $x = 0$ (in order to extrapolate to limit). If tableau of values shows faster convergence of higher-order formulae, then $|I_{j,k} - I_{j-1,k-1}| > |I_{j,k-1} - I_{j-1,k-1}| \gg |I_{j,k} - I_{j-1,k}|$.

Section 5.1.5

2. (a) Since (5.1.26) applies to quadratic pieces with C^1 continuity, $\int_{x_{k-1}}^{x_{k+1}}$
$$y(x)dx = h(\tfrac{1}{2}y_{k-1} + y_k + \tfrac{1}{2}y_{k+1}) - (h^2/12)[\pi'_{k+1}(x_{k+1}) - \pi'_k(x_{k-1})].$$
From (3.4.2), $\pi'_k(x) = \hat{y}_{k-1,k} - (\mu_k/h_k)(x_k + x_{k-1} - 2x)$, so that using (3.4.3), $\pi_{k+1}(x_{k+1}) - \pi'_k(x_{k-1}) = \hat{y}_{k,k+1} - \hat{y}_{k-1,k} + \mu_{k+1} + \mu_k = 2(\hat{y}_{k,k+1} - y_{k-1,k}) = 2\delta^2 y_k/h$, and so $\int_{x_{k-1}}^{x_{k+1}} y(x)dx = (h/3)(\hat{y}_{k-1} + 4y_k + y_{k+1})$.

   (b) Since $\Delta y_k = y_{k+1} - y_k$, therefore $h(y'_{2m+1} - y'_0) = [(2m+1)/m(m+1)]$
$$[m(y_0 + y_{2m+1}) + \sum_{k=1}^{m} (-1)^k(2m+1-2k)(y_k + y_{2m+1-k}). \text{ Thus, from}$$
(5.1.26), $\int_{x_0}^{x_{2m+1}} y(x)dx = h\sum_{k=0}^{m} w_k(f_k + f_{2m+1-k})$ where $w_0 = (4m+5)/[12(m+1)]$ and $w_k = 1 - (-1)^k(2m+1)(2m+1-2k)/[12m(m+1)]$. With $m = 1$: $w_0 = 3/8$, $w_1 = 9/8$. With $m = 2$: $w_0 = 13/36$, $w_1 = 29/24$, $w_2 = 67/72$. To check, assume $x_0 = 0$, $x_k = k$ (say) and $f_k = k^j$, giving correctly $\int_0^{x_{2m+1}} y(x)dx = (2m+1)^{j+1}/(j+1)$ for $j = 0$, 1, 2, 3 (and $m = 2$, but the result is true for any m).

3. (a) $n = 2$: $(b-a)^{-1}\int_a^b f(x)dx = (1/6)(y_0 + 4y_1 + y_2)$ — Simpson's rule; $n = 3$: $= (1/8)(y_0 + 3y_1 + 3y_2 + y_3)$ — three-eighths rule.

4. (a) (i) 2169 m/s² (ii) 2143 m/s² (iii) 2145 m/s².

   (b)
t (sec)	0	1	2	3	4	5	6	7	8
$a'(t)$(m/s³)	16.5	15.5	17	21	28.5	45.5	82.5	174	306
$a(t)$(m/s²)	150	166	181	200	223	257	314	422	662
$v(t)$	0	158	331	521	732	971	1254	1614	2145

   Values of $a'(t)$ at $t = 0$ and 8 secs follow from (5.1.28); elsewhere, $a'(t) = [a(t+1) - a(t-1)]/2$.

   (c) (i) 6561 m. (ii) 6611 m. (iii) 6604 m.

   (d) Values of $v'(t) = a(t)$, and $v(t)$ tabulated in part (b). Using (5.1.25),
t (sec)	0	1	2	3	4	5	6	7	8
$h(t)$ (m)	0	78	321	745	1370	2219	3327	4752	6611

5. (a) (i) Quadratic end-pieces imply $M_0 = M_1$, $M_{n-1} = M_n$ as in (3.4.10) and from stated problem, therefore $y'_0 + y'_1 = 2\hat{y}_{0,1}, y'_{n-1} + y'_n = 2\hat{y}_{n-1,n}$; i.e. if $h_k = h$, a constant, $h(y'_0 + y'_1) = 2\Delta y_0$, $h(y'_{n-1} + y'_n) = 2\nabla y_n$.
   (ii) The 'not-a-knot' condition implies from (3.4.11) that $h_2^2 y'_0 + (h_2^2 - h_1^2)y'_1 - h_1^2 y'_2 = 2(h_2^2 \hat{y}_{0,1} - h_1^2 \hat{y}_{1,2})$, and $h_n^2 y'_{n-2} + (h_n^2 - h_{n-1}^2)y'_{n-1} - h_{n-1}^2 y'_n = 2(h_n^2 \hat{y}_{n-2,n-1} - h_{n-1}^2 \hat{y}_{n-1,n})$; and if $h_k = h$, a constant then $h(y'_2 - y'_0) = 2\delta^2 y_1$, $h(y'_n - y'_{n-2}) = 2\delta^2 y_{n-1}$.

(b)

t (secs)	0	1	2	3	4	5	6	7	8	
$a'(t)(\text{m/s}^3)$	17.2	14.8	16.6	20.7	26.7	43.6	71.8	164.1	315.9	condition (i)
	18.8	14.4	16.8	20.6	26.8	43.2	73.4	158.3	337.4	condition (ii)

At $t = 8$ secs, from (5.1.26), $v(t) = 2144$ m/s, condition (i);
$= 2142$ m/s, condition (ii).

(c) $a'(0) = 16.9$ m/s^3 $a'(8) = 315.3$ m/s^3 condition (i)
$= 19.3$ m/s^3 $= 336.9$ m/s^3 condition (ii)

Section 5.2.1

1. (a) Required formula has symmetric form: $(1/2) \int_{-1}^{+1} f(x) \, dx = \lambda_0 [f(x_0) + f(-x_0)] + (1 - 2\lambda_0) f(0)$. Since formula is correct for $f(x) \equiv x^2$ and $f(x) \equiv x^4$: $(1/2) \int_{-1}^{+1} x^2 \, dx = 1/3 = 2\lambda_0 x_0^2$ and $(1/2) \int_{-1}^{+1} x^4 dx = 1/5 = 2\lambda_0 x_0^4$, whence $x_0^2 = 3/5$ (i.e., $x_0 = -\sqrt{3/5}$) and $\lambda_0 = 5/18$.

(b) $(1/2) \int_{-1}^{+1} f(x) dx = \lambda_1 [f(x_1) + f(-x_1)] + (½ - \lambda_1)[f(-1) + f(1)]$, and as above $1/3 = 1 - 2\lambda_1 (1 - x_1^2)$ and $1/5 = 1 - 2\lambda_1 (1 - x_1^4)$; thus $(1 + x_1^2) = [2\lambda_1 (1 - x_1^4)]/[2\lambda_1 (1 - x_1^2)] = (4/5)/(2/3) = 6/5$. Whence $x_1^2 = 1/5$ (i.e., $x_1 = -1/\sqrt{5}$) and $\lambda_1 = 5/12$.

(c) $(1/2) \int_{-1}^{+1} f(x) = (1/4)[f(x_0) + f(x_1) + f(-x) + f(-x)]$, and with $f(x) \equiv x^2$ and x^4 as above, $(2/3) = x_0^2 + x_1^2$ and $(2/5) = x_0^4 + x_1^4$. Squaring each side of the first equation and subtracting the second, $2x_0^2 x_1^2 = 2/45$, so that $(2/3) = x_k^2 + 1/(45 x_k^2)$ with $k = 0$ or 1. Thus $x_k^2 = (5 \pm 2\sqrt{5})/15$, giving $x_k = \pm 0.1876, \pm 0.7947$.

2. (a) $c = 4/175$. (b) $c = -16/525$ (c) $c = 16/945$.

3. Write $\theta = (\pi/4)(1 + x)$ and $f(x) = \sin[(\pi/4)(1 + x)]$: then required integral is of form (5.2.3) with $n = 2$. Now estimate is $(5/18)[f(-\sqrt{3/5}) + f(\sqrt{3/5})] + (4/9)f(0) = 0.63662494$ with error -5.17×10^{-6}. In (5.2.5), since $f^{(6)}(\xi) = -(\pi/4)^6 \sin[(\pi/4)(1 + \xi)]$, one finds $\xi = -0.024$ to match this error.

4. (a) Using (2.5.7) and neglecting E_0, show that $[2^{n+1}(n + 1)!]^4/[(2n + 3)! (2n + 2)!] \to (\pi/2)$ as $n \to \infty$.

(b) Required to show that $B_m/2^{m(m-3)} < \pi/2^{2m+1}$ i.e., $B_m < \pi \, 2^{m(m-5)-1}$. But from (2.5.6), $B_m/B_{m-1} = m(2m - 1)/(2\pi^2)$ whereas $2^{m(m-5)}/2^{(m-1)(m-6)} = 4^{m-3}$, so that this inequality must be satisfied for sufficiently large m. (It is satisfied for $m \geqslant 4$.)

5. $P_2(x) = (3x^2 - 1)$, $P_3(x) = x(5x^2 - 3)/2$, $P_4(x) = (35x^4 - 30x^2 + 3)/8$, and zeros of $P_4(x)$ are at $x^2 = (15 \pm 2\sqrt{30})/35$, i.e., $\pm 0.861136, \pm 0.339981$. Further, $P_5(x) = x(63x^4 - 70x^2 + 15)/8$, and $P_5'(x) = 3(126x^4 - 70x^2 + 5)/8$, and since for the Lobatto formula, the points of evaluation are at roots of $(1 - x^2)P_5'(x)$, these are at $x^2 = 1$, $(35 \pm \sqrt{595})/126$ i.e., at $x = \pm 1, \pm 0.686564, \pm 0.290147$.

6. Running totals of evaluations are 2, 5, 9, 13, 19, 25 for Gaussian quadrature and 3, 7, 15, 31, 63 for the Romberg method: former is clearly more efficient.

Section 5.2.2

1. The 3 fixed points are those of the Gauss–Legendre formula $x_0 = -\sqrt{3/5}$, $x_1 = 0$, $x_2 = \sqrt{3/5}$. Let zeros of $x^4 + 2ax^2 + b$ be the positions of added points. Then from (5.2.8) $\int_{-1}^{+1} x^{r+1} (x^2 - 3/5)(x^4 + 2ax^2 + b)\,dx = 0$ for $r = 0, 1, 2, 3$. These equations are inevitably satisfied for $r = 0$ and 2, since the integrand is then an odd function of x; the equations for $r = 1$ and 3 give $a = -5/9$, and $105 + 220a + 99b = 0$, i.e., $b = 155/891$. Zeros of $x^4 + 2ax^2 + b$ are consequently at $x^2 = (55 \pm 2\sqrt{330})/99$ i.e., $x = \pm 0.960491$, ± 0.434244, are positions of added points.

2. (a) Since there are no fixed points in Gaussian quadrature, optimal selection of $(n + 1)$ points of integrand evaluation would be the zeros of $G_{n+1}(x)$ where, from (5.2.8) $\int_{-1}^{+1} x^r G_{n+1}(x)\,dx = 0$ for $r = 0, 1, \ldots n$. But since $P_0(x)$, $P_1(x)$, . . ., $P_n(x)$ are of genuine degree $0, 1, \ldots n$ they form a basis of any polynomial of degree $\leqslant n$. Thus the equations (5.2.8) can be rewritten as $\int_{-1}^{+1} P_r(x) G_{n+1}(x)\,dx = 0$ for $r = 0, 1, \ldots n$. This is satisfied by $G_{n+1}(x) \equiv P_{n+1}(x)$, by definition.

   (b) $q_m(x)$ can be expressed as a linear combination of $p_0(x)$, $p_1(x)$, . . . $p_m(x)$.

   (c) $q_m(x) p_n(x)$ does not change sign in $[a, b]$, therefore $\int_a^b w(x) q_m(x) p_n(x)\,dx \neq 0$, which is impossible since this integral is zero by part (b).

3. (a) Follows from orthogonality condition of problem 2(b).

   (b) Integrating by parts, $\int_{-1}^{+1} x^2 (1 - x^r) P'_n(x)\,dx = \int_{-1}^{+1} [(r + 2) x^2 - r] x^{r-1} P_n(x)\,dx$ which is zero for $r = 0, 1, \ldots (n - 2)$.

4. (a) The shape function $s_{nk}(x_0, x_1, \ldots x_n; x) = P_{n+1}(x)/[(x - x_k)P'_{n+1}(x_k)]$ and result follows as in (5.1.2).

   (b) For example, expand $p_n(x)$ about $x = x_0$ as a Taylor polynomial.

   (c) $\int_{-1}^{+1} s_{nk}(x)[s_{nk}(x) - 1]\,dx = \int_{-1}^{+1} P_{n+1}(x)q_{n-1}(x)\,dx = 0$, since $q_{n-1}(x) = [P_{n+1}(x) - P'_{n+1}(x_k)(x - x_k)](x - x_k)^{-2}/[P'_{n+1}(x_k)]^2$ is by part (a) a polynomial of degree $(n - 1)$ and the integral vanishes by virtue of the orthogonality relation of problem 2(b). Thus $\int_{-1}^{+1} s_{nk}^2(x)\,dx = \int_{-1}^{+1} s_{nk}(x)\,dx$.

5. $|E_G^{(n)}| = |(1/2) \int_{-1}^{+1} f(x)\,dx - \sum_{k=0}^{n} \lambda_k f_k| \leqslant |M_1| + |M_2| + |M_3|$, say, where $M_1 = (1/2) \int_{-1}^{+1} [f(x) - p_N(x)]\,dx$, $M_2 = (1/2) \int_{-1}^{+1} p_N(x)\,dx - \sum_{k=0}^{n} \lambda_k p_N(x_k)$, and $M_3 = \sum_{k=0}^{n} \lambda_k [p_n(x_k) - f_k]$. If $(2n - 1) \geqslant N$ then $M_2 = 0$, and by the Weierstrasse theorem, $|M_1| \leqslant (1/2) \int_{-1}^{+1} |f(x) - p_N(x)|\,dx \leqslant \epsilon$ and $|M_3| \leqslant \sum_{k=0}^{n} |\lambda_k| \epsilon = \epsilon$, since the λ_k are positive and sum to unity. Thus $|E_G^{(n)}| \leqslant 2\epsilon$ $\forall$ $n \geqslant [N(\epsilon) + 1]/2$.

6. Suppose $p_{2n+1}(x)$ is the polynomial of degree $\leqslant (2n+1)$ which is the Hermitian interpolate of $f(x)$ at the $(n+1)$ zeros of $P_{n+1}(x)$, i.e., at the sampling points of the Gauss-Legendre formula. Since this formula integrates such a polynomial precisely, therefore $E_G^{(n)} = \frac{1}{2}\int_{-1}^{+1}[f(x) - p_{2n+1}(x)]dx$. Substitute the remainder term of problem 1, section 3.7.1, in the integrand, with $\pi_n(x) = P_{n+1}(x)/\hat{P}_{n+1}$, and use the integral mean value theorem.

7. (a) Observe from (5.2.6) that $q_{i+1}(x) = \sum\limits_{m=0}^{i+1} q_{i+1,i-m+1}\, P_m(x) = (x - x_i)$

$$\sum_{m=0}^{i} q_{i,i-m}\, P_m(x) = \sum_{m=0}^{i} q_{i,i-m}\, [(m+1)P_{m+1}(x) - (2m+1)x_i P_m(x)$$

$+ mP_{m-1}(x)]/(2m+1)$ and match coefficients of $P_m(x)$.

(b) For example, supposing q_{-2} and q_{-1} are defined:

        for $j : = -2$ **step** 1 **until** n **do** $q_j : = 0$;

        $q_0 : = 1$;

        if $n \geqslant 0$ **then for** i . $= 0$ **step** 1 **until** n **do**

        begin $k : = 0$;

            for $j : = i + 1$ **step** -1 **until** 0 **do**

            begin $q_j : = (k+1)*q_{j-2}/(2*k+3) - x_i*q_{j-1} + k*q_j/(2*k-1)$;

               $k : = k + 1$

            end;

      end,

(c) $q_{n+1,1} = q_{n+1,3} = \ldots = 0$, since $q_{n+1}(x)$ is either an odd or even function of x.

(d) $q_{n+1}(x) = \sum\limits_{m=0}^{n+1} q_{n+1,n-m+1}P_m(x) = (x - x_k) \sum\limits_{m=0}^{n} q_{n-m}^{(k)}\, P_m(x)$ and

proceed as in part (a).

8. (a) $(1/2)\int_{-1}^{+1} s_{nk}(x)\, dx = (1/2) \sum\limits_{m=0}^{n} r_{n-m}^{(k)} \int_{-1}^{+1} P_m(x)\, dx = (1/2)\, r_n^{(k)}$, by

orthogonality principle.

(b) Here $s_{nk}(x) = q_n^{(k)}(x)/q_n^{(k)}(x_k)$, so that $r_n^{(k)} = q_n^{(k)}/ \prod\limits_{\substack{i=0 \\ i \neq k}}^{n} (x_k - x_i)$. Thus,

from problem 6(d), for example:

for $k : = 0$ **step** 1 **until** n **do**

begin $qb : = qa : = j : = 0$;

    for $m : = n + 1$ **step** -1 **until** 1 **do**

    begin $qc : = (2*m-1)*(q_j + x_k*qb - (m+1)*qa/(2*m+3))/m$;

        $j : = j + 1; qa : = qb; qb : = qc$

    end;

    for $j : = 0$ **step** 1 **until** n **do if** $j \neq k$ **then** $qc : = qc/(x_k - x_j)$;

    $l_k : = qc$

end

(c) $64/225$.

Section 5.2.3

1. Observe that $\phi'(t)$ is an even function of t, and use (5.2.13).

2. The largest representable magnitude < 1 is $(b^p - 1)b^{-p}$ and the smallest magnitude which rounds to unity is $[b^p - (1/2)] b^{-p}$. Thus $\phi(t)$ must be less than this, i.e., $2c|t|/(1 - t^2) < \text{argtanh}[1 - (b^{-p}/2)] = (1/2)\ln(4b^p - 1) \cong (p/2)\ln(b) + \ln(2) = \ell$, say. Thus $|t| \leqslant \ell/[c + (c^2 + \ell^2)^{1/2}] \cong 1 - c/\ell$ if $c/\ell \ll 1$. With $b = c = 2$, $p = 24$, $|t| < 0.8024$, and at this limit $1 - x^2 = b^{-p}$ so that $\phi'(t) = 51.84b^{-p} = 3.09 \times 10^{-6}$.

3. (a) Here $\phi'(t) = m(1 - t^2)^{-3/2}(1 - x^2)$. But $(1 - x) - 2 \exp[-2^{-1/2} m/(1 - t)^{1/2}]$ as $t \to 1$, so that $g(t) = f(x)\phi'(t)$ or $2^{\nu - 1/2}A\, m(1 - t)^{-3/2} \exp[-2^{-1/2}\nu m/(1 - t)^{1/2}]$ as $t \to 1$, and $g(1) = g'(1) = g''(1) = \ldots = 0$.

(c)

j	1, 15	2, 14	3, 13	4, 12	5, 11	6, 10	7, 9	8
$\pm x_j$	.9999989	.9997702	.9966993	.9804635	.9243451	.7750271	.4652200	0
$\phi'(t_j)/16$	.0000046	.0003971	.0034636	.0148923	.0456863	.1099811	.2005752	.25

4. Precise answer equals $\Gamma(1/4)p^{-1/4} = 3.6256099082p^{-1/4}$.

SECTION 5.3

1. If the span of the component segments are $h_0, h_1, h_2, \ldots$ where $\sum_{k=0}^{N} h_k = (b - a)$, then the magnitude of the errors in the component integrals over each segment are individually less than (say) ϵh_k and so less than $(b - a)\epsilon$ in total. Thus the error of the integral mean over $[a, b]$ is less than ϵ.

2. Points of integrand evaluation are 0, 1/32, 1/16, 3/32, 1/8, 3/16, 1/4, 9/32, 5/16, 11/32, 3/8, 7/16, 1/2, 3/4, 1. Result is 0.2125 ... with error of 3×10^{-4}.

3. (a) Result follows from Taylor's theorem: i.e., $f(x \pm h) = f(x) \pm hf'(x) + (h^2/2)f''(x) \pm (h^3/6)f'''(x) + \ldots$ neglecting terms of order h^4. Value of $\delta^4 f(x)$ is approximately $\delta^2 f''(x)h^2$, i.e., $f^{iv}(x)h^4$.
   (b) Error of integral mean is from (5.1.11) and Table 5.1 equal to $(2h)^4 f^{iv}(\xi)/2880$ where $\xi \in [x - 2h, x = 2h]$. Accepting $\delta^4 f$ as estimate of $f^{iv}(\xi)h^2$, therefore $\delta^4 f < 180 \, \epsilon/(b - a)$.
   (c) $|\delta^4 f| < (3k/2)|\delta^4 f - 2\delta^2 f|$.

4. If $f''(x)$ is regarded as a constant over $[x_1 - h, x_1 + h]$, and true integral mean is I, then $I = f_1 + \epsilon$, say, and also $I = \lambda(f_0 + f_2) + (1 - 2\lambda)f_1 + [2\lambda^3 + (1 - 2\lambda)^3]\epsilon$, where $\lambda = 1 - \sqrt{3/5}$ is the span of each of the outer segments as a fraction of the interval span $(2h)$. Eliminating ϵ, since $1 - 2\lambda^3 - (1 - 2\lambda)^3 = 6\lambda(1 - \lambda)^2$, $I = f_1 + (f_0 + f_2 - 2f_1)/[6(1 - \lambda)^2] = f_1 + (5/18)(f_0 + f_2 - 2f_1) = (5/18)(f_0 + f_2) + (4/9)f_0$.

5. (a) Segments extend between the four end-points $x_1 \pm [\xi \pm (1 - \xi)]h$ where $\xi = \sqrt{3/5}$, and the points of integrand evaluation are at $x_1 \pm [\xi + \sigma\xi(1 - \xi)]h$, $x_1 + \xi(2\xi - 1)h$, for $\sigma = -1, 0, 1$. Thus $-\alpha_0 = \alpha_8 = (2 - \xi)\xi = 0.94919333848$, $-\alpha_1 = \alpha_7 = \xi = 0.77459666924$, $-\alpha_2 = \alpha_6 = 0.6$, $-\alpha_3 = \alpha_5 = (2\xi - 1)\xi = 0.42540333076$, $\alpha_4 = 0$.

(b) As in problem 4, If $I = I^{(6)} + \epsilon$, and $f^{vi}(x)$ can be considered constant, then $I^{(8)} = \lambda(I_1 + I_3) + (1 - 2\lambda)I_2 + [2\lambda^7 + (1 - 2\lambda)^7]\epsilon$, where $\lambda = 1 - \epsilon$, and eliminating ϵ, $\mu_1 = \mu_3 = 0.22885870755$, $\mu_2 = 0.55761233527$, $\mu_0 = -0.01532975038$.

(c) Since the weights of the Gauss–Legendre 3-point rule are $5/18, 4/9, 5/18$, the corresponding λ coefficients for $I^{(8)}$ are $\lambda_0 = \lambda_2 = \lambda_6 = \lambda_8 = 5\mu_1/18$, $\lambda_1 = \lambda_7 = (8\mu_1 + 5\mu_0)/18$, $\lambda_4 = 4(\mu_2 + \mu_0)/9$ and so: $\epsilon_0 = \epsilon_8 = -0.18779825 \times 10^{-2}$, $\epsilon_1 = \epsilon_7 = 1.092751196 \times 10^{-2}$, $\epsilon_2 = \epsilon_6 = -2.372550288 \times 10^{-2}$, $\epsilon_3 = \epsilon_5 = 2.031671996 \times 10^{-2}$, $\epsilon_4 = -1.128149294 \times 10^{-2}$.

(d) True integral mean is $I = \pi/4$. $I - I^{(6)} = -6.27 \times 10^{-3}$, $I - I^{(8)} = -1.23 \times 10^{-4}$, $I - I^{(10)} = -4.16 \times 10^{-5}$, $I^{(8)} - I^{(6)} = -6.14 \times 10^{-3}$ (judged error of $I^{(6)}$), $I^{(10)} - I^{(8)} = -8.15 \times 10^{-5}$ (judged error of $I^{(8)}$).

SECTION 5.4

1. (a) Skew symmetry implies that $c_{jk} = -c_{kj}$ and $c_{jj} = 0$.

(b) Let $x = \xi \cos \alpha - \eta \sin \alpha$, $y = \xi \sin \alpha + \eta \cos \alpha$, where (ξ, η) are Cartesian coordinates of a frame of axes rotated through angle α. Then $y_j x_k - x_j y_k \cong (\xi_j \sin \alpha + \eta_j \cos \alpha)(\xi_k \cos \alpha - \eta_k \sin \alpha) - (\xi_j \cos \alpha - \eta_j \sin \alpha)(\xi_k \sin \alpha + \eta_k \cos \alpha) = (\eta_j \xi_k - \xi_j \eta_k)(\sin^2 \alpha + \cos^2 \alpha) = \eta_j \xi_k - \xi_j \eta_k$.

(c) By integration by parts $J = \int_\gamma y(s)x'(s)\,ds - \frac{1}{2}(x_n y_n - y_o x_o) = \frac{1}{2}(x_n y_n - y_o x_o) - \int_\gamma x(s)y'(s)\,ds$. Adding these two expressions and dividing by 2, quoted result follows. Writing $x(s) = r(\theta)\cos \theta$, and $y(s) = r(\theta)\sin \theta$, then $x'(s) = [r'(\theta)\cos \theta - r(\theta)\sin \theta]\,(d\theta/ds)$, and similarly $y'(s) = [r'(\theta)\sin \theta + r(\theta)\cos \theta]\,(d\theta/ds)$ so that $[y(s)x'(s) - y'(s)x(s)] = -r^2(\theta)\,(\sin^2 \theta + \cos^2 \theta)\,(d\theta/ds) = -r^2(\theta)d\theta/ds$.

2. (a) Since $A = (1/2)(y_j x_k - x_j y_k)$ is invariant under rotation of axes, it may be assumed without loss of generality that P_j lies on the y-axis (say), so that $x_j = 0$; then $OP_j = y_j$ is the length of one side, and x_k is the altitude of the opposite point P_k, so that area of triangle is equal to A.

(b) By the result of part (a), $\frac{1}{2}(y_o x_n - x_o y_n)$ is the triangular area $OP_o P_n$, whereas J is the area between the curve γ and the radial lines OP_o, OP_n, as in Fig. 5.6(b).

(c) Verification follows from part (b). Further, since area K is independent of the position of the origin, therefore $\sum_{j=0}^{n} \sum_{k=0}^{n} c'_{jk}(b + y_j)(a + x_k)$ must be independent of a and b. Thus $\sum\limits_{j=0}^{n} \sum\limits_{k=0}^{n} c'_{jk}\, y_j = \sum\limits_{j=0}^{n} \sum\limits_{k=0}^{n} c'_{jk}\, x_k = \sum\limits_{j=0}^{n}$

$\sum\limits_{k=0}^{n} c'_{jk} = 0$. Since x_k and y_j are arbitrary, therefore every row and

column of $\mathbf{c}'$ must sum to zero.

3. (b) Let OX_2 at right angles to OP_1 be the x-axis, where $X_2 P_2$ is parallel to OP_1: similarly let $X_0 P_0$ be parallel to OP_1. Since origin is at mid-point of $P_0 P_2$, therefore relative to the chosen axis system, $x_0 = -x_2$, $y_0 = -y_2$ and by Simpson's rule the area bounded by the curve $P_0 P_1 P_2$, the lines $X_0 P_0$ and $X_2 P_2$ and the x-axis, is $y_1 (x_2 - x_0)/3$; but this is equal to the area K between the arc $P_0 P_1 P_2$ and the line $P_0 P_2$. By the trapezoidal rule (or otherwise) the area of the triangle $P_0 P_1 P_2$ is $y_1 (x_2 - x_0)/2$.

  (c) From problem 1, § 5.1, correction term to $\int_{x_0}^{x_2} y(x)\, dx$ is $(x_2 - x_0)$ $(\hat{y}_{01} + \hat{y}_{12})\,(x_2 - 2x_1 + x_0)/6$ and in (5.4.7), $(x_2 - x_0)\,(\hat{y}_{01} + \hat{y}_{12})/2$ is replaced by $(y_2 - y_0)$, which is only true if $x_1 - x_0 = x_2 - x_1$ when the correction is in any event zero. The disparity is due to the different directions assumed for the axis of the parabola in the two interpolations.

4. (a) With arbitrary origin O, a double segment application of the trapezoidal rule estimates the area J between the arc $P_0 P_1 P_2$ and the lines OP_0, OP_2 as the sum of the triangular areas $OP_0 P_1$, $OP_1 P_2$ whereas single segment application estimates the area J as the triangle $OP_0 P_2$. The difference is therefore the triangular area $P_0 P_1 P_2$. If the single segment estimate is assumed to have 4 times the error of the double segment estimate, on eliminating the error the area is $\{4[\text{area}(OP_0 P_2) + \text{area} (P_0 P_1 P_2)] - \text{area}(OP_0 P_2)\}/3 = (4/3)\,\text{area}\,(P_0 P_1 P_2) + \text{area}\,(OP_0 P_2)$.

  (b) Interpretation of Selmer's rule follows from result of problem 2(b). Richardson extrapolation proceeds as in part (a) except that 3 segments are compared with one segment application of trapezoidal rule, and error is reduced by factor of 9 (not 4).

5. Taking $x_k = kh$ (say), then $I = \tfrac{1}{2}(x_n y_n - x_o y_o) + y^T C\, x$ becomes $I \cong$ $h\,[(3/8)y_0 + (7.6)y_1 + (23/24)y_2 + y_3 + y_4 + \ldots + (7/6)y_{n-1} + (3/8)y_n] =$ $h\,[\tfrac{1}{2}y_0 + y + y_2 + \ldots + y_{n-1} + \tfrac{1}{2}y_n + (1/24)\,(4y_1 - 3y_0 - y_2 + 4y_{n-1} - 3y_n - y_{n-2})]$, consistent with (5.1.29).

6. (a) Let $\theta_j = -2\pi j/n$, so that $x_j = r\cos(2\pi j/n)$, $y = -r\sin(2\pi j/n)$: then since

$$x_{j+k} - x_{j-k} = -2r\,\sin(2\pi j/n)\sin(2\pi k/n), \text{ therefore area is } 2r^2 \sum_{j=1}^{n}$$

$$\sin^2(2\pi j/n) \sum_{k=1}^{m} b_k^{(m)} \sin(2\pi k/n) = r^2 \sum_{j=1}^{n} [(1 - \cos(4\pi j/n)] \sum_{k=1}^{m} b_k^{(m)}$$

$$\sin(2\pi k/n) = r^2 n \sum_{k=1}^{m} b_k^{(m)} \sin(2\pi k/n) \text{ since } \sum_{j=1}^{n} \cos(4\pi j/n) = 0.$$

(b) $A_{1,8} = 4r^2 \sin(\pi/4) = 2\sqrt{2}r^2 = 2.828r^2$; $A_{2,8} = 2(4\sqrt{2}-1)r^2/3 = 3.104r^2$;
$A_{3,8} = 2(23\sqrt{2}-9)r^2/15 = 3.137r^2$; $A_{4,8} = 8(44\sqrt{2}-21)r^2/105 = 3.141r^2$.

(c) $A_{1,n} = (nr^2/2)\sin(2\pi/n) = (nr^2/2)[(2\pi/n) - (1/6)(2\pi/n)^3 + \ldots] = \pi r^2 [1 - 2\pi^2/(3n^2) + \ldots]$, error of order $1/n^2$. $A_{2,n} = (nr^2/12)$ $[8 \sin(2\pi/n) - \sin(4\pi/n)] = (nr^2/6)[4 - \cos(2\pi/n)] \sin(2\pi/n) = (r^2/3)$ $[1 - 2\pi^2/(3n^2) + 2\pi^4/(15n^2) - + \ldots] [3 + 2\pi^2/n^2 - 2\pi^4/(3n^4) + - \ldots]$ $= \pi r^2 [1 - 8\pi^4/(15n^4) + \ldots]$, error of order $1/n^4$.

7. (a) Trapezoidal rule ($m = 1$) gives 17.15×10^6 km^2; Bessel interpolation ($m = 2$) 17.51×10^6 km^2; and class C^2 interpolation ($m = 3$) 17.52×10^6 km^2.

   (b) Composite Brun's rule gives 17.76×10^6 km^2, and (5.4.8) gives 17.45×10^6 km^2. (Actual surface area of S. America is quoted as c.17.85×10^6 km^2, but much closer definition of the coast line would of course be needed to compute this.)

8. For instance:
   ```
   area : = 0;
   if read(x0, y0) then
       begin xlast : = 0; ylast : = 0;
           while read(x, y) do
               begin x : = x − x0; y : = y − y0;
                   area : = x ∗ ylast − y ∗ xlast + area
                   xlast : = x; ylast : = y
           end
   end;
   ```
   If the magnitude of the coordinates is larger than $b^{p/2}$ then the increments to area will lose precision by cancellation error. In that event the increment to area would be better calculated as $(x − x$last$)*y$last $+ (y − y$last$)*x$last, and values accumlated in double precision (or using an extra 'spill' register).

SECTION 5.5

1. (a) Since $\int_0^x \ln(1/x)\, dx = x[\ln(1/x) + 1] = x \ln(e/x)$, the error is $h \ln(e/h) - h(2/h) = h \ln(e/2)$.

   (b) Points of integrand estimation are at $x = x_j = (2j − 1)h/2$ and estimate is
   $$h \sum_{j=1}^{N} \ln\{2/[(2j − 1)h]\} = h \sum_{j=1}^{N} \ln\{4j/[2j(2j − 1)h]\} = h \ln[(4/h)^N N!/(2N)!],$$
   whereas correct value is $Nh \ln(e/Nh) = h \ln[(e/Nh)^N]$. Thus error is $h \ln[(e/4N)^N(2N)!/N!]$. But using (2.5.7), $(e/4N)^N(2N)!/N! \cong \sqrt{2}$.

   (c) The error is simply the difference between the errors of $\int_0^{Nh} \ln(1/x)\, dx$ and $\int_0^h \ln(1/x)\, dx$, that is $h \ln(\sqrt{2}) − h \ln(e/2) = (1/N)\ln(2^{3/2}/e)$. However, if ϵ is fixed, then the second derivative of the integrand being bounded independently of N, the error will be proportional to $(1/N^2)$.

2. (a) $x = t^2$, giving $2\int_0^1 \cos(t^2)\,dt$.

  (b) No singularity.

  (c) $x = 1 - t^2$, giving $2\int_0^1 (2 - t^2)^{1/2} t^2 \sin(\pi t^2)\,dt$.

  (d) Either partition integral into two components with singularities at $x = 0$ in one, and at $x = 1$ in the other, or use (say) $x = \sin^2(t)$, giving $2\int_0^{\pi/2} [1 + \sin^2(t)]^{-1/2} dt$.

3. (a) Integral becomes $(1/2)\int_0^1 [\exp(-2x) - 1]\, x^{-1}\, dx = -\int_0^1 x^{-1} \sinh(x)\exp(-x)\,dx$.

  (b) Integral can be partitioned as $\int_0^{\pi/2} x\ln[\sin(x)]\,dx + \int_{\pi/2}^{\pi} x\ln[\sin(x)]\,dx = \int_0^{\pi/2} x\ln[\sin(x)]\,dx + \int_0^{\pi/2} (\pi - x)\ln[\sin(x)]\,dx = \int_0^{\pi/2} (\pi - 2x)\ln[\sin(x)]\,dx$ and then integrating by parts, integral becomes $\int_0^{\pi/2} (\pi - x)x \cot(x)\,dx$, where $x\cot(x)$ is analytic in $[0, \pi/2]$.

4. Partitioning range, and integrating by parts, $\int_0^1 \exp(-x)\ln(x)\,dx = -\int_0^1 x^{-1}[1 - \exp(-x)]\,dx$; and $\int_1^{\infty} \exp(-x)\ln(x)\,dx = \int_1^{\infty} x^{-1} \exp(-x)\,dx = \int_0^1 t^{-1} \exp(-1/t)\,dt$, so that on adding the integral becomes $\int_0^1 [\exp(-1/x) + \exp(-x) - 1]\, x^{-1}\,dx$.

5. Let $p_n(x) = (c_0/2) + \sum_{j=1}^{n} c_j T_j(x)$ be the Tchebyshev nth degree interpolate of $f(x)$. Now $\int_{-1}^{+1} (1 - x^2)^{-1/2} T_j(x)\,dx = \int_0^{\pi} \cos(j\theta)\,d\theta = \pi\delta_{j0}$, so that $\int_{-1}^{+1} (1 - x^2)^{-1/2} p_n(x)\,dx = (\pi c_0/2) = [\pi/(n+1)] \sum_{k=0}^{n} f(x_k)$, from section 3.7.2, problem 3(c).

6. (a) Substituting $t = (x^2 + s)^{1/2}$ in the integrand, $\mathrm{erfc}(x) = \pi^{-1/2} \exp(-x^2)\int_0^{\infty} (x^2 + s)^{-1/2} \exp(-s)\,ds$ and applying the Gauss–Laguerre formula,

$$\mathrm{erfc}(x) = \pi^{-1/2} \exp(-x^2) \sum_{k=1}^{n} \lambda_k^{(n)} (x^2 + x_k)^{-1/2}.$$

  But $(x^2 + s)^{-1/2}$ behaves like $s^{-1/2}$ when s is large, and its approximation by a polynomial (implicit in the use of Gauss–Laguerre quadrature) would be inappropriate.

  (b) Substitute $t = x/s^{1/2}$ transform integral to $\mathrm{erfc}(x) = \pi^{1/2} x \exp(-x^2)\int_0^1 s^{-3/2} \exp[-x^2(1 - s)/s]\,ds$.

7. (a) Relation follows by noting that $\cos(m\pi) = (-1)^m$ and $\sin(m\pi) = 0$, and series development by repeated applications of this relation, with $R_N = (-1)^N (m\pi)^{-2N} \int_0^1 \sin(m\pi x) f^{(2N)}(x)\,dx$.

  (b) Here $f(x) = (1/2)[(2 - x)^{-1} - (2 + x)^{-1}]$ so that $f^{(k-1)}(x) = [(2 - x)^{-k} + (-1)^k (2 + x)^{-k}](k - 1)!/2$. Thus with $m = 10$, $(-1)^{m-1} f^{(2k)}(1) + f^{(2k)}(0) = -(1 - 3^{-2k-1})(2k)!/2$, and series is semi-convergent since term-ratio magnitude $|t_{k+1}/t_k|$ exceeds 1 for $(k + 1)(2k + 1) > (10\pi)^2$ (i.e., $k \geqslant 22$). Since $f^{(2N)}(x) > 0$ for $0 \leqslant x \leqslant 1$, the truncation error is equal to $(-1)^N (m\pi)^{-2N} \sin(m\pi\xi)\int_0^1 f^{(2N)}(x)\,dx$ for some $\xi \in (0, 1)$, so

that its magnitude is $|R_N| < (10\pi)^{-2N} | f^{(2N+1)}(1) - f^{(2N-1)}(0)| <$
$(10\pi)^{-2N}(2N - 1)!/2$. Thus $|R_3| < 6.2 \times 10^{-8}$: therefore with $N = 3$,

$$\int_0^1 [x/(4 - x^2)] \sin(10\pi x) \, dx \cong \tfrac{1}{2} \sum_{k=0}^{2} (- 1)^{k+1}(1 - 3^{-2k-1}) (2k)!/$$
$$(10\pi)^{2k+1} = - 0.0105797 \text{ (actual value} = - 0.0105796517...).$$

8. (a) By de l'Hopital's theorem, $h/8 \to [g'(\xi)f'(\xi) - f(\xi)g''(\xi)]/g'(\xi)$ as $x \to \xi$.
   (b) Placing (say) $f(x) \equiv \sin(x)$, $g(x) = x \sin(x) - \cos(x)$, in result of part (a),
        $\int_0^{\pi/2} (x - \cot x)^{-1} \, dx = [\ln(\pi/2) + \int_0^{\pi/2} (x - \xi^2 \tan x)(1 - x \tan x)^{-1} \, dx]/$
        $(\xi^2 + 2)$, with $\xi = \cot\xi = 0.86033358902...$ (leading easily to value of
        $0.2901640048...$).

9. (b) Partition integral at $x = 3$ (say). Integral from $x = 1$ to $x = 3$ equals
        1.445 and (semi-convergent) series of part (a) gives 0.284 for the integral
        over the remaining range. Answer = 1.73.

List of Principle Notation

ALPHABETIC CHARACTERS

$\arcsin(x)$, $\arccos(x)$, &c.	inverse trigonometric functions.
$\operatorname{arg\,sinh}(x)$, $\operatorname{arg\,cosh}(x)$, &c.	inverse hyperbolic functions.
$B_1, B_2, \dots, B_n, \dots$	Bernoulli numbers (p. 58).
$C^n[a,b]$	space of functions which are n times differentiable in the interval $[a,b]$.
C^n	class of a plane curve (p. 129).
ceil (x)	smallest integer not less than x.
Ey_k	$= y_{k+1}$, shift operator (p. 99).
$E^n y_k$	$= y_{k+n}$, repeated shift operator (p. 99).
$E_1(x)$	exponential integral (p. 287).
entier (x)	largest integer not greater than x.
erf (x)	error function (p. 60).
erfc (x)	complementary error function (p. 55).
exp (x)	$= e^x$, exponential function.
h_k	$= x_k - x_{k-1}$.
$\displaystyle\mathop{K}_{j=m}^{n} p_j \downarrow q_j$	$= p_m/(q_m + \mathop{K}\limits_{j=m+1}^{n} p_j \downarrow q_j)$, if $m \leqslant n$; $= 0$ if $m = n+1$; continued fraction (p. 64).
$\displaystyle\mathop{K'}_{j=m}^{n} p_j \downarrow q_j$	$= 1/(q_m + \mathop{K}\limits_{j=m+1}^{n} p_j \downarrow q_j)$, continued fraction (p. 65).
$\ln(x)$	natural logarithm of x.
$O(\epsilon)$	a value M (say,) such that $|M/\epsilon|$ is bounded (above) in some neighbourhood of $\epsilon = 0$.
$P_n(x)$	Legendre polynomial of degree n (p. 215).
$p_n(x)$	polynomial of degree not greater than n.
sgn (x)	$= x/|x|$ if $x \neq 0$; $= 0$ if $x = 0$; signum function.
sqrt (x)	$= \sqrt{x}$, positive square root of x, $x \geqslant 0$.

$T_n(x)$	Tchebyshev polynomial of first kind and of degree n (p. 135).
$U_n(x)$	Tchebyshev polynomial of second kind and of degree n (p. 322).
$\Gamma(x)$	gamma function (p. 40).
Δy_k	$= y_{k+1} - y_k$, forward difference operator (p. 99).
$\Delta^n y_k$	$= \Delta(\Delta^{n-1} y_k)$, repeated forward difference operator (p. 99).
δy_k	$= y_{k+\frac{1}{2}} - y_{k-\frac{1}{2}}$, central difference operator (p. 100).
$\delta^n y_k$	$= \delta(\delta^{n-1} y_k)$, repeated central difference operator (p. 100).
ϵ	machine unit (p. 19); *also* any small quantity.
$\displaystyle\prod_{j=m}^{n} \alpha_j$	$= \alpha_m \alpha_{m+1} \ldots \alpha_n$ if $m \leqslant n$; $= 1$ if $m = n + 1$; continued product.
$\displaystyle\sum_{j=m}^{n} t_j$	$= t_m + t_{m+1} + \ldots + t_n$ if $m \leqslant n$; $= 0$ if $m = n + 1$; series sum.
∇y_k	$= y_k - y_{k-1}$, backward difference operator (p. 100).
$\nabla^n y_k$	$= \nabla(\nabla^{n-1} y_k)$, repeated backward difference operator (p. 100).

SYMBOLS

$[a, b]$	closed interval of x (say) such that $a \leqslant x \leqslant b$.
(a, b)	open interval of x (say) such that $a < x < b$.
(x, y)	cartesian coordinates of point in a plane.
$y_{i,j,\ldots m,n}$	polynomial interpolating (x_k, y_k) for $k = i, j, \ldots, m, n$.
$y_{i,j,\ldots m,n}$	$= (\hat{y}_{i,j,\ldots m} - \hat{y}_{j,\ldots m,n})/(x_i - x_n)$, divided difference ($\hat{y}_i \equiv y_i$).
$\hat{p}_n(x), \hat{p}_n$	coefficient of x^n in polynomial $p_n(x)$.
$f'(x), f''(x), \ldots f^{(n)}(x)$	first, second, ... nth derivatives of function $f(x)$. ($f^{(0)}(x) \equiv f(x)$.)
$f'_k, f''_k, \ldots f_k^{(n)}$	first, second, ... nth derivatives of function $f(x)$ at $x = x_k$. ($f_k^{(0)} \equiv f_k$.)
$n!$	$= \Gamma(n+1) = \displaystyle\prod_{k=1}^{n} k$, $(n \geqslant 0$, integer); fractorial function.
$\lvert x \rvert$	magnitude of x.
$\lvert \mathbf{S} \rvert$	cardinality of set $\mathbf{S}$.
$(x)_n$	$= \Gamma(x+n)/\Gamma(x)$, Pochhammer's symbol (p. 284).

$\tilde{x}, \tilde{f}(x), \tilde{f}(\tilde{x})$ approximate (computed) values of x, $f(x)$,
                                                    $f(\tilde{x})$.

$\tilde{\omega}$ approximate (computed) arithmetic operation ω.

$\simeq, \cong$ 'approximate' equals and equivalence symbols.

$x * y$ $= xy$, multiplication.

$x \uparrow n$ $= x^n$, exponentiation.

Index